Conference Proceedings of the Society for Experimental Mechanics Series

Series Editor

Kristin B. Zimmerman, Ph.D.
Society for Experimental Mechanics
Bethel, CT, USA

More information about this series at http://www.springer.com/series/8922

Dan Casem • Leslie Lamberson • Jamie Kimberley
Editors

Dynamic Behavior of Materials, Volume 1

Proceedings of the 2016 Annual Conference on Experimental and Applied Mechanics

Springer

Editors
Dan Casem
Army Research Laboratory
Aberdeen, MD, USA

Leslie Lamberson
Drexel University
Philadelphia, PA, USA

Jamie Kimberley
New Mexico Inst. of Mining & Technology
Socorro, NM, USA

ISSN 2191-5644 ISSN 2191-5652 (electronic)
Conference Proceedings of the Society for Experimental Mechanics Series
ISBN 978-3-319-41131-6 ISBN 978-3-319-41132-3 (eBook)
DOI 10.1007/978-3-319-41132-3

Library of Congress Control Number: 2016952246

Printed on acid-free paper

This Springer imprint is published by Springer Nature
The registered company is Springer International Publishing AG Switzerland

Preface

Dynamic Behavior of Materials represents one of ten volumes of technical papers presented at the 2016 SEM Annual Conference & Exposition on Experimental and Applied Mechanics organized by the Society for Experimental Mechanics and held in Orlando, FL, on June 6–9, 2016. The complete proceedings also includes volumes on: *Challenges in Mechanics of Time-Dependent Materials*; *Advancement of Optical Methods in Experimental Mechanics*; *Experimental and Applied Mechanics*; *Micro and Nanomechanics*; *Mechanics of Biological Systems and Materials*; *Mechanics of Composite & Multifunctional Materials; Fracture, Fatigue, Failure and Damage Evolution*; *Residual Stress, Thermomechanics & Infrared Imaging, Hybrid Techniques and Inverse Problems*; and *Joining Technologies for Composites and Dissimilar Materials.*

Each collection presents early findings from experimental and computational investigations on an important area within Experimental Mechanics. Dynamic Behavior of Materials is one of these areas.

The Dynamic Behavior of Materials track was initiated in 2005 and reflects our efforts to bring together researchers interested in the dynamic behavior of materials and structures, and provide a forum to facilitate technical interaction and exchange. In the past years, this track has represented an ever growing area of broad interest to the SEM community, as evidenced by the increased number of papers and attendance.

The contributed papers span numerous technical divisions within SEM, which may be of interest not only to the dynamic behavior of materials community but also to the traditional mechanics of materials community.

The track organizers thank the authors, presenters, organizers, and session chairs for their participation, support, and contribution to this track. We are grateful to the SEM TD chairs for co-sponsoring and/or co-organizing the sessions in this track. They would also like to acknowledge the SEM support staff for their devoted efforts in accommodating the large number of paper submissions this year, making the 2016 Dynamic Behavior of Materials Track successful.

Aberdeen, MD — Dan Casem
Philadelphia, PA — Leslie Lamberson
Socorro, NM — Jamie Kimberley

Contents

Contributors

Stephen Alexander TKC Global Solutions LLC, Herndon, VA, USA

Khalid Alshibli Department of Civil and Environmental Engineering, University of Tennessee, Knoxville, TN, USA

Yiannis Andreopoulos Department of Mechanical Engineering, The City College of New York, New York, NY, USA

D. Andrews U.S. Army Research Laboratory, Aberdeen, MD, USA

Mark Andrews U.S. Army Research Laboratory, Aberdeen Proving Ground, Aberdeen, MD, USA

Brady Aydelotte U.S. Army Research Laboratory, Aberdeen Proving Ground, Aberdeen, MD, USA

Erich D. Bain U.S. Army Research Laboratory, Aberdeen Proving Ground, Aberdeen, MD, USA

Radhakrishnan Balu U.S. Army Research Laboratory, Materials and Manufacturing Sciences Division, Aberdeen Proving Ground, Aberdeen, MD, USA

Todd D. Beaudet U.S. Army Research Laboratory, Materials and Manufacturing Sciences Division, Aberdeen Proving Ground, Aberdeen, MD, USA

J. Berthe Onera—The French Aerospace Lab, Lille, France

Stephan Bless Tandon School of Engineering, New York University, Brooklyn, NY, USA

Eric Bodenchak University of Nevada, Las Vegas, Las Vegas, NV, USA

Matthew Boswell University of Nevada, Las Vegas, Las Vegas, NV, USA

Steven E. Boyd U.S. Army Research Laboratory, Aberdeen Proving Ground, Aberdeen, MD, USA

M. Brieu Centrale Lille/LML, Villeneuve d'Ascq, France

Geoffrey H. Campbell Materials Science Division, Lawrence Livermore National Laboratory, Livermore, CA, USA

Dan Casem Army Research Laboratory, Aberdeen, MD, USA

J.E. Catalano RDRL-WMM-F, Aberdeen Proving Ground, Aberdeen, MD, USA

Satish Chaparala Corning Incorporated, Corning, NY, USA

Austin Daykin National Security Technologies, LLC, North Las Vegas, NV, USA

Edward Daykin National Security Technologies, LLC, North Las Vegas, NV, USA

Camilo Daza Tandon School of Engineering, New York University, Brooklyn, NY, USA

Ann Mae DiLeonardi US Army Research Laboratory, Aberdeen Proving Ground, Aberdeen, MD, USA

R. Doney U.S. Army Research Laboratory, Aberdeen Proving Ground, Aberdeen, MD, USA

Andrew Druckrey Department of Civil and Environmental Engineering, University of Tennessee, Knoxville, TN, USA

Robert M. Elder U.S. Army Research Laboratory, Aberdeen Proving Ground, Aberdeen, MD, USA

B. Erzar CEA/DAM/GRAMAT, Gramat, France

Ali Fahad Fahem Department of Mechanical Engineering, University of South Carolina, Columbia, SC, USA

Department of Mechanical Engineering, University of Al-Qadisiyah, Al-Diwaniyah, Qadisiyah Province, Iraq

P. Forquin Soils Solids Structures Risks (3SR) Laboratory, Grenoble-Alps University, Grenoble Cedex 9, France

C.G. Fountzoulas U.S. Army Research Laboratory, WMRD, RDRL-WMM-B, Aberdeen, MD, USA

Zach Fussell National Security Technologies, LLC, North Las Vegas, NV, USA

S. Gangireddy MML, NIST, Materials Science and Engineering Division, Gaithersburg, MD, USA

Antonio Garcia Department of Mechanical Engineering, New Mexico Institute of Mining and Technology, Socorro, NM, USA

E. Gauch Naval Undersea Warfare Center, Division Newport, Newport, RI, USA

Amos Gilat Scott Laboratory, Department of Mechanical and Aerospace Engineering, The Ohio State University, Columbus, OH, USA

John W. Gillespie Jr. Center for Composite Materials, University of Delaware, Newark, DE, USA

Department of Mechanical Engineering, University of Delaware, Newark, DE, USA

Department of Materials Science and Engineering, University of Delaware, Newark, DE, USA

Department of Civil and Environmental Engineering, University of Delaware, Newark, DE, USA

Michael D. Grapes Department of Materials Science and Engineering, Johns Hopkins University, Baltimore, MD, USA

Thomas Graves National Security Technologies, LLC, New Mexico Operations, Los Alamos, NM, USA

C. Allan Gunnarsson US Army Research Laboratory, Aberdeen Proving Ground, Aberdeen, MD, USA

Sachin Gupta Dynamic Photo Mechanics Laboratory, Department of Mechanical, Industrial and Systems Engineering, University of Rhode Island, Kingston, RI, USA

Bazle Z. (Gama) Haque Center for Composite Materials, University of Delaware, Newark, DE, USA

Cameron Hawkins National Security Technologies, LLC, North Las Vegas, NV, USA

Michael Heika National Security Technologies, LLC, North Las Vegas, NV, USA

Mark D. Hindenlang U.S. Army Research Laboratory, Aberdeen Proving Ground, Aberdeen, MD, USA

Robert Hixson National Security Technologies, LLC, New Mexico Operations, Los Alamos, NM, USA

Zhenxing Hu Department of Mechanical Engineering, The University of Texas at Dallas, Richardson, TX, USA

Emily Huskins Department of Mechanical Engineering, United States Naval Academy, Annapolis, MD, USA

Magued Iskander Tandon School of Engineering, New York University, Brooklyn, NY, USA

Douglas Jahnke Department of Mechanical Engineering, The City College of New York, New York, NY, USA

P. Jannotti U.S. Army Research Laboratory, Aberdeen Proving Ground, Aberdeen, MD, USA

Richard Jennings University of Nevada, Las Vegas, Las Vegas, NV, USA

L.J. Kecskes U. S. Army Research Laboratory, Aberdeen Proving Ground, Aberdeen, MD, USA

Michael Keefe Center for Composite Materials, University of Delaware, Newark, DE, USA

Department of Mechanical Engineering, University of Delaware, Newark, DE, USA

Christopher T. Key Applied Physical Sciences, Groton, CT, USA

Addis Kidane Department of Mechanical Engineering, University of South Carolina, Columbia, SC, USA

J.H. Kim Materials Measurement Laboratory, National Institute of Standards and Technology, Gaithersburg, MD, USA

Jamie Kimberley New Mexico Institute of Mining and Technology, Socorro, NM, USA

Energetic Materials Testing and Research Center, New Mexico Institute of Mining and Technology, Socorro, NM, USA

E.M. Klier RDRL-WMM-F, Aberdeen Proving Ground, Aberdeen, MD, USA

Daniel B. Knorr Jr. U.S. Army Research Laboratory, Aberdeen Proving Ground, Aberdeen, MD, USA

Behrad Koohbor Department of Mechanical Engineering, University of South Carolina, Columbia, SC, USA

Veli-Tapani Kuokkala Department of Materials Science, Tampere University of Technology, Tampere, Finland

James M. LeBlanc Naval Undersea Warfare Center, Division Newport, Newport, RI, USA

Joseph L. Lenhart U.S. Army Research Laboratory, Aberdeen Proving Ground, Aberdeen, MD, USA

Guojing Li Department of Mechanical Engineering, Michigan State University, East Lansing, MI, USA

Tian Li Materials Science Division, Lawrence Livermore National Laboratory, Livermore, CA, USA

Jonathan P. Ligda Weapons and Materials Research Directorate, Army Research Laboratory, Aberdeen Proving Ground, Aberdeen, MD, USA

Dahsin Liu Department of Mechanical Engineering, Michigan State University, East Lansing, MI, USA

Jeffrey T. Lloyd Weapons and Materials Research Directorate, Army Research Lab, Aberdeen Proving Ground, Aberdeen, MD, USA

Tyler R. Long U.S. Army Research Laboratory, Aberdeen Proving Ground, Aberdeen, MD, USA

Nicholas J. Lorenzo Weapons and Materials Research Directorate, Army Research Laboratory, Aberdeen Proving Ground, Aberdeen, MD, USA

Hongbing Lu Department of Mechanical Engineering, The University of Texas at Dallas, Richardson, TX, USA

Wei-Yang Lu Sandia National Laboratories, Livermore, CA, USA

Huiyang Luo Department of Mechanical Engineering, The University of Texas at Dallas, Richardson, TX, USA

Kevin A. Masser U.S. Army Research Laboratory, Aberdeen Proving Ground, Aberdeen, MD, USA

S.P. Mates Materials Measurement Laboratory, National Institute of Standards and Technology, Gaithersburg, MD, USA

Helio Matos Dynamic Photo Mechanics Laboratory, Department of Mechanical, Industrial and Systems Engineering, University of Rhode Island, Kingston, RI, USA

Melissa Matthes University of Nevada, Las Vegas, Las Vegas, NV, USA

Christopher S. Meredith Weapons and Materials Research Directorate, Army Research Lab, Aberdeen Proving Ground, Aberdeen, MD, USA

Christopher S. Meyer U.S. Army Research Laboratory, ATTN: RDRL-WML-H, Proving Ground, Aberdeen, MD, USA

Paul Moy Army Research Laboratory, Aberdeen Proving Ground, Aberdeen, MD, USA

John A. O'Neill Army Research Laboratory, Aberdeen Proving Ground, Aberdeen, MD, USA

Brendan O'Toole University of Nevada, Las Vegas, Las Vegas, NV, USA

S.B. Park Department of Mechanical Engineering, State University of New York at Binghamton, Binghamton, NY, USA

Ronak Patel Department of Mechanical Engineering, University of South Carolina, Columbia, SC, USA

Thomas Plaisted Weapons and Materials Research Directorate, U.S. Army Research Laboratory, Aberdeen Proving Ground, Aberdeen, MD, USA

G. Portemont Onera—The French Aerospace Lab, Lille, France

V.B.S. Rajendra Prasad Vasavi College of Engineering, Ibrahimbagh, Hyderabad, Telangana, India

Vahid Azadeh Ranjbar Department of Mechanical Engineering, The City College of New York, New York, NY, USA

G. Venkata Rao Vasavi College of Engineering, Ibrahimbagh, Hyderabad, Telangana, India

R.L. Rhorer Engineering Laboratory, National Institute of Standards and Technology, Gaithersburg, MD, USA

Adam D. Richardson U.S. Army Research Laboratory, Aberdeen Proving Ground, Aberdeen, MD, USA

Shawoon Roy University of Nevada, Las Vegas, Las Vegas, NV, USA

Brett Sanborn Oak Ridge Institute for Science and Education, Oak Ridge, TN, USA

Sandia National Laboratory, Albuquerque, NM, USA

Emil Sandoz-Rosado U.S. Army Research Laboratory, Materials and Manufacturing Sciences Division, Aberdeen Proving Ground, Aberdeen, MD, USA

T. Sano U. S. Army Research Laboratory, Aberdeen Proving Ground, Aberdeen, MD, USA

Melissa K. Santala Materials Science Division, Lawrence Livermore National Laboratory, Livermore, CA, USA

Dan Schleh Liuman Technologies, Lansing, MI, USA

J. Schneider SAFRAN Snecma, Moissy-Crammayel, France

Brian E. Schuster Weapons and Materials Research Directorate, Army Research Laboratory, Aberdeen Proving Ground, Aberdeen, MD, USA

Jeremy D. Seidt Scott Laboratory, Department of Mechanical and Aerospace Engineering, The Ohio State University, Columbus, OH, USA

M. Shafiq Department of Mechanical and Aerospace Engineering, University of Florida, Gainesville, FL, USA

C. Shillings Naval Undersea Warfare Center, Division Newport, Newport, RI, USA

Arun Shukla Dynamic Photo Mechanics Laboratory, Department of Mechanical, Industrial and Systems Engineering, University of Rhode Island, Kingston, RI, USA

Charandeep Singh Department of Mechanical Engineering, State University of New York at Binghamton, Binghamton, NY, USA

Timothy W. Sirk U.S. Army Research Laboratory, Aberdeen Proving Ground, Aberdeen, MD, USA

Jarrod L. Smith Scott Laboratory, Department of Mechanical Engineering, The Ohio State University, Columbus, OH, USA

Subramani Sockalingam Center for Composite Materials, University of Delaware, Newark, DE, USA

William A. Spurgeon U.S. Army Research Laboratory, Aberdeen Proving Ground, Aberdeen, MD, USA

G. Subhash Department of Mechanical and Aerospace Engineering, University of Florida, Gainesville, FL, USA

Eduardo Suescun-Florez Tandon School of Engineering, New York University, Brooklyn, NY, USA

Balamurugan M. Sundaram Department of Mechanical Engineering, Auburn University, Auburn, AL, USA

Hareesh V. Tippur Department of Mechanical Engineering, Auburn University, Auburn, AL, USA

Mohamed Trabia University of Nevada, Las Vegas, Las Vegas, NV, USA

N. Tran SAFRAN Snecma, Moissy-Crammayel, France

Thomas Voisin Department of Materials Science and Engineering, Johns Hopkins University, Baltimore, MD, USA

T.R. Walter U. S. Army Research Laboratory, Aberdeen Proving Ground, Aberdeen, MD, USA

Tusit Weerasooriya US Army Research Laboratory, Aberdeen Proving Ground, Aberdeen, MD, USA

Timothy P. Weihs Department of Materials Science and Engineering, Johns Hopkins University, Baltimore, MD, USA

Eric D. Wetzel U.S. Army Research Laboratory, Materials and Manufacturing Sciences Division, Aberdeen Proving Ground, Aberdeen, MD, USA

C.L. Williams U. S. Army Research Laboratory, Aberdeen Proving Ground, Aberdeen, MD, USA

Tingge Xu Department of Mechanical Engineering, The University of Texas at Dallas, Richardson, TX, USA

Peiyu Yang Scott Laboratory, Department of Mechanical and Aerospace Engineering, The Ohio State University, Columbus, OH, USA

Jian H. Yu U.S. Army Research Laboratory, Aberdeen Proving Ground, Aberdeen, MD, USA

Yong Zhang Department of Mechanical Engineering, Johns Hopkins University, Baltimore, MD, USA

J.L. Zinszner Soils Solids Structures Risks (3SR) Laboratory, Grenoble-Alps University, Grenoble Cedex 9, France

CEA/DAM/GRAMAT, Gramat, France

Chapter 1
Atomistic Simulation of a Two-Dimensional Polymer Tougher Than Graphene

Emil Sandoz-Rosado, Todd D. Beaudet, Radhakrishnan Balu, and Eric D. Wetzel

Abstract A graphene/polyethylene hybrid 2D polymer, "graphylene", exhibits a higher theoretical fracture toughness than graphene, while remaining 2× stiffer and 9× stronger than Kevlar®, per mass. Within the base structure of graphylene, the sp^3-bonded polyethylene linkages provide compliance for ductile fracture, while the benzene rings contribute to high stiffness and strength. Combining stiff and compliant units to achieve enhanced mechanical performance demonstrates the potential of designing 2D materials at the molecular level.

1.1 Introduction

The extraordinary in-plane stiffness and intrinsic strength of graphene [1] in its pristine state have made it a desirable candidate as a structural material. Chemical vapor deposition of large-area graphene has been refined [2] to the point that grain boundaries of graphene approach the breaking strength of perfect crystalline graphene [3], a phenomenon that has been supported by atomistic simulations [4]. Graphene has the theoretical potential to enable ballistic barriers that have 10–100× less weight than barriers composed of Kevlar with the same ballistic limit [5], and has also demonstrated a specific kinetic energy of penetration an order of magnitude greater than steel and 2–3× greater than Kevlar, as measured by microscale ballistic experiments [6]. However, because graphene is a network of very stiff sp^2 bonds, it is highly resistant to fracture initiation but, once formed, a crack will propagate in a brittle manner [7, 8]. This brittle behavior may limit graphene's potential as a structural engineering material, as local failure due to a flaw or stress concentration is likely to trigger a sudden and catastrophic global failure.

To demonstrate a two-dimensional (2D) material with a more ductile fracture response compared to graphene, we propose a new family of 2D polymer which we refer to as "graphylene." This 2D covalent polymer network can be conceptually described as a graphene/polyethylene hybrid comprising benzene rings linked by short polyethylene chains. These short polyethylene links give graphylene in-plane stiffness and strength values that are somewhat lower than graphene. However, we demonstrate that the flexibility of the sp^3 bonded carbon atoms in the polyethylene chains leads to ductile fracture propagation behavior, with significantly higher energy required to propagate cracks relative to graphene.

2D polymers in the form of hexagonal carbon rings connected by linear carbon links have been recently described. Graphyne [9] and related allotropes [10] are composed strictly of extremely stiff carbon–carbon double and triple bonds, likely leading to brittle behavior. Graphane [11] adds single hydrogen bonds to each carbon atom in graphene, resulting in a hexagonal network of sp^3 bonds. Studies have also examined carbon allotropes that are randomly hydrogen functionalized [12]. Stiffness and strength in these graphene-like polymers have been incompletely reported, while fracture has not been directly studied in any of these systems. The polyethylene links in the graphylene 2D polymer structure should add compliance mechanisms that are not present in graphene, graphyne, graphane, or other 2D carbon allotropes, allowing for local ductility and greater flaw tolerance.

E. Sandoz-Rosado • T.D. Beaudet • R. Balu • E.D. Wetzel (✉)
U.S. Army Research Laboratory, Materials and Manufacturing Sciences Division,
Bldg. 4600, Aberdeen Proving Ground, Aberdeen, MD 21005, USA
e-mail: eric.d.wetzel2.civ@mail.mil

D. Casem et al. (eds.), *Dynamic Behavior of Materials, Volume 1*, Conference Proceedings
of the Society for Experimental Mechanics Series, DOI 10.1007/978-3-319-41132-3_1

1.2 Establishing the Structure of Graphylene

We have broadly considered the graphylene-*n* (GrE-*n*) family of 2D polymers, where *n* indicates the characteristic length of the polyethylene (PE) chains in terms of the number of methylene bridge units (–CH_2–) between each nearest neighbor C_6 ring. A complete description of the various GrE-*n* configurations considered is provided in the Supplemental Information, with the results summarized as follows. Since the system is composed of more rigid graphene units (C_6 rings) and softer polyethylene $(CH_2)_n$ units, we hypothesize that global stiffness and strength will reduce, while fracture toughness will increase, with increasing *n*. Simulations of various GrE-*n* polymers revealed that odd *n* configurations were more difficult to solve for converged stable states compared to even *n* configurations, primarily due to symmetry challenges that required larger unit cells and computational domain sizes. Therefore, although odd *n* graphylene systems are likely to be physically realizable, only even *n* systems were considered for further study. Of the possible $n = 2, 4, 6$, etc. configurations, GrE-2 polymers were selected as our system of interest as the closest comparison to graphene.

1.3 Elastic Modulii of Graphylene

First-principles DFT uniaxial stress calculations for both the first and second C_6 nearest-neighbor (1NN and 2NN, respectively) directions of GrE-2 were performed to determine its strength and stiffness, and to benchmark subsequent molecular dynamics (MD) predictions.

The elastic modulii were extracted from this stress–strain response during uniaxial tensile testing according to the following relationship:

$$\sigma_{2D} = E_{2D}\epsilon + D_{2D}\epsilon^2 \tag{1.1}$$

where σ_{2D} is stress, ϵ is strain, and E_{2D} and D_{2D} are the first and second order elastic modulii respectively [5]. Because graphylene is being studied as an atomically-thin sheet, the stress and moduli values are expressed in units of [N/m]. In-plane elastic modulii are calculated by curve fitting the DFT data with Eq. (1.1) to find E_{2D} and D_{2D} for the 1NN and 2NN lattice directions.

Analogous uniaxial tensile calculations were performed using classical MD modeling via the LAMMPS software package, which will also be used for subsequent fracture studies. A single sheet of GrE-2 with dimensions 21.5 nm $\times$ 25.5 nm (Fig. 1.1a) held at 0 K was stretched by fixing one edge and displacing the opposing edge at a constant rate of 0.1 nm/ps for a strain rate of 0.5 ns^{-1} which is of the same order as strain rates for other graphene fracture simulation studies [13, 14]. To achieve uniaxial stress, the boundaries normal to the direction of stretching are periodic, but permitted to relax (Fig. 1.1a).

In-plane elastic modulii, and stress and strain to failure determined by DFT are reported in Table 1.1 for both directions of crystalline symmetry. Figure 1.1b shows the comparison between DFT and MD for tensile simulations. The results show that graphylene-2 is somewhat anisotropic, with the 1NN and 2NN directions having linear elastic modulii values of $E_{2D} = 78.9$ N/m and $E_{2D} = 97.6$ N/m respectively. The linear elastic modulus values are over three times smaller than that of pristine graphene, which is reported to be 340 N/m by numerous theoretical and experimental studies [5], but is still over 2$\times$ and 5$\times$ higher, per mass, than high performance engineering materials such as Kevlar and titanium, respectively [5] (based on an initial GrE-2 areal density of 6.53×10^{-7} kg/m^2). The non-linear stress–strain response in GrE-2 manifests at smaller strains than graphene, making it a nonlinear elastic material.

The elastic response of GrE-2 predicted by the DFT and MD simulations are in good agreement (Fig. 1.1a), indicating that the REBO potential is suitable for modeling graphylene polymers. However, the ultimate stress to failure of GrE-2 predicted by MD is significantly lower than that predicted by DFT. This We attribute the discrepancy in failure points to the interaction cutoff imposed on the REBO potential [4, 7, 11] to prevent aphysical carbon-carbon scissioning that is a result of computational artifacts in the potential's switching function [15]. While the MD simulations of GrE-2 may fail prematurely they will not contain any of the strain-hardening and high breaking strengths typical to REBO simulations without cutoffs. Therefore, the following MD simulations characterizing fracture toughness can be considered conservative, and that true intrinsic toughness values of graphylene may be even higher than our predictions.

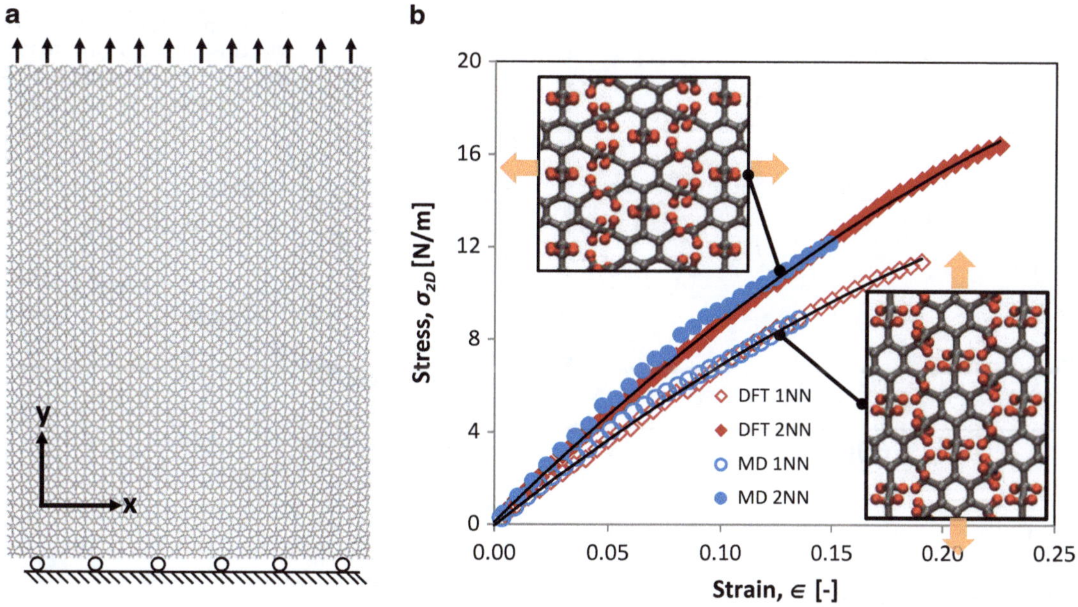

Fig. 1.1 (**a**) MD domain and boundary conditions for modeling GrE-2 under uniaxial tension. Hydrogen atoms are not depicted for ease of viewing. (**b**) Predicted nominal stress vs. nominal strain for uniaxial tension simulations of GrE-2 to failure using both DFT and MD in 1NN and 2NN directions. Inlays depict deformed lattices (from MD simulations) in both directions at 12.5 % strain. Solid lines depict curve fits of Eq. (1.1) to the DFT data

Table 1.1 In-mechanical properties of GrE-2 determined from DFT

	E_{2D} [N/m]	D_{2D} [N/m]	σ_f [N/m]	ϵ_f [−]
First NN	78.9	−94.7	11.4	0.19
Second NN	97.6	−107.1	16.4	0.23

First and second nearest neighbor directions are denoted as 1NN and 2NN respectively

1.4 Initiation of Mode-I Crack Growth in GrE-2 Graphylene

The intrinsic ability of a material to resist the initiation of crack growth from a pre-existing crack is quantified in terms of fracture toughness. There are three metrics we use to quantify the fracture behavior of graphylene: (i) critical fracture energy (quasi-static), (ii) flaw-tolerance (quasi-static) and (iii) fracture energy release rate (dynamic). To characterize the critical fracture energy of graphylene, we performed MD simulations of fracture of a domain of GrE-2 with a pre-existing crack and benchmarked it with graphene under the same conditions. The crack domain, depicted in Fig. 1.2a, is 60 nm by 30 nm which corresponds to an aspect ratio acceptable for modeling crack growth [16] and matches the dimensions of previous MD simulations of graphene fracture [7]. The domain was permitted to relax in the direction perpendicular to strain.

A pre-crack of length a_0 was created in the GrE-2 and graphene domains and strain was applied at a constant rate in the direction perpendicular to the pre-crack. This configuration is consistent with a mode-I crack, the results of which can be seen in Fig. 1.2b. The crack width was 0.3 nm and the crack tip radii were 0.15 nm for consistency. To quantify fracture toughness we adopt a 2D formulation of the quasi-static critical fracture energy of a Griffith crack which is formulated assuming linear elasticity appropriate for the small (less than 5 %) strain in our crack models:

$$G_{Ic,2D} = \frac{K_{Ic,2D}^2}{E_{2D}} = \frac{\sigma_{c,2D}^2}{2E_{2D}} \pi a_0 \tag{1.2}$$

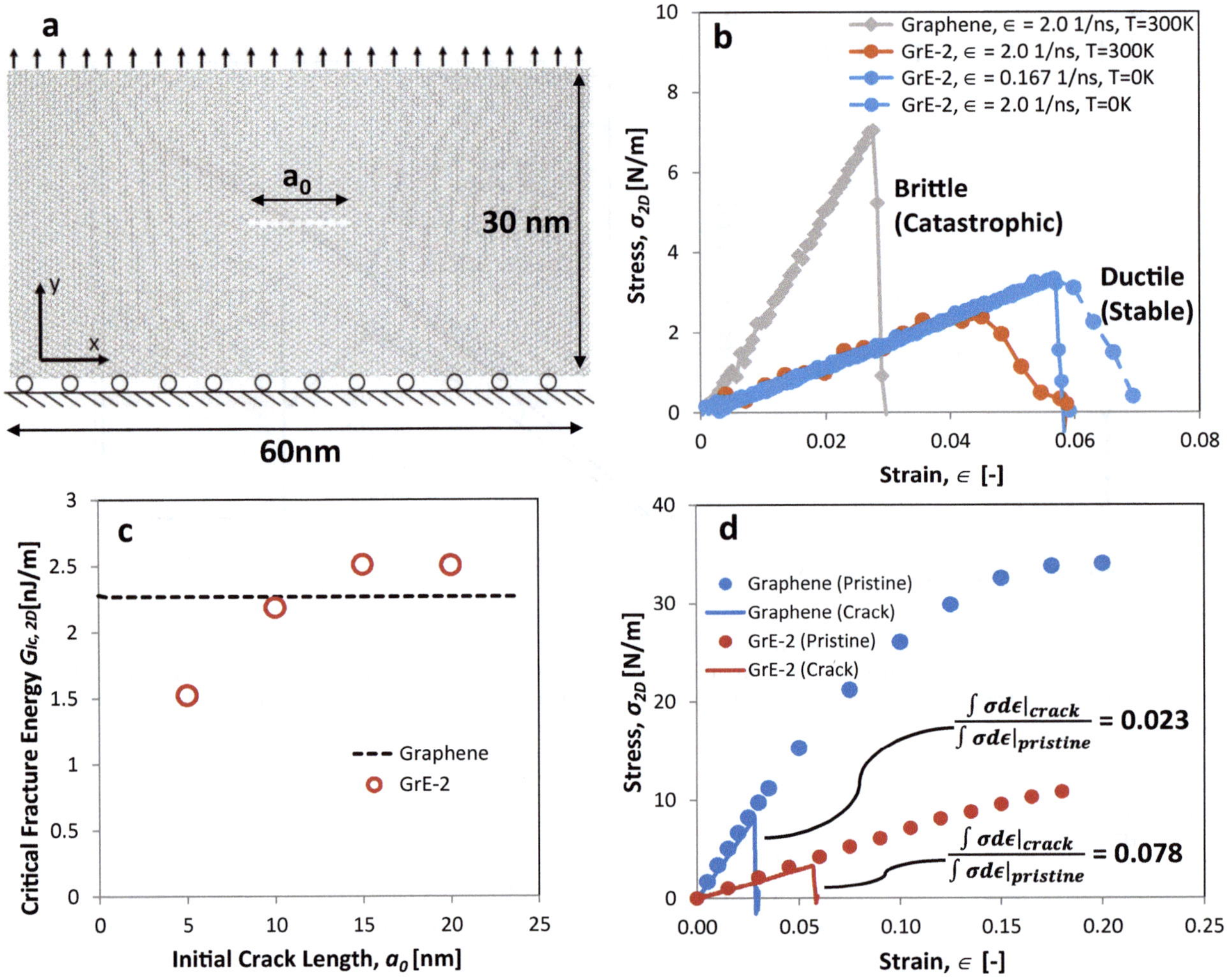

Fig. 1.2 (**a**) Crack domain of GrE-2 to examine fracture toughness, (**b**) nominal stress vs. nominal strain of graphene and a GrE-2 Griffith crack propagating perpendicular to the 1NN direction at various conditions ($a_0 = 10$ nm), (**c**) the critical fracture energy of graphene and GrE-2 at varying initial crack lengths initialized at 0 K, and (**d**) comparison in stress–strain response between pristine and cracked graphene and GrE-2 at 0 K demonstrating superior flaw tolerance of GrE-2

where $G_{Ic,2D}$ is the 2D mode-I critical fracture energy, $K_{Ic,2D}$ is the critical 2D mode-I stress concentration factor and $\sigma_{c,2D}$ is the 2D critical fracture stress. The critical fracture energy was first determined for graphene with a pre-crack length of a_0 of 10 nm, resulting in a value of $G_{Ic,2D} = 2.28$ nJ/m, in agreement with the critical fracture energy of graphene reported by a previous study, $G_{Ic,2D} = 2.33$ nJ/m [7]. This value was found to be constant across all pre-crack lengths for the same domain dimensions, a signature of brittle material response. A parametric study of pre-crack length was then performed for GrE-2, and the critical fracture energy can be seen in Fig. 1.2c compared to that of graphene. Immediately it is clear that GrE-2, unlike graphene, has a critical fracture energy that is dependent on pre-crack length, a signature of ductile fracture response. Furthermore, the critical fracture energy of GrE-2, ($G_{Ic,2D} = 2.51$ nJ/m at $a_0 = 20$ nm) rapidly exceeds that of graphene ($G_{Ic,2D} = 2.28$ nJ/m across all pre-crack lengths) as pre-crack length increases. While GrE-2 shows initial crack growth at a lower critical stress than graphene, this initiation event occurs at a significantly higher strain than graphene.

Another metric for flaw tolerance is the ratio of the strain energy of the flawed material relative to the strain energy of the pristine material [17]. The simulation results show that GrE-2 exhibits three times the value of this metric compared to graphene (Fig. 1.2d). In other words, pre-existing defects in graphene will deteriorate the toughness of the material twice as much as identical pre-existing defects in graphylene.

1.5 Conclusions

The simulations in this study suggest that GrE-2 graphylene exhibits features consistent with ductile fracture propagation, in contrast to the brittle fracture that has been predicted and observed for graphene. Combined with its very high stiffness and strength relative to current engineering materials, graphylene possesses a unique set of mechanical properties that could enable engineering structures with unprecedented performance. Although not presented here, simulations of the transport and electronic properties of graphylene are also underway.

The present GrE-2 graphylene polymer is a useful model system, in particular considering the maturity of DFT and MD bond potential functions for the comprising benzene and polyethylene subcomponents of the structure. However, a nearly limitless range of 2D polymers can be imagined, providing ample opportunity for further study and improvement in properties. The practical realization of these materials will require parallel mechanical modeling and synthesis efforts to identify systems with useful properties that can also be readily fabricated.

References

1. Lee, C., et al.: Measurements of the elastic properties and intrinsic strength of monolayer graphene. Science **321**(5887), 385–388 (2008)
2. Petrone, N., et al.: Chemical vapor deposition-derived graphene with electrical performance of exfoliated graphene. Nano Lett. **12**(6), 2751–2756 (2012)
3. Lee, G.-H., et al.: High-strength chemical-vapor–deposited graphene and grain boundaries. Science **340**(6136), 1073–1076 (2013)
4. Grantab, R., Shenoy, V.B., Ruoff, R.S.: Anomalous strength characteristics of tilt grain boundaries in graphene. Science **330**(6006), 946–948 (2010) (Copyright 2010, The Institution of Engineering and Technology)
5. Wetzel, E.D., Balu, R., Beaudet, T.D.: A theoretical consideration of the ballistic response of continuous graphene membranes. J. Mech. Phys. Solids **82**, 23–31 (2015)
6. Lee, J.-H., et al.: Dynamic mechanical behavior of multilayer graphene via supersonic projectile penetration. Science **346**(6213), 1092–1096 (2014)
7. Zhang, P., et al.: Fracture toughness of graphene. Nat. Commun. **5**, 3782 (2014)
8. Hwangbo, Y., et al.: Fracture characteristics of monolayer CVD-graphene. Sci. Rep. **4**, 4439 (2014)
9. Cranford, S.W., Buehler, M.J.: Mechanical properties of graphyne. Carbon **49**(13), 4111–4121 (2011)
10. Enyashin, A.N., Ivanovskii, A.L.: Graphene allotropes. Phys. Status Solidi B **248**(8), 1879–1883 (2011)
11. Pei, Q., Zhang, Y., Shenoy, V.: A molecular dynamics study of the mechanical properties of hydrogen functionalized graphene. Carbon **48**(3), 898–904 (2010)
12. Li, Y., et al.: Mechanical properties of hydrogen functionalized graphene allotropes. Comput. Mater. Sci. **83**, 212–216 (2014)
13. Zhang, Z., Wang, X., Lee, J.D.: An atomistic methodology of energy release rate for graphene at nanoscale. J. Appl. Phys. **115**(11), 114314 (2014)
14. Le, M.-Q., Batra, R.C.: Single-edge crack growth in graphene sheets under tension. Comput. Mater. Sci. **69**, 381–388 (2013)
15. Pastewka, L., et al.: Describing bond-breaking processes by reactive potentials: importance of an environment-dependent interaction range. Phys. Rev. B **78**(16), 161402 (2008)
16. Buehler, M.J., Gao, H.: Modeling dynamic fracture using large-scale atomistic simulations. In: Shukla, A. (ed.) Dynamic Fracture Mechanics, p. 1. World Scientific, Singapore, (2006)
17. Markus, J.B., et al.: Cracking and adhesion at small scales: atomistic and continuum studies of flaw tolerant nanostructures. Model. Simul. Mater. Sci. Eng. **14**(5), 799 (2006)

Chapter 2
Transverse Compression Response of Ultra-High Molecular Weight Polyethylene Single Fibers

Subramani Sockalingam, John W. Gillespie Jr., Michael Keefe, Dan Casem, and Tusit Weerasooriya

Abstract This work reports on the experimental quasi static transverse compression response of ultra-high molecular weight polyethylene (UHMWPE) Dyneema SK76 single fibers. The experimental nominal stress-strain response of single fibers exhibits nonlinear inelastic behavior under transverse compression with negligible strain recovery during unloading. Scanning electron microscopy (SEM) reveals the presence of significant voids along the length of the virgin and compressed fibers. The inelastic behavior is attributed to the microstructural damage within the fiber. The compressed fiber cross sectional area is found to increase to a maximum of 1.83 times the original area at 46 % applied nominal strains. The true stress strain behavior is determined by removing the geometric nonlinearity due to the growing contact area. The transverse compression experiments serve as validation experiments for fibril-length scale models.

Keywords UHMWPE • Ballistic impact • Transverse compression • Finite element analysis (FEA) • Constitutive model

2.1 Introduction

Ultra-high molecular weight polyethylene (UHMWPE) fibers are used in personnel protection ballistic impact applications [1] in the form flexible textile fabrics and laminated composites. UHMWPE fiber is made up of extremely long chains of polyethylene (monomer unit >250,000 per molecule) with a hierarchy of sizes and exhibits a fibrillar structure. Macro-fibrils [2] consist of bundles of micro-fibrils which in turn are composed of bundles of nano-fibrils. These fibers lend themselves to such applications due to their superior specific axial tensile strength and specific modulus. The fibers experience multi-axial loading [3] including axial tension, axial compression, transverse compression and transverse shear during impact. While axial specific toughness and longitudinal wave speed are important fiber properties contributing to the ballistic performance [4], the role of transverse properties and multi-axial loading during impact is not well understood. In this work we investigate the quasi static (QS) transverse deformation behavior of UHMWPE Dyneema SK76 single fibers.

S. Sockalingam (✉)
Center for Composite Materials, University of Delaware, Newark, DE, USA
e-mail: sockalsi@udel.edu

J.W. Gillespie Jr.
Center for Composite Materials, University of Delaware, Newark, DE, USA

Department of Mechanical Engineering, University of Delaware, Newark, DE, USA

Department of Materials Science and Engineering, University of Delaware, Newark, DE, USA

Department of Civil and Environmental Engineering, University of Delaware, Newark, DE, USA

M. Keefe
Center for Composite Materials, University of Delaware, Newark, DE, USA

Department of Mechanical Engineering, University of Delaware, Newark, DE, USA

D. Casem • T. Weerasooriya
Army Research Laboratory, Aberdeen, MD, USA

D. Casem et al. (eds.), *Dynamic Behavior of Materials, Volume 1*, Conference Proceedings of the Society for Experimental Mechanics Series, DOI 10.1007/978-3-319-41132-3_2

Fig. 2.1 Schematic of single fiber transverse compression

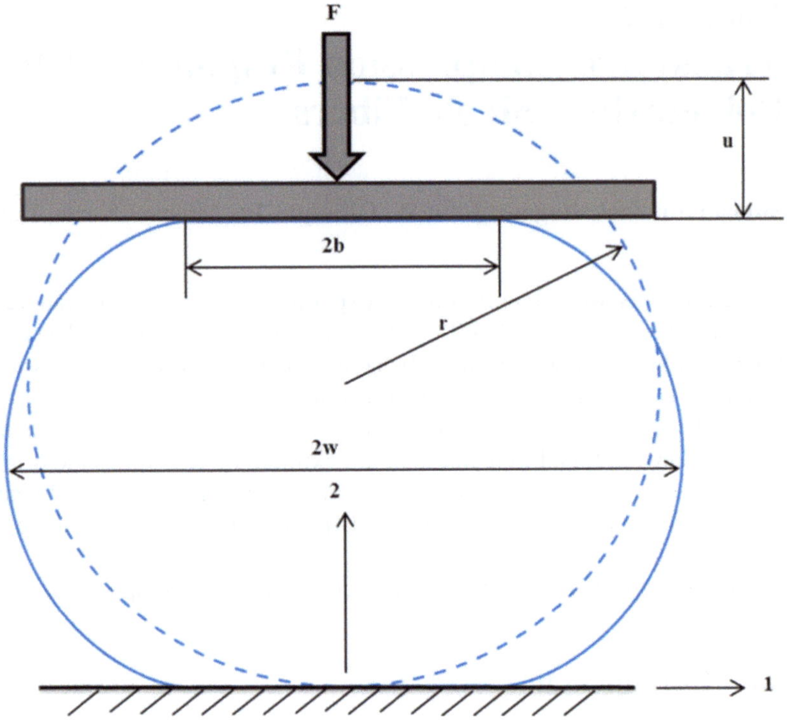

2.2 Experimental Set up

The QS experimental set up involves compressing a single fiber between rigid parallel platens (Sapphire substrates) in plane strain conditions as shown in the schematic in Fig. 2.1. The compressive load per unit length (F), platen displacement (u) and compressed width ($2w$) of the fiber are measured in real time. Contact width ($2b$) is measured post-test at different applied load levels. The fibers are compressed at 0.1 μm/s (average strain rate of 0.0059 s^{-1}). A more detailed explanation of the experimental set up is reported in [5].

2.3 Results and Discussion

The QS experimental nominal stress $\overline{\sigma} = \frac{F}{d}$ nominal strain $\overline{\varepsilon} = \frac{u}{d}$ (d is undeformed fiber diameter) response due to monotonic and cyclic loading is shown in Fig. 2.2. The fibers exhibit a nonlinear inelastic response. The nonlinearity is attributed to both geometric stiffening due to growing contact area and material softening. After each load-unload cycle residual strains are measured.

Figure 2.3 shows scanning electron microscopy (SEM) images of fibers subjected to different levels of maximum nominal strains. The undeformed fiber shown in Fig. 2.3a indicates the presence of significant void like features on the surface and along the length of the fiber. Void like features are also observed on the surface of the compressed fibers.

The normalized contact width (b/r) and compressed width (w/r) growth is shown in Fig. 2.4. At maximum load the compressed width has increased by a factor of 4.5. It is also seen both 2b and 2w plateau at higher load levels to approximately four and a half fiber diameters. Negligible elastic recovery or reduction in width growth is observed during unloading.

The elastic modulus in transverse compression is determined using an analytical solution based on the Hertzian contact for a transversely isotropic fiber given by eqs. 2.1 [6] and 2.2 [7]. The experimental measurements are fitted into eqs. 2.1 and 2.2 to as shown in Fig. 2.5a using the properties in Table 2.1(v31 and v12 in Table 2.1 are assumed values). The analytical solution compares well with the experimental results until about 2 % nominal strains where material softening and damage may be on setting. A transverse modulus of 2.37 GPa and a Hertzian elastic limit of 2 % is determined.

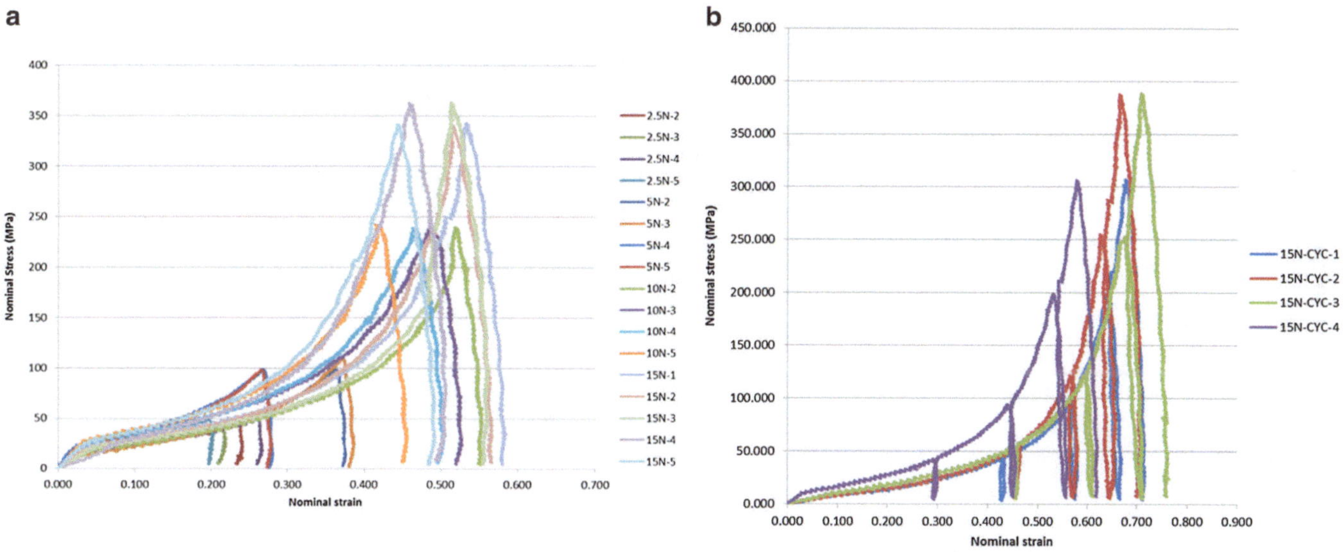

Fig. 2.2 QS experimental (**a**) one load unload cycle on different fibers (**b**) multiple load unload cycle on the same fiber

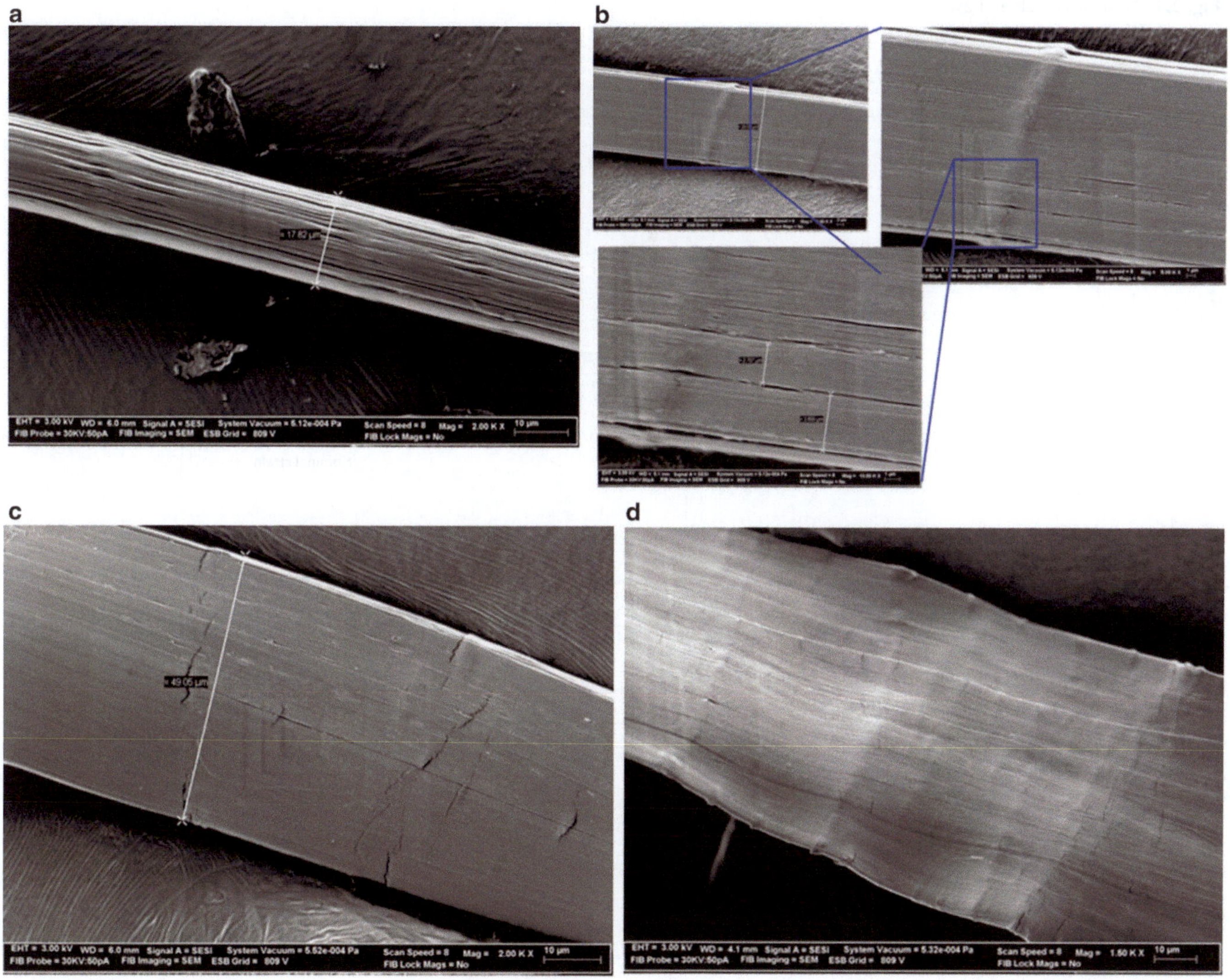

Fig. 2.3 SEM images of deformed fibers at nominal strain levels (**a**) 0 % (**b**) 22 % (**c**) 51 % (**d**) 71 %

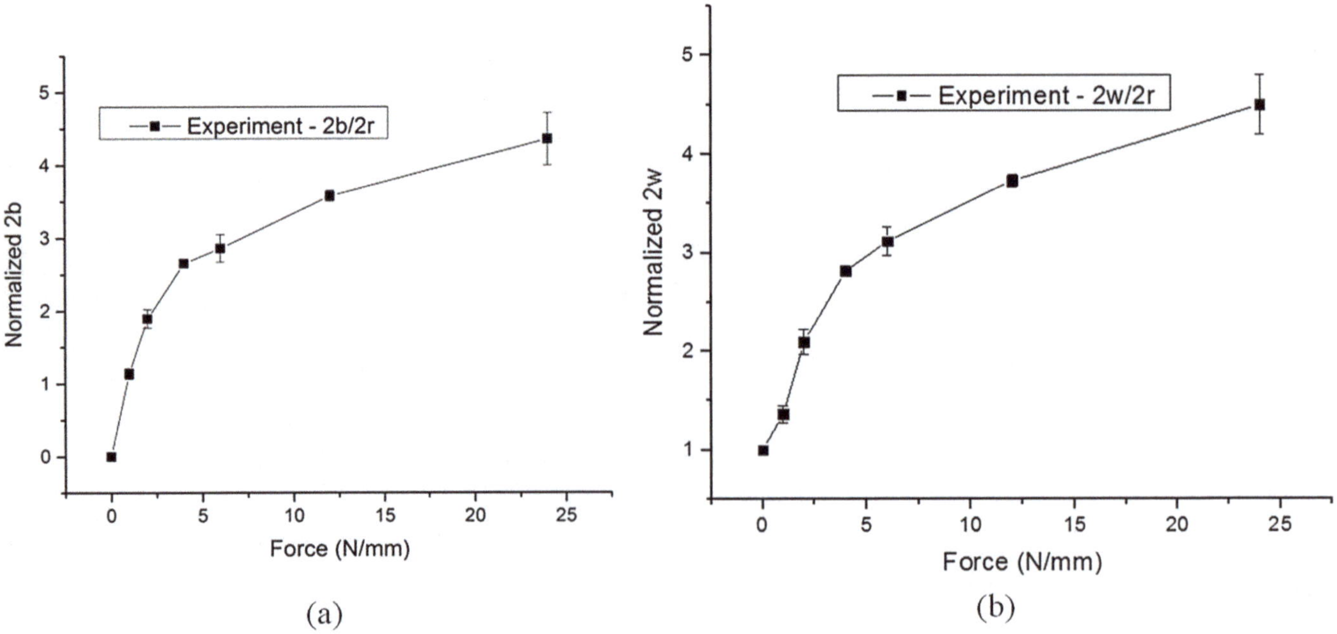

Fig. 2.4 Normalized 2b and 2w

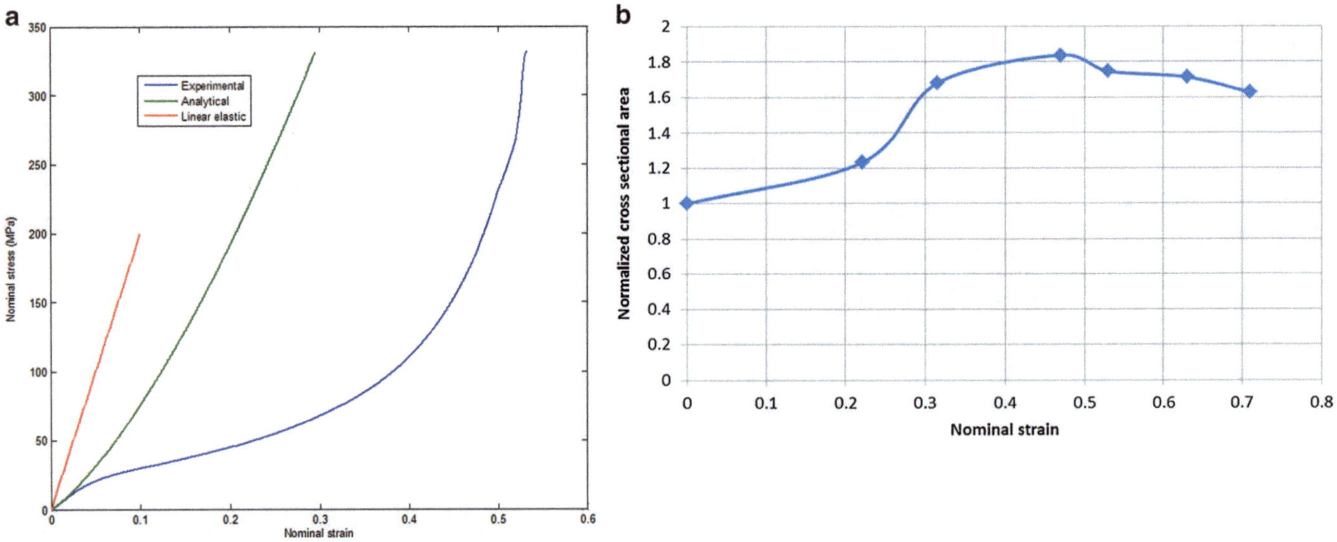

Fig. 2.5 (**a**) Nominal stress strain (**b**) normalized cross sectional area

Table 2.1 Transversely isotropic properties of Dyneema SK76

ρ (g/cm^3)	D (μm)	E_3 (GPa)	ν_{31}	ν_{12}	E_1 (GPa)	Hertzian elastic limit	True elastic limit
0.97	17.0	116.0	0.60	0.40	2.37	2 %	0.5 %

$$u = \frac{4F}{\pi}\left(s_{11} - \frac{s_{13}^2}{s_{33}}\right)\left[0.19 + \sinh^{-1}\left({}^{r}\!/_{b}\right)\right] \tag{2.1}$$

$$b^2 = \frac{4Fr}{\pi}\left(\frac{1}{E_1} - \frac{\nu_{31}^2}{E_3}\right) \tag{2.2}$$

where,
F—load per unit length along the fiber longitudinal direction
u—platen displacement
L—length of the compressed fiber
r—radius of the fiber
b—contact half width
E_1—Young's modulus in 1–2 plane of transverse isotropy
E_3—longitudinal Young's modulus in the fiber direction
w—compressed half width

$$s_{11} = \frac{1}{E_1} \quad s_{12} = -\frac{\nu_{12}}{E_1} \quad s_{13} = -\frac{\nu_{13}}{E_1} \quad s_{33} = \frac{1}{E_3}$$

The intrinsic material behavior is isolated from the total response by defining true stress as $\sigma = \frac{F}{2w_{eff}}$ where the effective width $w_{eff} = \frac{w+b}{2}$. The original fiber cross section is assumed to be a circle and the compressed fiber cross section is assumed to be a rectangle. The compressed cross sectional area is estimated using $2w_{eff} \times (2r - u)$. The normalized compressed apparent cross sectional area is found to increase to a maximum of 1.83 times the original area at 46 % nominal strain and then plateau at higher strain levels as shown in Fig. 2.5b. A more detailed study is required to understand the increase in cross sectional area.

The true strain is obtained by equating the internal energy to the external work done given by eq. 2.3. The true strains are computed incrementally using trapezoidal rule to evaluate the integrals in eq. 2.3.

$$\int_{V_0}^{V}\int_{0}^{\varepsilon_{eff}} \sigma_{eff}\, d\varepsilon_{eff}\, dV = \int_{0}^{u} F\, du \tag{2.3}$$

The true stress strain response obtained by imposing a constant volume along with a linear elastic behavior with a stiffness of 2.37 GPa is shown in Fig. 2.6. The true elastic limit of 0.5 % is identified as the point at which the instantaneous stiffness is lower than the elastic stiffness of 2.37 GPa.

The single fiber transverse compression experiment is modeled using a quarter symmetric finite element (FE) model shown in Fig. 2.6b in the commercial FE code LS-DYNA. The fiber is modeled as a nonlinear inelastic continuum material with a user defined constitutive model [5] UMAT. The yield stress-effective plastic strain required as input to the model is determined by subtracting the elastic portion of the true stress-strain curve (Fig. 2.6a). The UMAT force displacement predictions are much better correlation to the experimental results compared to a linear elastic behavior with stiffness of 2.37 GPa. However the model under predicts contact area compared to the experiments as shown in Fig. 2.6d. This deviation may be attributed to the increase in fiber cross sectional area observed in the experiments which is a topic for future studies.

2.4 Conclusions

This paper presented the quasi-static transverse compression response of UHMWPE Dyneema SK76 ballistic fibers. The fibers exhibit nonlinear inelastic behavior under large compressive strains. The true stress strain behavior of the fiber is determined by removing the geometric nonlinearity using the measured contact area. The true elastic limit of these fibers in transverse compression is found to be 0.5 % and the Hertzian elastic limit is found to be 2 %. The numerical FE model

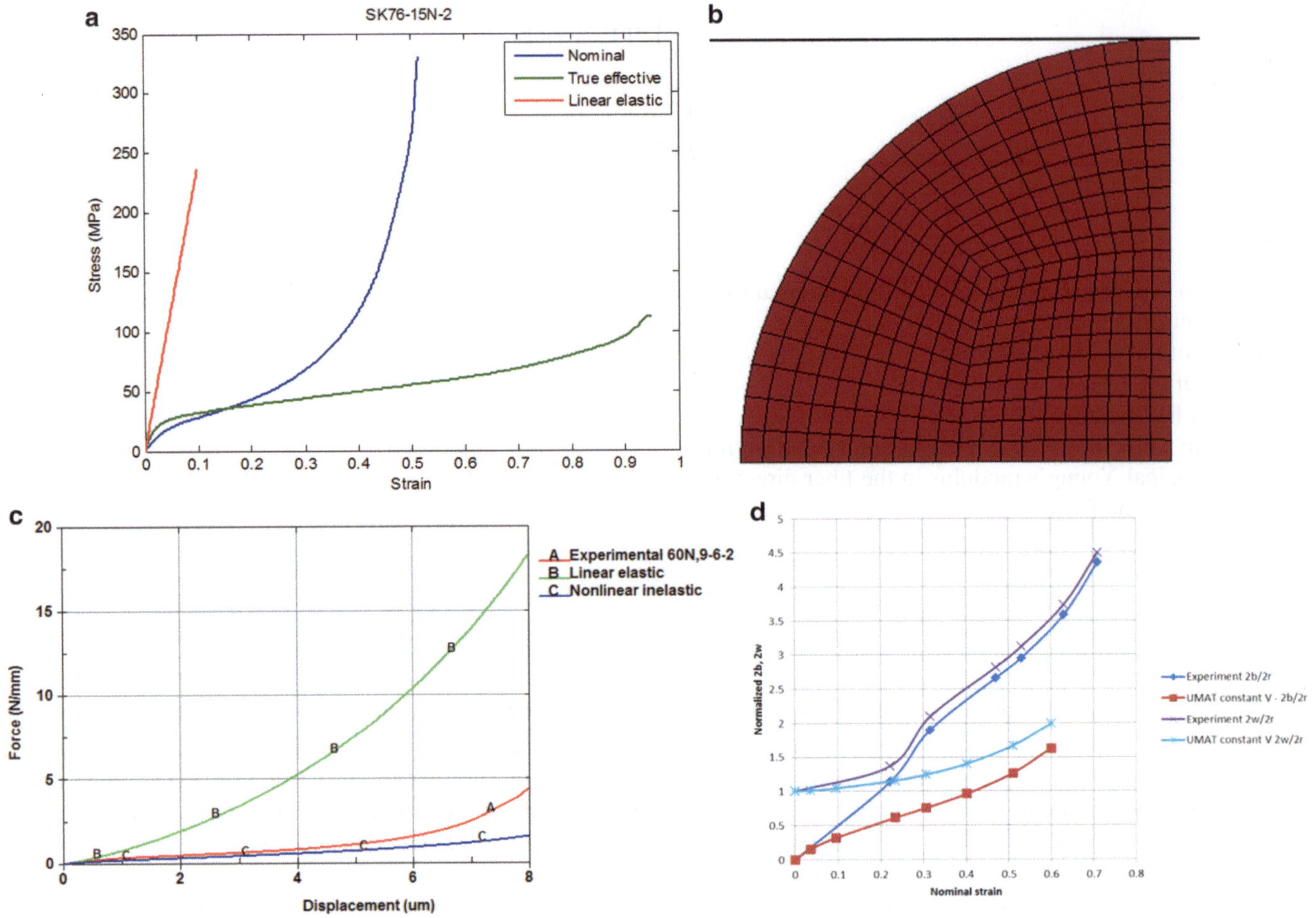

Fig. 2.6 (**a**) True stress strain behavior (**b**) FE model (**c**) comparison of model and experiments (**d**) normalized contact and compressed width

predictions using true stress-strain curve is found to be in a much better correlation to the experimental force displacement response compared to a linear elastic assumption. Both virgin and compressed fibers are found to possess void like features on the fiber surface. Further investigation is required to better understand the increase in apparent fiber cross sectional area during transverse compression.

Acknowledgements Research was sponsored by the Army Research Laboratory and was accomplished under Cooperative Agreement Number W911NF-12-2-0022. The views and conclusions contained in this document are those of the authors and should not be interpreted as representing the official policies, either expressed or implied, of the Army Research Laboratory or the U.S. Government. The U.S. Government is authorized to reproduce and distribute reprints for Government purposes notwithstanding any copyright notation herein.

References

1. Krishnan, K., Sockalingam, S., Bansal, S., Rajan, S.: Numerical simulation of ceramic composite armor subjected to ballistic impact. Compos. Part B **41**(8), 583–593 (2010)
2. McDaniel, P.B., Deitzel, J.M., Gillespie, J.W.: Structural hierarchy and surface morphology of highly drawn ultra high molecular weight polyethylene fibers studied by atomic force microscopy and wide angle X-ray diffraction. Polymer **69**, 148–158 (2015)
3. Sockalingam, S., Gillespie Jr., J.W., Keefe, M.: Dynamic modeling of Kevlar KM2 single fiber subjected to transverse impact. Int. J. Solids Struct. **67–68**, 297–310 (2015)
4. Cunniff, P.M.: Dimensionless parameters for optimization of textile-based body armor systems. In: Proceedings of the 18th International Symposium on Ballistics, November, San Antonio, pp. 1303–1310. Technomic, Lancaster, PA (1999)

5. Sockalingam, S., Bremble, R., Gillespie Jr., J.W., Keefe, M.: Transverse compression behavior of Kevlar KM2 single fiber. Compos. A: Appl. Sci. Manuf. **81**, 271–281 (2016)
6. Jawad, S.A., Ward, I.M.: The transverse compression of oriented nylon and polyethylene extrudates. J. Mater. Sci. **13**(7), 1381–1387 (1978)
7. Hadley, D.W., Ward, I.M., Ward, J.: The transverse compression of anisotropic fibre monofilaments. Proc. R. Soc. Lond. A. Math. Phys. Sci. **285**(1401), 275–286 (1965)

Chapter 3
Morphology and Mechanics of the Young Minipig Cranium

Stephen Alexander, C. Allan Gunnarsson, Ann Mae DiLeonardi, and Tusit Weerasooriya

Abstract The Göttingen miniature pig is a useful surrogate to understand mechanisms of traumatic brain injury (TBI) in the human. However, the mechanical response of the minipig skull has not been previously reported. In this study, cranial samples were extracted from the skulls of adolescent minipigs (six months of age, average weight of 13.8 kg). The microstructure was first characterized using high-resolution μCT. A highly gradient structure was observed, with the bone volume fraction (BVF) almost doubling in the through-thickness (depth) dimension. These specimens were then mechanically loaded in quasi-static compression. The surface strain distribution along the loading direction was measured during the experiments using digital image correlation (DIC). Depth-dependent moduli were derived from the measured DIC strains rather than machine displacement, due to the large gradient in morphology. An elasticity-morphology relationship from literature was extended to represent the modulus variation in the functionally gradient skull structure (BVF), by calibrating the relationship with the experimentally derived local moduli. The model enables the prediction of local moduli based solely on the morphological parameter BVF measured with μCT, and also provided an estimation of the modulus of the bony phase of the cranium.

Keywords Cranial bone mechanical properties • Fabric tensor • Mean intercept length • Minipig skull morphology • Skull mechanics

3.1 Introduction

The Göttingen minipig is used as a surrogate in impact experiments designed to better understand the mechanisms by which mechanical loading induces traumatic brain injury (TBI). However, cranial mechanics and morphology must be understood in order to accurately scale any quantitative results, such as injury thresholds, from non-human TBI experiments to the human anatomy.

Both the mechanical response and the morphology of the skull have been previously studied for the human [1, 2] and large-breed pig [3]. Mature human cranial bone is a sandwich structure composed of three layers spanning the through-thickness axis, from the brain-most side of the skull to the skin-most side. The outer and inner tables are composed of dense cortical bone, while the middle layer, the diploë, is porous trabecular bone. On the other hand, studies have reported a random orientation of bone across the plane transverse to the through-thickness axis [4]. Importantly, skull morphology changes dramatically during maturation, developing from a single layer of dense bone [3].

However, the size and porosity of the minipig skull is known to differ significantly from that of larger breeds [5], requiring independent characterization. In this study, micro-computed tomography (μCT) was used to measure spatial variation in porosity and to quantify the orientation of the bone phase within the structure. Specimens were also loaded at a quasi-static rate in the through-thickness (depth) direction, normal to the outer surface of the skull. The surface strain field of the specimen was captured during compression and was used to derive depth-dependent moduli. These moduli were related to the morphology by optimizing parameters of a mechanical model. The tissue modulus of pure bone was also approximated as part of the modeling procedure.

S. Alexander (✉)
TKC Global Solutions LLC, 13873 Park Center Drive, Herndon, VA 20171, USA
e-mail: stephen.l.alexander18.ctr@mail.mil

C.A. Gunnarsson • A.M. DiLeonardi • T. Weerasooriya
US Army Research Laboratory, Aberdeen Proving Ground, Aberdeen, MD 21005, USA

D. Casem et al. (eds.), *Dynamic Behavior of Materials, Volume 1*, Conference Proceedings of the Society for Experimental Mechanics Series, DOI 10.1007/978-3-319-41132-3_3

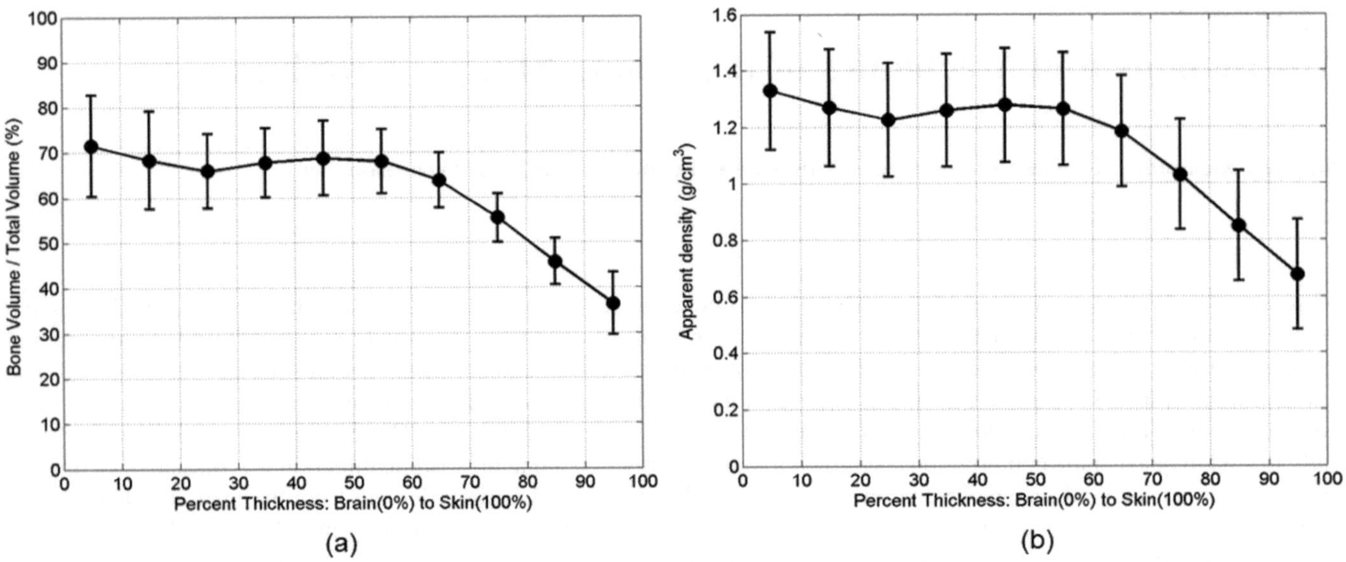

Fig. 3.1 The average bone volume fraction (**a**) and apparent density (**b**) plotted across the through-thickness axis. Percent thickness ranges from the brain-most side of the bone at 0 % to the skin-most side at 100 %. Error bars are standard deviations

3.2 Specimen Collection and Morphological Analysis

Cubical specimens (coupons), of side length 4–6 mm, were extracted from the skulls of three adolescent Göttingen minipigs (average age and weight: 26.8 weeks, 13.9 kg), following procedures previously described [6]. A total of 14 coupon specimens were imaged prior to mechanical testing using a μCT scanner (Skyscan 1172, Bruker microCT) at 62 kV and 161 mA. The size of the isotropic voxel ranged from 2.82 to 2.95 μm. The μCT image stack sampled the thickness (depth) dimension of the skull. Each image represented a cross-sectional slice of the specimen, transverse to the depth dimension and with a thickness of one voxel. The image plane will also be referred to as the transverse plane.

Morphological analysis was performed with the CTAn software, provided by the CT manufacturer. The ratio of bone area to total area of each image (depth) was assigned as the value of the bone volume fraction (f_{bv}) at that depth. The apparent density ρ_a was calculated by multiplying the bone volume fraction by the average density of pure bone, ρ_b, $\rho_a = f_{bv}\rho_b$. Figure 3.1 shows the change of bone volume fraction and apparent density as a function of depth.

The spatial orientation of the bone phase within each specimen was quantified from the μCT data using the mean intercept length (MIL) tensor [7]. In general, the principal axes of the MIL tensor correspond to the principal axes of the bone phase, averaged over the volume of interest (VOI). The degree of anisotropy (DA) was calculated as the ratio of the largest to the smallest eigenvalues of the MIL tensor.

The orientation analysis was performed on a sub-population of four specimens. The average bone orientation was measured by calculating the MIL tensor for the largest possible spherical VOI which could be extracted from the specimen. The resulting DA for the four specimens ranged from 1.20 to 1.45, much lower than the DA reported for bones from other anatomical sites. For example, Mulder et al. reported the human calcaneus to have a DA of 1.99 $\pm$ 0.29 [8]. The comparatively low DA supports the approximation of the bone phase of the skull specimens as lacking a dominant orientation.

3.3 Compression Loading and Mechanical Parameters

Unconfined, quasi-static compressive loading was applied to a sub-population of four cranial specimens (coupons). Compression was in the depth direction at an average nominal strain rate of 0.001/s. One of the surfaces perpendicular to the loading platen was speckled for digital image correlation (DIC), and a CCD camera captured the deformation of the speckled plane during loading.

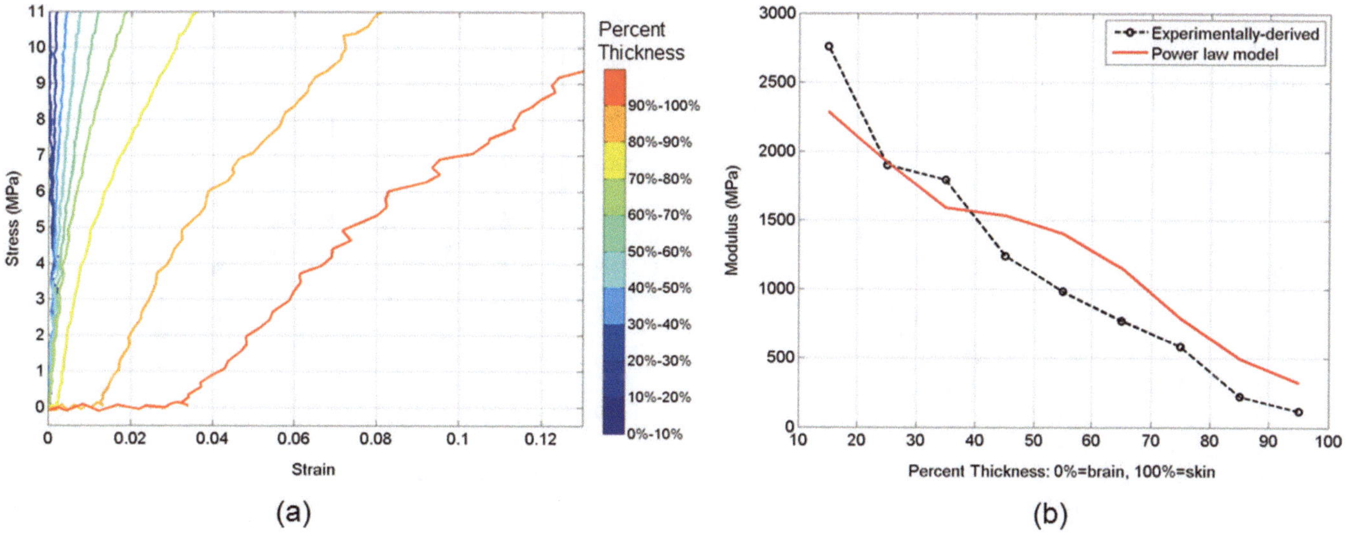

Fig. 3.2 An example of localized mechanical properties for Spec A. (**a**) The stress-strain response of each layer of Spec A, as calculated from the experimental results. Line colors correspond to the thickness percentage of the layer, as shown in the legend on right, in which a thickness of 0 % is the brain-most side and 100 % is the skin-most side. (**b**) The modulus of each layer calculated from the localized stress-strain response, together with model results

The depth-dependent stress-strain response was calculated from the experimental results of the four specimens. For each specimen, the depth dimension was divided into ten layers of equal thickness. The specimen was idealized as a parallelepiped composed of the ten layers stacked in series. The strain field over the speckled plane was calculated for each time point of the experiment. The average strain of a specific layer, ϵ_{yy}^{r}, was defined as the average strain over the specific $x - y$ space of the layer. The compressive stress of each layer as a function of time was calculated using the iso-stress approximation. Under this assumption, the stress in each layer was obtained directly from far-field measurements (load cell). Figure 3.2a shows an example stress-strain response for each of the layers of Spec A. The elastic moduli of each layer, E^{r}, were calculated from a linear regression of the initial portion of the layer's stress-strain response. These moduli were taken at layer-specific strain values roughly between $0.001 < \epsilon_{yy}^{r} < 0.01$ (0.1– 1% strain). Those layers for which an initial, linear stress-strain response was not discernible were not included in further analysis.

3.4 Relating Morphology and Mechanics

The extent to which the morphology of the layer could predict the layer's experimentally-derived modulus was investigated using data from the compression experiments. All morphological (e.g. density) and mechanical (e.g. strain) fields were assumed to be constant within each layer. The experimentally-measured modulus, E^{r}, of each layer, r, within a given specimen was assumed to be related to the tissue modulus of pure bone, E_0, scaled by a factor of the bone volume fraction of the layer,

$$E^{r} = E_0 (f_{bv}^{r})^{k}. \tag{3.1}$$

In Eq. 3.1, ${f_{bv}}^{r}$, refers to the average bone volume fraction over the spatial area of the layer r. The relationship between the apparent modulus of the porous structure and the modulus of pure bone, scaled by a power of the bone volume fraction, has been well-established [9]. However, these power-law relationships have primarily been used to relate the far-field modulus to the specimen-averaged bone volume fraction. The present methodology extends this method to calculate localized moduli within each specimen, based on spatial changes in morphology.

Equation 3.1 requires two parameters to be optimized, E_0 and the scaling parameter, k. They were optimized by fitting to the layer-specific moduli calculated from the experimental results, and the resulting relationship was

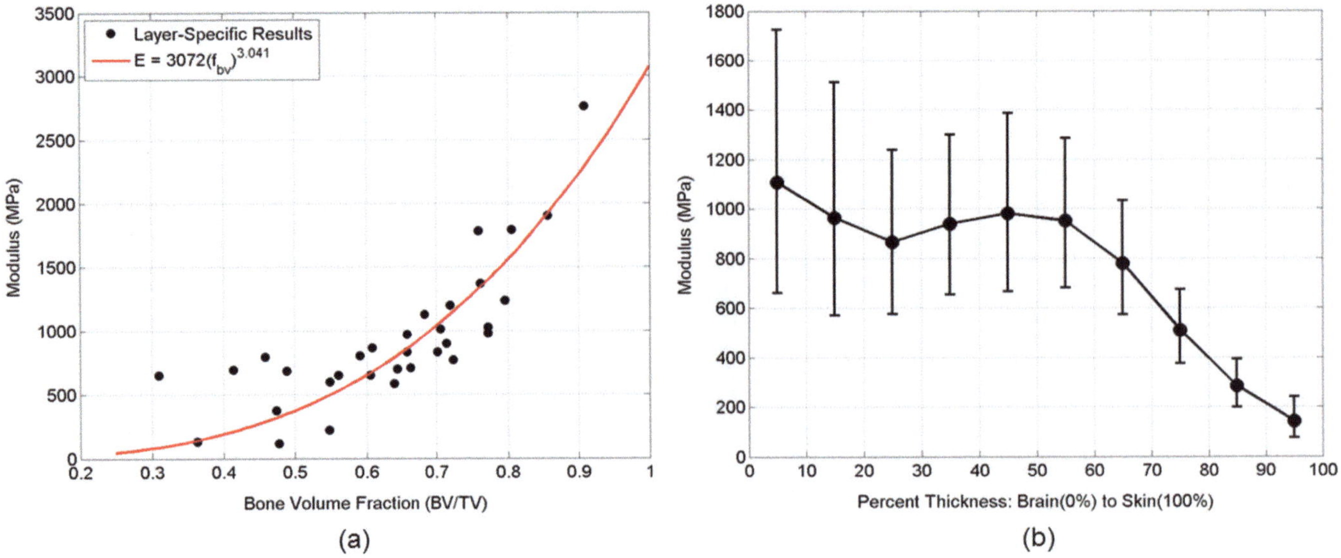

Fig. 3.3 Model results and prediction. (**a**) The modulus of each layer, calculated from the compression experiments, is plotted as a function of the layer's bone volume fraction, as measured from the μCT results. The red line is the model fit. (**b**) Model prediction for the change of modulus as a function of depth for all 14 specimens. Error bars indicate standard deviation of the bone volume fraction of each layer (Fig. 3.1a)

$$E = 3072\left(f_{bv}^{r}\right)^{3.041}. \tag{3.2}$$

Figure 3.2b shows an example of the model fit for one of the specimens, Spec A. Figure 3.3a shows the data from all specimens used in optimization, plotting the modulus of each of the layers as a function of the layer's averaged bone volume fraction.

After optimization, Eq. 3.2 was then applied to the average change of bone volume fraction as calculated for all 14 specimens, previously shown in Fig. 3.1b. Figure 3.3b shows the resulting prediction of the change of modulus with depth. The brain-surface portion of the bone was predicted to be, on average, approximately five times stiffer than the skin-surface.

3.5 Conclusions

The cranial bone morphology of the adolescent Göttingen minipig was quantified using μCT imaging with a resolution on the order of 3 μm. The volume fraction of the bone and its orientation within the skull were calculated from the images. Adolescent minipig skull specimens had denser bone near the brain, which gradually transitioned to more porous bone near the skin. This arrangement is the inverse of what has been previously reported in immature large-breed pigs and humans [3].

The compressive mechanical response of cranial bone was also measured. A relationship between localized bone volume fraction and modulus was developed using a small subset of specimens. One of the fundamental assumptions of the modeling technique was the absence of significant mechanical anisotropy. A preliminary investigation of the extent to which the structure was randomly arranged relied on MIL analysis of the μCT dataset and indicated a low degree of structural anisotropy.

The results of the isotropic model were applied to the larger group of specimens to predict the modulus-depth dependence. The model parameter corresponding to the bone tissue modulus was found to be 3.1 GPa. Researchers have previously approximated the tissue modulus of trabecular (highly porous) bone as 12 GPa, based on the range of results from experimental studies which directly measured this modulus [1]. The slightly lower value found for the minipig in this study likely arises from age effects. The isotropic model scaled the tissue modulus by the value of the bone volume fraction with an exponential factor of 3.04. This value aligns with previous studies, which show the scaling parameter to range from 2 to 3 [2].

This study was limited in the assumptions employed in the mechanical model, the small number of specimens used in parameter optimization, and the lack of an independent validation study. Perhaps most importantly, the model did not account for the spatial variation of bone orientation. Further investigation of the influence of the structural anisotropy on the mechanical results, and any conclusions as to the presence of mechanical anisotropy, would require other methods which could capture local variations in the morphology with greater precision.

Disclaimer The research reported in this document was performed in connection with contract/instrument W911-QX-14-C0016 with the U.S. Army Research Laboratory. The views and conclusions contained in this document are those of TKC Global Inc. and the U.S. Army Research Laboratory. Citation of manufacturer's or trade names does not constitute an official endorsement or approval of the use thereof. The U.S. Government is authorized to reproduce and distribute reprints for Government purposes notwithstanding any copyright notation hereon.

References

1. Gibson, L.J., Ashby, M.F.: Cellular solids: Structure and properties. Cambridge University Press, Cambridge (1997)
2. Mow, V.C., Huiskes, R.: Basic Orthopaedic Biomechanics & Mechano-Biology. Lippincott Williams & Wilkins, Philadelphia (2005)
3. Margulies, S.S., Thibault, K.L.: Infant skull and suture properties: Measurements and implications for mechanisms of pediatric brain injury. J. Biomech. Eng. **122**(4), 364–371 (2000)
4. McElhaney, J.H., Fogle, J.L., Melvin, J.W., Haynes, R.R., Roberts, V.L., Alem, N.M.: Mechanical properties of cranial bone. J. Biomech. **3**(5), 495–511 (1970)
5. Sauleau, P., Lapouble, E., Val-Laillet, D., Malbert, C.H.: The pig model in brain imaging and neurosurgery. Animal **3**(8), 1138–1151 (2009)
6. DiLeonardi, A.M., Gunnarsson, C.A., Sanborn, B., Weerasooriya, T.: Orientation dependent mechanical response and quantification of cranial bone of the Göttingen pig. In Proceedings of the Conference: SEM 2015 Annual Conference & Exposition on Experimental & Applied Mechanics, Riverside, CA (2015)
7. Harrigan, T., Mann, R.: Characterization of microstructural anisotropy in orthotropic materials using a second rank tensor. J. Mater. Sci. **19**(3), 761–767 (1984)
8. Souzanchi, M.F., Palacio-Mancheno, P., Borisov, Y.A., Cardoso, L., Cowin, S.C.: Microarchitecture and bone quality in the human calcaneus: Local variations of fabric anisotropy. J. Bone Miner. Res. **27**(12), 2562–2572 (2012)
9. Helgason, B., Perilli, E., Schileo, E., Taddei, F., Brynjólfsson, S., Viceconti, M.: Mathematical relationships between bone density and mechanical properties: A literature review. Clin. Biomech. **23**(2), 135–146 (2008)

Chapter 4
Dynamic Characterization of Nitronic 30, 40 and 50 Series Stainless Steels by Numerical Analysis

C.G. Fountzoulas, E.M. Klier, and J.E. Catalano

Abstract Nitronic® stainless steels are austenitic, high-strength, corrosion-resistant products that provide higher-performance alternatives to many conventional 300 and 400 Series stainless steels. They provide excellent mechanical properties at sub-zero and elevated temperatures, impact resistance at low temperatures and, superb resistance to high- temperature oxidation while they retain good weldability and fabricability. The microstructure of Nitronic® 30, 40 and 50 series of 25 and 35 % cold worked, thermally aged at 750 C for 8 h, and annealed coupons was characterized by metallography, X-ray diffraction (XRD) and scanning electron microscopy. The coupons etching of each series needed specific attention and it resulted in an improved procedure. Nitronic® 50 steels remain non-magnetic even after extensive cold work. Nitronic 50 metallographic analysis showed the existence of cementite and enlarged grain boundaries after thermal ageing. The austenite, martensite and ε phases were recognized by XRD. The dynamic behavior of these alloys was studied by modeling and simulation using the commercial software ANSYS/AUTODYN and was compared to simulation results of projectile made of 4340 steel on known materials models. The correlation of the microstructure to the alloy properties and the dynamic behavior of these alloys will be presented in detail.

Keywords Nitronic 30, 40, 50 stainless steels • Dynamic characterization • Numerical analysis • Modeling

4.1 Introduction

Austenitic stainless steels, also known as γ-iron, are the most common of stainless steels. Austenitic stainless steels are non-magnetic and they are comprised of austenite, chromium and nickel. They are formable and weldable, and they can be successfully used in a wide range of temperatures, from cryogenic temperatures to the high temperatures of furnaces and jet engines. They are nonmagnetic because γ-iron is a microstructure formed at temperatures higher than 727 °C, the eutectoid reaction temperature. Austenite can be produced at room temperature by increasing the amount of chromium from 16 to 25 %. Moreover, nitrogen in solution contributes to their high corrosion resistance. Addition of nickel helps the stabilizing of their austenitic structure and it would be used even more widely, if it were not for the cost of nickel [1].

Ferritic steels contain less than 0.10 % carbon and they are magnetic. The fact that they cannot be hardened via heat treatment and do not weld to a high standard limits the use of these metals somewhat, but they are still suitable for a wide range of applications.

Austenitic steels are the most common type of stainless steel, accounting for up to 70 % of all stainless steel production. Its versatility is in largely due to the fact that it can be formed and welded with successful results.

Martensitic steels share some characteristics with ferritic, but boasts higher levels of carbon, up to a full 1 %. This means that they can be tempered and hardened and thus they are highly useful in situations where the strength of the steel is more important than its resistance to corrosion.

Duplex steels are a combination of ferritic and austenitic steels, a structure which renders duplex steel stronger than both. Addition of elements such as aluminum, copper and niobium make these steels extremely strong. They can be machined and

C.G. Fountzoulas (✉)
U.S. Army Research Laboratory, WMRD, RDRL-WMM-B, Aberdeen, MD, USA
e-mail: constantine.fountzoulas.civ@mail.mil

E.M. Klier • J.E. Catalano
RDRL-WMM-F, Aberdeen Proving Ground, Aberdeen, MD 21005-5069, USA

D. Casem et al. (eds.), *Dynamic Behavior of Materials, Volume 1*, Conference Proceedings of the Society for Experimental Mechanics Series, DOI 10.1007/978-3-319-41132-3_4

Table 4.1 Composition of Nitronic steels and 4340

	30	40	50	60	4340
Carbon	0.03 max.	0.08	0.06	0.10	0.37–0.43
Manganese	7.0–9.0	3–10.00	8	8	0.6–0.8
Phosphorus	0.040 max.		4.00–6.00		0.035
Sulfur	0.030 max.	0.030	0.030		0.04
Silicon	1.00 max.	1.00	1.00	4	0.15–0.30
Chromium	15.0–17.0	19–21.50	17	17	0.7–0.9
Nickel	1.5–3.0	5.50–7.50	11.50–13.50	8.5	1.65–2.00
Nitrogen	0.15–0.30	0.15–0.40	0.20–0.30	0.13	
Copper	1.00 max.				
Molybdenum			1.50–3.00		0.2–0.3
Columbium/Niobium			0.10–0.30		
Vanadium			0.10–0.30		

The remaining element if iron [3–7]

Table 4.2 Properties of Nitronic and 4340 steels [3–7]

	ρ		E		UTS		YS		El%	Hardness
Nitronic 30	0.284 lb/in^3	7.862 g/cm^3	28 × 10^3 Ksi	193 GPa	108 Ksi	749 MPa	49 Ksi	339 MPa	56	B89
Nitronic 40	0.285 lb/in^3	7.89 g/cm^3	28 × 10^3 Ksi	193 GPa	95 Ksi	655 × 10^3 MPa	48 Ksi	330 MPa	35	B100
Nitronic 50	0.285 lb/in^3	7.89 g/cm^3	28 × 10^3 Ksi	193 GPa	121 Ksi	834 MPa	57 Ksi	393 MPa	46	B91
4340	0.284 lb/in^3	7.86 g/cm^3	29 × 10^3 Ksi	200 GPa	108 Ksi	745 MPa	68.2 Ksi	470 MPa	22	B95

worked into a wide variety of shapes without becoming distorted and, in terms of corrosion, have the same resistance levels as austenitic steels [1].

Nitronic stainless steel alloys contain nitrogen and are completely austenitic. Nitronic steels do not become magnetic when they are cold worked. Nitronic stainless steel contains nitrogen, which makes it stronger and also more resistant to corrosion. Nickel is also a primary ingredient, improving resistance to corrosion and overall workability. Due to their composition, Nitronic steel is not only one of the strongest steel, but it is temperature resistant and continuously non-magnetic as well [2].

This versatility is reflected in the fact that there are actually over 150 different grades of stainless steel, with fifteen of them being the ones most commonly used. Table 4.1 shows the stoichiometry of Nitronic and 4340 steels. Table 4.2 shows a summary of mechanical properties of Nitronic steels and 4340 steel.

4.1.1 Grades of Nitronic Stainless Steel [2]

- *Nitronic 30* is highly resistant to both corrosion and abrasion and it is also resistant to impact, and as such it can be used for heavy-duty jobs. It is one of the toughest and most durable steels and it is used for coal buckets, ore separators, food processing equipment, sewage treatment, and truck or bus frames [3].
- *Nitronic 40*, like other Nitronic steels, remains austenitic and retains its toughness all the way down to—253 °C, making this type of stainless steel perfect for cryogenic applications. Therefore it is used in many different applications such as, for aircraft as ducting and bellows systems, fasteners, clamps, and flanges. It is also resistant to oxidation at high temperatures [4].
- *Nitronic 50* stainless steel is also highly resistant to corrosion and is also very strong. However, when annealed, it has nearly twice its normal yield strength. It is used for pumps, fasteners, cables, wire cloth, boat and marine hardware, springs, and photographic equipment [5].
- *Nitronic 60* is a high quality steel of lower cost than other alloys. Nitronic 60 stainless steel provides resistance to oxidation at high temperatures and resistance to impact at lower temperatures. It is ideal for valve seats, trim and stems, as well as nuts and bolts, screening, pins, roller bearings, and pump components, bridge pins and other fastening systems [6].

Table 4.3 Specimen details: cold rolling and heat treatment

Process	N30	N40	N50
Annealed	x		x
25 % cold work (CW)	x	x	x
35 % cold work (CW)	x		x
25 % CW-annealed @ 750 °C 8 h	x	x	x
35 % CW-annealed @ 750 °C 8 h	x		x
As-received, ½ hardness (1/2H)	x		

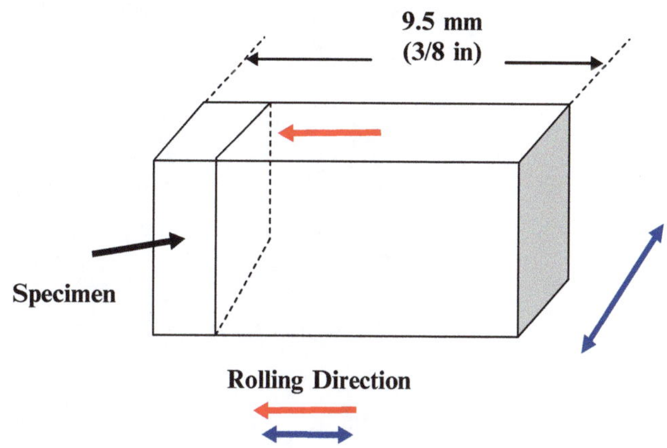

Fig. 4.1 Specimen preparation for metallographic observation

4.1.2 Experimental Details

Selected specimens of as –received and cold rolled Nitronic 30, 40 and 50 steels of full and 1/2H (hardness) were used for this study (Table 4.3). Specimens from the cold rolled Nitronic steels were sectioned in directions parallel and perpendicular to the rolling direction (Fig. 4.1). The etchants used for metallographic study were in-house developed chemical solutions. They were consisted of Aqua Regia and Glyceregia, a variation of Aqua Regia with 10 ml glycerol. The etching time was determined by trial error until the microstructure was clearly observed in the microscope. Figures 4.2–4.5 show microphotographs of all Nitronic steels examined using a metallographic microscope. The XRD analysis was conducted with the aid of a Rigaku Mini Flex II, Desktop Diffractometer using CuKα X-rays, $\lambda = 0.154$ nm (Fig. 4.6). Figure 4.7 shows the Austenitic steel XRD pattern after cold rolling [8].

4.2 Results

4.2.1 Modeling and Simulations

An initial numerical study of identical geometry projectiles consisted of Nitronic 30, 40, 50 and 4340 steels impacting same targets were conducted by 2D modeling and simulation using the commercial software package ANSYS/AUTODYN. The targets were consisted of borofloat glass strike faces, polyurethane, glass ceramic and polycarbonate. The only available in the AUTODYN library and open literature equation of state (EOS) and material failure and strength models was of the 4340 steel. As EOS and material failure and strength models of the Nitronic steels we used the existed in the library models of the 4340 steel modified according to their properties. For the Nitronic and 4340 steels linear equation EOS, Johnson Cook (JC) material strength model and no failure were used: polynomial EOS, and Johnson-Holmquist 2 (JH2) strength and failure models were used for borofloat glass and glass ceramic; shock EOS, piecewise JC and plastic strength and failure models were used for the polycarbonate; and linear and elastic and principal stress strength and failure material models were used for the polyurethane. All materials were discretized by SPH solver with particle size of 0.2 mm for increased modeling accuracy. The impact velocity was set to 914 m/s. The exit velocity of the Nitronic projectiles and the damage of the target

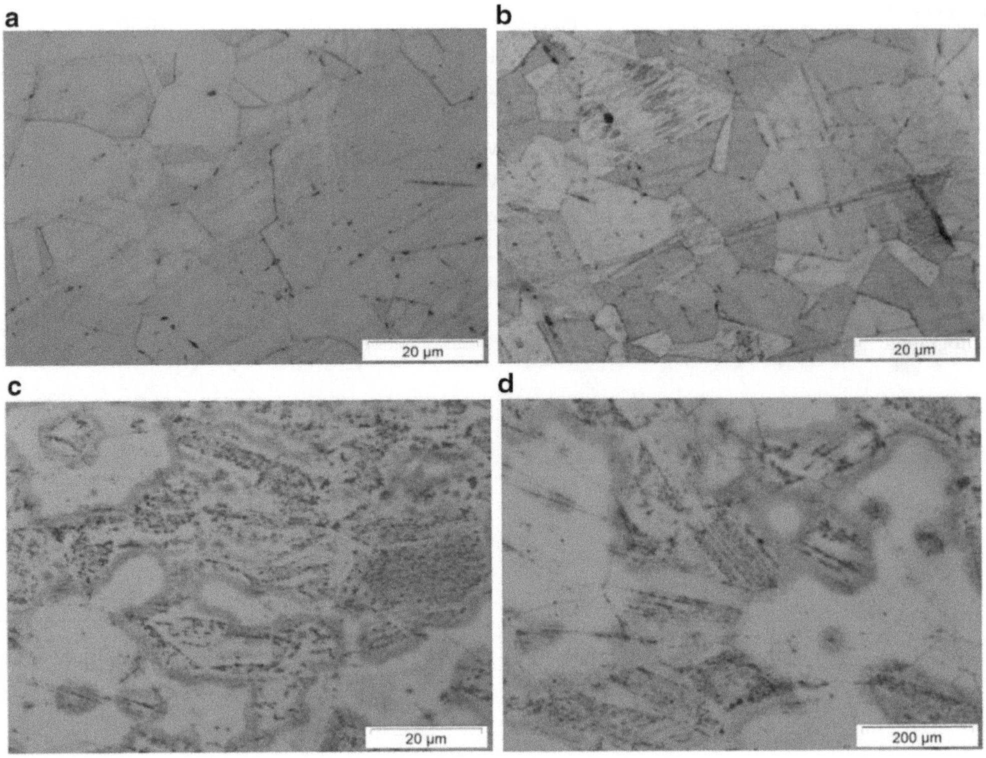

Fig. 4.2 Nitronic 30 metallography (**a**) Annealed; (**b**) Annealed 750 °C 8 h (parallel); (**c**) Annealed 750 °C 8 h (perpendicular) and (**d**) 25 % Cold Work (CW) (Parallel)

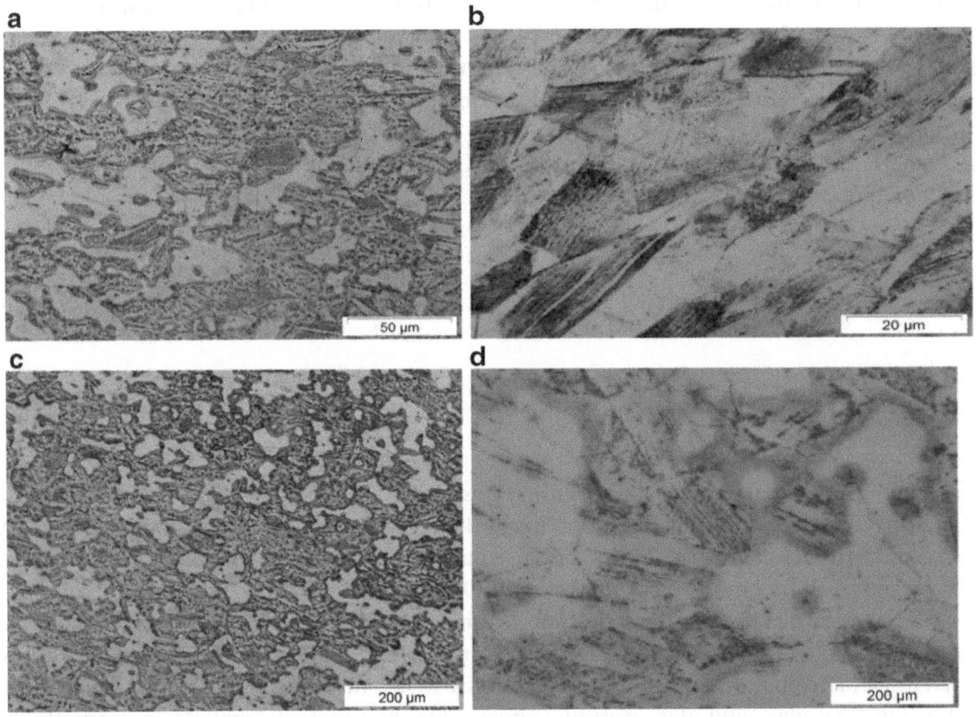

Fig. 4.3 Nitronic 30 metallography (**a**) 25 % CW—Annealed—750 °C-8 h (Perpendicular); (**b**) 25 % CW—Annealed—750 °C-8 h (Parallel); (**c**) 25 % CW (Perpendicular) and (**d**) 25 % CW (Parallel)

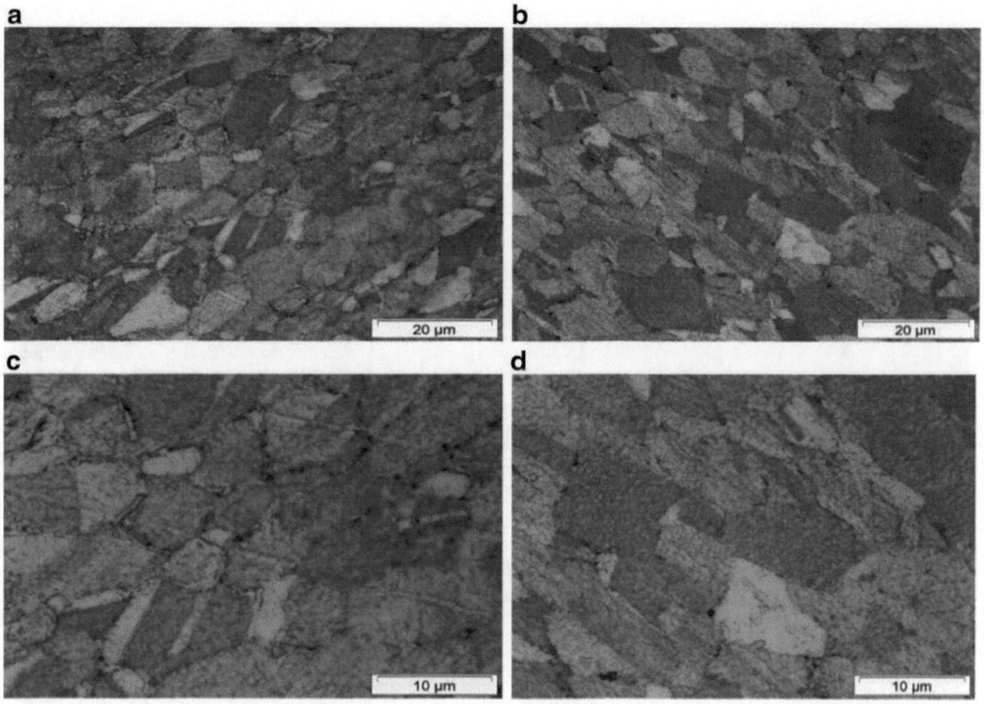

Fig. 4.4 Nitronic 30 metallography (**a**) 25 % CW—Annealed—700 °C-8 h (Perpendicular); (**b**) 25 % CW—Annealed—700 °C-8 h (Parallel); (**c**) 25 % CW—Annealed—700 °C-8 h (Perpendicular) and (**d**) 25 % CW—Annealed—700 °C-8 h (Parallel)

after 0.2 ms simulation were compared to the corresponding simulated results of the 4340 steel impact. The simulated overall damage of the targets appears to be similar (Fig. 4.8). The simulated exit velocities are close with the N40 steel to be smaller compared to N30, N50 and 4340 steels (Fig. 4.9).

4.3 Conclusions

The evolution of the microstructure of extruded coupons of Nitronic 30, 40 and 50 stainless steels was studied by metallography:

- In parallel and perpendicular directions to the rolling direction
- And as a function of cold work and annealing
- The Nitronic alloys are very effectively strengthened by CW
- These alloys surprisingly retain much of this CW strength after exposure to relatively high temperatures for relatively long times.

  Precipitation of cementite was observed in the grain boundaries in the annealed coupons
  XRD analysis showed that for

- Nitronic30 Stainless Steels: No new phases were observed

– All alloys showed the same pattern with the strongest diffraction peak observed at diffraction angle 19.08° ($2\theta = 38.16°$).
– However the as-received alloy, with half hardness showed different pattern with the strongest XRD peak at 22.11° ($2\theta = 44.22°$) and some different peaks than the CW Nitronic steels.

- Nitronic 50 Stainless Steels: No new phases were observed.

– All alloys showed the same pattern with the strongest diffraction peak observed at diffraction angle 19.08° ($2\theta = 38.16°$).
– The 35 % cold work, annealed at 750 °C for 8 h coupon showed an additional peak at 37.5° ($2\theta = 43°$).

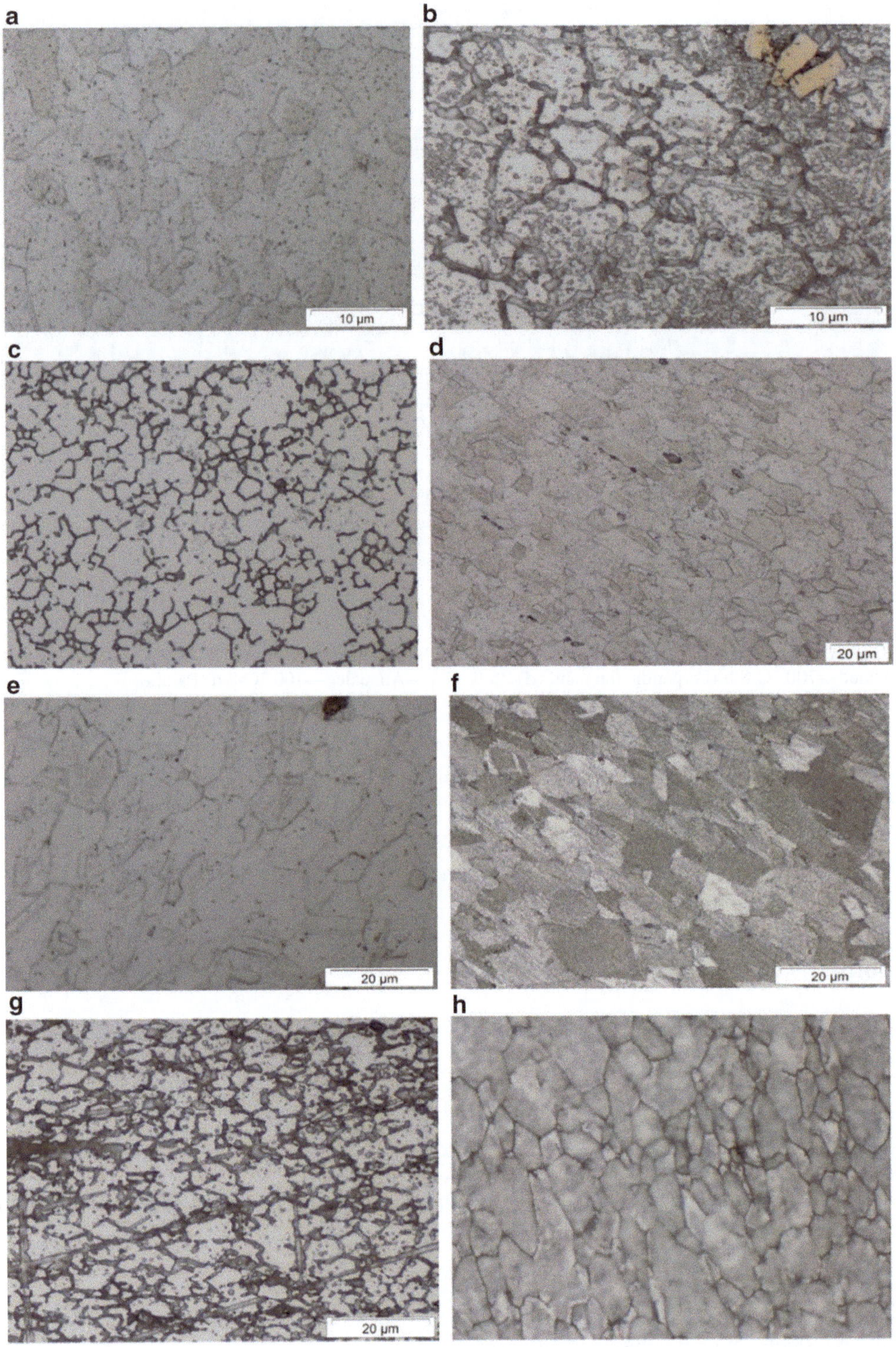

Fig. 4.5 Nitronic 30 metallography (**a**) Annealed; (**b**) Annealed—750 °C-8 h (Parallel); (**c**) Annealed—750 °C-8 h (Perpendicular); (**d**) 25 % CW (Parallel); (**e**) 25 % CW—(Perpendicular); (**f**) 25 % CW—Annealed—700 °C-8 h; (**g**) 25 % CW—Annealed—700 °C-8 h (Perpendicular) and (**h**) 25 % CW—Annealed—700 °C-8 h (Parallel)

Initial modeling showed that the overall damage of the targets caused by all steels projectiles chosen were similar. However, the exit velocity of projectile made of 4340 steel was 200 m/s compared to the 54, 52 and 48 m/s of N30 N40 and N50 steels.

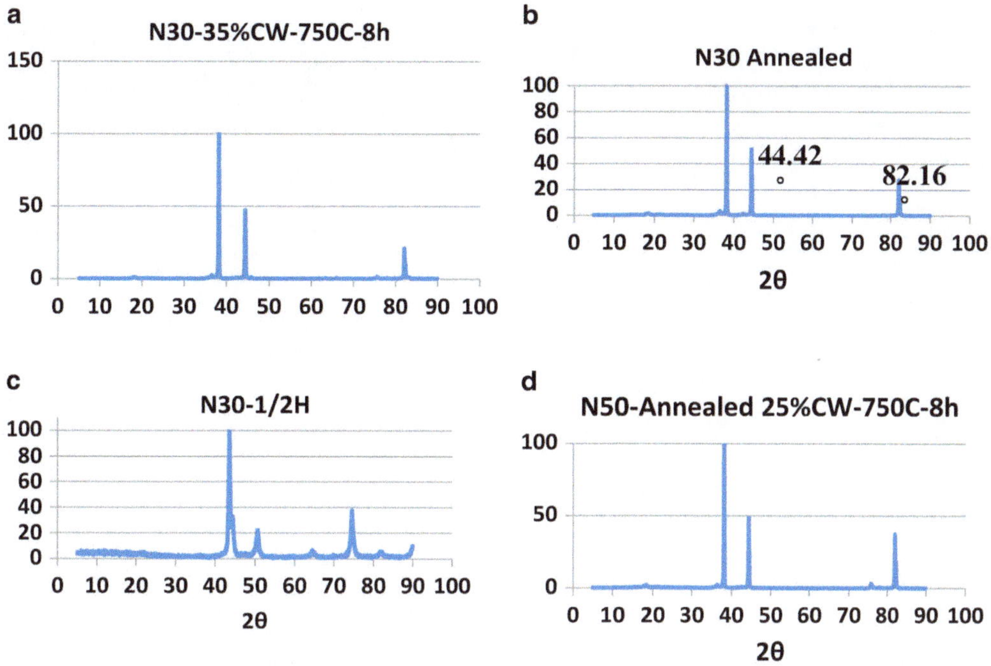

Fig. 4.6 XRD pattern of selected Nitronic steels specimens

Fig. 4.7 X-ray diffraction patterns of X5CrNi18-8 steel after 43 % of plastic deformation with and heat treatment at 500 °C

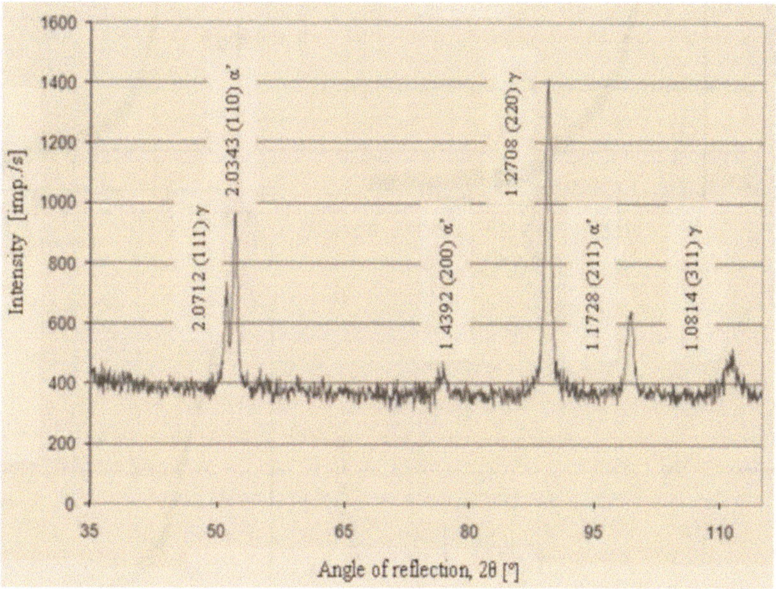

Further analysis will be conducted to completely recognize the constituents of the phases of these alloys using SEM and TEM and modeling using various targets consisted of other materials. Additional experimentation is need for production of accurate EOS and strength and failure material models.

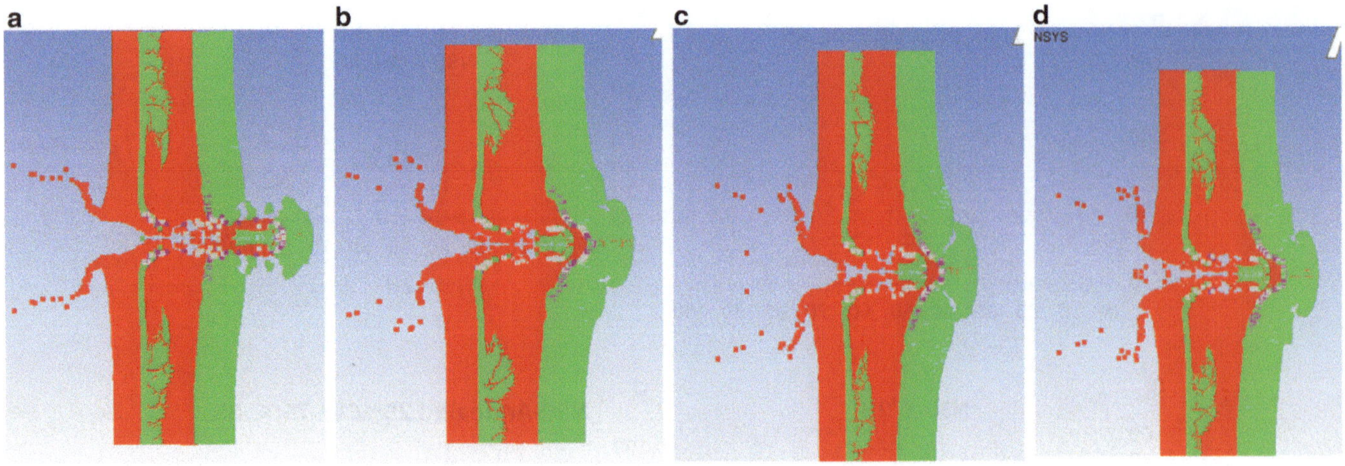

Fig. 4.8 Damage developed in the target by (**a**) 4340; (**b**) N30; (**c**) N40; and (**d**) N50 impactors respectively

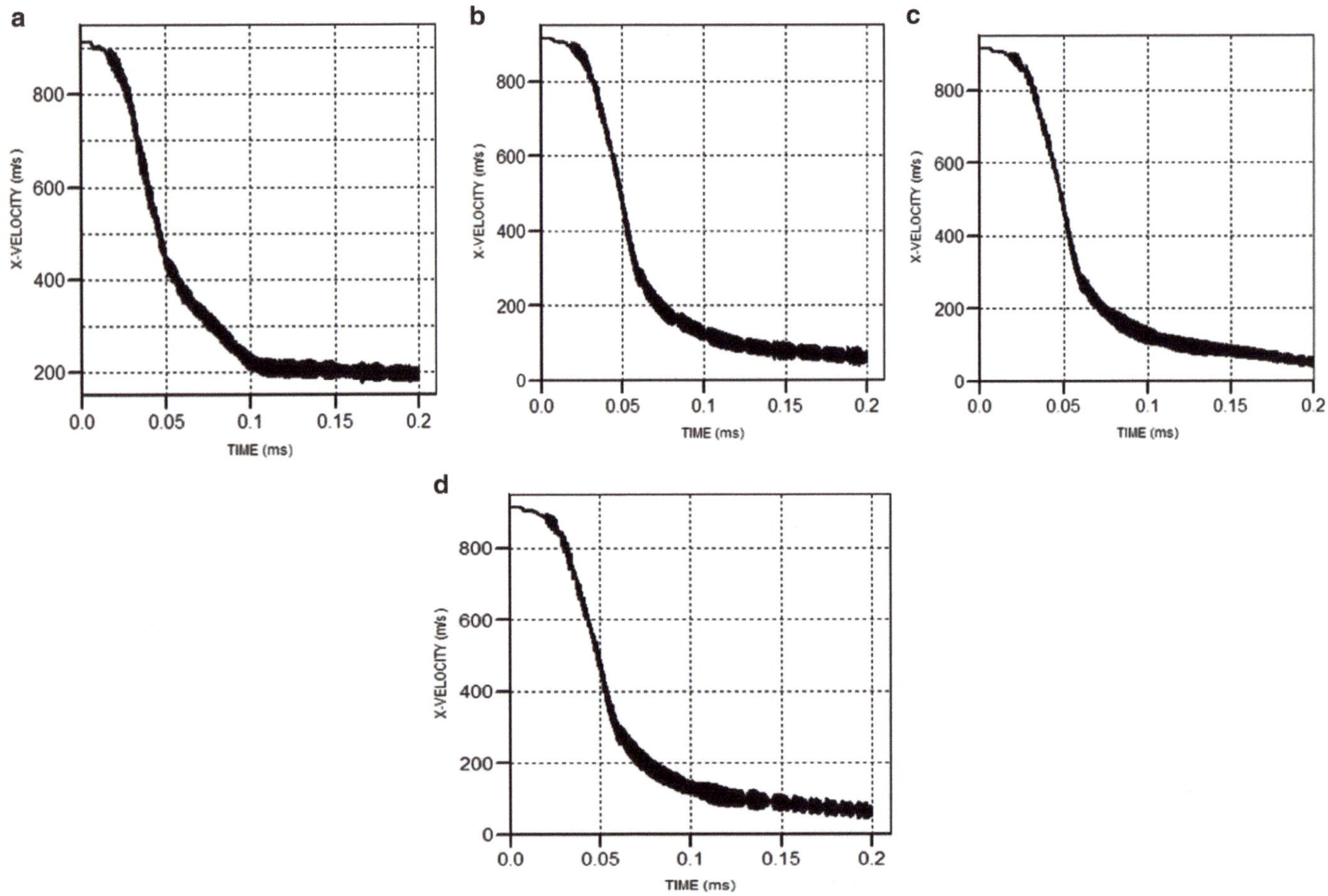

Fig. 4.9 Exit velocities of (**a**) 4340 steel-200 m/s; (**b**) N30-54 m/s; (**c**) N40-51 m/s; and (**d**) N50-58 m/s

References

1. www.asminternational.org
2. http://blog.bergsen.com/2013/09/19/nitronic-stainless-steel/
3. http://www.hpalloy.com/Alloys/descriptions/NITRONIC30.aspx
4. http://www.hpalloy.com/Alloys/descriptions/NITRONIC40.html
5. http://www.hpalloy.com/Alloys/descriptions/NITRONIC50.aspx
6. http://www.hpalloy.com/Alloys/descriptions/NITRONIC60.aspx
7. http://www.azom.com/article.aspx?ArticleID=6772
8. Kurc-Lisiecka, A., Kalinowska-Ozgowicz, E.: Structure and mechanical properties of austenitic steel after cold rolling. JAMME **44**(2), 148–153 (2011)

Chapter 5
Mechanical Response of T800/F3900 Composite at Various Strain Rates

Peiyu Yang, Jeremy D. Seidt, and Amos Gilat

Abstract Composite materials are becoming more and more popular in the automotive and aerospace industries. Many models exist that can describe the plastic deformation and failure response of composites, however, a new orthotropic material model (MAT_213 in LS-DYNA) has been developed to improve the predictive accuracy of numerical simulations of dynamic structural events involving composite materials. A test series was devised to calibrate this model for T800/F3900, a strengthened epoxy carbon-fiber reinforced polymer. In this paper, the mechanical response of T800/F3900 is studied experimentally under different loading conditions. Tests include in plane uniaxial compression, in plane transverse tension, compression and out of plane tension, compression at various strain rates. High rate tension and compression tests are conducted using the split Hopkinson bar technique, while static tests are conducted on a hydraulic load frame. In plane and out of plane deformations are studied using Digital Image Correlation (DIC).

Keywords Composite material • T800/F3900 • Digital image correlation • Split Hopkinson Bar

5.1 Introduction

Composite materials are made from two or more constituent materials with different material properties. T800/F3900 is a strengthened epoxy carbon-fiber reinforced polymer which is strong in fiber direction but a brittle material. Since the material model (MAT_213) has already been established in LS-DYNA Goldberg et al. (Journal of Aerospace Engineering 04015083, 2015), the verification of the material response of T800/F3900 by experiments becomes necessary. Dynamic and quasi-static material responses are studied from tension and compression tests. The strain rates that are used for testing both tension and compression tests are 10^{-3}, 1 and 10^{3}, respectively. Compression and tension Split Hopkinson pressure Bar are applied for tension and compression tests with strain rate of 1000 1/s and MTS hydraulic load frame is used for tests with strain rates of 0.001 1/s and 1000 1/s. 2D and 3D DIC are used for calculating the change of strain during the tests.

5.2 Experiments

Tension and compression tests are performed on T800/F3900 to study the material response under strain rates of 0.001 1/s, 1 1/s and 1000 1/s. The 90° orientation tension specimen geometry is shown in Fig. 5.1. The gage section is designed to be 0.2 in. in length and 0.4 in. in width. The thickness of the specimen is designed to be 0.125 in.

To do strain rates of 0.001 1/s and 1 1/s tension tests, the tension specimen is glued into adapters by using Loctite Ablestik 2106 T Bipax adhesive and the assembly is pinned into universal joint to prevent any bending that may happen on the specimen. The assembly with universal joint is shown in Fig. 5.2a. Two Point Grey GLZ-41C6M-C cameras with Schneider 35 mm lenses are used in the tension tests with strain rate of 0.001 1/s using frame rate of 15 frames per second. 3D DIC is used to measure the full field strain change in the gage section of the specimen. For strain rate of 1 1/s tension tests, specimen is patterned in both gage section and side surface. Two Vision Research Phantom V7.0G cameras are employed to observe both surfaces, which is shown in Fig. 5.2b. 2D DIC is used on each surface to analyze strain change.

P. Yang (✉) • J.D. Seidt • A. Gilat
Scott Laboratory, Department of Mechanical and Aerospace Engineering, The Ohio State University, 201 W 19th avenue, Columbus, OH 43210, USA
e-mail: yang.1948@osu.edu

D. Casem et al. (eds.), *Dynamic Behavior of Materials, Volume 1*, Conference Proceedings of the Society for Experimental Mechanics Series, DOI 10.1007/978-3-319-41132-3_5

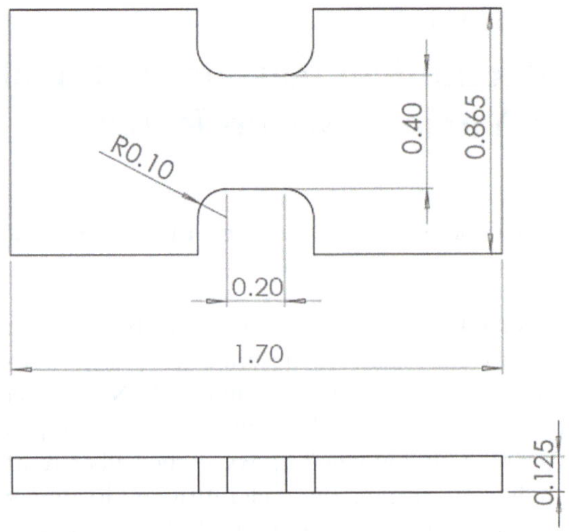

Fig. 5.1 Tension specimen geometry (units in inch)

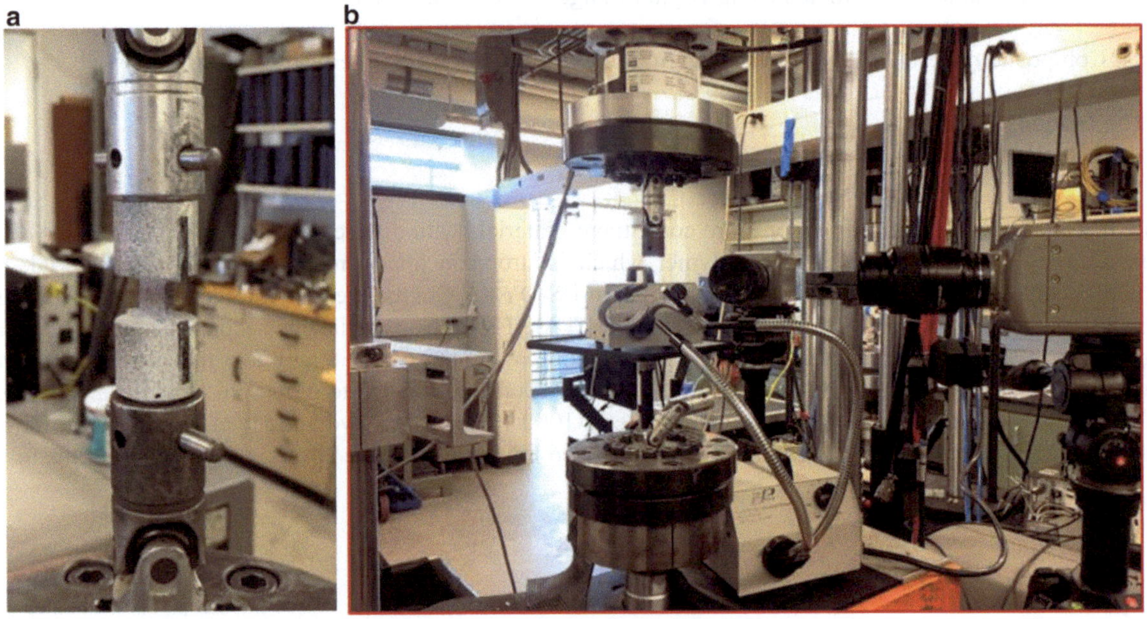

Fig. 5.2 (**a**) Tension specimen with adapters pinned into universal joint, (**b**) Strain rate of 1 1/s tension tests with two Phantom MC2 cameras

A point, a virtual extensometer and a rectangle are placed on the surface of specimen in DIC to see the uniformity of strain on the specimen (Fig. 5.3).

Figure 5.4 shows the geometry of compression specimen. The specimen has a square cross section with 0.125 in. long on each side and the height is designed to be 0.25 in.

For strain rate of 0.001 1/s and 1 1/s compression tests, specimen is placed between two tungsten carbide plates with a square notch on each side (Fig. 5.5a) which fit the square cross section of the specimen (Fig. 5.5b). The reason to cut the square notch is to prevent specimen from moving during the test.

Since the material is the assembly of laminates, delamination may happen when it is being compressed, which will cause inconsistent strain change from DIC. Therefore, four Point Grey cameras with 35 mm lenses are used to observe all four surfaces of the specimen. The layout of the setup is shown in Fig. 5.6. 2D DIC is applied to measure strain from all four surfaces. Similar as those in tension tests, strain from a point, a virtual extensometer and a rectangle are measured (Fig. 5.7).

For strain rate of 1000 1/s compression test, Split Hopkinson compression bar is applied. The Split Hopkinson compression bar consists a strike bar, an incident bar and a transmitted bar which are made of titanium (Fig. 5.8a). The strike bar is fired by compressed air and loading wave is generated from the strike bar. After the strike bar hitting on the incident bar, the wave travels through the incident bar, specimen and transmitted bar. Before the test starts, air is filled into the tank on a certain level. The level of pressure depends on the strain rate of the test. Two surfaces are patterned and

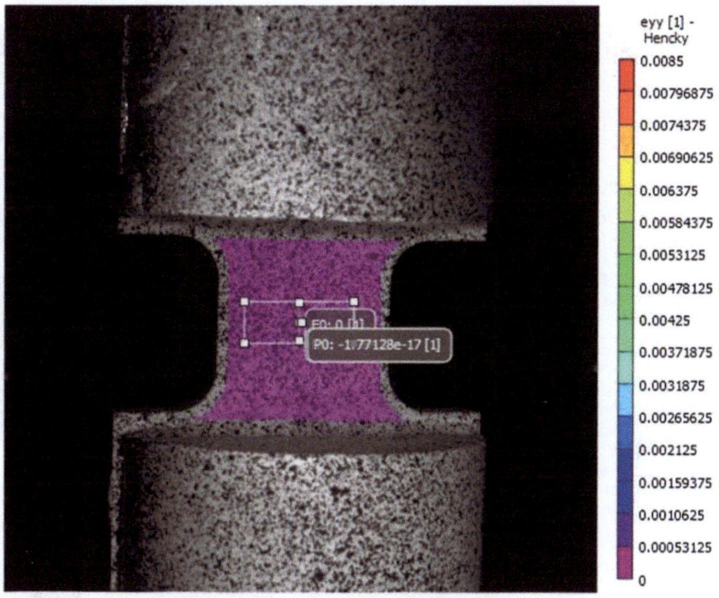

Fig. 5.3 DIC from one of tension tests

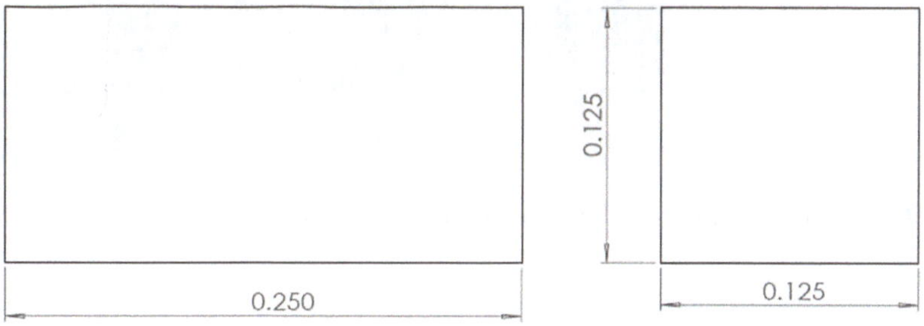

Fig. 5.4 Compression specimen geometry (units in inch)

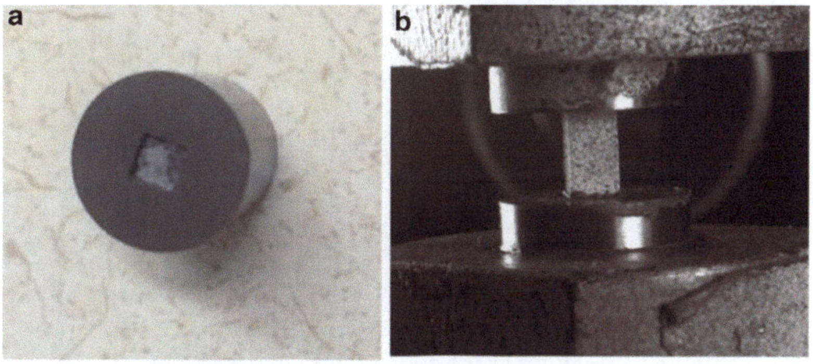

Fig. 5.5 (**a**) Tungsten carbide with a square notch, (**b**) Specimen placed between two tungsten carbide plates

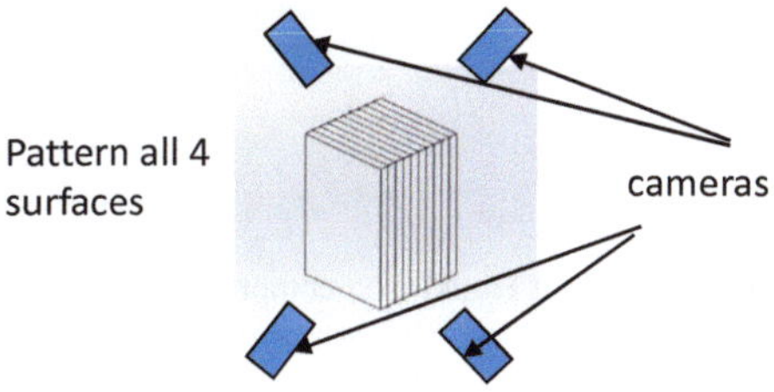

Fig. 5.6 Test setup for strain rate of 0.001 1/s and 1 1/s compression tests

Fig. 5.7 (**a**) DIC from one of the specimen surfaces, (**b**) Specimen placed on Split Hopkinson compression bar and (**c**) 2D DIC for the specimen

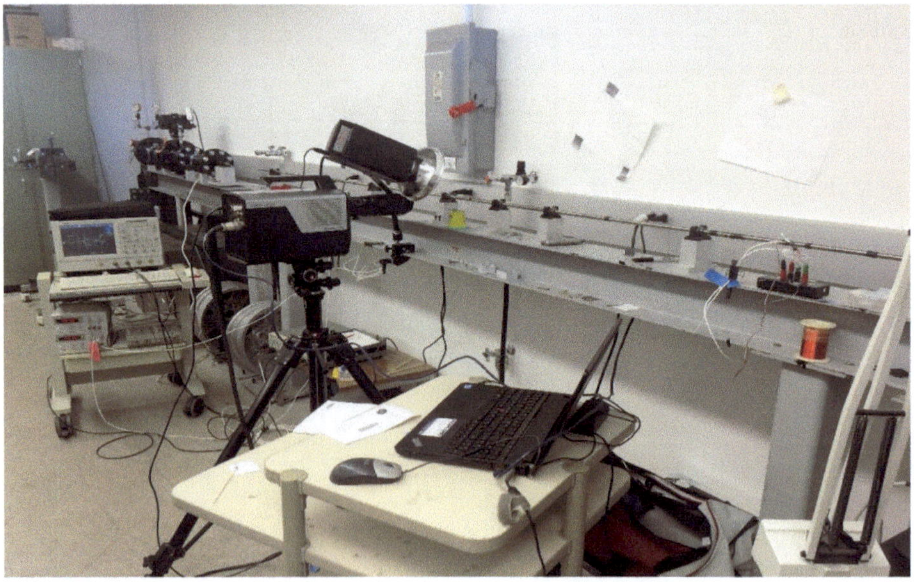

Fig. 5.8 Split Hopkinson compression bar

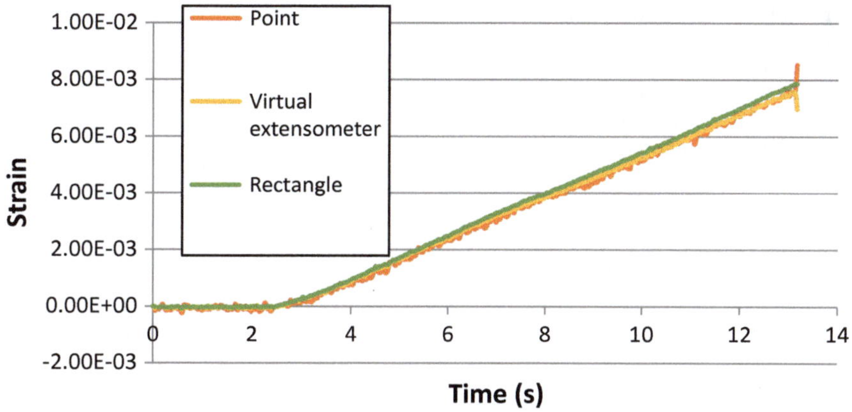

Fig. 5.9 Strain vs. time

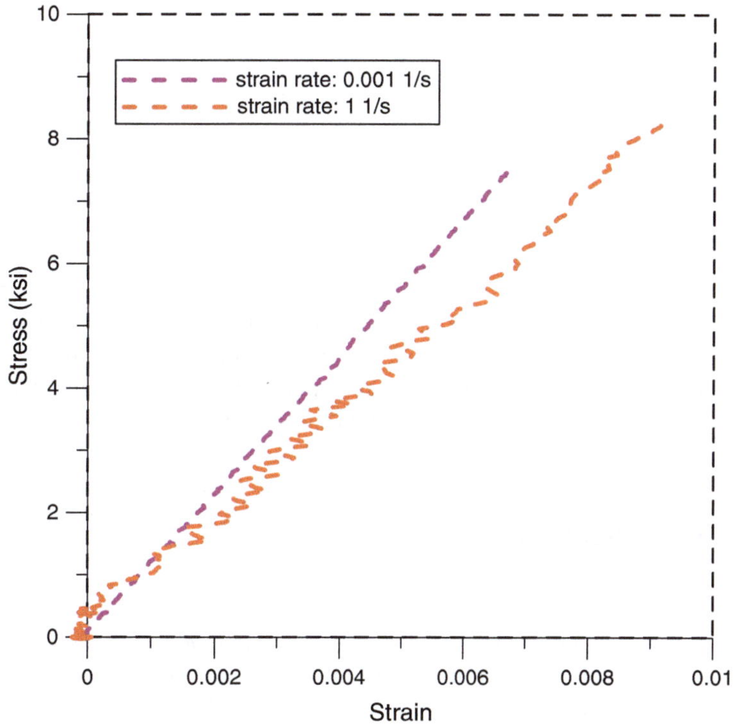

Fig. 5.10 Stress vs. strain curve from 90° tension tests

Fig. 5.8b shows how the specimen placed on Split Hopkinson compression bar. One Shimadzu HPV-*X*2 camera is applied in this test and it records the entire test procedure at 5,000,000 frames per second. Strain is calculated by using 2D DIC. Same as compression tests with strain rate of 0.001 1/s and 1 1/s, strain is calculated from a virtual extensometer, a point and an area on both surfaces (Fig. 5.8c).

5.3 Results

Recall that three ways are measured in DIC for tension tests (Fig. 5.3), Figure 5.8 shows the representative curve of strain vs. time from the results of tension test of 90° orientation with strain rate of 0.001 1/s. Three curves matched up well indicated that strain was changing uniformly when the specimen was being stretched. Strain went up to 0.75 % when the specimen broke (Fig. 5.9).

Figure 5.10 shows the stress vs. strain curve from tension tests of 90° orientation specimen with strain rates of 0.001 1/s and 1 1/s. For strain rate of 0.001 1/s tension test result, peak stress is about 0.5 ksi lower than the peak stress from strain rate 1 of 1/s tension test and strain at failure is 0.2 % lower. In Fig. 5.8, strain data is measured from virtual extensometer.

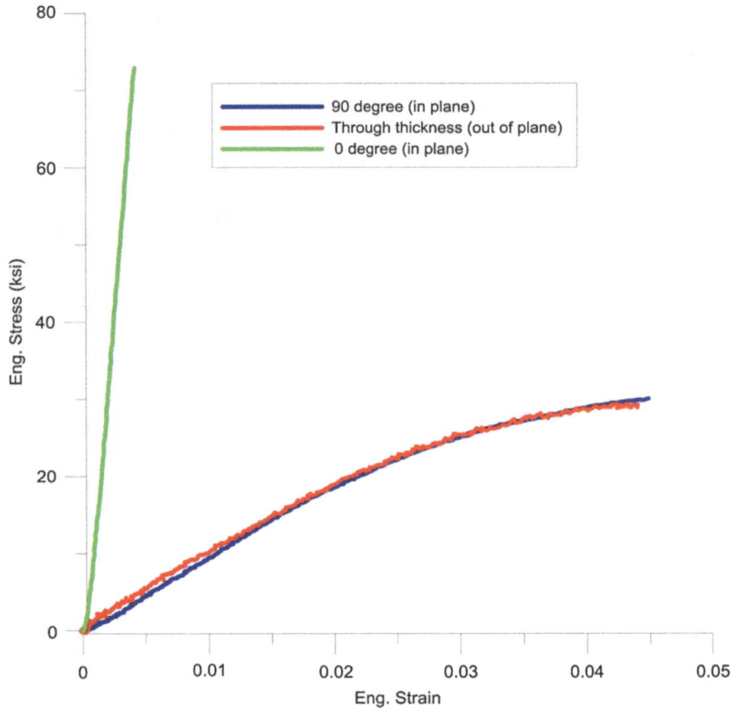

Fig. 5.11 Stress vs. Strain data from compression tests [Engineering stress and strain data are supposed to be negative. All the data shown in Fig. 5.11 were multiplied by −1 for convenience.]

Stress and strain data from 0° (in plane), 90° (in plane) and through thickness (out of plane) compression tests with strain rate of 0.001 1/s are shown in Fig. 5.11. The data indicates that T800/F3900 in 90° in plane and through thickness out of plane are showing similar material response in strain rate of 0.001 1/s compression tests. The material in 0° orientation is much more brittle (0.4 % of strain at failure) comparing with other two orientations (4.4 % of strain at failure).

5.4 Conclusion

Stress at failure from 90° tension tests (Fig. 5.10) is similar to material properties data provided by Philip [1]. Figure 5.10 also shows that the tensile response of material in 90° orientation is sensitive to strain rate. Peak stress from 90° compression test agree with the data provided Philip as well. Compressive material response is similar in all transverse fiber direction (including in plane and out of plane transverse direction).

Acknowledgement The research was supported by the USA Federal Aviation Administration.

References

1. Goldberg, R.K., Carney, K.S., Dubois, P., Hoffarth, C., Harrington, J., Rajan, S., Blankenhorn, G.: Development of an orthotropic elasto-plastic generalized composite material model suitable for impact problems. J. Aerosp. Eng. 04015083 (2015)
2. Bogert, P., Satyanarayana, A., Chunchu, P.: Comparison of damage path predictions for composite laminates by explicit and standard finite element analysis tools. In: 47th AIAA/ASME/ASCE/AHS/ASC Structures, Structural Dynamics, and Materials Conference (14th AIAA/ ASME/AHS Adaptive Structures Conference) 7th (2006). (n. pag. Web)

Chapter 6
Full-Field Temperature and Strain Measurement in Dynamic Tension Tests on SS 304

Jarrod L. Smith, Veli-Tapani Kuokkala, Jeremy D. Seidt, and Amos Gilat

Abstract The thermomechanical response of 304-stainless steel tension specimens to a range of strain rates from $7 \times 10^{-3}\ s^{-1}$ to $2600\ s^{-1}$ was investigated. Quasi-static tests (7×10^{-3} to $0.8\ s^{-1}$) were completed on a hydraulic load frame, intermediate tests ($200\ s^{-1}$) were performed with a modified pressure bar, and high strain rate tests ($2600\ s^{-1}$) on a split Hopkinson pressure bar. Full-field infrared thermography and strain measurements were recorded during each test. Infrared measurements were taken using the Telops FAST-IR 1000 infrared camera at rates up to 30,0000 frames per second. 2D-DIC was used to compute strain from simultaneously recorded visible images taken at rates up to 90,000 frames per second. Max temperatures of 290 °C were recorded in the necking region of a uniaxial specimen at a strain rate of $2600\ s^{-1}$. These measurements can be used to investigate the transition of isothermal deformation to adiabatic deformation and to determine the portion of plastic work converted to heat at each strain rate.

Key words Infrared thermography • Digital image correlation • Split Hopkinson bar • Thermomechanical • High strain rate

6.1 Introduction

Inelastic deformation behavior of all metals is known to be affected by strain rate and temperature. The effects of strain rate and temperature are coupled because a significant portion of the energy required for deformation is also converted to heat during the process which raises the temperature of the material. At low strain rates, plastic deformation takes place in virtually isothermal conditions because as the heat is generated in the material there is sufficient time for the heat to dissipate into the surroundings. At higher strain rates the temperature in the material reaches higher temperatures because the event occurs in period of time that is too short for the heat to transfer from the material. To observe this behavior, material specimens are normally tested in simple tension tests utilizing a load frame for quasi-static tests and Split Hopkinson Bar for higher strain rate tests. The force on the specimen can be measured by utilizing a load cell for quasi-static tests and by analyzing the elastic waves in the Split Hopkinson Bar. The strains on the surface of the specimen can be determined utilizing Digital Image Coorelation (DIC). DIC is implemented by painting the specimen surface with a black and white speckle pattern and capturing the derformation of the surface with high speed cameras. The images are then processed with software that determines the displacements of the speckles during deformation and then calculates the strains based off of the displacements. Due to recent advancements in Infrared Thermography cameras it is also possible to measure the temperature on the surface of the specimen at high speeds. Simultaneous full-field measurement of the strain and temperature has been completed by many reasearchers at low strain rates. For example, Saai et al. [1] studied the thermomechanical behavior of Al bi-crystals with tensile tests at the strain rate of ca. $10^{-2}\ s^{-1}$ making use of simultaneous DIC and IR measurements. In their experiments, the infrared and visible cameras were observing the same specimen area, which was first coated with black paint and then sprayed with white paint to form a random pattern for DIC. They estimated that the emissivity of the paint (s) was 0.96, i.e., very close to that of a black body. The IR images were recorded during the tensile tests at 20 frames per second (fps) with a resolution of 320 × 240 pixels. Oliferuk et al. [2] used infrared thermography and visible imaging to determine the energy storage rate in the area of strain localization in an austenitic stainless steel similar to AISI 304 L. In their tests, the mean value of strain rate was $6.6 \times 10^{-1}\ s^{-1}$ and the imaging frame rate of both visible and IR cameras was 538 fps. For local strain determinations, graphite dot markers were painted on one surface of the specimen, while the opposite surface on the IR camera side was covered with soot with estimated emissivity of 0.95.

J.L. Smith (✉) • J.D. Seidt • A. Gilat
Scott Laboratory, Department of Mechanical Engineering, The Ohio State University, 201 W 19th Ave, Columbus, OH 43210, USA
e-mail: jasmith116@gmail.com

V.-T. Kuokkala
Department of Materials Science, Tampere University of Technology, P.O.B. 589, Tampere 33101, Finland

D. Casem et al. (eds.), *Dynamic Behavior of Materials, Volume 1*, Conference Proceedings of the Society for Experimental Mechanics Series, DOI 10.1007/978-3-319-41132-3_6

At high strain rates (above 100 s^{-1}) the temperature measurement during deformation experiments has been mostly done by using infrared detectors. In their studies on adiabatic shear band formation, Hartley et al. [3] and Marchand and Duffy [4] measured the temperature distribution in the gage section of a specimen deforming in shear at a strain rate of the order of 1000 s^{-1} in a torsional SHB apparatus using ten and twelve IR detectors. Mason et al. [5] used an array of eight detectors to measure the temperature increase in the specimen during a compression SHB experiment at a strain rate of 3000 s^{-1}. Noble and Harding [6] were the first to measure the temperature of a specimen during a tensile SHB test. Using a scanning thermal camera, a maximum temperature in the range of 240–310 °C was measured in the tests on iron at a nominal strain rate of 1600 s^{-1}. A common feature to most of the published studies related to the simultaneous use of full-field thermal imaging and digital image correlation is that the applied strain rates have been relatively low, in most of the cases well below 1 s^{-1}, which means that the required framing speeds of the IR cameras have been rather moderate, mostly well below 1000 fps. In many cases the temperature variations to be detected have been relatively small, in some cases only fractions of a kelvin.

In the present paper, simultaneous full-field strain and temperature measurements, using 2D DIC and a high-speed IR camera, respectively, are made in tensile tests at low (7×10^{-3} to 0.8 s^{-1}), intermediate (200 s^{-1}) and high (2600 s^{-1}) strain rates. Three different mechanical testing methods together with high speed visible and infrared cameras are used. The results show the evolution of strain and temperature during the uniform deformation in the early part of the tests and in the necking region during the localization. Practical and theoretical issues related to the emissivity of the specimens, calibration of infrared cameras, as well as the current limitations of the infrared techniques, such as the limited temperature range of measurements at a constant exposure time are discussed.

6.2 Experimental Procedures and Techniques

Tensile tests on flat thin stainless steel 304 specimens are conducted at various strain rates ranging from 7×10^{-3} s^{-1} to 3000 s^{-1}. Full-field deformation and full-field temperature are measured simultaneously and synchronously during the tests. The deformation is measured using DIC on one side of the specimen, and the temperature is measured on the opposite side with an IR camera. Since the specimen is thin, it is assumed that the deformation and temperature are uniform through the thickness. Three different testing methods are used as shown in Fig. 6.1.

Quasi-static tests at the strain rates of 7×10^{-3}, 0.08, and 0.8 s^{-1} are done using a servo-hydraulic MTS load frame. Tests at the high strain rates of 200 and 3000 s^{-1} are completed using an intermediate strain rate apparatus and a Split Hopkinson Bar (SHB), respectively. The SHB consists of two titanium 6AL-4V 15.9 mm diameter bars (incident and transmitter). The specimen is attached between the bars and is loaded with a tensile wave that is generated in the incident bar by shooting a striker tube that impacts a flange at the far end of the incident bar. The force in the specimen is determined from the wave that is transmitted to the transmitter bar. Tests at the strain rate of 200 s^{-1} are done with an intermediate strain rate apparatus. The apparatus consists of a fast hydraulic actuator and a long (40 m) transmitter bar. Its operation is similar to the SHB apparatus except that the hydraulic actuator replaces the incident bar. The specimen is attached between the end of the transmitter bar and the actuator and is loaded by a force applied by the actuator, which can move at a speed of up to 2 m/s. Once the specimen is loaded, an elastic tensile wave propagates through the transmitter bar to the free end and reflects back toward the end where the specimen is attached. The force in the specimen is determined by measuring the wave in the transmitter bar.

6.2.1 *Specimen Geometry and Material*

Tests at the high strain rates require specimens with a short gage length, and in order to eliminate possible effects of the specimen geometry on the results, the gage section geometry of the specimens used in the quasi-static tests is the same. A drawing of the specimen used in the high strain rate tests is shown in Fig. 6.2.

The specimens in the quasi-static tests are the same, except that the flanges are longer such that they can be held by hydraulic grips. The specimens are made of austenitic stainless steel AISI 304 L. This material exhibits high ductility, strong capability of strain hardening, and low thermal conductivity, which all together lead to notable internal heating of the material already at relatively low strain rates. The specimens were also sandblasted to increase the emissivity of the surface so the temperature of the surface could be measured at lower temperatures and exposure times.

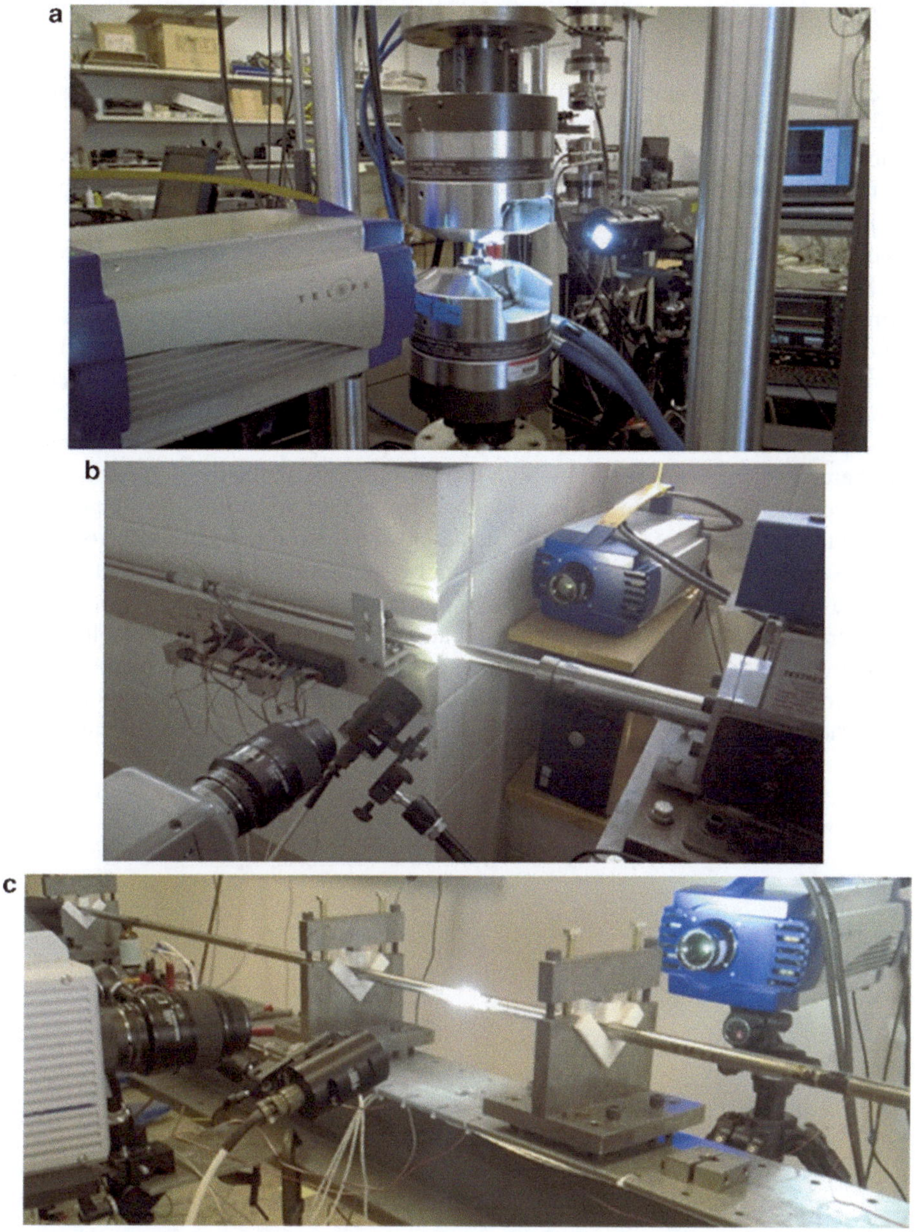

Fig. 6.1 Experimental setup for tensile testing at low (**a**), intermediate (**b**), and high (**c**) strain rate

Fig. 6.2 Specimen geometry

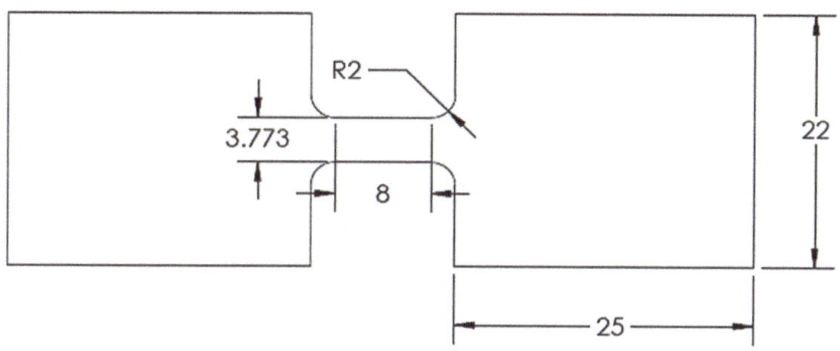

6.2.2 *Full-Field Thermal Imaging and Calibration*

Thermal image capture is done with Telops FAST-IR 1000 high speed infrared camera, capable of 256 × 320 pixel resolution at speeds up to 1000 fps and a reduced pixel resolution at speeds up to 30,000 fps. The exposure time of the IR camera varied in the current tests from 100 μs down to 5 μs. With decreasing exposure time the IR camera can register higher surface temperatures, but at the same time the lowest measurable temperature increases so that the usable temperature range at each constant exposure time is ca. 250 K, as seen in Fig. 6.3.

The resolution, speed and exposure time for all cameras used in the various strain rate tests are listed in Tables 6.1 and 6.2.

To synchronize the Phantom visible camera and the IR camera, they are externally driven by a dual-channel Keysight 33,500 series waveform generator, which enables running the cameras at different but phase-synchronized frequencies. In practice, the visual camera typically is run at a frequency which is the same or a multiple of the IR camera frequency. The waveform generator sends TTL level timing pulses to both cameras in an externally triggered burst mode. In the low strain rate tests with the MTS frame, the external trigger to the waveform generator is sent through the digital output port of the controller unit 100 ms before the actuator starts to move. The data acquisition of the MTS controller starts at the same time as the trigger is sent, and therefore the images taken by both cameras as well as the load cell and LVDT signals of the load frame have a common recorded time base. In the intermediate strain rate apparatus the trigger signal is generated by a proximity sensor detecting the movement of the hydraulic actuator. The force signal obtained from the strain gage bridges is also synchronized with the cameras. The time of the force signal is shifted in order to account for the time it takes for the wave to propagate from the specimen end to the strain gage bridge location on the transmitter bar. With the Split Hopkinson Bar, Photron SA 1.1 camera is used instead of the Phantom camera because of its better resolution at the required higher frame rates. Since the Photron camera cannot be clocked externally, the synchronization with the IR camera is done by using one channel of the waveform generator to run the infrared camera and using the other channel to trigger the autonomous image acquisition of the Photron camera at a preset frequency (normally a multiple of the IR camera's framing frequency).

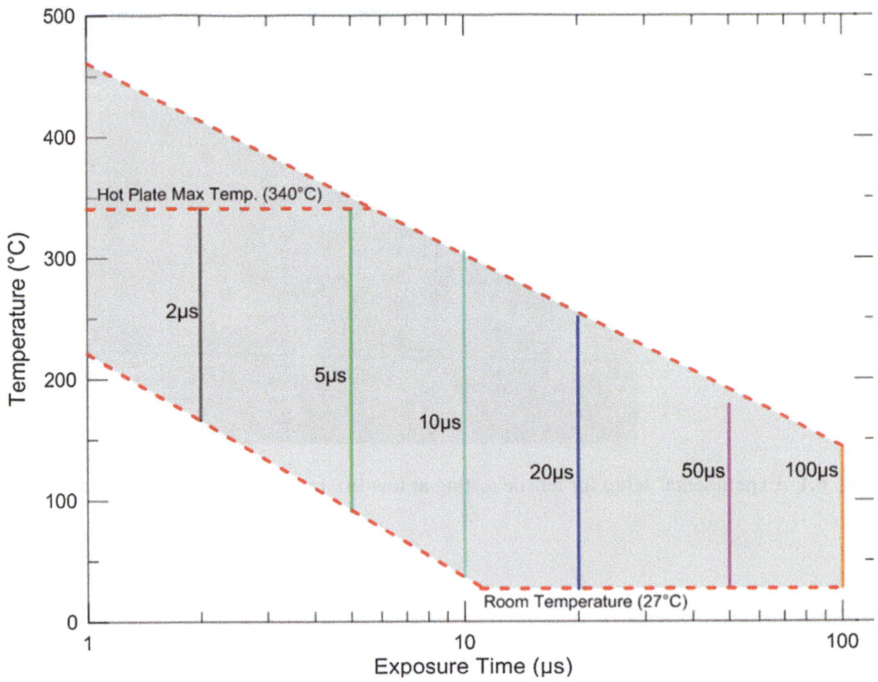

Fig. 6.3 Exposure time versus measurable temperature range for the Telops camera

Table 6.1 Visual cameras with resolutions and frame rates used in the various tests

Strain rate (s^{-1})	Camera	Pixel resolution	Frame rate (fps)
0.01	Vision research phantom v7.3	800 by 600	5
0.1	Vision research phantom v7.3	800 by 600	70
1	Vision research phantom v7.3	800 by 600	500
200	Vision research phantom v7.3	512 by 128	20,000
3000	Photron SA 1.1	384 by 128	90,000

Table 6.2 Resolution and frame rate used by the thermal camera in the various tests

Strain rate (s^{-1})	Pixel resolution	Frame rate (fps)	Exposure time (μs)
0.007	256 × 320	5	100
0.08	256 × 320	70	50
0.8	256 × 320	500	10
200	64 × 128	10,000	10
2600	24 × 64	30,000	10

After adjustment of the time shift between the two cameras' clock signals, the Telops and Photron cameras are synchronized with high accuracy. The signal from the stain gages on the incident bar of the SHB apparatus is used as a trigger signal for waveform generator.

The infrared camera used in this work can record the temperature data as so-called radiometric temperatures or as non-uniformity corrected photon counts for each pixel. Since the radiometric temperature method assumes that the emissivity of the object is unity, i.e., it is a black body, the temperature readings obtained for stainless steel specimens will be too low because of its lower emissivity. Emissivity values for various substances can be found in the literature, but since the actual emissivity depends on many factors, such as the surface finish of the object, a calibration with real samples is often the only way to obtain reliable temperature values. As stated in the Introduction, in many previous works some kind of a coating, such as soot or spray paint, has been applied on the specimen surface to increase its emissivity close to that of a black body. The downside of this approach is that the coating tends to peel off, crack, or thin too much especially in the necking area of the specimen at large strains, which increases the uncertainty of the obtained temperature readings. In this work, calibration measurements with several specimens were conducted in the following manner: thermocouple(s) were attached to the surface of a specimen, which was heated on a hot plate in 10–20 K steps and the true temperature, indicated by the thermocouple(s), was recorded together with the IR camera reading obtained as an average of an area of reasonable size. To examine the effect of plastic deformation on the emissivity the calibration was done with tested specimens which have areas with different amount of plastic deformation. The necking area is highly deformed and appears a little rougher compared to the surface outside the gage section which is not deformed plastically. In the calibration procedure, IR camera readings at all temperature steps from room temperature up to ca. 350 °C were recorded from the deformed sections of tested specimens using exposure times of 100, 50, 20, 10, 5 and 2 μs. As Fig. 6.3 shows, at each exposure time the temperature difference between starvation and saturation of the IR detector is ca. 250 K. From the calibration measurements, polynomial fits were formed separately for each exposure time, which were then used to convert the radiometric temperatures given by the IR camera to true surface temperatures of the specimen.

6.3 Experimental Results

Results are presented for the tension tests conducted at strain rates from 0.007 to 2600 s^{-1}. Engineering stress strain curves from all the tests are displayed in Fig. 6.4. The strain to failure ranges from 0.65 at the lowest strain rate to 0.4. From these results it is evident that there is extensive strain rate hardening at the higher strain rates, leading to an increase in yield strength and a decrease in ductility.

Figures 6.5, 6.6, 6.7, 6.8, 6.9 present the strain and temperature along the specimen at different time intervals for each strain rate test. Each figure shows the true axial strain (measured by DIC) along the center line of the specimen at different times and the temperature (measured with the IR camera) at the same location and time. At the lowest strain rate of 0.007 s^{-1}, the strain and temperature increase almost uniformly across the specimen until necking is initiated. Once, necking is initiated the strain and temperature of the specimen increase around a point in the specimen near 6 mm. Similarly, all of the other specimens at higher strain rates react in the same manner where strain and temperature increase uniformly until a necking region is developed and then all strain and temperature increase localizes in that area. The temperature observed in the lowest strain rate test of 0.007 s^{-1} was the lowest of all the tests with a peak value close to 105 °C. An increase of nearly 70 °C in the lowest strain rate tests shows that the deformation of the specimens at all strain rates will not be isothermal. All other tests realized a peak temperature value close to 250 °C which leads one to believe that above strain rates of 0.007 s^{-1}, all heating in the specimens due to deformation effects are nearly adiabatic. The temperature of the tests above 0.007 s^{-1}, is also near the peak value for each of the exposure times used in those tests which can lead to increased error in the temperature reading and saturation of the sensor. In the case of the 2600 s^{-1} strain rate test, it is obvious that the sensor became saturated with photons and produced erroneous results leading to the dip in temperature in the localization region in Fig. 6.9. In future tests more attention will need to be placed on the selection of exposure time for the emissivity of the tension test specimen.

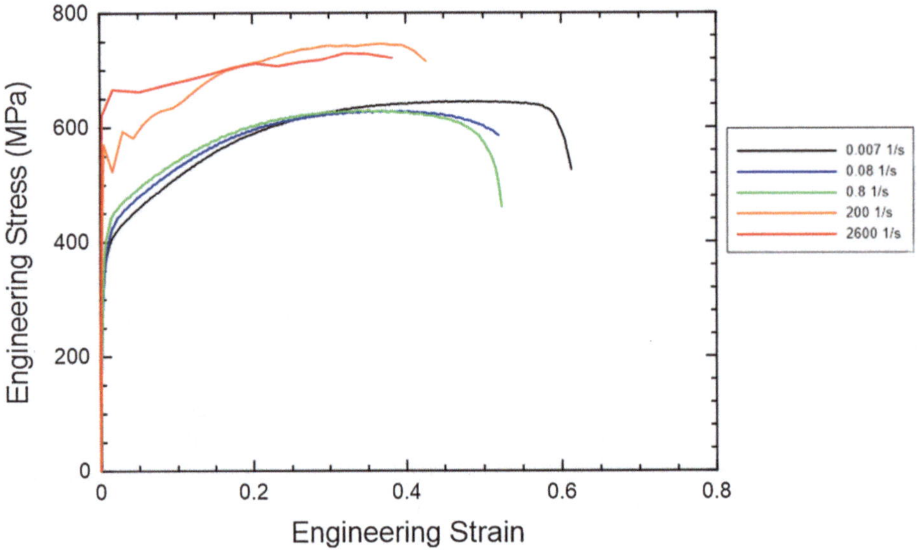

Fig. 6.4 Engineering stress strain curves of AISI 304 stainless steel at various strain rates

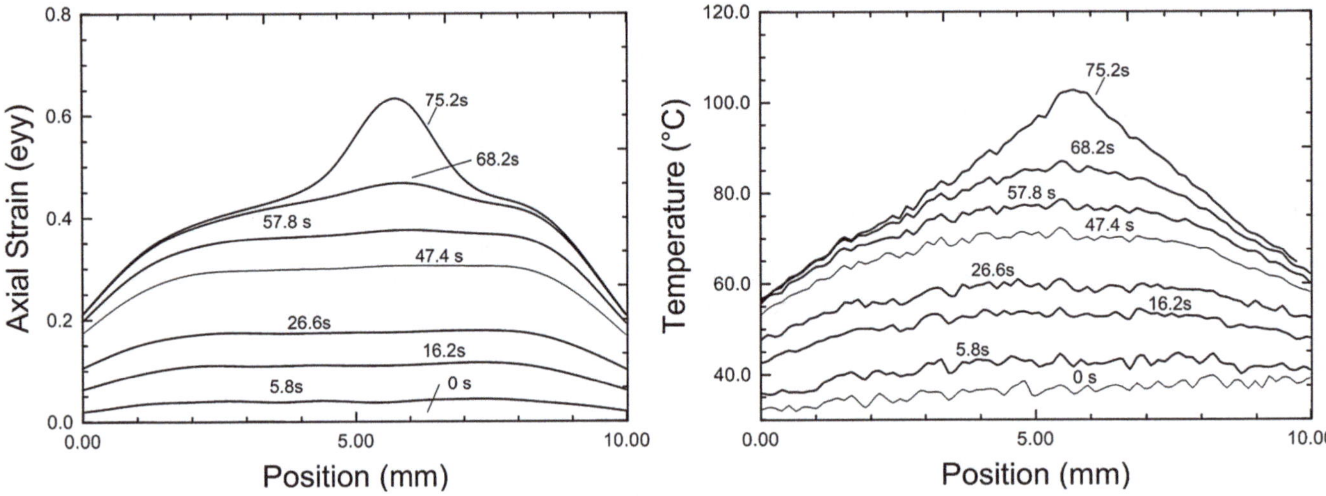

Fig. 6.5 Strain and temperature along the specimen at a strain rate of 0.007

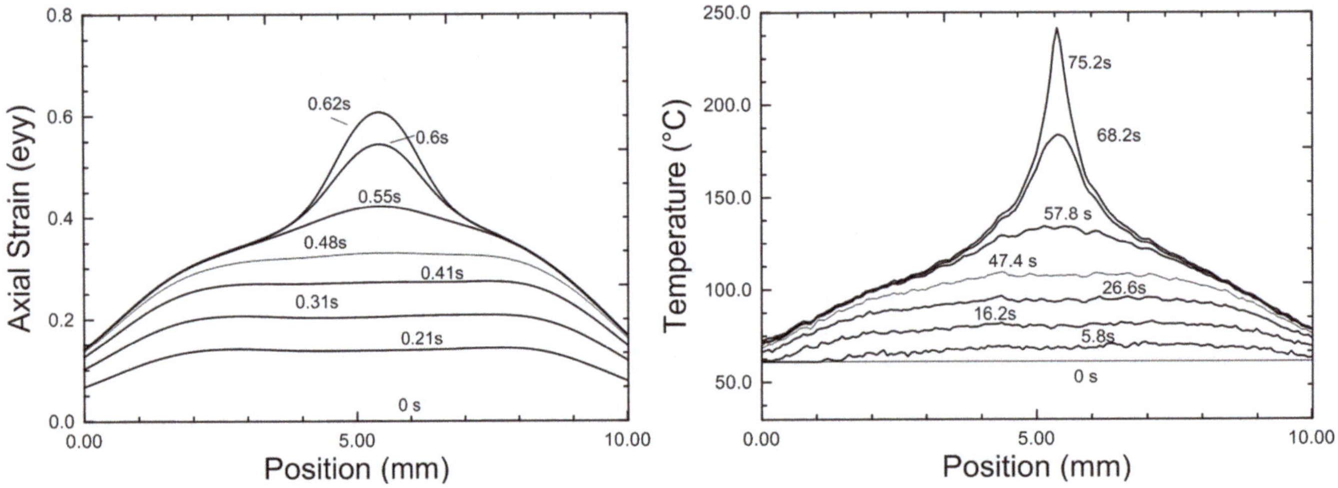

Fig. 6.6 Strain and temperature along the specimen at different intervals at a strain rate of 0.08

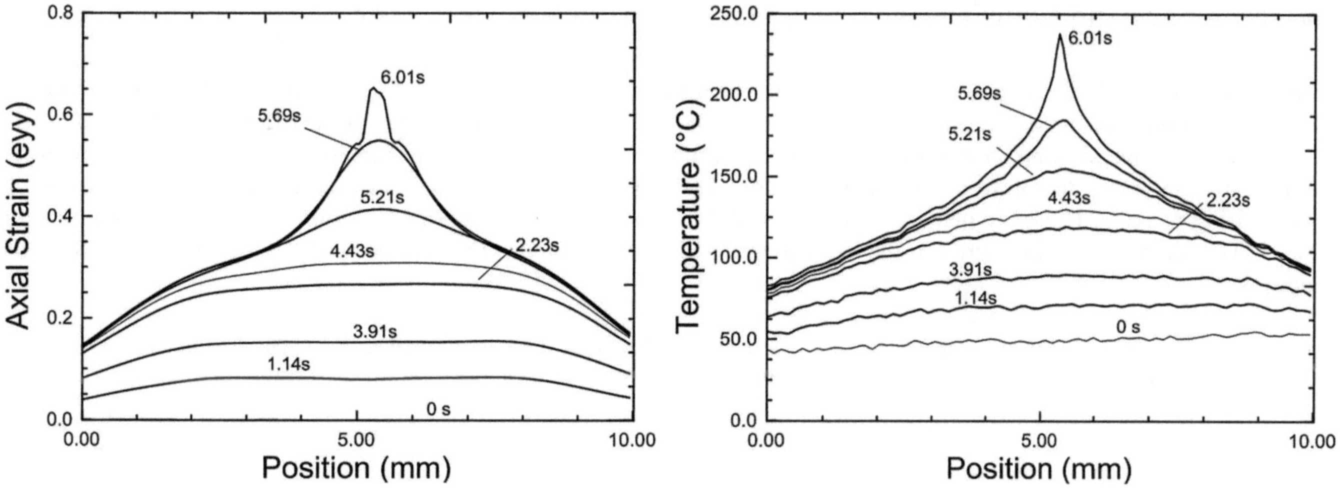

Fig. 6.7 Strain and temperature along the specimen at different intervals at a strain rate of 0.8

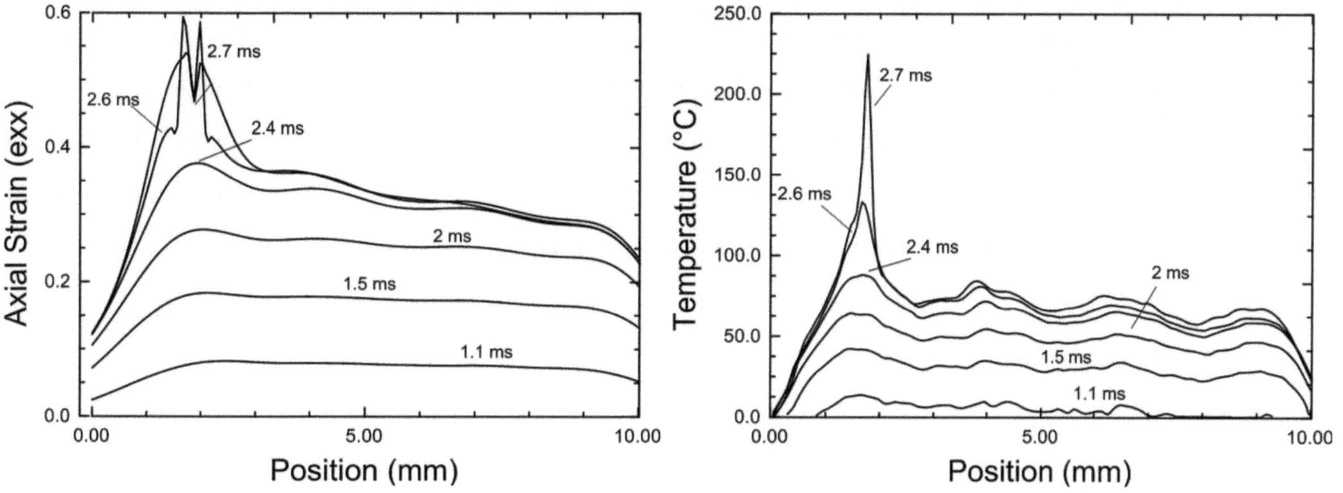

Fig. 6.8 Strain and temperature along the specimen at different intervals at a strain rate of 200

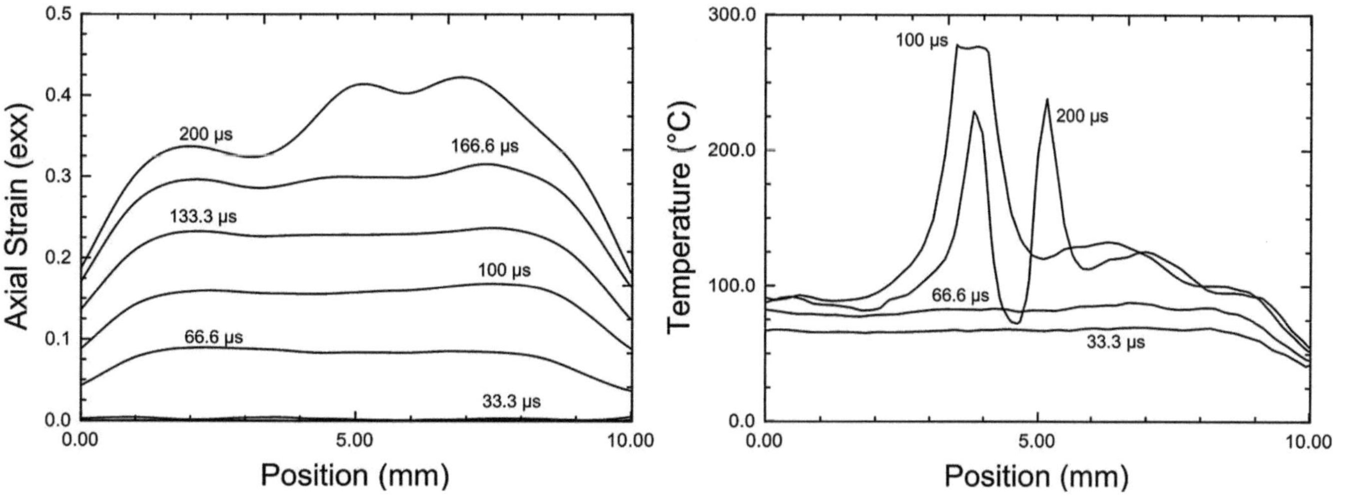

Fig. 6.9 Strain and temperature along the specimen at different intervals at a strain rate of 2600

6.4 Summary and Conclusions

Simultaneous full-field deformation and temperature measurements on the surface of the specimen have been done in tension tests conducted at nominal strain rates of 0.007, 0.08, 0.8, 200, and 2600 s^{-1}. The results show a modest and nearly uniform temperature increase in the test at the lowest strain rate of 0.01 s^{-1}. At all higher strain rates there is uniform temperature rise during the uniform deformation before necking and faster localized deformation and heating in the neck region during necking. Higher temperatures are measured in the test at strain rate of 1 s^{-1} than in the tests at strain rate of 0.1 s^{-1}. The temperatures measured in the tests at strain rates of 200 and 3000 are nearly the same indicating adiabatic conditions at these rates.

Acknowledgement The research was supported by the U.S.A. Federal Aviation Administration, Grant No. 11-G-004. The authors are grateful to Mr. William Emmerling, and Dr. Chip Queitzsch for their support and involvement.

References

1. Saai, A., Louche, H., Tabourot, L., Change, H.J.: Experimental and numerical study of the thermo-mechanical behavior of Al bi-crystal in tension using full field measurements and micromechanical modeling. Mech. Mater. **42**(3), 275–292 (2010)
2. Oliferuk, W., Maj, M., Zembrycki, K.: Determination of the energy storage rate distribution in the area of strain localization using infrared and visible imaging. Exp. Mech. **55**(4), 753–760 (2013)
3. Hartley, K., Duffy, J., Hawley, R.: Measurement of the temperature profile during shear band formation in steels deforming at high strain rates. J. Mech. Phys. Solids **35**(3), 283–301 (1987)
4. Marchand, A., Duffy, J.: An experimental study of the formation process of adiabatic shear bands in a structural steel. J. Mech. Phys. Solids **36** (3), 251–283 (1988)
5. Mason, J., Rosakis, A., Ravichandran, G.: On the strain and strain rate dependence of the fraction of plastic work converted to heat: an experimental study using high speed infrared detectors and the Kolsky bar. Mech. Mater. **17**(2–3), 135–145 (1994)
6. Noble, J.P., Harding, J.: Temperature measurement in the tensile Hopkinson bar test. Meas. Sci. Technol. **5**(9), 1163–1171 (1994)

Chapter 7
Dynamic Fracture Response of a Synthetic Cortical Bone Simulant

Thomas Plaisted, Allan Gunnarsson, Brett Sanborn, and Tusit Weerasooriya

Abstract This work characterizes the fracture response of a composite material designed to mimic the response of human cortical bone. We have identified additive manufacturing, more generally known as 3-D printing, as a means of reproducing the curvature, variation in thickness, and gradient in porosity characteristic of the human bone between the cortical and trabecular regions. As the base material for developing bone surrogates via additive manufacturing, we evaluate a photocurable polymer with a high loading of ceramic particulate reinforcement that is compatible with stereolithographic additive (SLA) manufacturing. Specimens were printed in two orientations to measure fracture response perpendicular and parallel to the direction of deposition of the layer-by-layer manufacturing process. Mode I fracture behavior of the material was measured in four point bending configuration at high rate via modified split Hopkinson pressure bar for both orientations. In this paper, the fracture behavior of the bone simulant are presented and are compared to the mode I fracture behavior of human cortical bone perpendicular to the long axis of the human femur characterized under the same conditions.

Keywords Tissue simulant • Bio-mimicry • Fracture toughness • Cortical bone

7.1 Introduction

Dynamic loading to the human body often results in bone fracture. The study and prevention of these injuries requires bio-fidelic representation of bone in order to understand the kinematics of injury and make accurate assessments of protection schemes. Post Mortem Human Surrogates (PMHS) may be used in these studies, although the bone tissue frequently comes from older donors and may not represent the mechanical characteristics of the general population, particularly if the target population is that of a typical athlete or soldier. Moreover, there can be large variations in the mechanical response of the tissue depending on the age and health of the donor. Developing a synthetic material to mimic the mechanical response of human bone alleviates a number of these concerns. The response of a surrogate material can be tailored to represent a particular age population and specific bone within the human skeleton.

Human bone is a complex hierarchical composite material with variable mechanical properties throughout the different regions of the body. The basic constituents of human bone remain the same however: collagen fibers, hydroxyapatite mineral, and water. Each of these constituents are present in the hard outer cortical layer as well as the porous inner trabecular region of the bones. Long bones such as the femur exhibit increased strength and stiffness along the primary loading direction, whereas cranial bones are less stiff and exhibit transverse isotropy in directions tangent to the outer layer due to the absence of a predominant loading direction on the cranium [1].

Additive manufacturing offers an attractive means to reproduce the complex features such as the shape, curvature and gradient in porosity of human bones. The SLA manufacturing technique builds material by cyclically raising and lowering a platform on which the parts are fabricated in a bath of photo-sensitive liquid resin. With each cycle, a blade is passed across the platform to create a uniform layer of resin. The resin layer is exposed to an ultra-violet laser that traces a cross-section of the desired geometry, thereby selectively curing the resin into a specific pattern. The process is repeated as the platform is lowered into the resin and the next layer is cured on top of the previous layer.

T. Plaisted (✉) • A. Gunnarsson • T. Weerasooriya
Weapons and Materials Research Directorate, U.S. Army Research Laboratory, Aberdeen Proving Ground, Aberdeen, MD 21005-5069, USA
e-mail: thomas.a.plaisted.civ@mail.mil

B. Sanborn
Oak Ridge Institute for Science and Education, P.O. Box 117, Oak Ridge, TN 37831-0117, USA

Sandia National Laboratory, Albuquerque, NM 87185, USA

D. Casem et al. (eds.), *Dynamic Behavior of Materials, Volume 1*, Conference Proceedings of the Society for Experimental Mechanics Series, DOI 10.1007/978-3-319-41132-3_7

The work reported here characterizes the mode I fracture behavior of a polymer material that was produced by additive manufacturing to mimic human cortical bone. Results are compared to the mode I fracture response of cortical bone extracted from the human femur and characterized by the same experimental technique.

7.2 Experimental

7.2.1 *Materials and Fabrication*

A photosensitive polymer with a high loading of ceramic particulate reinforcement was chosen for its similar density and anticipated mechanical properties relative to the human cranial cortical bone [2]. Furthermore it was chosen for its compatibility with the SLA manufacturing process. Fracture specimens were printed via the SLA manufacturing process into rectangular beams with slightly oversized dimensions and were subsequently sanded to achieve the nominal final dimension of 50 mm × 4 mm × 3 mm, per guidelines in ASTM C1421-10 [3]. The planar, non-porous specimens were manufactured in two orientations: *flat* type specimens were made such that layer-by-layer deposition of the SLA process coincided with the 4 mm dimension of the specimen geometry, and *edge* type specimens in which the deposition coincided with the 3 mm dimension. Each layer was deposited with a nominal thickness of about 80 μm.

Specimens were then notched to a length of approximately 1.9 mm and the thickness approximately 200 μm. The notch was made such that the crack would propagate perpendicular to the layer deposition and for *edge* type specimen and parallel to layer deposition in *flat* type specimen (Fig. 7.1).

7.2.2 *Testing*

Four-point bending experiments on the notched beam specimens were conducted at dynamic loading rates using an aluminum Split Hopkinson Pressure Bar (SHPB) with specially designed loading fixtures. The loading fixtures consisted of a four point bending fixture, having two steel pins with a center to center distance of 10 mm on the incident side of the specimen, and two steel support pins with a center to center distance of 20 mm on the opposite side. The diameter of the loading and support pins were 2.4 mm as shown in Fig. 7.2.

The incident and transmission bars of the SHPB had a diameter of 31.75 mm and a length of 3.7 m. Wave shaping of the input pulse was employed to ensure dynamic equilibrium and a constant loading rate such that the incident bar end velocity was approximately 550 mm/s. The incident and transmission bars were instrumented with embedded quartz transducers to achieve high signal with little noise for the low loads inherent in fracture experiments on low impedance materials. Further details on the SHPB technique may be found here [5, 6]. A schematic of the SHPB experimental setup is shown in Fig. 7.3. A total of 7 *flat* printed specimen and eight *edge* printed specimen were tested under dynamic conditions.

The fracture toughness measured by the four point bending method specified in ASTM C1421-10 was calculated according to:

$$K_{I_{PB}} = f\left[\frac{P_{max}(S_0 - S_1)10^{-6}}{BW^{\frac{3}{2}}}\right]\left[\frac{3\left[\frac{a}{w}\right]^{\frac{1}{2}}}{2\left[1-\frac{a}{w}\right]^{\frac{3}{2}}}\right] \tag{7.1}$$

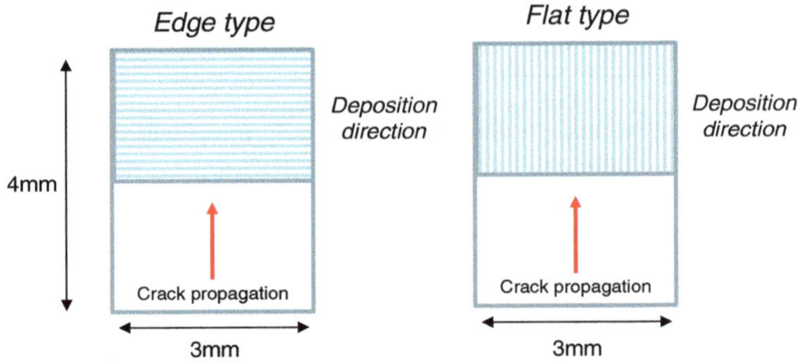

Fig. 7.1 Direction of crack propagation with respect to deposition direction for edge and flat type specimens

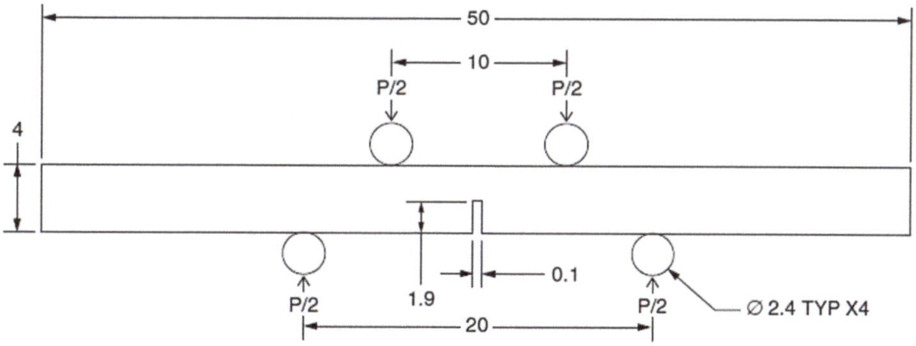

Fig. 7.2 Schematic of specimen and four point bending test fixture geometry [4]

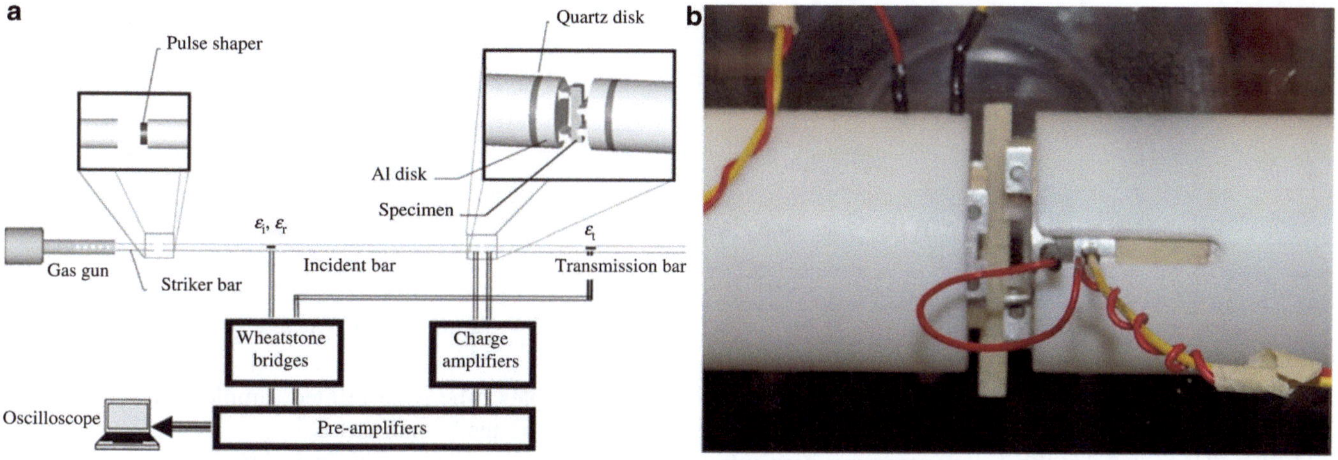

Fig. 7.3 Schematic (*left*) and photograph (*right*) of the SHPB experimental setup used for dynamic testing [4]

with

$$f = 1.9887 - 1.326\frac{a}{w} - \frac{\left\{3.49 - 0.68\left[\frac{a}{w}\right] + 1.35\left[\frac{a}{w}\right]^2\right\}\left[\frac{a}{w}\right]\left\{1 - \frac{a}{w}\right\}}{\left\{1 + \left[\frac{a}{w}\right]^2\right\}} \tag{7.2}$$

where K_{IPB} (MPa*√m) is the fracture toughness of a brittle four-point bending beam specimen with a notched pre-crack, P_{max} is the measured peak axial force (N) by the transmission quartz transducer, B is the specimen width (m) and W the specimen height (m), a is the notch or crack length (m), S_0 is the distance between the two supporting points on the notch side of the specimen (m), and S_1 is the distance between the two loading points on the loading side of the specimen (m). This equation applies to a specimen in dynamic equilibrium having a crack length to height ratio [a/w] between 0.35 and 0.60.

7.3 Results and Discussion

The initiation fracture toughness, K_{IPB}, for the *edge* printed and *flat* printed samples are provided in Table 7.1 as well as +/− one standard deviation, compared to the average of cortical bone from the femurs of three human donors (age 36, 43 and 50 years) reported previously [4].

These results are also plotted in Fig. 7.4 with respect to the dynamic loading rate, expressed as fracture toughness rate in units of MPa*√m*s^{-1}.

The results indicate that the dynamic initiation fracture toughness for the *flat* and *edge* samples both fell within the range of measured values for that of femoral cortical bone. The average fracture toughness was lower than the average of the

Table 7.1 Fracture toughness of donor tissue and cortical simulant material measured at quasi-static and dynamic rates

	Fracture toughness (MPa*√m)		
FT rate (MPa*√m/s)	Donor average	Simulant-flat	Simulant-edge
~10 K–50 K	5.7 ± 1.5	4.2 ± 0.4	4.9 ± 0.7

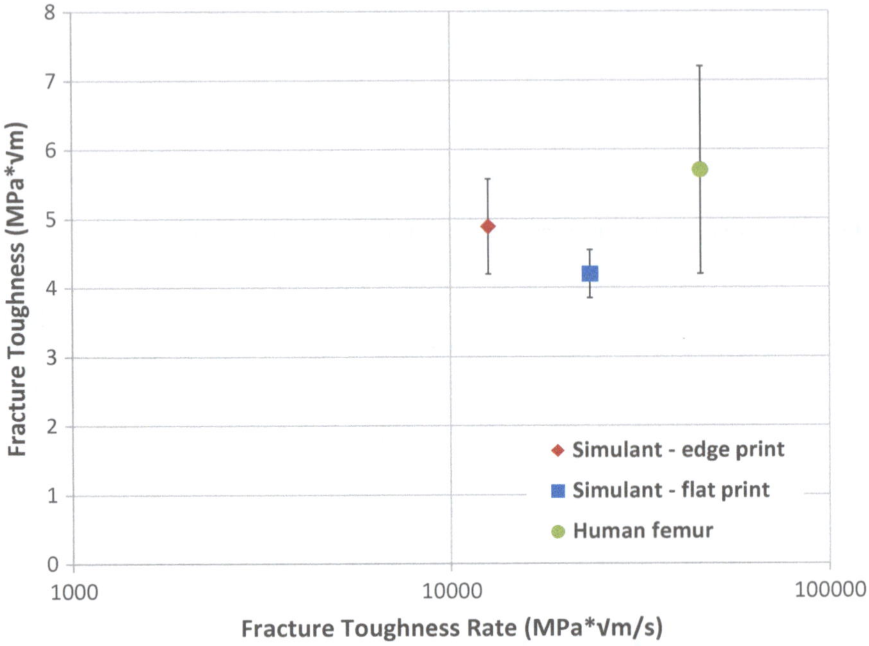

Fig. 7.4 Fracture toughness of simulant material printed in edge and flat orientation compared to that of human femur bone [4]

donors by about 14 % for the *flat* samples and 26 % for the *edge* samples. The *edge* printed samples showed a higher variation than the *flat* but the variation was still well below that observed in the human tissue.

It should be noted that the donor specimens were extracted such that the long axis of the beam specimen was parallel to the long axis of the femur. In this orientation, osteons within the bone lie perpendicular to the direction of crack propagation. Osteons are long cylindrical structures that house the blood supply, with nominal diameters around 300 μm, and run approximately along the axial direction of long bones. Fracture toughness in this direction is known to be significantly higher than that parallel to the osteons due to toughening mechanisms such as crack tip bridging and crack path deflection along the mineral rich and brittle interface of the osteon [7–10].

The simulant material studied here however does not possess a microstructural feature to mimic the fiber-like properties and resulting anisotropy due to the osteons. The particulate reinforced resin is largely isotropic in its liquid state. Any subsequent anisotropy is attributed to the conditions under which it was processed, including deposition induced anisotropy. In this respect the *edge* and *flat* samples differed in terms of the orientation of the layers relative to the direction of crack propagation. The *flat* samples exhibited a lower fracture toughness, where the crack was propagating parallel to the direction of deposition, which may indicate a weaker material along the inter-layer bond line. Further analysis of the crack initiation point and observation of the crack shape within each specimen is ongoing, as well additional studies to exploit this anisotropic feature to achieve closer correlation to human bone.

7.4 Summary

This research has evaluated a composite material consisting of a particulate reinforced polymer to serve as a surrogate material for human cortical bone. Fracture specimens were produced by SLA additive manufacturing and tested in 4 point bending using the SHPB technique to measure their mode I fracture response. The fracture toughness of the material measured parallel to the deposition direction (*flat* specimen) was 4.2 ± 0.4 MPa*√m, and that perpendicular to the deposition direction (*edge* specimen) was 4.9 ± 0.7 MPa*√m. Both average toughness values are lower than the average fracture toughness of human femur cortical bone (perpendicular to the osteonal direction) measured using the same technique, at 5.7 ± 1.5 MPa*√m, but within the standard deviation. In its current form, the material may be considered approximately close in fracture response to the femur. However it may even be closer to the fracture toughness of bones with less osteon induced anisotropy, such as the cranial bone, which is the subject of future characterization.

Acknowledgments This research was supported in part by an appointment to the Postgraduate Research Participation Program at the U.S. Army Research Laboratory administered by the Oak Ridge Institute for Science and Education through an interagency agreement between the U.S. Department of Energy and USARL.

References

1. McElhaney, J.H., Fogle, J.L., Melvin, J.W., Haynes, R.R., Roberts, V.L., Alem, N.M.: Mechanical properties of cranial bone. J. Biomech. **3**(5), 497–511 (1970)
2. Plaisted, T.A., Gardner, J.M., Gair, J.L.: ARL-RP-0552: characterization of a composite material to mimic human cranial bone. US Army Research Laboratory, Aberdeen Proving Ground (2015)
3. ASTM International: Standard test methods for determination of fracture toughness of advanced ceramics at ambient temperature, p. 31. ASTM International, West Conshohocken (2010). *C1421–10*
4. Gunnarsson, C.A., Sanborn, B., Foster, M., Moy, P., Weerasooriya, T.: Initiation fracture toughness of human cortical bone as a function of loading rate. In: Chalivendra, V., Song, B., Casem, D. (eds.) Dynamic behavior of materials, vol. 1, pp. 45–56. Springer, New York (2013)
5. Weerasooriya, T., Moy, P., Casem, D., Cheng, M., Chen, W.: A four-point bend technique to determine dynamic fracture toughness of ceramics. J. Am. Ceram. Soc. **89**(3), 990–995 (2006)
6. Casem, D., Weerasooriya, T., Moy, P.: Inertial effects of quartz force transducers embedded in a split Hopkinson pressure bar. Exp. Mech. **45** (4), 368–376 (2005)
7. Koester, K.J., Ager, J.W., Ritchie, R.O.: The true toughness of human cortical bone measured with realistically short cracks. Nat. Mater. **7**(8), 672–677 (2008)
8. Nalla, R.K., Kinney, J.H., Ritchie, R.O.: Mechanistic fracture criteria for the failure of human cortical bone. Nat. Mater. **2**(3), 164–168 (2003)
9. Yeni, Y.N., Norman, T.L.: Calculation of porosity and osteonal cement line effects on the effective fracture toughness of cortical bone in longitudinal crack growth. J. Biomed. Mater. Res. **51**(3), 504–509 (2000)
10. Weerasooriya, T., Sanborn, B., Gunnarsson, C.A., Foster, M.: Orientation dependent compressive response of human femoral cortical bone as a function of strain rate. J. Dynam. Behav. Mater. **2**(1), 74–90 (2016)

References

Chapter 8
Fracture Response of Cross-Linked Epoxy Resins at High Loading Rate as a Function of Glass Transition Temperature

John A. O'Neill, C. Allan Gunnarsson, Paul Moy, Kevin A. Masser, Joseph L. Lenhart, and Tusit Weerasooriya

Abstract The failure behavior of cross-linked polymer epoxies with different glass transition temperatures (Tg) was investigated under Mode I fracture at high loading rate using a novel experimental method with in situ observation of the fracture process. By varying the monomer choices, the properties of the epoxies can be tailored to achieve greater resistance to cracking and higher impact toughness. For these experiments, a unique four-point bending specimen was used. High rate experiments were conducted on a modified split Hopkinson pressure bar with pulse-shaping. High speed digital imaging was used to visualize failure initiation. The images were also used with digital image correlation to optically measure the crack opening displacement and crack propagation velocity. The experimental results were used to calculate the energy required to initiate fracture at high loading rate. The results indicate that the critical energy required to initiate fracture at high loading rate was higher for epoxies with lower Tg values, up to an optimum Tg. This dependence of critical energy on the Tg of the epoxy was similar to that which has been previously measured for the epoxy's impact resistance. In this paper, the experimental methods and results are discussed.

Keywords High Strain Rate • Glass Transition Temperature • Epoxy • Fracture Response • Mode I Fracture • Hopkinson Bar

8.1 Introduction

Polymer epoxies are frequently used in many applications to structurally bond components of similar or dissimilar materials. Some of the many uses can be found in the form of fiber reinforced polymers (FRP) for the aerospace and automotive industries, sporting goods, and marine crafts. When exposed to sudden impacts, these structural components and bonds are subjected to high loading rates. Typically, the failure of the resin matrix in FRPs is the dominating factor leading to interply delamination. Therefore, it is important to understand how these epoxies behave and fail at high loading rate. High loading rates are especially important as they are characteristic of blast and ballistic impact.

The fracture response of polymer epoxies has been investigated by many researchers. At quasi-static loading rates, studies have been performed to understand crack initiation, the effects of molecular weight between cross-links, and the fracture behavior of epoxy systems using various tougheners. Specimen geometries based on various ASTM standards have been developed for both three and four point bend specimen geometries [1–6].

Experimental techniques for dynamic fracture behavior have been developed and can be summarized into three categories: high rate bending, high rate tension, and dynamic wedging. Many different techniques exist; some of these have been summarized previously by Jiang et al. [7]. Previous studies have shown that there are two important criteria for valid and useful high rate fracture experiments: a state of dynamic equilibrium of the specimen, and a constant loading rate of the specimen during the experiment. Both of these criteria can be obtained by shaping the input loading pulse [8–14]. Additionally, low impedance materials, such as pre-cracked beams and polymer materials, frequently cause traditional strain gauge measurement techniques to be unusable. Other researchers, such as Chen et al. [9], have solved this problem by embedding quartz based loading devices in the loading train, which allows for direct measurement of dynamic equilibrium and loading rate. Weerasooriya et al. developed an experimental method to measure fracture toughness under four-point

Certain commercial equipment, instruments, or materials are identified in this paper in order to specify the experimental procedure adequately. Such identification is not intended to imply recommendation or endorsement by the Army Research Laboratory, nor is it intended to imply that the materials or equipment identified are necessarily the best available for the purpose.

J.A. O'Neill • C.A. Gunnarsson • P. Moy • K.A. Masser • J.L. Lenhart • T. Weerasooriya (✉)
Army Research Laboratory, Aberdeen Proving Ground, Aberdeen, MD 21005, USA
e-mail: tusit.weerasooriya.civ@mail.mil

D. Casem et al. (eds.), *Dynamic Behavior of Materials, Volume 1*, Conference Proceedings of the Society for Experimental Mechanics Series, DOI 10.1007/978-3-319-41132-3_8

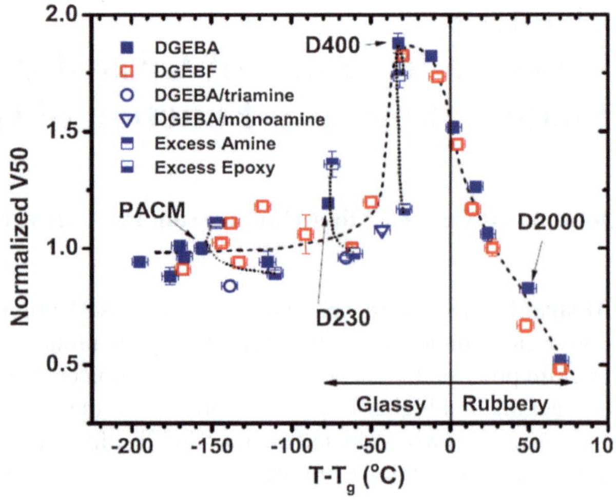

Fig. 8.1 Normalized impact resistance (V50) data for a series of epoxy networks as a function of reduced temperature [1]

bend at high loading rates, and used it to measure the dynamic fracture toughness of SiCN ceramic [10] and PMMA [11]. Gunnarsson et al. [12] further developed this technique to measure the fracture toughness of human cortical bone, integrating ultra-high speed imaging and crack tip strain analysis. Syn et al. [13] studied the surface morphology effect on the fracture behavior along dissimilar material interfaces (in this case, aluminum and epoxy) using a wing-like specimen. Weerasooriya et al. [14] used a similar method to study the dynamic failure behavior of adhesive bond interfaces. Both Weerasooriya et al. and Syn et al. showed an increase in failure load and energy rate for fracture initiation as loading rate increased from low to high rates.

Previous investigations have demonstrated that the proximity of the testing temperature to the glass transition temperature strongly influences the ballistic performance of amine-cured epoxy networks [15, 16]. Epoxy/amine mixtures well into the glassy state (DGEBA/PACM) or well into the rubbery state (DGEBA/D2000) perform relatively poorly, failing in a brittle manner in the case of DGEBA/PACM, or by complete projectile penetration of the rubbery material in the case of DGEBA/D2000. This is shown more clearly in Fig. 8.1. Epoxy/amine mixtures such as DGEBA/D400, whose glass transition temperature is close to the ballistic measurement temperature, fail through a combination of modes, exhibiting plastic deformation and radial and cone cracking. The observation that being close to the glass transition improves ballistic performance has also been observed by other researchers [17–20], and is likely a general trend of polymer networks.

In this investigation, the high rate fracture behavior of three amine-cured epoxy networks is characterized by conducting Mode I fracture experiments on epoxy (DGEBA-based) specimens with varying Tg. Epoxy (DGEBA) and amine mixtures were chosen such that they were well below (DGEBA/PACM), below (DGEBA/D230), or near (DGEBA/D400) the glass transition temperature, corresponding to a range of poor to relatively good ballistic performance. A fracture specimen was used similar to previous work [13, 14]. The specimen's loading surfaces, the nose and the two wing-tips, were designed to be curved instead of flat. This modification stiffened the loading nose of the specimen to reduce the potential of the specimen bending when loaded. It is important that the specimens do not bend, and that all energy absorbed is used to initiate failure and crack propagation. The effectiveness of the curved surface design in eliminating the bending problem in the specimens was verified using ultra high speed digital image correlation (DIC). An additional benefit is the ability to use flat-faced loading fixtures instead of pin-based loading fixtures. This allows for use of the incident and transmission bar ends to load the specimen, instead of special fixtures. Directly using the bar ends for specimen loading reduces experimental complexity, and improves wave transmission by removing changes in impedance in the loading train. All experiments were conducted at dynamic loading rate using an aluminum split pressure Hopkinson bar (SHPB) system. High sensitivity semi-conductor strain gages were used to accurately measure the bar strain histories and stresses at the loading interfaces to verify dynamic equilibrium. The experimental results were used to quantify the Tg dependence of DGEBA based polymer system fracture properties, such as crack propagation velocity (CPV), failure load, and failure energy. This research will provide a quantitative fracture-resistance experimental technique for evaluation of epoxy systems as a function of Tg. Future work will be performed to relate the failure load to the initiation fracture toughness.

8.2 Experimental

8.2.1 Materials

Diglycidyl ether of bisphenol A (DGEBA), i.e., EPON 825 (average molecular weight ~355 g/mol), was obtained from Miller-Stephenson [7] and was used as the epoxy. Three curing agents were used with the DGEBA: polypropylene oxide based-Jeffamine diamines with a vendor-specified molecular weight of 230 g/mol (D230) and 400 g/mol (D400) were provided by Huntsman. Diamine curing agent, 4,4'-methylenebis(cyclohexylamine) (PACM), was provided by Air Products. PACM is an unmodified cycloaliphatic amine commonly used in filament winding and wet layup laminating for fabrication of fiber-reinforced polymers, where high fracture toughness and better corrosive properties are necessary. All epoxies and curing agents were used as received without further purification. All formulations were stoichiometric mixtures of the epoxy and diamine curing agent. Sample designations are based on the amine curing agent used, e.g. D230 refers to a stoichiometric mixture of DGEBA and D230. The chemical structure of the epoxy and curing agents used in this study are shown in Fig. 8.2.

The high rate fracture specimen geometry is shown in Fig. 8.3 (all dimensions mm). Silicone molds were prepared by casting Dow Xiameter brand silicone RTV onto approximately 50 steel sample blanks. After curing, the sample blanks were removed, revealing high fidelity cavities. The molds were coated with several layers of mold release to aide sample removal. For all sample preparations, epoxies and curing agents were preheated to 60 °C. Each epoxy and diamine was then mixed and stirred vigorously at 60°C for 5 min, poured into silicone molds, and degassed under vacuum. Every epoxy mixture appeared uniformly mixed, with no phase separation visible. All epoxies were cured under a nitrogen purge with a cure cycle of 80°C for 2 h, 150°C for 8 h, and 200°C for 2 h. This cure cycle has been shown to result in networks where no uncured functional groups can be detected.

The complexity of the fracture geometry prevented the use of an aluminum mold to direct cast the epoxy into the fracture geometry. This is due to the difficultly in extracting the epoxy from an aluminum mold without damaging the sample even if

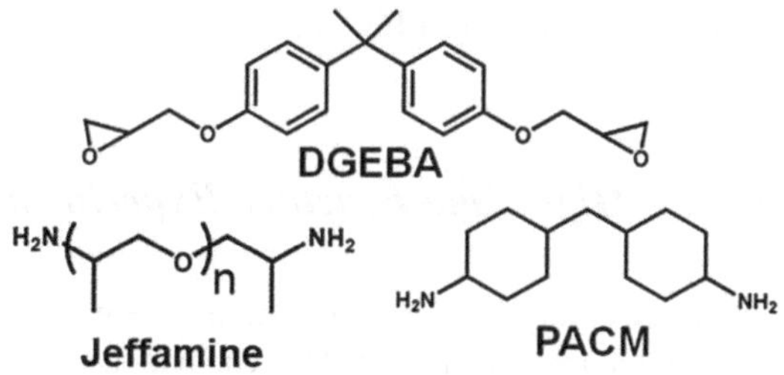

Fig. 8.2 Chemical structures of DGEBA, the general structure of the Jeffamine series (D230, D400) and the chemical structure of the PACM diamine

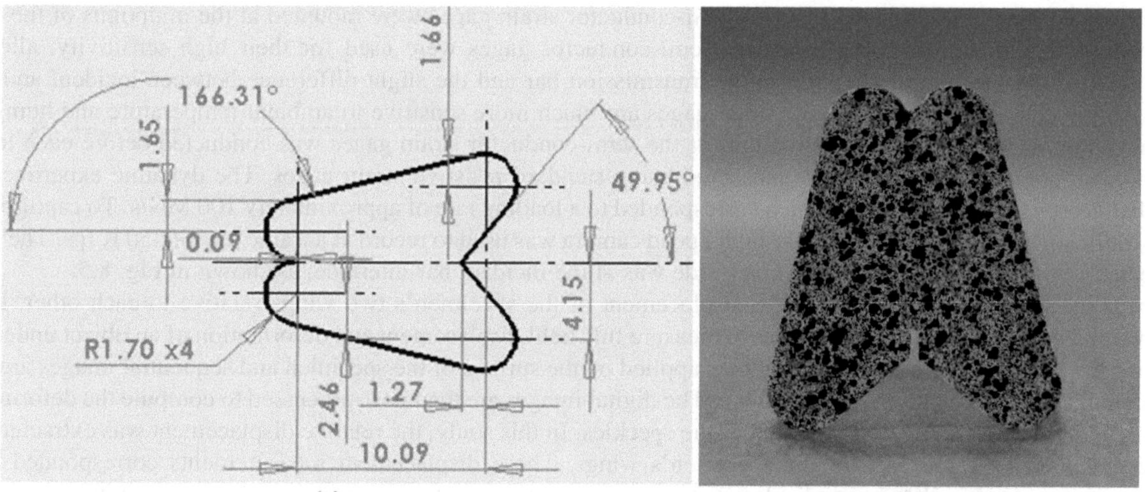

Fig. 8.3 (**a**) Fracture specimen geometry and (**b**) typical specimen with notch and DIC speckle pattern

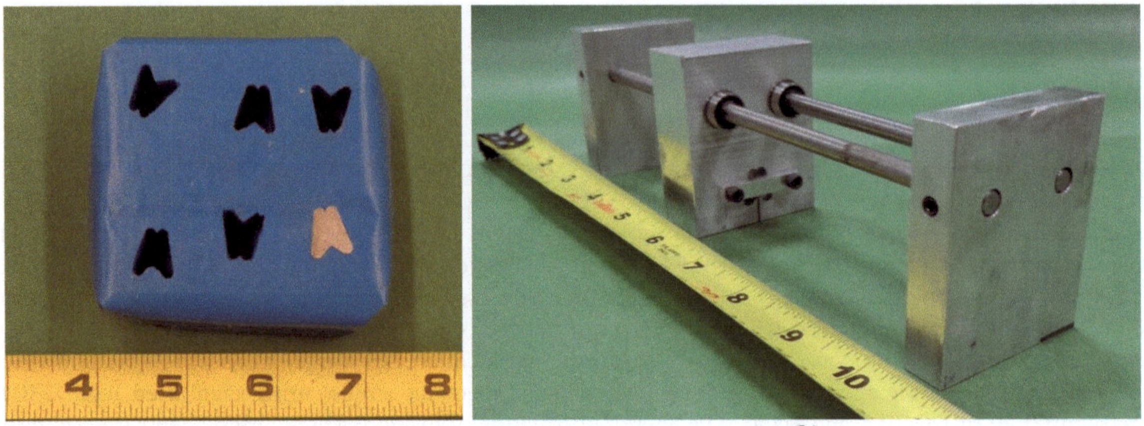

(a) (b)

Fig. 8.4 (**a**) Silicon mold used to mold the epoxy and (**b**) custom notching fixture

mold release is applied. Instead, a flexible silicon mold was created. To fabricate the mold, fracture specimens with the geometry as shown in Fig. 8.3 were machined from 25.4 mm thick aluminum and cast in silicon. The flexibility of the silicone enabled the replicate aluminum specimens to be removed from the silicone mold once the mold was cured. An example of the silicone mold is shown in Fig. 8.4. The epoxy mixture could then be poured into the silicon mold and cured.

Once the epoxy specimens were finished curing, they were pressed out of the mold and milled to a thickness of 23.25 mm using a table mill. This process was necessary to remove the uneven material present near the mold surfaces after curing and to create flat and parallel surfaces. With the specimens at the proper length, a notch was cut with a 300 μm width diamond saw blade. The present study did not investigate any other pre-crack profile other than the notch created from the saw blade. A custom fixture (Fig. 8.4b) for specimen notching was designed to create a constant 1 mm deep notch into the center of each specimen as shown in Fig. 8.3b.

8.2.2 *High Rate Fracture Experiments*

Dynamic epoxy fracture experiments were conducted on a modified SHPB. The working principle of this experimental setup is well documented [7–10, 21]. The bars used for the incident and transmission bar in the test setup were made from Al 7075 with a diameter of 31.8 mm and a length of 3.66 m. A gas gun propelled Al 7075 striker with a length of 0.61 m was used to generate a stress pulse in the incident bar. Pulse shaping was used to shape the input pulse to achieve the necessary dynamic equilibrium and a near constant loading rate. Semi-conductor strain gages were mounted at the midpoints of the input and output bars to record the bar stress histories. Semi-conductor gages were used for their high sensitivity, allowing for measurement of the low amplitude pulse in the transmission bar and the slight difference between incident and reflected pulse in the incident bar. Semi-conductor strain gages are much more sensitive to ambient temperature and humidity than traditional strain gages; therefore a calibration for the semi-conductor strain gages was conducted before each test series, using a known compressive strain pulse measured using standard resistive strain gages. The dynamic experiments had a relative bar velocity of about 5.0 m/s, which corresponded to a loading rate of approximately 100 MN/s. To capture the onset of crack initiation and propagation, an ultra-high-speed camera was used to record at a frame rate of 250 K fps. The specimen was oriented in the SHPB so that the pre-crack side was at the incident bar interface, as shown in Fig. 8.5.

DIC was used to accurately measure the displacement of the specimen's two wings relative to each other. DIC is an optical method which uses sequential images to measure full-field displacement and deformation of an object under load. To summarize the DIC process, a speckled pattern is applied on the surface of the specimen and sequential images are captured using a digital camera throughout an experiment. The digital images are then post-processed to compute the deformation and displacement field by tracking the movement of the speckles. In this study, the relative displacement was extracted between two selected points, one on each of the specimen's wings. These displacement measurements corresponded to digital extensometers measuring the perpendicular distances across the crack, and were used to obtain the crack opening displacement (COD) as a function of time during the experiment. COD is the distance that the two halves of the specimen separate away from each other as the crack progagates along the specimen length.

Fig. 8.5 Mode I epoxy fracture specimen loaded in SHPB (speckle surface faces away, towards camera)

8.3 Results and Discussion

8.3.1 Crack Propagation Velocity

DIC was used to calculate the COD at various locations along the specimen using digital extensometers, similar to bonded crack detection gages. These extensometers and their locations are shown in Fig. 8.6a. The COD at each of these locations was extracted and plotted together (Fig. 8.6b). By visually inspecting the images for crack tip arrival at an extensometer location, and comparing this to the corresponding extensometer's COD, it was determined that the crack had reached the extensometer location when the COD reached 0.5 pixels. The time at which each extensometer reached 0.5 pixel was extracted, and compared to the location of the extensometer along the specimen length to extract a time-crack tip location relationship.

The digital extensometer results indicated that the crack traveled at a linear rate (constant velocity) along the specimen. As an example, Fig. 8.7 shows the crack location as a function of time for a single PACM specimen under a high loading rate experiment. The five data points correspond to the five digital extensometers used for this experiment; the time value of each data point is the amount of time it required for each extensometer to measure a length increase (COD) of 0.5 pixels. The crack tip location followed a linear trend and the slope of the linear fit to the points indicates the crack tip propagation velocity (CPV), in this case 234.4 m/s.

8.3.2 Failure Load and Energy

The failure load is simply the maximum force that the specimen experienced during the experiment. Future work will relate this failure load to the fracture toughness (K_I) of the material. The failure energy is the energy required to initiate crack propagation from the original precrack tip. Failure energy was calculated using Eq. 8.1 by integrating load history with respect to displacement, from zero displacement to the displacement corresponding to crack growth.

$$FE = \int_{x_o}^{x_{crack\ growth}} P(x)\,dx \tag{8.1}$$

Table 8.1 shows a summary of the average results of the Mode I fracture DGEBA epoxy experiments at the dynamic loading rate, including plus/minus statistics representing plus/minus one standard deviation of the data. Figure 8.8 shows these results graphically as a function of the reduced temperature for (a) crack propagation velocity, (b) crack initiation load, and (c) failure energy. The reduced temperature (T−Tg) is simply the material Tg subtracted from the material temperature at time of testing (here, for all experiments, room temperature, or 23 °C). In each plot, the black dots represent the average value for each loading rate. The results show that all three fracture parameters for DGEBA epoxy increased when the Tg

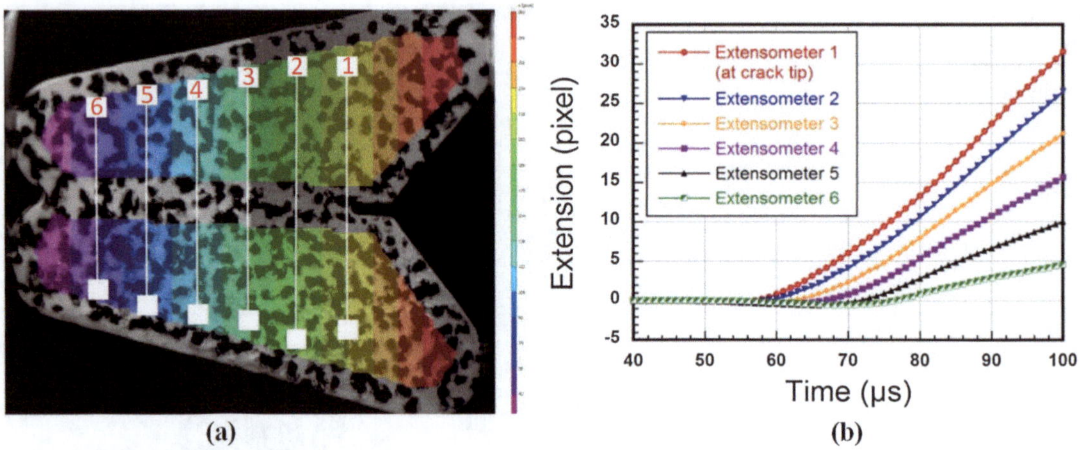

Fig. 8.6 (**a**) Location of digital extensometers across specimen and (**b**) COD of extensometers vs time

Fig. 8.7 Example of crack location vs time for a PACM specimen under high-rate loading

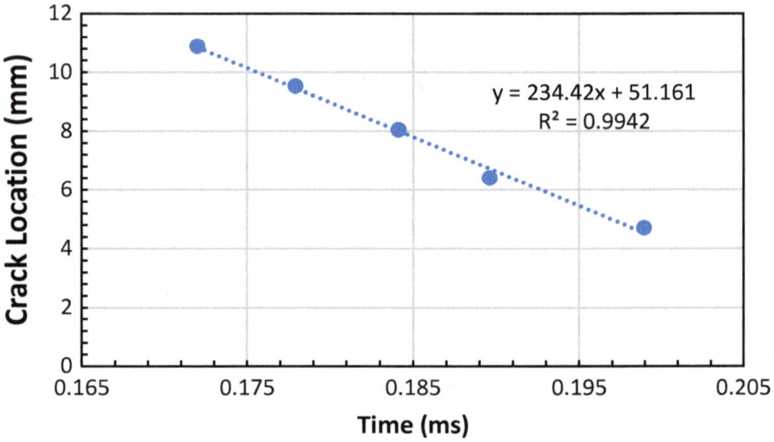

Table 8.1 Summary of results for Mode I fracture experiments of DGEBA-based epoxies at high-loading rate

Cure agent	Tg (°C)	T−Tg (°C)	Crack propagation velocity (m/s)	Failure load (kN)	Failure energy (J)
PACM	160	−137	234.47 ± 16.65	9.101 ± 0.596	1.502 ± 0.357
D230	90	−67	302.79 ± 47.62	10.916 ± 1.224	2.373 ± 0.786
D400	45	−22	291.96 ± 43.43	10.485 ± 1.788	2.989 ± 1.409

decreased from 160 to 90 °C, corresponding to a change in cure agent from PACM to D230. However, both CPV and failure load decreased slightly when the Tg decreased from 90 to 45 °C, with the cure agent changed from D230 to D400,. The failure energy showed a clear trend, linearly increasing as the T−Tg increased (Tg decreased). The failure energy was greatest for the DGEBA formulation that utilized D400 as the cure agent, which represents a Tg for the epoxy that is closest to the apparent ideal Tg for impact protection.

8.4 Conclusions

Mode I fracture experiments were conducted at high loading rate on DGEBA epoxy based systems using a unique fracture specimen geometry. The Tg of the epoxy system was varied by altering the curing agent, so that the fracture properties of the DGEBA epoxy could be studied as a function of Tg. High speed imaging was utilized to accurately deterimine the timing of crack initiation/failure, and DIC was used to measure the crack propagation velocity. The results show that the measured

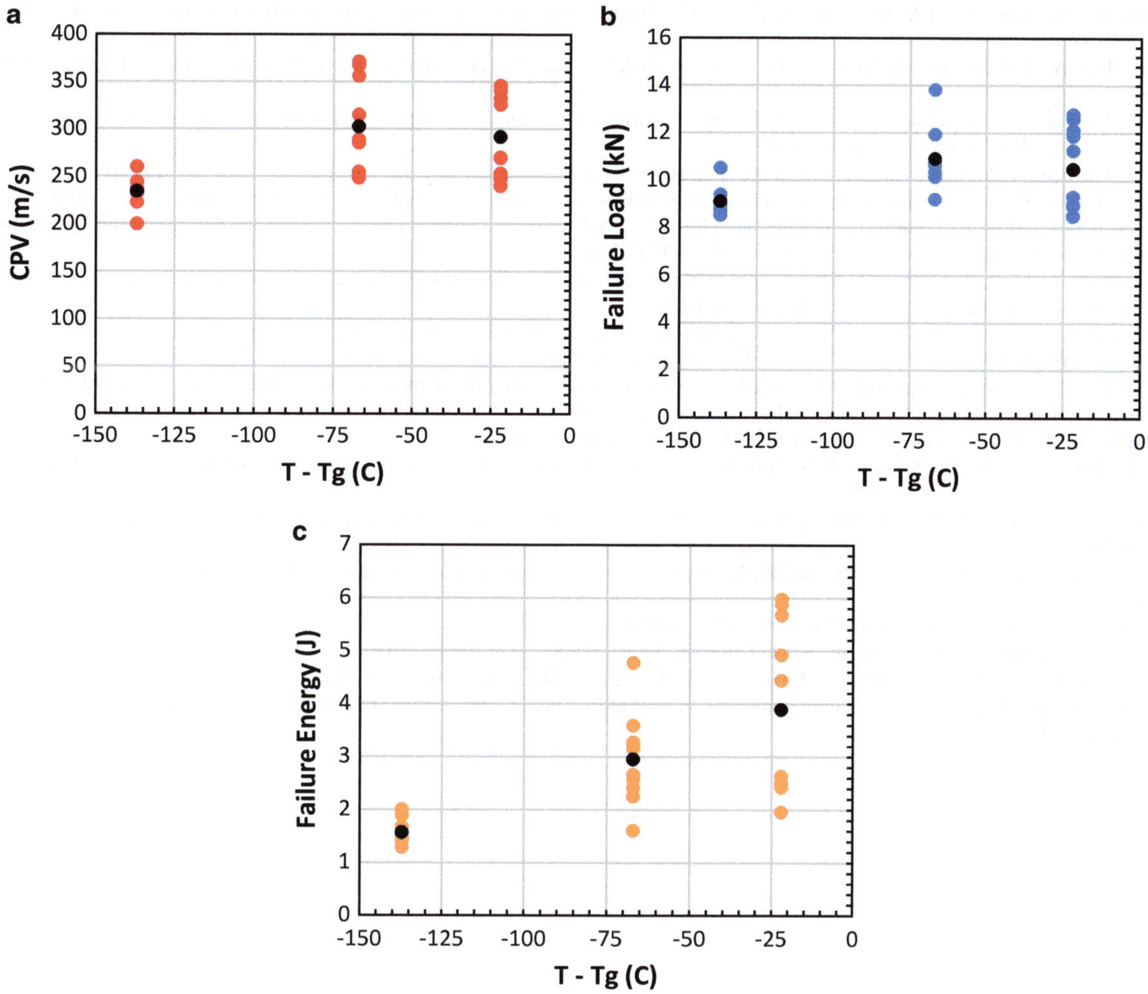

Fig. 8.8 Relationship between the curing agent glass transition temperature and (**a**) crack propagation velocity, (**b**) failure load, and (**c**) failure energy

critical fracture parameters for DGEBA epoxy increase when the cure agent was altered from PACM to D230 (Tg decreased from 160 to 90 °C). However, both CPV and failure load decreased slightly as the cure agent was again changed from D230 to D400, with the Tg decreasing from 90 to 45 °C. On the other hand, there was a clear linear trend of increasing failure energy as the reduced temperature (T−Tg) increased, corresponding to decreasing epoxy Tg. The failure energy was greatest for the D400-DGEBA formulation, which represents a reduced temperature (T−Tg) closest to the point which has demonstrated maximum impact resistance. Future studies will investigate the effect of moving the cure agent Tg to the rubbery region, other epoxy systems with dispersed tougheners, changing the cross-link density, and molecular weight. Additionally, the rate dependence of the fracture properties shown here will be investigated for lower loading rates.

References

1. Cho, K., Lee, D., Park, C.E., Huh, W.: Effect of molecular weight between crosslinks on fracture behaviour of diallylterephthalate resins. Polymer **37**(5), 813–817 (1996)
2. Kang, B.U., Jho, J.Y., Kim, J., Lee, S., Park, M., Lim, S.: Effect of molecular weight between crosslinks on the fracture behavior of rubber-toughened epoxy adhesives. J. Appl. Polym. Sci. **79**, 38–48 (2001)
3. Hussain, M., Nakahira, A., Nishijima, S., Niihara, K.: Fracture behavior and fracture toughness of particulate filled epoxy composites. Mater. Lett. **27**, 21–25 (1996)
4. Mostovoy, S., Ripling, E.J.: Fracture toughness of an epoxy system. J. Appl. Polym. Sci. **10**, 1351–1371 (1966)

5. Wakako, A., Kentaro, N., Tadaharu, A., Akihiko, Y.: Fracture toughness for mixed mode I/II of epoxy resin. Acta Mater. **53**(3), 869–875 (2005)
6. Cho, K., Huh, W.: Effect of molecular weight between crosslinks on fracture behaviour of diallylterephthalate resins. Polymer **37**(5), 813–817 (1996)
7. Jiang, F., Vecchio, K.: Hopkinson Bar loaded fracture experimental technique: A critical review of dynamic fracture toughness tests. Appl. Mech. Rev. **62** (2009). Transactions of the ASME
8. Wu, X.J., Gorham, D.A.: Stress equilibrium in the split Hopkinson pressure Bar test. J. Phys. IV **7**(C3), 91–96 (1997)
9. Chen, W., Lu, F., Zhou, B.: A quartz crystal imbedded split Hopkinson bar for soft materials. Exp. Mech. **40**(1), 1–6 (2000)
10. Weerasooriya, T., Moy, P., Casem, D., Cheng, M., Chen, W.: A four point bending load technique for determination of dynamic fracture toughness for ceramics. J. Am. Ceram. Soc. **89**(3), 990–995 (2006)
11. Weerasooriya, T., Moy, P., Casem, D., Cheng, M., Chen, W.: Fracture toughness for PMMA as a function of loading rate. Proceedings of the 2006 Society of Experimental Mechanics Annual Conference
12. Gunnarsson, C.A., Sanborn, B., Foster, M., Moy, P., Weerasooriya, T.: Initiation fracture toughness for human cortical bone as a function of loading rate. Proceedings of the 2012 Society of Experimental Mechanics Annual Conference
13. Syn, C., Chen, W.: Surface morphology effects on high-rate fracture of an aluminum epoxy interface. J. Compos. Mater. **42**, 1639–1658 (2008)
14. Weerasooriya, T., Gunnarsson, C.A., Jensen, R., Chen, W.: Strength and failure energy for adhesive interfaces as a function of loading rate. Proceedings of the 2011 Society of Experimental Mechanics Annual Conference
15. Knorr Jr., D.B., Yu, J.H., Richardson, A.D., Hindenlang, M.D., McAninch, I.M., La Scala, J.J., Lenhart, J.L.: Polymer **53**(25), 5917–5923 (2012)
16. Masser, K.A., Knorr Jr., D.B., Hindenlang, M.D., Yu, J.H., Richardson, A.D., Strawhecker, K.E., Beyer, F.L., Lenhart, J.L.: Polymer **58**, 96–106 (2015)
17. Bogoslovov, R.B., Roland, C.M., Gamache, R.M.: Impact-induced glass transition in elastomeric coatings. Appl. Phys. Lett. **90**, 221910 (2007)
18. Roland, C.M.: Rubber Chem. Technol. **79**(3), 429–459 (2006)
19. Roland, C.M., Fragiadakis, D., Gamache, R.M.: Compos. Struct. **92**(5), 1059–1064 (2010)
20. Roland, C.M., Fragiadakis, D., Gamache, R.M., Casalini, R.: Philos. Mag. **93**(5), 468–477 (2013)
21. Davies, E.D.H., Hunter, S.C.: The dynamic compression testing of solids by the method of the split Hopkinson pressure bar. J. Mech. Phys. Solids **11**(3), 155–179 (1963)

Chapter 9
Measurement of Dynamic Response Parameters of an Underdamped System

Charandeep Singh, Satish Chaparala, and S.B. Park

Abstract The dynamic behavior of a system is highly influenced by the energy dissipation due to its damping. To clearly understand dynamic behavior, an accurate measurement of damping is very important. One of the most commonly used methods for evaluating damping coefficients is the measurement of decay of free vibrations. The purpose of this study is to introduce a new experimental technique, a pluck test coupled with High speed Digital Image Correlation (DIC) [1] technique, to analyze the dynamic response of underdamped systems. It is further validated comparing the results from this test with those of obtained from the existing techniques such as using laser vibrometer and by using a general purpose single-axis accelerometer. Using the pluck test, an initial displacement is introduced into the system and the response with time is measured using DIC. High speed DIC employs two synchronized high speed digital cameras. Images captured at a high frame rate by high speed cameras are analyzed using 3D DIC software and therefore, a full field dynamic response of the test subject is captured The parameters such as natural frequency and damping ratio are obtained from the measured time history of the displacement using logarithmic decrement method [2]. The test vehicles used in this study to implement this technique are different liquid crystal display (LCD) modules used in commercially available smart phones and Corning® Gorilla® Glass that is used as cover glass in the same.

Keywords Pluck test • Digital image correlation (DIC) • Finite element modelc (FEM) • Underdamped system • Laser vibrometer

9.1 Introduction

DIC is a non-contact optical deformation measurement technique that is capable of measuring both in plane and out of plane deformation of a test subject. It provides full field measurement up to sub pixel accuracy [3–8]. DIC correlates the digital images of a test subject captured at a reference stage and subsequent deformed stage. It tracks the gray scale intensity of speckle patterns on the surface of the test subject during and after deformation. In the reference image, based of gray scale of the pixel, numerical value between 0 and 225 is assigned to each pixel. 0 is complete black and 225 is complete white. A certain number of pixels is bundled up and form a subset. By choosing subsets from reference image and searching for an optimal match throughout the deformed images, a displacement field is obtained.

Laser Doppler Vibrometer (LDV) is another precision optical measurement technique to determine vibration velocity and displacement at a certain point. It significantly extends the measurement capabilities as compared to traditionally used surface contacting vibration sensors such as accelerometers and strain gauges [9]. Like DIC it is non-contact measurement technique and its fundamental principle of operation is based on the Doppler effect, sensing the frequency shift of back scattered light from a vibrating body. Many case of engineering interest like light weight testing sample or rotation specimen, a physical contact from accelerometer can cause significant error in results [10]. In cases like these, non-contact techniques like high speed DIC and Laser vibrometer are very useful.

The purpose of this work is to introduce new experimental technique, which is pluck test coupled with high speed Digital Image Correlation to measure dynamic decay in underdamped systems. A pluck test involves applying a displacement initial condition (rather than a velocity initial condition, as in impulsive tests) to a suitably mounted test object and measuring its subsequent response at various locations as it executes free vibrations. By properly selecting the locations and the

C. Singh • S.B. Park (✉)
Department of Mechanical Engineering, State University of New York at Binghamton, Binghamton, NY 13902, USA
e-mail: sbpark@binghamton.edu

S. Chaparala
Corning Incorporated, Corning, NY 14831, USA

D. Casem et al. (eds.), *Dynamic Behavior of Materials, Volume 1*, Conference Proceedings of the Society for Experimental Mechanics Series, DOI 10.1007/978-3-319-41132-3_9

magnitudes of the initial displacements, it is sometimes possible to excite various modes of vibration, provided that these modes are reasonably uncoupled. In this study, this technique is implemented on several commercially available LCD modules used in smartphones and Corning® Gorilla® Glass used as cover glass in several smart phones to measure natural frequency and damping ratio. The results are compared with those obtained from laser vibrometer and single axis accelerometer. The pluck test is the most common test that uses the initial-displacement method. A schematic diagram of the test set-up is shown in Fig. 9.2. The test object is initially deflected by pulling it manually and releasing it. When the test panel is suddenly released, it undergoes free vibrations about its static equilibrium position. The response is measured using DIC for several locations of the test object and analyzed to obtain the required parameters.

Sometimes it is difficult to excite a single mode when trying to search for natural frequency. If two natural frequencies are close together, modal interactions of the two frequencies invariably will be present in the response measurements. Because of the closeness of the frequencies, the response curve can show a beat phenomenon and that makes it difficult to determine damping by the logarithmic-decrement method. It is difficult to distinguish between decay caused by damping and rapid drop-off caused by beating. In this case, one of the frequency components must be filtered out, using a very narrowband-pass filter, before computing damping.

9.2 Analytical Model

9.2.1 Equation of Motion

Equation 9.1 is the equation of motion of a simple spring mass system with damping written in terms of damping ratio, ξ and undamped natural frequency, ω_n where $\ddot{x}$ represents the dynamic response.

$$\ddot{x} + 2\xi\omega_n\dot{x} + \omega_n^2 x = 0 \tag{9.1}$$

Solution obtained is as follows:

$$x(t) = e^{-\xi\omega_n t}\left[a_1 e^{i\omega_n t\sqrt{1-\xi^2}} + a_2 e^{i\omega_n t\sqrt{1-\xi^2}}\right] \tag{9.2}$$

Using Euler's formula Eq. 9.2 can be written in more convenient form:

$$x(t) = Ae^{-\xi\omega_n t}\, Sin\,[\omega_d t + \phi] \tag{9.3}$$

Where A and ϕ are the amplitude and phase of the response respectively. Damped frequency is expressed in terms of natural frequency and damping ratio is $\omega_d = \omega_n\sqrt{1-\zeta^2}$.

9.2.2 Determination of Damping Ratio Fig. 9.1

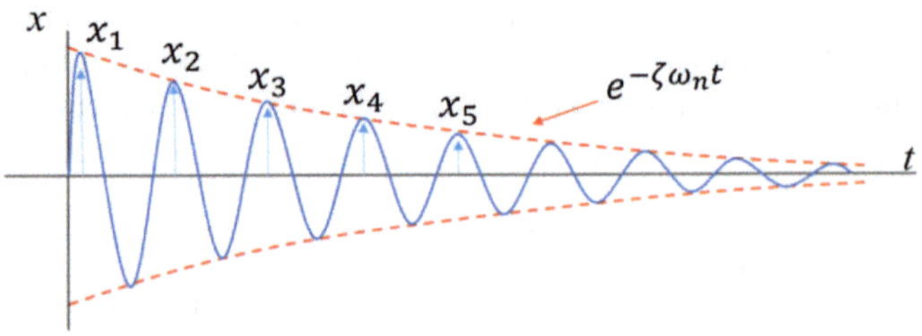

Fig. 9.1 Deformation decay in case of underdamped system with respect to time

$$\frac{x_1}{x_2} = \frac{Ae^{-\xi\omega_n T1} sin\,(\omega_d T_1 + \phi)}{Ae^{-\xi\omega_n T2} sin\,(\omega_d T_2 + \phi)} \tag{9.4}$$

$$\frac{x_1}{x_2} = \frac{e^{-\xi\omega_n T1}}{e^{-\xi\omega_n\left(T_1 + \frac{2\pi}{\omega_d}\right)}} = e^{2\pi\frac{\omega_n}{\omega_d}} \tag{9.5}$$

$$\ln\left(\frac{x_1}{x_{n+1}}\right) = \ln\left(\frac{x_1}{x_2}\right) + \ln\left(\frac{x_2}{x_3}\right) .. + \ln\left(\frac{x_n}{x_{n+1}}\right) = n\delta \tag{9.6}$$

$$\delta = \frac{1}{n}\ln\left(\frac{x_1}{x_{n+1}}\right) \tag{9.7}$$

$$\xi = \frac{\delta}{\sqrt{(2\pi)^2 + \delta^2}} \tag{9.8}$$

9.3 Experimenatal Study

9.3.1 Pluck Test

In this study six test vehicles are used: they are LCD from commercially available four different smart phones (Display Stack A, B, C and D) as well as Corning Gorilla single glass and laminate glass. Laminate glass consists of two individual glasses bonded using Optically Clear Adhesive (OCA). The bottom edge of the specimen is constrained in a vise grip as shown in schematic for all in Fig. 9.2. Sample is manually displaced from top edge and dynamic decay is observed by three techniques.

9.3.2 High Speed Digital Image Correlation

9.3.2.1 Specimen Preparation and Calibration

A thin layer of speckle pattern is spray painted on to the surface of the sample. Spray paint is applied such that random, high contrast black and white dots cover the surface of the sample. The cameras are calibrated to a field of view (FOV) suitable for the size of the sample so that maximum number of pixels is used by image of the sample, resulting in less noise and a more accurate result. Pictures of the calibration panel are sequentially captured at appropriate distance and orientation. A photogrammetry process known as bundle adjustment is used to create an accurate relationship between two cameras.

9.3.2.2 High-Speed Measurement

When the calibration process completes, calibration panel is removed and sample is introduced. The schematic in Fig. 9.2a shows the experimental set up consisting of two high speed cameras, halogen lights and sample which is held in vise grip. Halogen lights of power 650 W allows cameras to have minimum exposure setting and higher frames per second capability. To avoid heating of cameras and sample, lights are only used during measurement. In this study 3000 frames per second were acquired. Cameras are triggered and sample is manually plucked in the middle. Dynamic decay is captured by cameras at resolution of 1024 × 1024 pixels. Images from high speed cameras are imported in DIC. DIC traces the movement of the speckle pattern on the sample, by comparing the speckle pattern on deformed image with reference image, in plane and out of plane deformation is calculated. Experimental evaluation of dynamic decay is presented in Fig. 9.3.

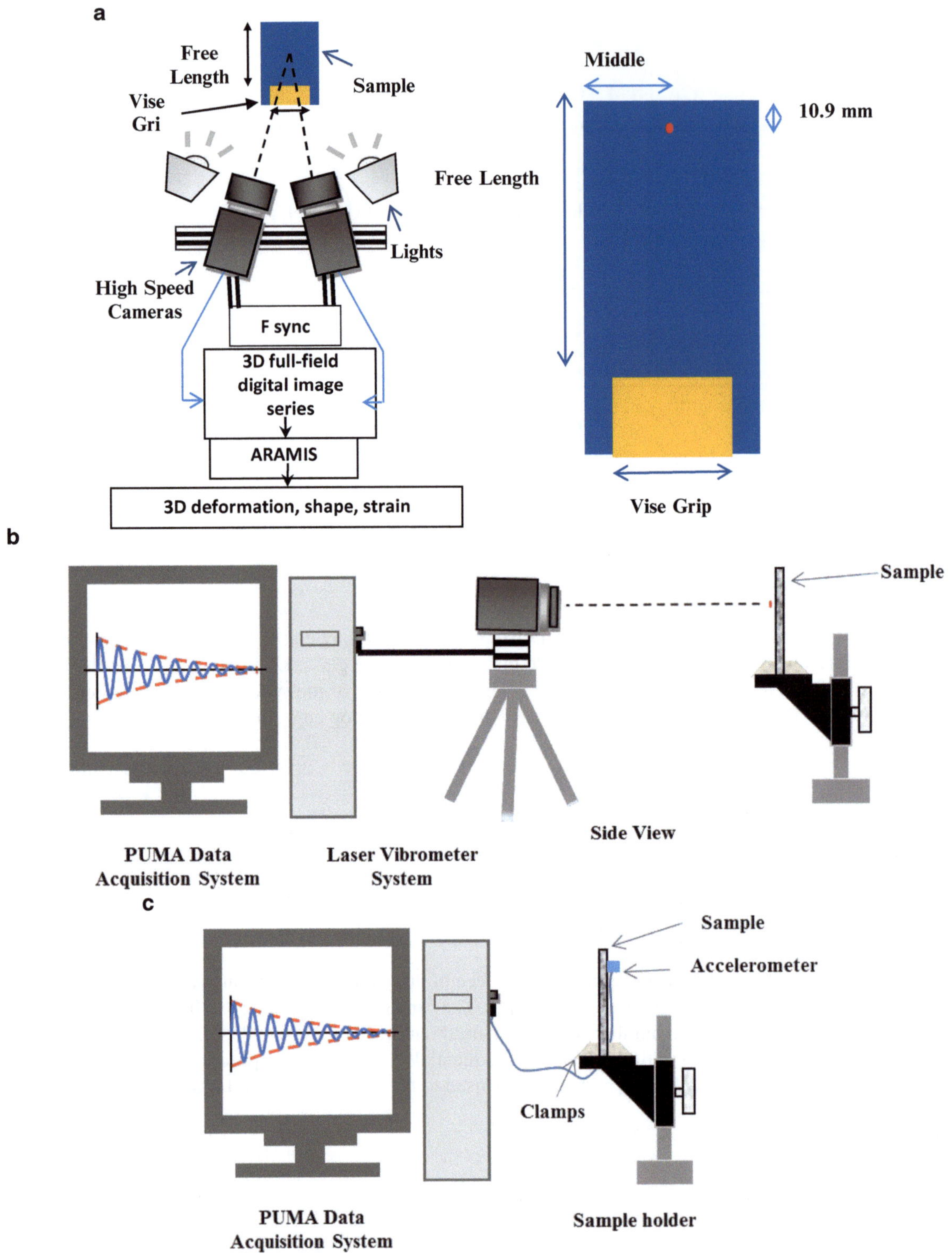

Fig. 9.2 Experimental set up

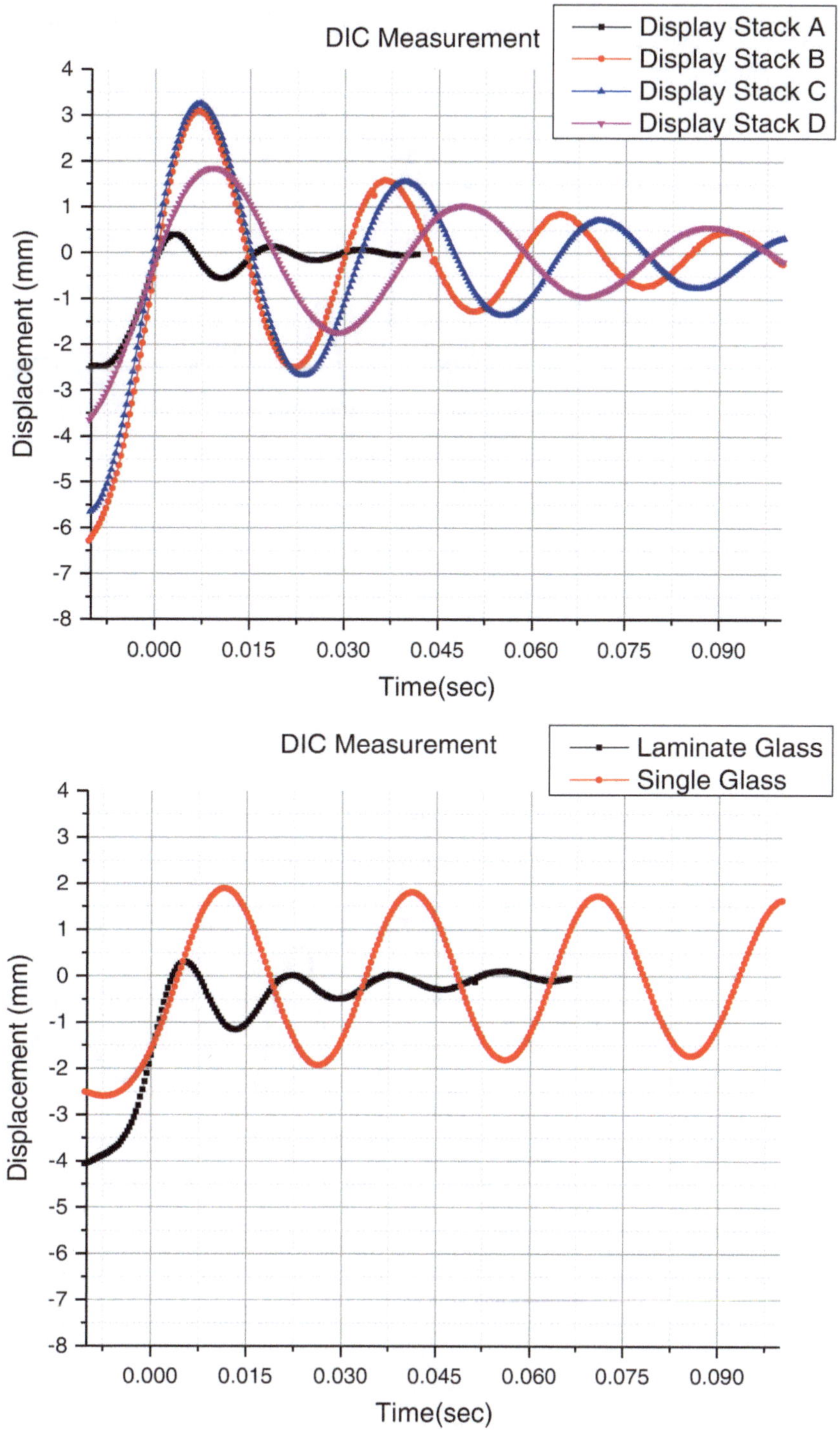

Fig. 9.3 Experimental measurement of dynamic decay using Digital Image Correlation technique

9.3.3 Laser Vibrometer

Laser vibrometry is another non-contact technique used extensively to measure the vibration of solid surfaces due to its un-intrusive nature and high spatial resolution. It allow measurements with high frequency bandwidth up to 20 MHz, velocity range of ±30 m/s, displacement resolution of about 8 nm and velocity resolution of 0.5 μm/s [11]. In this study to verify the DIC measurement, the laser vibrometer is used to measure the dynamic decay of the test subject. Boundary conditions under r both DIC and the laser vibrometer were help equal. The sample is held in the vise grips and laser is pointed at the middle of the sample at distance 10.9 mm from the top as shown in Fig. 9.2a. The schematic in Fig. 9.2b shows the experimental set up. The PUMA data acquisition system is used to obtain result from laser vibrometer. Result from all seven different samples are shown in Fig. 9.4.

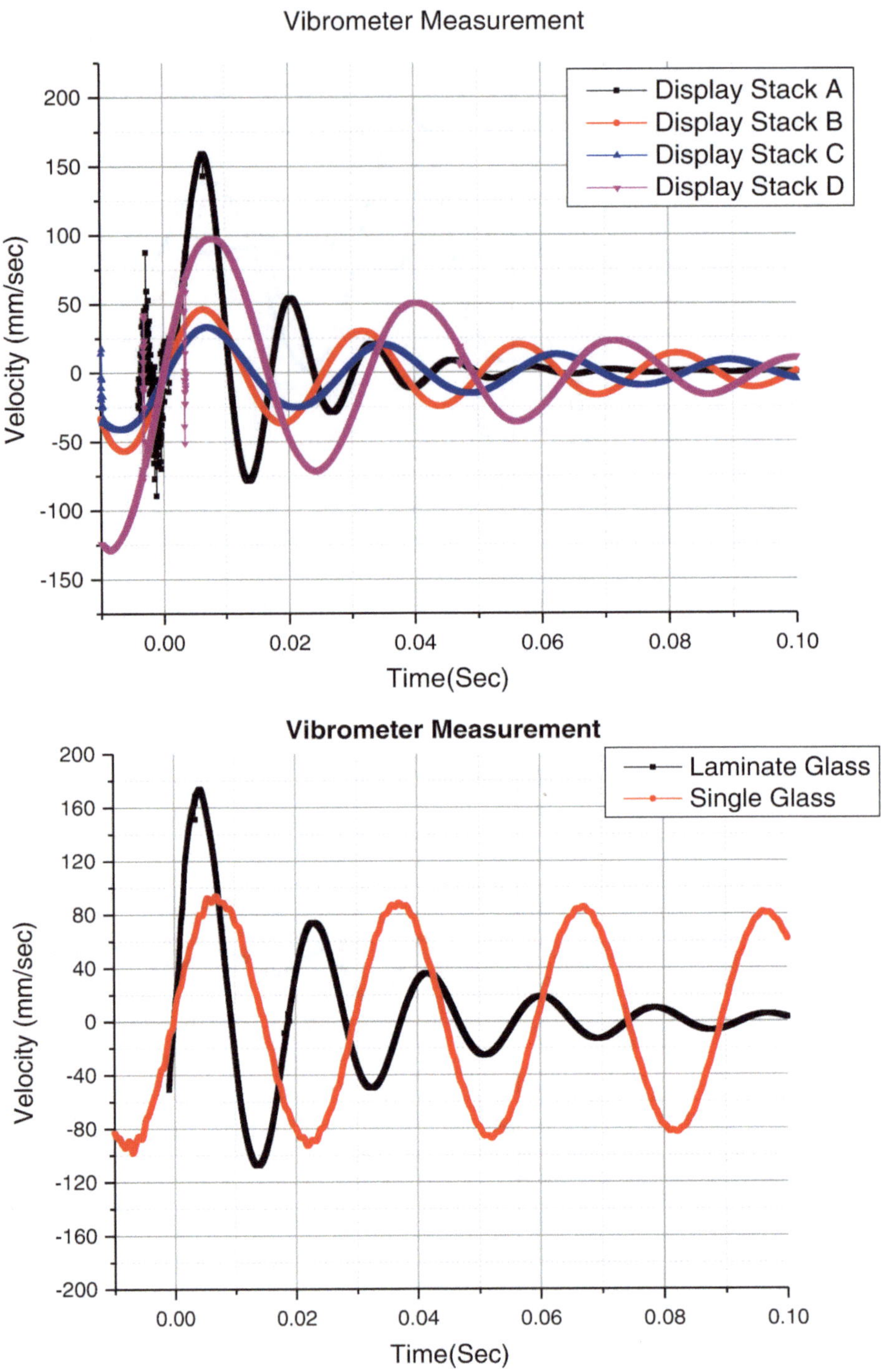

Fig. 9.4 Experimental measurement of dynamic decay using laser vibrometer

9.3.4 Single Axis Accelerometer

A single axis accelerometer is used to validate the results obtained from DIC. An accelerometer with sensitivity of 5 mV/g and weighing just 0.2 g is used to avoid the effect of its weight on experimental measurement. The schematic in Fig. 9.5 shows the experimental set up. Boundary conditions for all the samples are kept similar to DIC and laser vibrometer measurement. The accelerometer is adhered to at the middle on the sample 10.9 mm from the top with the help of beeswax. A PUMA data acquisition system is used to obtain results from accelerometer.

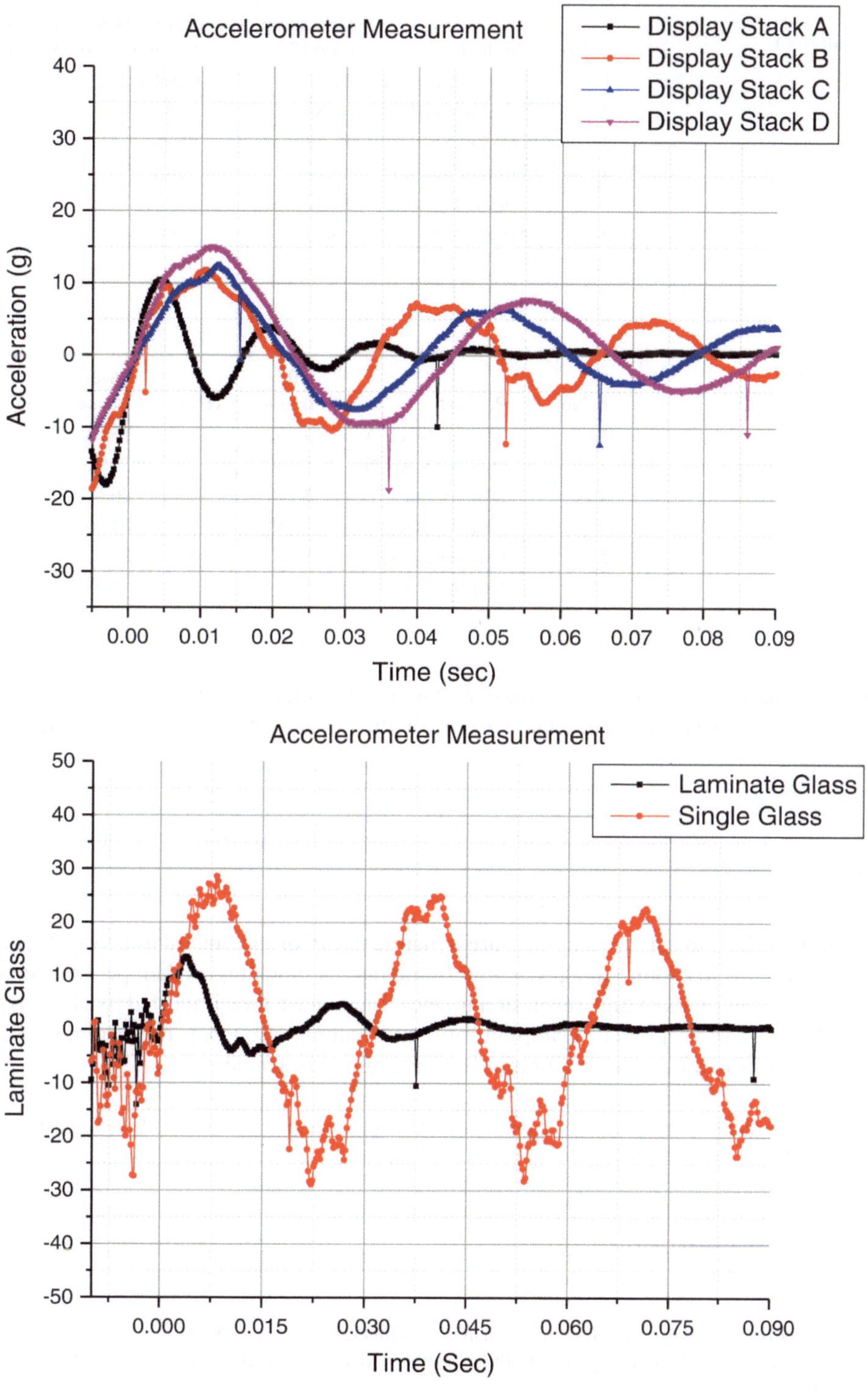

Fig. 9.5 Experimental measurement of dynamic decay using single axis accelerometer

9.4 Results and Discussion

Dynamic decay of each specimen is presented in the Figs. 9.3, 9.4, and 9.5 for DIC, the laser vibrometer and single axis accelerometer respectively. For high-speed DIC measurement images captured at 3000 frames per second provide great resolution for out of plane deformation. Dynamic decay of each specimen is clearly measured. Similarly, Laser vibrometer provides smooth dynamic decay measurement. The single axis accelerometer, however, is noise prone and for light weight

Table 9.1 Comparison of Damping ratio measurement from three techniques

S. No.	Product	DIC damping ratio (%)	Laser vibrometer damping ratio (%)	Accelerometer damping ration (%)
1	Display Stack A	14.2	13.98	11.04
2	Display Stack B	9.89	6.0	6.24
3	Display Stack C	10.67	6.65	7.9
4	Display Stack D	9.2	8.82	8.88
5	Laminate glass	12.42	10.87	13.26
6	Single glass	0.79	0.66	1.35

Table 9.2 Comparison of natural frequency measurement from three techniques

S. No.	Product	DIC frequency (Hz)	Laser vibrometer frequency (Hz)	Accelerometer frequency (Hz)
1	Display Stack A	67.62	75.51	67.97
2	Display Stack B	36.46	39.46	31.44
3	Display Stack C	30.79	36.5	26.98
4	Display Stack D	25.19	31.08	23.31
5	Laminate glass	59.67	53.97	52.11
6	Single glass	33.7	33.51	31.60

samples like single glass, the effect of additional mass of the accelerometer is seen to cause error in the measurement. Damping ratio and natural frequency are calculated from all the three experimental techniques and are compared in Tables 9.1 and 9.2.

9.5 Conclusion

In this study, the capability of DIC to capture and measure dynamic decay of the underdamped system is highlighted. DIC is coupled with high speed cameras to capture images at high frame rate. A pluck test is then performed and dynamic decay of the specimens are captured in hundreds of high resolution images. Images are then imported to a DIC system and out of plane deformation decay with time can be plotted. Damping ratio and natural frequency measurement from DIC are in good agreement with the results obtained from laser vibrometer and single axis accelerometer.

Acknowledgment The work presented in this paper is sponsored by Corning Incorporated. The author would like to thank Dr. Quang Su, Binghamton University for help with laser vibrometer.

References

1. Sharpe, W.N.: Springer handbook of experimental solid mechanics. Springer, New York (2008)
2. Tweten, D.J., Ballard, Z., Mann, B.P.: Minimizing error in the logarithmic decrement method through uncertainty propagation. J. Sound Vib. **333**(13), 2804–2811 (2014)
3. Sánchez-Arévalo, F.M., Pulos, G.: Use of digital image correlation to determine the mechanical behavior of materials. Mater Charact **59**(11), 1572–1579 (2008)
4. Pan, B., Qian, K., Xie, H., Asundi, A.: Two-dimensional digital image correlation for in-plane displacement and strain measurement: a review. Meas. Sci. Technol. **20**(6), 62001 (2009)
5. Lyons, J.S., Liu, J., Sutton, M.A.: High-temperature deformation measurements using digital-image correlation. Exp. Mech. **36**(1), 64–70 (1996)
6. Kwak, J.B., Lee, D.G., Nguyen, T., Park, S.: Deformation and strain measurement of flip-chip solder bump under in-situ thermal loading. In: ASME 2009 International Mechanical Engineering Congress and Exposition, pp. 311–318 (2009)
7. Bing, P., Hui-min, X., Tao, H., Asundi, A.: Measurement of coefficient of thermal expansion of films using digital image correlation method. Polym. Test. **28**(1), 75–83 (2009)
8. GOM ARAMIS User's Manual

9. Tomasini, E.P., Revel, G.M., Castellini, P.: Laser based measurement. In: Encyclopaedia of vibration, pp. 699–710. Academic, London (2001)
10. Pickering, C.J.D., Halliwell, N.A., Wilmshurst, T.H.: The laser vibrometer: a portable instrument. J. Sound Vib. **107**(3), 471–485 (1986)
11. Castellini, P., Martarelli, M., Tomasini, E.P.: Laser Doppler vibrometry: development of advanced solutions answering to technology's needs. Mech. Syst. Signal. Process. **20**(6), 1265–1285 (2006)

[illegible]
[illegible]
[illegible]
Mech. Syst. Signal Process. 20(6), 1206–1235 (2006)

Chapter 10
Dynamic Penetration and Bifurcation of a Crack at an Interface in a Transparent Bi-Layer: Effect of Impact Velocity

Balamurugan M. Sundaram and Hareesh V. Tippur

Abstract Dynamic fracture behavior of layered PMMA sheets is studied using transmission-mode Digital Gradient Sensing (DGS) technique. DGS is a relatively new optical method that exploits elasto-optic effects exhibited by transparent solids allowing a direct quantification of two orthogonal in-plane stress gradients simultaneously and hence crack tip parameters when used to study fracture mechanics problems. The current work builds on authors' previous two reports on this topic. Interfacial trapping, bifurcation and mixed-mode penetration into the second layer of a dynamically growing mode-I crack in the first layer encountering a normally oriented interface in a bi-layered configuration was reported in the first report [1]. In the second, the role of the location of a weak interface relative to the initial crack tip within the given geometry of the specimen was studied and interfacial penetration vs. bifurcation mechanisms was demonstrated and analyzed [2]. The current work focuses on the effect of impact velocity and the resulting loading rate on crack branching/penetration phenomenon when the mode-I crack encounters a normally oriented interface. In this ongoing work, a select location of interface relative to the initial crack tip is re-examined by varying the impact velocity. Using DGS for visualization and quantification, fracture mechanisms associated with crack growth are explained for the bi-layered system.

Keywords Digital gradient sensing • Transparent material • Layered material • Dynamic fracture • Crack path selection

10.1 Introduction

Many works reported in the literature focus on crack-interface interactions involving a growing crack and an inclined interface [3], resulting in a dominant penetrated crack in the subsequent layer following interfacial crack growth. Yet, relatively few reports exist on a more common and practical scenario of growing cracks interacting with normally oriented interfaces [4]. Hence, this series of studies investigate the fracture response of an elastically homogeneous PMMA bi-layer with a discrete plane of weakness. The current work builds on authors' previous two reports [1, 2] on this topic. Interfacial trapping, bifurcation and mixed-mode penetration into the second layer of a dynamically growing mode-I crack in the first layer encountering a normally oriented interface in a bi-layered configuration was reported in the first report. In the second, the role of the location of a weak interface relative to the initial crack tip within the given geometry of the specimen was studied and interfacial penetration vs. bifurcation mechanisms was demonstrated and analyzed.

The current work focusses on the effect of striker impact velocity and the resulting loading rate on crack branching vs. penetration phenomenon when the mode-I crack encounters a normally oriented interface. The goal here is to study the fracture response when dynamically growing cracks of different velocities and stress intensity factor reach an interface oriented perpendicular to the crack growth direction. The stress intensity factors are evaluated by using an optical method called Digital Gradient Sensing (DGS) in conjunction with high-speed photography. In this preliminary work two striker velocities of 13 m/s and 17 m/s are investigated.

10.2 Specimen Geometry and Sample Preparation

As explained previously in Ref. [1], the bi-layered specimen with interface normally oriented to the prospective crack growth direction was prepared by bonding two cast PMMA sheets using a commercially available acrylic adhesive Weldon 16. A rectangular layered material system of approximate dimension 6″ × 4″ and thickness × 0.33″ was thus obtained.

B.M. Sundaram • H.V. Tippur (✉)
Department of Mechanical Engineering, Auburn University, Auburn, AL 36849, USA
e-mail: htippur@eng.auburn.edu

D. Casem et al. (eds.), *Dynamic Behavior of Materials, Volume 1*, Conference Proceedings of the Society for Experimental Mechanics Series, DOI 10.1007/978-3-319-41132-3_10

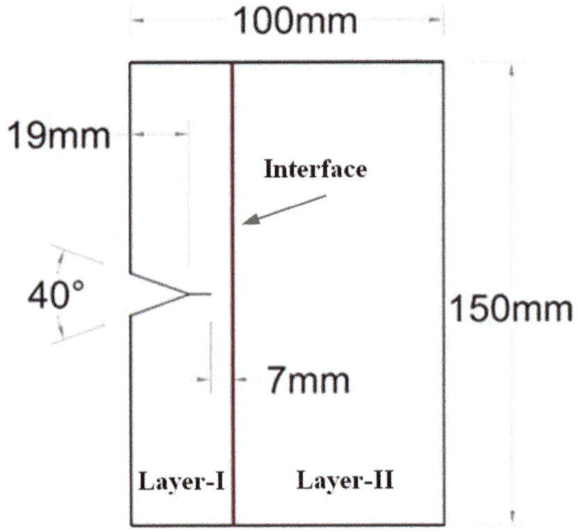

Fig. 10.1 Bi-layered PMMA with a normally oriented interface at 7 mm from the extended crack tip

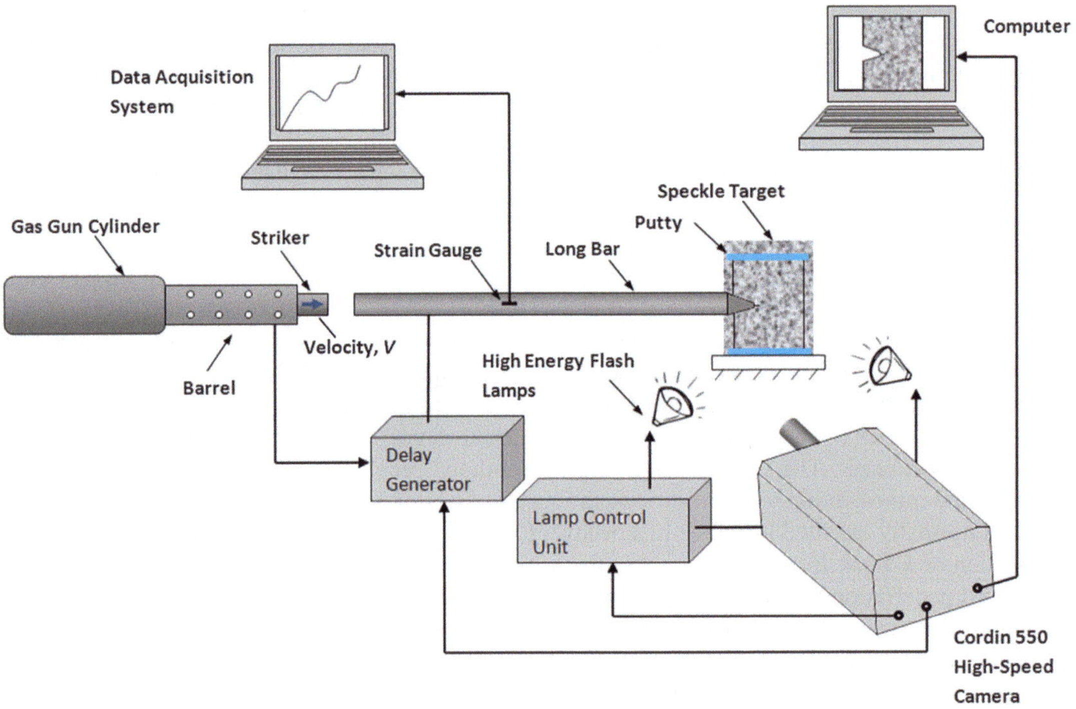

Fig. 10.2 Schematic of the experimental setup

The interfacial fracture toughness of interface (100 μm thick) was [1] approximately 50 % of the virgin PMMA (~1.3 MPa√m). The interface was located at 7 mm from the extended crack tip emanating from a 40° V-notch cut into the first layer. Figure 10.1 shows the specimen geometry used in this work. Further details of the sample preparation have been previously reported by the authors in Ref. [1].

10.3 Experimental Details

The schematic of the experimental setup used for dynamic fracture tests is shown in Fig. 10.2. A Hopkinson pressure bar was used to load the sample. A 6 ft long, 1-in. diameter long-bar with a wedge shaped tip was held against the unconstrained specimen with a matching wedge shaped notch. A 12-in. long, 1-in. diameter striker held inside the barrel of a gas-gun was used to impact the long-bar. Both the long-bar and the striker were of the same diameter and made of aluminum 7075 to eliminate the impedance mismatch between them. The striker was launched towards the long-bar at two different velocities

(V) of 13 m/ and 17 m/s by using two different air pressures (10 psi and 15 psi, respectively) in the gas-gun for this study. When the striker contacted the long-bar, a compressive stress wave propagated along the bar before transmission into the specimen in contact along the faces of the V-notch producing a dynamic wedge loading.

A target plate with random black and white speckles was placed behind the specimen at a fixed distance (~25 mm) to measure deformations using DGS [5]. The speckle pattern was photographed through the specimen using a Cordin-550 ultrahigh-speed digital camera with 32 sensors. Before loading the specimen, a set of 32 reference images were recorded at a rate of ~200,000 frames per second at a resolution of 1000 × 1000 pixels per image. When the striker contacted the long-bar, a trigger signal initiated recording of a second set of 32 images (at same framing rate) of the fracture event. Thus each reference image had a corresponding deformed image. The corresponding reference and deformed images from same sensor were matched. The corresponding sets of images were subsequently correlated to obtain in-plane displacements, using which angular deflections of light rays were evaluated at 2-D array of spatial positions in the field of view.

10.4 Crack Path Selection

In Figs. 10.3a, b the V-notch was extended using a diamond impregnated blade to form the initial crack tip. The crack growth occurs from left to right in these photographs. There is a single crack initiated at the crack tip in layer-I which is predominantly mode-I in nature. In Fig. 10.3a it can be seen that for a striker impact velocity of 13 m/s, the initiated single crack enters layer-II as a single mode-I crack. On the other hand, in Fig. 10.3b, for a striker velocity of 17 m/s, a complex growth patterns involving crack bifurcation and interfacial crack growth followed by mixed-mode penetrated cracks in layer-II can be seen. The two daughter cracks produced in layer-II travel symmetrically (relative to the specimen configuration) throughout the fracture event. It can be seen that the distance travelled by the two daughter cracks in the interface is ~10 mm and the angle of emergence is 22°.

Figure 10.4 shows plots of crack velocity histories for both the cases. The open circle represents the crack velocity when the crack enters the interface and the dotted circle represents the crack velocity when the crack exits the interface. It can be seen that crack velocity as the crack enters the interface for a striker velocity of 13 m/s is lower (~268 m/s) than that for a striker velocity of 17 m/s (~357 m/s). From this, one could infer that the velocity of the crack when it impinges on the interface is a key factor in the subsequent interfacial bifurcation, and penetration/branching mechanisms observed. It should also be noted that the crack velocity observed in layer-II for the former (~265 m/s) is lower compared to the latter (~360 m/s) at later stages of fracture event.

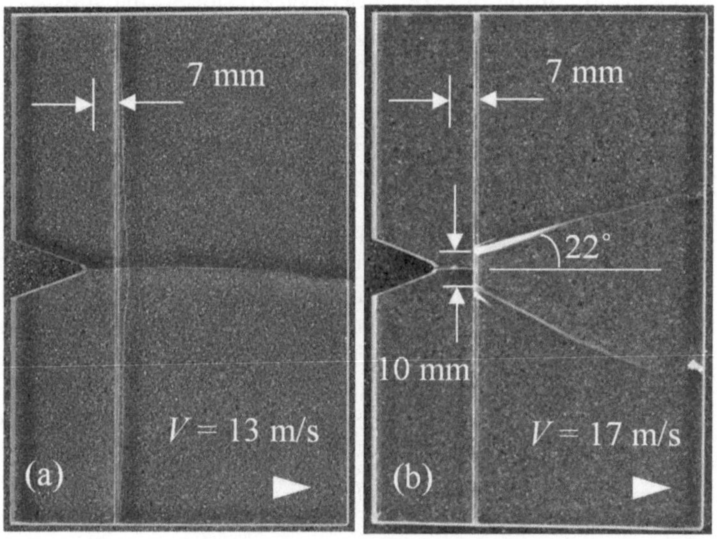

Fig. 10.3 Photographs of fractured specimens showing crack path differences for a striker impact velocity, V, of (**a**) 13 m/s and (**b**) 17 m/s, respectively. *Arrowheads* indicate crack growth direction

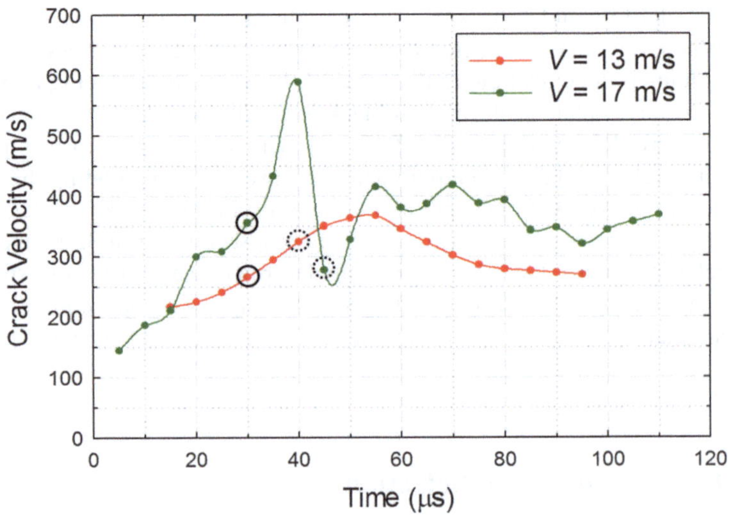

Fig. 10.4 Crack velocity histories for different striker velocities

10.5 Optical Measurements

The recorded reference and deformed images were correlated using ARAMIS™ to obtain contours of constant angular deflections near a propagating crack tip. The details are avoided here for brevity. The resulting contours in Figs. 10.5 and 10.6 represent the angular deflections of light rays for two different configurations and represent the stress gradients [5] in two mutually perpendicular directions. Here only few representative plots at two time instants are shown ($t = 0$ μs represents the time at which the crack initiates). The vertical dotted line represents the interface in these images. For a striker velocity of 13 m/s (Fig. 10.5), a single crack is observed in the second layer, whereas, for a striker velocity of 17 m/s, (Fig. 10.6), two symmetrically propagating daughter cracks can be observed in the second layer.

10.6 Work-In-Progress

The equation for obtaining stress intensity factors from the measured data were formulated during the previous study on dynamic crack propagation reported in Ref. [1]. The SIF histories for both the cases are currently being evaluated. Subsequently, using the SIF and velocity histories, the stresses experienced by the interface as the crack approaches the interface will be back calculated. This would shed some light on the prevailing crack penetration/bifurcation mechanism.

10.7 Conclusion

The results indicate that the impact velocity of the striker and hence the crack loading rate can alter the crack growth morphology along the interface and the second layer of an elastically homogeneous bi-layered brittle solid with a plane of weakness oriented normal to the initial crack growth direction. When the specimen was impacted with lower striker velocity (13 m/s) a single crack penetrated the interface and the second layer sequentially. Whereas, when a higher impact velocity (17 m/s) was used, a single crack branched into two successful interfacial daughter cracks which kink into layer-II. The factors affecting these dramatically different fracture patterns are incident crack velocity and stress intensity factor when impingement on the interface occurs. The velocity histories show that the velocity of the crack when it approached the interface for the two cases were different. Significantly higher crack velocity was observed in the latter compared to the former. Further study to investigate the details of these mechanisms is currently underway.

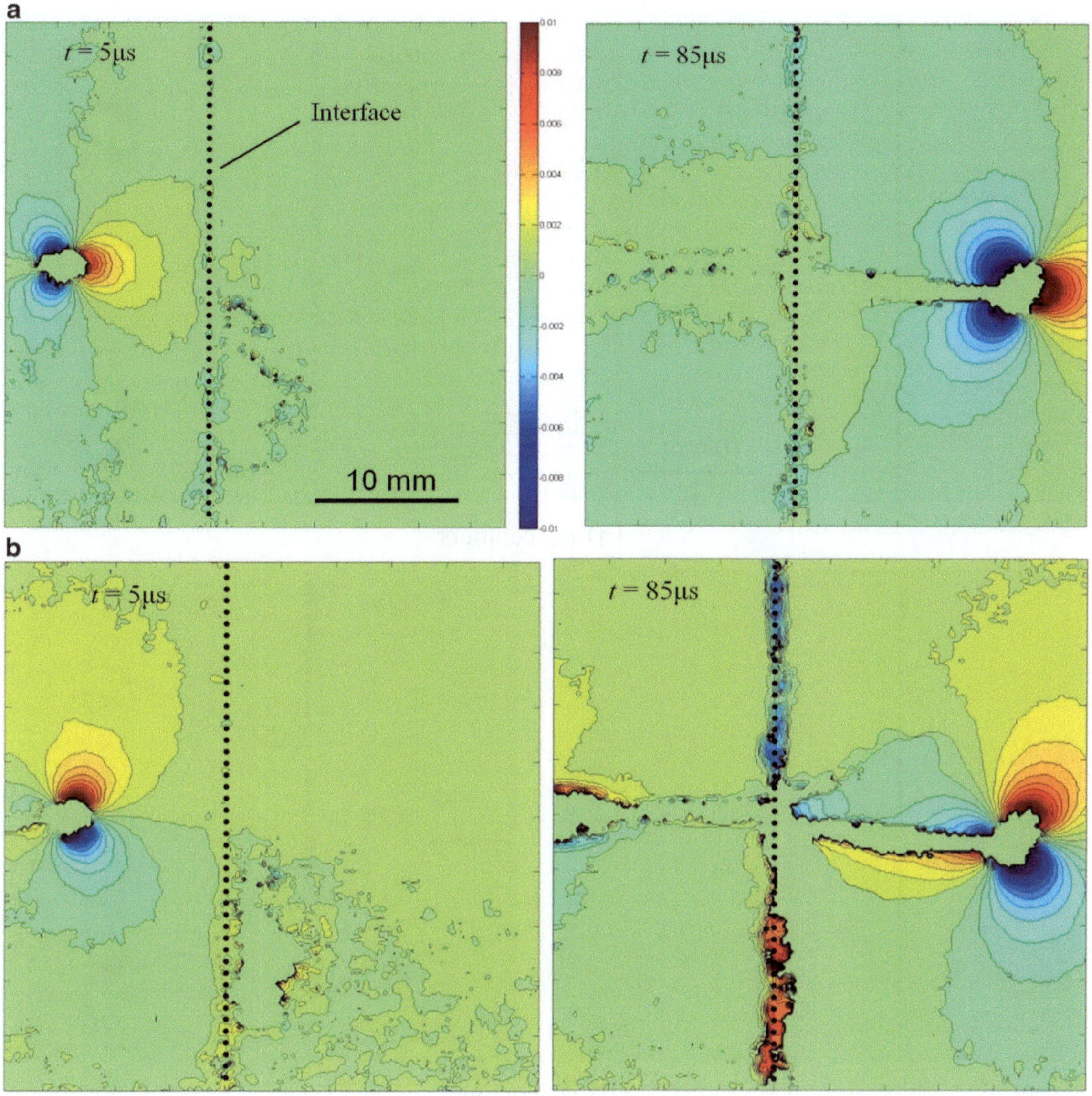

Fig. 10.5 Angular deflection contour plots proportional to stress-gradients in the x- and y-directions for a striker velocity of 13 m/s. (**a**) ϕ_x contours, (**b**) ϕ_y contours

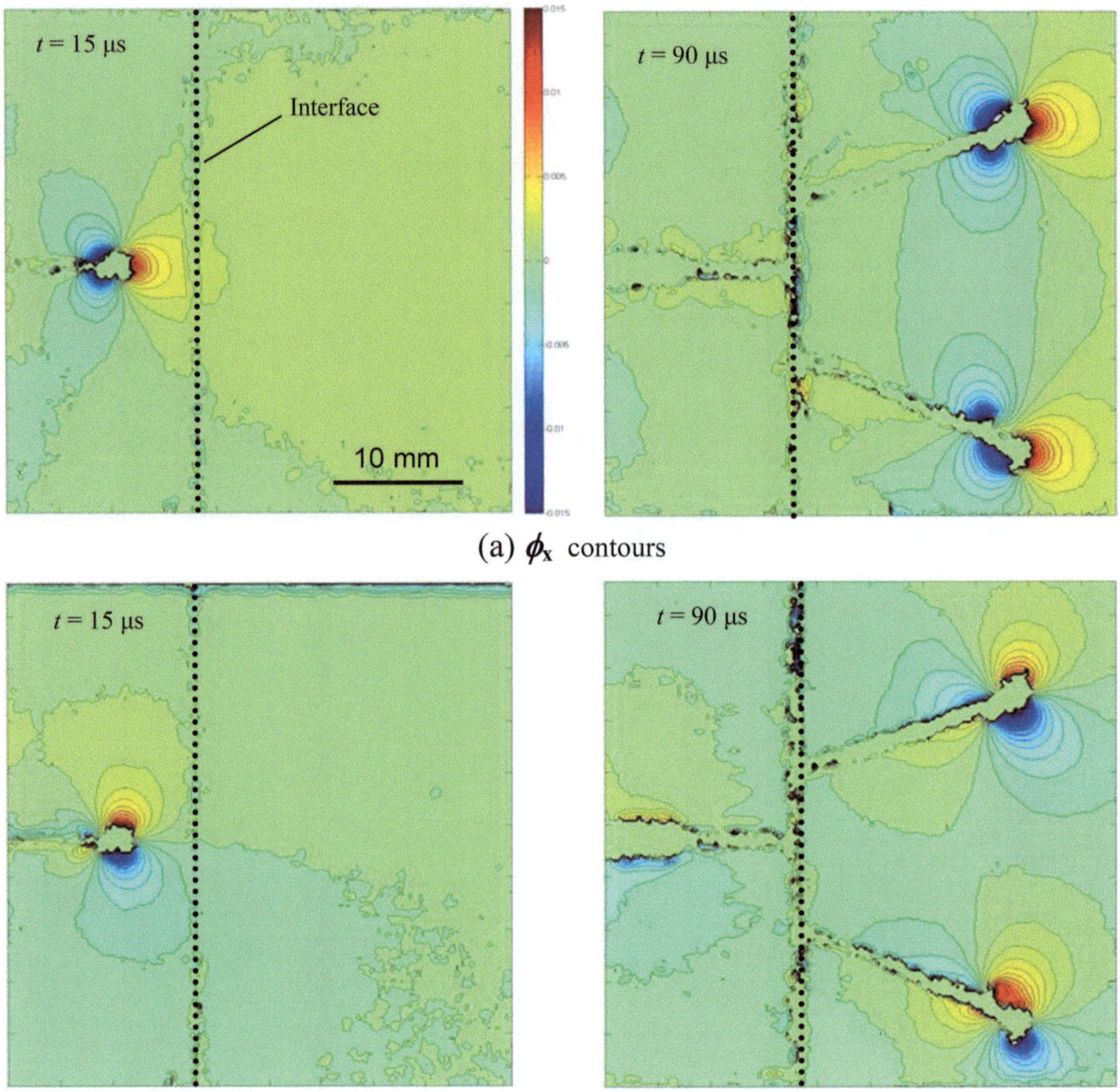

(b) ϕ_y contours

Fig. 10.6 Plots of angular deflection contours proportional to stress-gradients in the *x*- and *y*-directions for a striker velocity of 17 m/s. (**a**) ϕ_x contours, (**b**) ϕ_y contours

Acknowledgment The authors would like to thank the U.S. Army Research Office for supporting this research through grant W911NF-12-1-0317.

References

1. Sundaram, B.M., Tippur, H.V.: Dynamic Crack growth normal to an interface in Bi-layered materials: an experimental study using digital gradient sensing technique. Exp. Mech. **56**, 37–57 (2016)
2. Sundaram, B.M., Tippur, H.V.: Dynamic crack propagation in layered transparent materials studied using digital gradient sensing method: Part II. In: Proceedings of 2015 SEM Annual Conference, Costa mesa, CA, 2015
3. Xu, L.R., Huang, Y.Y., Rosakis, A.J.: Dynamic crack deflection and penetration at interfaces in homogeneous materials: experimental studies and model predictions. J. mech. phys. solids **51**(3), 461–486 (2003)
4. Park, H., Chen, W.: Experimental investigation on dynamic crack propagating perpendicular through interface in glass. J. Appl. Mech. **78**(5), 051013 (2011)
5. Periasamy, C., Tippur, H.V.: A full-field digital gradient sensing method for evaluating stress gradients in transparent solids. Appl. Optics **51** (12), 2088–2097 (2012)

Chapter 11
Influence of Loading Rate on Fracture Strength of Individual Sand Particles

Andrew Druckrey, Dan Casem, Khalid Alshibli, and Emily Huskins

Abstract Dynamic loading on granular materials, such as impact, blast, or projectile penetration, can impose large inter-particle forces to cause significant particle fracture within individual particles. Extensive research has been conducted at different strain rates on granular media mass, but very little has been published to investigate the influence of strain or loading rate on individual particles. Therefore, a gap in the knowledge base is present since comprehensive multi-scale modeling of granular material begins at the micro (particle) scale. In this paper, individual natural sand particles are compressed to fracture at loading rates of 0.2 mm/min, 2.25 m/s, and 14.5 m/s using quasi-static unconfined compression and unconfined mini-Kolsky bar techniques. Fracture loads are compared for various "types" of particles within the natural sand, and compared to conventional quasi-static failure definitions for particles. Particles exhibited loading rate dependence when comparing Weibull characteristic tensile strength with loading rate.

Keywords Loading rate • Sand particle fracture • Weibull statistics • Kolsky bar • Fragmentation

11.1 Introduction

It is well-known that fracture and comminution of particles affect the constitutive and deformation behavior of granular material during quasi-static loading as well as high-rate loading conditions. During blast loading, energy dissipates partially through sand particle fracture and comminution [1, 2]. Fracture of individual particles influences the dynamic behavior of the ejecta resulting from buried explosives [3] and therefore affect the loading applied to an object such as an armored vehicle or structures close to the explosive source. Particle fracture may occur in other high strain rate loadings on granular materials such as projectile penetration into sand [4–6], drilling and mining applications, and impact. Assessing dynamic properties of fracture at the particle level is an essential step for better understanding of the behavior of bulk granular materials that experience high strain rate loading and plays a critical role in improving design against such dynamic forces.

Constitutive behavior of granular materials is highly influenced by particle scale properties, including crushing behavior. For example, McDowell, Bolton [7] updated the relative density index proposed by Bolton [8] to include the mean tensile strength of particles because it provides a more micromechanical insight into the dilatant behavior of crushable soils. Also, it is well-known that during quasi-static loading of granular materials particles align into force chains that resist the majority of the external loads, and particles within those chains fracture first when the continuum is exposed to high stresses [9–12]. Many approaches have been proposed to quantify particle fracture. For example, Cavarretta, O'Sullivan [13] used a micromechanical approach to describe the compression of individual particles and developed two simple analytical models that incorporated frictional sliding, bulk particle compression, and fragmentation based on geometry and kinematic degradation of stiffness. More recently, Alshibli et al. [14] used 3D X-ray diffraction to measure strains within individual silica particles in compression and calculated lattice strains and stresses within individual particles.

A more common quantification of particle fracture has been described and used in the literature thoroughly using Weibull statistics (e.g. [15–17]). The characteristic tensile strength (σ_{0f}), as determined through Weibull statistics, is the particle stress where 37 % of particles survive. The Weibull modulus (m) defines the variability in particle crushing strength. Variability increases with decreasing modulus. These parameters quantify the strength of a set of particles as well as the

A. Druckrey (✉) • K. Alshibli
Department of Civil & Environmental Engineering, University of Tennessee, 325 John D. Tickle Bldg., Knoxville, TN 37916-2313, USA
e-mail: adruckre@vols.utk.edu

D. Casem
Army Research Laboratory, Aberdeen, MD 21005, USA

E. Huskins
Department of Mechanical Engineering, United States Naval Academy, 590 Holloway Road, Annapolis, MD 21402-5042, USA

D. Casem et al. (eds.), *Dynamic Behavior of Materials, Volume 1*, Conference Proceedings of the Society for Experimental Mechanics Series, DOI 10.1007/978-3-319-41132-3_11

variability, and adequately describe overall fracture strength. They account for individual particle variabilities such as potential flaws within a particle.

High rate testing of individual sand particles provides the initial particle-scale information that can be used in dynamic modeling across the scales. In particular, high strain-rate tests can determine strain-rate dependence of particle fracture. Strain-rate dependence of particle fracture can in turn be used to calibrate and validate high fidelity particle-scale poly-ellipsoidal discrete element models (DEM) that include particle fracture, which in turn can be used to develop lower fidelity soil blast models at larger scale to replace costly and time consuming full scale testing [18]. Few studies have been published to investigate the behavior of single particle failure under dynamic loading. Recently, Parab et al. [19] began closing this knowledge gap by using a confined Kolsky bar to dynamically compress particles during in-situ high speed phase contrast imaging (PCI). Two confined approximately spherical Ottawa sand particles were placed in a hollow cylindrical aluminum housing and were subjected to dynamic load while 2D radiographs and load versus time data were collected. Visual assessments of damage propagation were conducted and when particles were dry, one of the two particles had extensive interfacial cracking followed by pulverization. Wet sand particles broke into large fragments followed by pulverization.

This paper investigates the effects of loading rate on particle fracture. Particles were tested at three rates, 0.2 mm/s, 2.25 m/s and 14.5 m/s using quasi-static and dynamic testing methods. For the 0.2 mm/s and 2.25 m/s rates, particles were imaged using 3D X-ray computed tomography before testing and fragments were collected so that post-mortem imaging could be conducted. 3D X-ray imaging aids in the analysis of fracture mechanisms such as type of failure and failure surface. Fragments could not be collected for the experiments conducted at 14.5 m/s because particles completely pulverized.

11.2 Materials and Experimental Setup

A natural sand from a quarry in Longmont, Colorado (Colorado Materials, Inc.), known as mason sand, was used in this study. It is not a typical "research" sand which is generally uniform and has well-defined properties. Mason sand is composed of mineral groups that generally correspond to the color of the particle as revealed by X-ray powder diffraction. Silica particles, mainly tridymite, are colorless to white (defined in this paper as clear) and have relatively solid microstructure. White particles are highly weathered and porous. Black particles are determined to be ferrihydrite, mica or iron, and potassium feldspar particles were generally red. ASTM 20-30 Ottawa sand particles were also used to provide a baseline. ASTM 20-30 sand is composed of relatively rounded silica particles that are well defined. Experiments with rates of 0.2 mm/min and 14.5 m/s were conducted on the ASTM 20-30 particles. All particles chosen in this paper have a size range of 0.5 mm to 1.5 mm.

Quasi-static (0.2 mm/min) experiments were conducted using the apparatus described in [20] during in-situ X-ray imaging. Particles were placed between two loading plates and diametrically loaded until major failure of the particle was observed by the loading curve and X-ray images. In quasi-static experiments, major failure was through the splitting of the particle into two or more fragments, ignoring minor asperity fracture.

A Kolsky bar setup (Fig. 11.1) was used for the 2.25 m/s experiments. Particles were individually placed between the incident (508 mm long by 4.76 mm diameter) and transmitter (152.5 mm by 1.5875 mm diameter) bars and a striker was launched by pneumatic pressure into the incident bar. A wave pulse was transmitted through the incident bar, measured by the strain gauges, and imposed onto the particle. Waves reflected off the loading end of the incident bar and measured again by the strain gauges. The wave pulse loaded the particle and was transmitted to the transmitter bar, and pulses in the transmitter bar were measured by a normal displacement interferometer through a series of beam splitters and mirrors. Fragments were then collected for X-ray imaging.

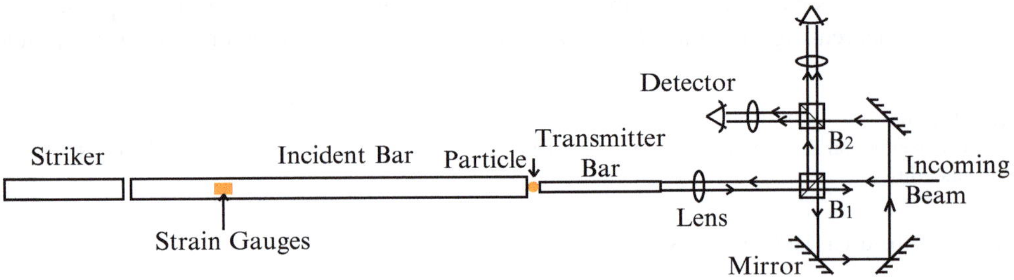

Fig. 11.1 Schematic of mini-Kolsky bar setup with NDI to measure displacement of transmitter bar due to transmitted pulse (after [22])

A similar Kolsky bar setup was used for the 14.5 m/s experiments. However, a transverse displacement interferometer (TDI) was setup on a grating on the incident bar (37.87 mm long by 1.61 mm diameter) in place of the incident bar strain gauges. Transmitter bar in this experimental setup was 37.92 mm long by 1.61 mm diameter and both bars were made of tungsten carbide (WC). WC bars were used so that the bars would not be damaged when crushing particles. Optics were used for sensitivity and to eliminate the need for strain gauges on the small bars. Particles exposed to 14.5 m/s loading were completely pulverized and fragments could not be collected for post-mortem imaging.

11.3 Results and Discussion

Loading curves were obtained for particles at each of the loading rates, and Fig. 11.2 shows an example of the experimental results (velocity measurements were not obtained for 0.2 mm/min experiments). Figure 11.2a displays the loading curve for a quasi-static loading rate. Figure 11.2b displays an example of the loading curve for an experiment conducted at an average of 2.25 m/s, as well as the velocity curve during the duration of the experiment. Figure 11.2c displays an example of a loading curve for an experiment conducted at an average of 14.5 m/s along with the velocity curve during the experiment. Velocity curves were relatively consistent during the higher loading rate experiments. The force required to cause major fracture or pulverization of each particle, F_f, is denoted for each of the experiments.

Single particle fracture strength is defined in this paper as the major failure of the particle under force F_f as $\sigma_f = {}^{F_f}/_{d^2}$, where d^2 is the distance between loading plates reflecting the particle diameter in the loading direction. Weibull [21] proposed a statistical distribution function that has widely been used to describe particle fracture such that: $P_s = exp\left[-\left({}^{\sigma_f}/_{\sigma_{0f}}\right)^{m_f}\right]$, where P_s is the probability of survival for a particular particle ($P_s = \frac{\#\ of\ particles\ crushing\ at\ \sigma \geq \sigma_f}{total\ \#\ of\ particles}$). The Weibull modulus ($m_f$) and the characteristic tensile stress at which the probability of survival is 37 % (σ_{0f}) can be determined from the plot of ln(σ_f) versus $\left[\ln\left({}^{1}/_{P_s}\right)\right]$. The Weibull modulus is the slope of the best fit line to the data and the characteristic tensile stress is the intersection of the best fit line with the x axis. Figure 11.3 shows the results for ASTM 20-30 stand under quasi-static loading. The Weibull modulus for these particles is 3.011 and the characteristic tensile stress corresponding to 37 % probability of survival, where the best fit line crosses the x axis, is 141.6 MPa. Overall, the linear fit is good, having a relatively high coefficient of determination.

Characteristic tensile stress corresponding to 37 % probability of survival and Weibull moduli for ASTM 20-30 and Mason sands are presented in Table 11.1. The ASTM 20-30 sand particles, tested at 0.2 mm/min and 14.5 m/s, have significantly larger characteristic tensile failure stresses than any of the Mason sand particles. This is because Mason sand particles are angular in shape. The Weibull moduli for all Mason sand particle types are fairly low and variable for all loading rates because there was significant variability in fracture tensile stresses. Of the Mason sand particles during quasi-static testing (0.2 mm/min rate), the clear silica particles had the largest characteristic tensile stress at failure, implying that at these rates the clear particles had the highest strength, followed by the black, red, and white particles. At the intermediate loading rate (2.25 m/s), the clear silica particles again has the largest characteristic tensile stress followed by red, white, and

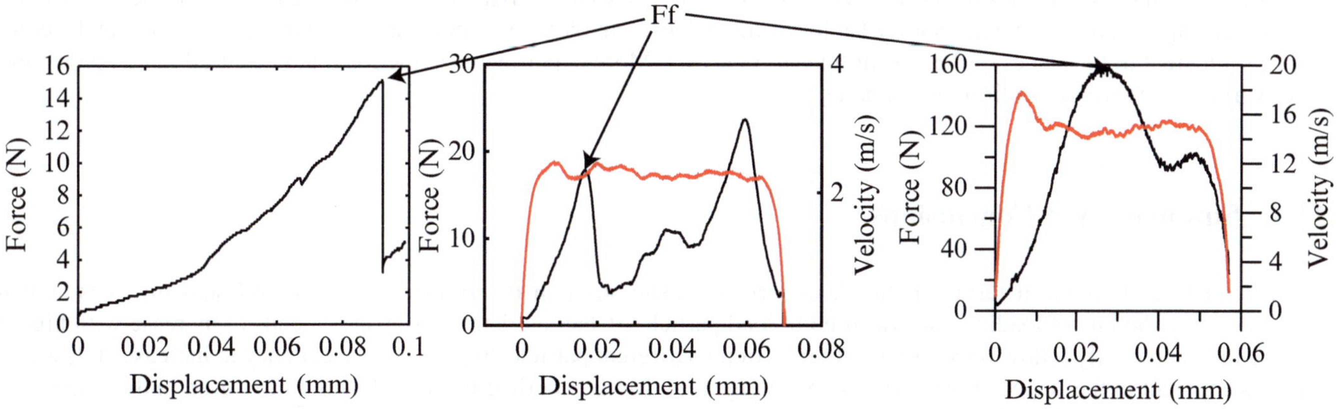

Fig. 11.2 Experimental loading curves for each loading rate. (**a**) 0.2 mm/min, (**b**) 2.25 m/s, (**c**) 14.5 m/s

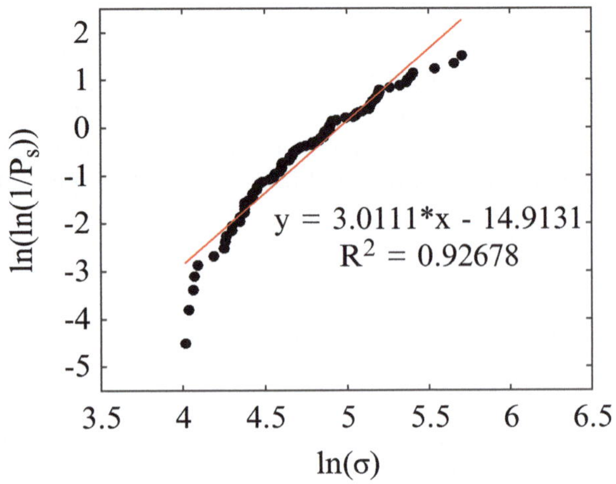

Fig. 11.3 Weibull probability distribution for individual ASTM 20-30 silica sand particles

Table 11.1 Results from Weibull analysis for particles

Loading rate	0.2 mm/min		2.25 m/s		14.5 m/s	
Material	σ_{0f} (Mpa)	*m*	σ_{0f} (Mpa)	*m*	σ_{0f} (Mpa)	*m*
ASTM 20-30	141.6	3.011	n/a	n/a	313.2	2.458
Mason sand						
Black	61.8	1.368	28.7	5.173	113.1	2.312
Clear	72.8	1.358	77.2	1.239	94.3	2.798
Red	37.9	2.523	58.6	1.559	106.9	1.174
White	23.9	0.820	31.0	0.745	117.8	1.375

black particles. At the highest loading rate (14.5 m/s), white particles have the largest characteristic tensile stress of the Mason sand.

To illustrate rate effects on particle fracture, the loading rate versus characteristic tensile stress at failure for all particle types is shown in Fig. 11.4. The overall trend for all particle groups is that higher applied loading rates result in larger characteristic tensile strengths. The linear relationship demonstrate the increase in characteristic tensile strength with increase in loading rate represent the degree of rate dependency and most have a coefficient of determination close to one. Black particles (Fig. 11.4a) have a slightly worse fit because of the low characteristic tensile stress observed at the 2 m/s loading rate. Of the Mason sand particles, white particles exhibit the largest increase in characteristic tensile stress with higher loading rates, while the clear silica particles exhibit the lowest. Although experiments with only two loading rates were conducted for the ASTM 20-30 sand, a high increase in characteristic tensile stress was observed between the 0.2 mm/min and 14.5 m/s experiments and the increase was significantly larger than any of the Mason sand particle groups.

Particle fracture strength, in terms of the characteristic tensile strength determined from Weibull statistics, has shown dependency on the rate of loading that was used to load particles. Characteristic tensile stress tends to increase with higher loading rate, apparently following a somewhat linear trend. Different particle "types" have a tendency to increase at different rates, as apparent in the slopes of the best fit lines in Fig. 11.4. More experiments at the loading rates used in this paper and other various rates would confirm these findings.

11.4 Summary and Conclusions

Particle fracture strength, in terms of the characteristic tensile strength determined from Weibull statistics, has shown dependency on the rate of loading that was used to load particles. Characteristic tensile stress tends to increase with higher loading rate, apparently following a somewhat linear trend. Different particle "types" have a tendency to increase at different rates, as apparent in the slopes of the best fit lines in Fig. 11.4. More experiments at the loading rates used in this paper and other various rates would confirm these findings.

Particle tensile strengths were calculated based on the fracture force required for bulk failure of the particle, and stresses were analyzed using Weibull statistics. Tensile stress as determined in this paper may not be the best way to describe particle

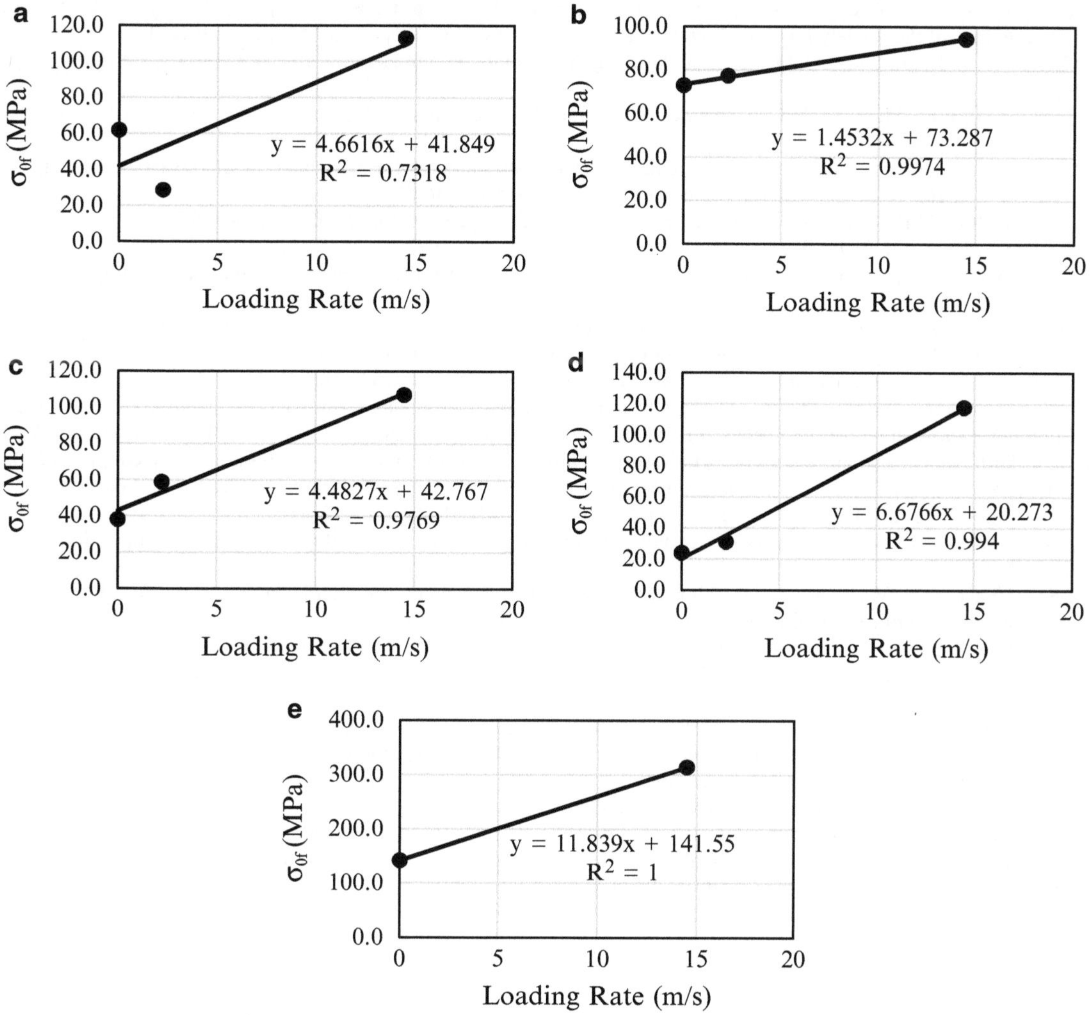

Fig. 11.4 Loading rate versus characteristic tensile strength for each of the sand types. (**a**) *Black* particles, (**b**) *Clear* particles, (**c**) *Red* particles, (**d**) *White* particles, (**e**) ASTM 20-30 particles

stress, especially if a particle is non-spherical and angular, but has shown good correlation with particle fracture in the literature. Weibull statistics parameter of characteristic tensile strength, which is the tensile strength probability of particle survival is 37 %, was determined for each of the particle types and loading rates. All particle types exhibited an increase of characteristic tensile strength with increase in loading rate, suggesting that particle fracture is rate dependent. The degree of rate dependency on particle fracture is variable with particle type, and experiments at more rates could determine whether or not there is any curvature in the degree of rate dependency.

Acknowledgment This material is funded by the Office of Naval Research (ONR) grant No. N00014-11-1-0691.

References

1. Børvik, T., Olovsson, L., Hanssen, A.G., Dharmasena, K.P., Hansson, H., Wadley, H.N.G.: A discrete particle approach to simulate the combined effect of blast and sand impact loading of steel plates. J. Mech. Phys. Solids **59**(5), 940–958 (2011). doi:10.1016/j.jmps.2011.03.004
2. Braslau, D.: Partitioning of energy in hypervelocity impact against loose sand targets. J. Geophys. Res. **75**(20), 3987–3999 (1970). doi:10.1029/JB075i020p03987
3. Regueiro, R., Pak, R., McCartney, J., Sture, S., Yan, B., Duan, Z., Svoboda, J., Mun, W., Vasilyev, O., Kasimov, N., Brown-Dymkoski, E., Hansen, C., Li, S., Ren, B., Alshibli, K., Druckrey, A., Lu, H., Luo, H., Brannon, R., Bonifasi-Lista, C., Yarahmadi, A., Ghodrati, E., Colovos,

J.: ONR MURI project on soil blast modeling and simulation. In: Song, B., Casem, D., Kimberley, J. (eds.) Dynamic Behavior of Materials. Conference Proceedings of the Society for Experimental Mechanics Series, Vol. 1, pp. 341–353. Springer (2014)
4. Cole, R.P.: Ballistic penetration of a sandbagged redoubt using silica sand and pulverized rubber of various grain sizes. University of South Florida (2010)
5. Allen, W.A., Mayfield, E.B., Morrison, H.L.: Dynamics of a projectile penetrating sand. J. Appl. Phys. **28**(3), 370–376 (1957). doi:10.1063/1.1722750
6. Cooper, W., Breaux, B.: Grain fracture in rapid particulate media deformation and a particulate media research roadmap from the PMEE workshops. Int. J. Fract. **162**(1–2), 137–150 (2010). doi:10.1007/s10704-010-9467-8
7. McDowell, G.R., Bolton, M.D.: On the micromechanics of crushable aggregates. Geotechnique **48**, 667–679 (1998)
8. Bolton, M.D.: The strength and dilatancy of sands. Geotechnique **36**, 65–78 (1986)
9. Peters, J.F., Muthuswamy, M., Wibowo, J., Tordesillas, A.: Characterization of force chains in granular material. Phys. Rev. E **72**(4), 041307 (2005)
10. Wang, Z., Yang, X., Chen, Q., Zhang, Y., Zhao, Y.: Study of the contact forces and grain size distribution during grain crushing. In: Multimedia Technology (ICMT), 2011 International Conference on, 26–28 July 2011, pp. 2617–2622 (2011)
11. Cil, M.B., Alshibli, K.A.: 3D evolution of sand fracture under 1D compression. Géotechnique (2014)
12. Cooper, W.: Communication of stresses by chains of grains in high-speed particulate media impacts. In: Proulx, T. (ed.) Dynamic Behavior of Materials. Conference Proceedings of the Society for Experimental Mechanics Series, Vol. 1, pp. 99–107. Springer, New York (2011)
13. Cavarretta, I., O'Sullivan, C.: The mechanics of rigid irregular particles subject to uniaxial compression. Geotechnique **62**(8), 11 (2012)
14. Alshibli, K., Cil, M., Kenesei, P., Lienert, U.: Strain tensor determination of compressed individual silica sand particles using high-energy synchrotron diffraction. Granul. Matter. **15**(5), 517–530 (2013). doi:10.1007/s10035-013-0424-x
15. McDowell, G.R., Amon, A.: The application of Weibull statistics to the fracture of soil particles. Soil. Found. **40**(5), 133–141 (2000). doi:10.3208/sandf.40.5_133
16. Nakata, Y., Kato, Y., Hyodo, M., Hyde, A.F.L., Murata, H.: One-dimensional compression behaviour of uniformly graded sand related to single particle crushing strength. J. Jpn. Geotech. Soc. soil. found. **41**(2), 39–51 (2001)
17. Cil, M.B., Alshibli, K.A.: 3D assessment of fracture of sand particles using discrete element method. Géotechnique Letters (2012)
18. Antoun, T., Herbold, E., Johnson, S.: Dynamic behavior of sand: Annual Report FY 11. In., vol. LLNL-TR-539077. Lawrence Livermore National Laboratory (2012)
19. Parab, N.D., Claus, B., Hudspeth, M.C., Black, J.T., Mondal, A., Sun, J., Fezzaa, K., Xiao, X., Luo, S.N., Chen, W.: Experimental assessment of fracture of individual sand particles at different loading rates. Int. J. Impact. Eng. **68**, 8–14 (2014). doi:10.1016/j.ijimpeng.2014.01.003
20. Druckrey, A.M., Alshibli, K.A.: 3D finite element modeling of sand particle fracture based on in situ X-Ray synchrotron imaging. Int. J. Numer. Anal. Methods Geomech. **40**(1), 105–116 (2015). doi:10.1002/nag.2396
21. Weibull, W.: A statistical distribution function of wide applicability. J. Appl. Mech. **18**(3), 293–297 (1951)
22. Casem, D.T., Grunschel, S.E., Schuster, B.E.: Normal and transverse displacement interferometers applied to small diameter kolsky bars. Exp. Mech. **52**(2), 173–184 (2012). doi:10.1007/s11340-011-9524-x

Chapter 12
Arrested Compression Tests on Two Types of Sand

Eduardo Suescun-Florez, Stephan Bless, Magued Iskander, and Camilo Daza

Abstract Silica sand and quartz sand were subjected to uniaxial loading and unloading at rates of 0.1/s and 0.0001/s. The particle size distribution was measured, and found to be significantly altered when peak strains were 10 % or greater. The loading modulus for silica sand was bilinear, and suggestive of elastic-plastic behavior, where the plastic part is due to void closure. On unloading, the modulus is close to the loading "elastic" value. Coral sand is softer than silica sand on loading, and the modulus is almost constant and much less than for silica sand. Both types of sand are recovered with a higher density than can be obtained with the starting particle mix. This suggests particles have crushed and filled some of the voids. Indeed, reduction of mean particle size is verified from post-test analysis. Coral sand, which has the greater reduction in void content, also exhibits increased particle breakup.

Keywords Silica sand • Coral sand • Confined compression • Crushing • Porosity

12.1 Introduction

Particle breakage is important for understanding the constitutive mechanisms in sand [1]. Breakage progresses in a number of stages starting with damage, followed by crushing, and possibly reaching complete pulverization of the grains which is also known as comminution. At all stages, particle breakage influences the macroscopic stress–strain behavior and permeability of sand especially at high pressures. [2, 3]. The role of particle breakage on the constitutive behavior of sand has been investigated by different means including breakage indices ([4, 5]. Other studies have correlated vertical stresses with pore volume reduction [6]. At high vertical stresses the behavior may be described using parallel normal compression and critical state lines, both of which are related to the degree of particle damage [7].

The main goal of this paper was to assess the evolution of particle damage of two natural sands subjected to one dimensional compression loading. Strain-controlled tests were arrested between 5 and 30 % strain. Evolution of crushing was assessed when post-tests samples were subjected to dynamic imaging techniques.

12.2 Materials Tested

Two sands were selected for investigation, a silica sand and a coral sand. The two sands were selected mainly for the difference in their grain strength, which is attributed to the different mineralogical compositions. The physical properties of the tested sands are summarized below. More information of these two materials can be found in [8].

Silica sand, was named #1-Dry by its distributor US Silica. Particles are sub-rounded to sub-angular with medium to high sphericity having diameters ranging from 10 to 800 μm, and aspect ratio values between 0.6 and 0.75. The mean particle size (D50) was found to be 340 μm and the coefficient of uniformity was determined as 1.59. The material is poorly graded and classified as SP according to ASTM's Unified Soil Classification System (USCS). The specific gravity of the grains was 2.66. Maximum and minimum densities were determined following standards [9, 10], and were found to be 1.84 and 1.43 g/cm^3 respectively, corresponding to porosity of 30 % and 45 %, respectively.

Coral sand labeled with the commercial name Carib Sea was purchased from an aquarium store. It was crushed in a grinder and separated into different grain size ranges. These particles are sub-rounded and less angular than those of the

E. Suescun-Florez • S. Bless (✉) • M. Iskander • C. Daza
Tandon School of Engineering, New York University, 6 Metrotech Center, Brooklyn, NY 11201, USA
e-mail: sbless@nyu.edu

D. Casem et al. (eds.), *Dynamic Behavior of Materials, Volume 1*, Conference Proceedings of the Society for Experimental Mechanics Series, DOI 10.1007/978-3-319-41132-3_12

silica sand. Particles size varies from 10 to 1500 μm exhibiting aspect ratios around 0.70. The mean particle size (D50) was found to be 615 μm and the coefficient of uniformity was 1.92. The material is poorly graded and is classified as SP according to the USCS. Specific gravity of the grains is 2.65, and its maximum and minimum densities are 1.81 and 1.55 g/cm^3 respectively; corresponding porosities were 32 % and 42 % respectively.

12.3 Arrested Testing Technique

Uniaxial compression tests were conducted on cylindrical samples of dry sand having a diameter 76.2 mm and thickness of 12.7 mm. corresponding to an aspect ratio of six. Chambers were lubricated before being filled with sand to minimize side friction on the cell boundary. Samples were prepared with an approximate relative density of 85 %, corresponding to initial porosities of 32 % and 33 % for silica and coral sand. The compression was arrested at 5, 15, or 25 % strain, and then samples were unloaded.

Once uniaxial tests were complete, all material was retrieved, and particle size and shape analysis were performed on the entire sample using a QuicPic dynamic image particle analyzer. The particle analyzer provides a particle detection limit of 10 μm.

12.4 Stress-Strain Behavior of Sands

Results from uniaxial compression tests arrested at 5, 15, and 20–25 % strain for silica sand are shown in Fig. 12.1, and coral sand in Fig. 12.2. Solid lines correspond to high strain rate (0.1/s) whereas dashed lines show the behavior at low strain rate (0.0001/s).

Since the deformation is uniaxial, the strain is equal to the change specific volume divided by the initial specific volume. Hence density is related to strain by

$$\rho = \rho_0/(1 - \varepsilon) \tag{12.1}$$

Since the moduli of void-free polycrystalline quartz (e.g. quartzite) or coral skeletons are orders of magnitude higher than the modulus of these sands, we can assume all of the change in density is due to removal of porosity, n, hence

$$n = 1 - \rho/\rho_{00} \tag{12.2}$$

where ρ_0 and ρ_{00} are initial test density and grain density, respectively. Figure 12.3 shows stress vs. porosity for the 20 % peak strain case at the higher strain rate for both materials.

For silica sand, the loading modulus appears to be bilinear and increases with strain rate. Unloading moduli are near constant and have less rate dependence. It is interesting that there is substantial increase in porosity during unloading. After the load/unload cycle the density has increase to about 2 g/cm^3. In other uniaxial experiments [8] the modulus also decreased with strain however the transition between initial and final modulus was difficult to resolve. In other words, there are three behaviors in the load-unload curves for silica sand which probably correspond to different micromechanical deformation

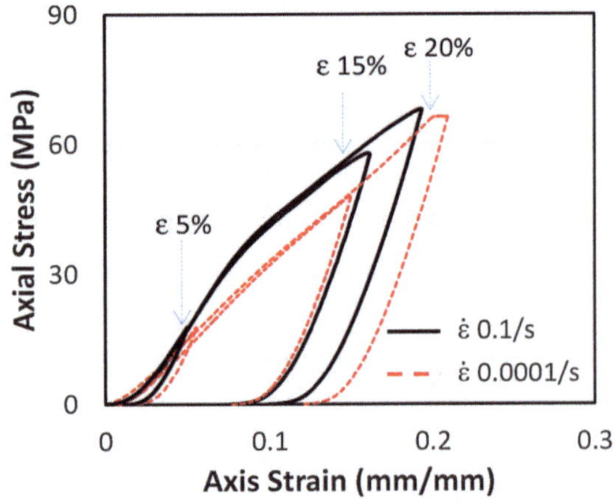

Fig. 12.1 Stress-strain data from arrested tests in silica sand

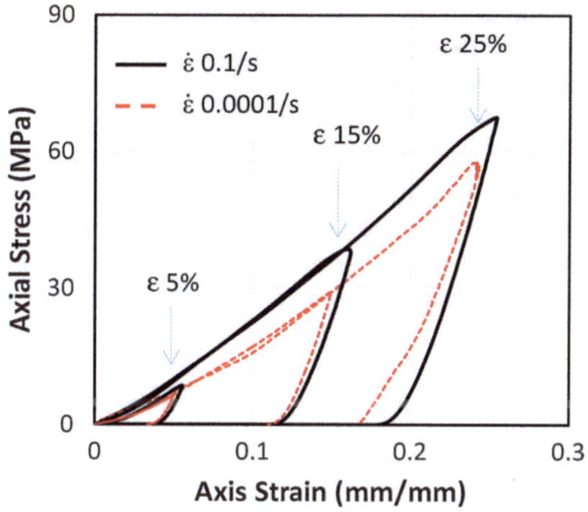

Fig. 12.2 Stress–strain data from arrested tests in coral sand

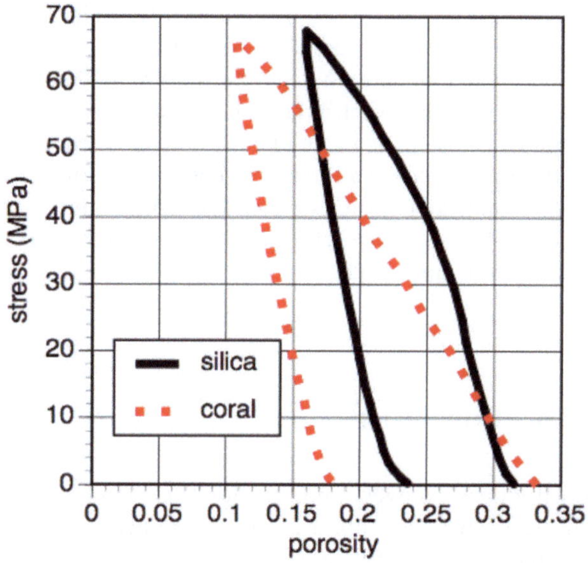

Fig. 12.3 Stress-porosity data from silica sand and coral sand tests unloaded from 20 % strain

mechanisms. After a short ramp due to seating, the stress increases linearly with strain, with a modulus of about 550 MPa at the higher strain rate. At about 9 % strain the modulus softens but thereafter remains nearly constant at about 280 MPa. Unloading occurs with a nearly linear modulus which is commensurate with that of the initial loading, about 1100 MPa, except at the end.

At the peak strain of each test, the secant modulus (0 to peak stress) increases about 10 % as the strain rate is increased by three orders of magnitude. This rate dependence of the loading modulus is consistent with that observed previously in [8], although the modulus values there are slightly stiffer than seen in these tests.

The only reversible process active in sand is presumably elastic deformation of grains as they shift position. This mechanism is probably responsible for the initial modulus and the unloading modulus. Inelastic mechanisms could include frictional sliding, irreversible closing of voids, and crushing to produce small particles that can fill the gaps between larger particles. The porosity after the load/unload cycle from 20 % strain is 24 %, which is less than the lowest porosity that could be observed in the starting material. This indicates that some material has been crushed, which was verified (see below) with the particle size analyzer.

The stress/strain curve appears to be very similar to an elastic-plastic shock compression curve, which is also for 1-D strain. If we apply that interpretation to these data, then the initial loading represents elastic compression in which the modulus is $K + 4G/3$, and the higher-stress part is volumetric compression without additional shear strain, for which the modulus is K. The hypothesis that the upper modulus is associated with volumetric collapse is supported by the observation that the difference in porosity between the initial and unloaded sample is about equal to the change in porosity during the "plastic" part of the compression curve. However, whether or not this is correct for silica sand should still be considered speculative.

For coral sand, loading takes place with a continuously stiffening modulus. Unloading takes place with a nearly constant modulus that does not vary with strain. The unloading modulus is much stiffer than the loading modulus. Loading moduli stiffen with strain rate, but unloading moduli do not. The unloading modulus, between about 50 and 20 MPa of axial stress, is 1000 MPa. This is remarkably close to the unloading modulus of quartz sand. It seems that the unloading moduli of these granular materials has little to do with the intrinsic properties of the grains themselves. Given that the unloading modulus is so much greater than the loading modulus, it appears that most of the loading behavior must be inelastic in coral sand. Because of the large hysteresis between the loading and unloading moduli, substantially greater porosity is removed by the load/unload cycle in coral sand than for silica sand. As was the case with silica sand, the final porosity is less than the porosity of the initial material at its maximum density; therefore, some void volume must be filled with crushed material.

12.5 Particle Damage

The material recovered from arrested tests was subjected to size analysis via dynamic imaging techniques. Results are shown in Fig. 12.4 for both types of sand. At 5 % strain there is little change in the particle size distribution. Reduction in particle sizes is evident at 10 and 20 % peak strain.

The degree of particle size reduction from 10 to 20 % peak strain was higher at the higher strain rate. Surprisingly, it was also observed that when either sand fractures, particles preserve similar aspect ratios to that of the parent material. The effect is more consistent as the applied strain increases.

Particle damage was greater in coral sand. For example, consider the initial median particle size. After 20 % strain 85 % of the coral sand particles were less than the initial median size, whereas in silica sand, only 70 % were smaller than the median. The small particles can fit between the large ones, which explains why the porosity decreases was greater in coral sand than in silica sand.

12.6 Conclusions

Load-unload-recovery compression tests on sand display characteristics that are specific to particle composition, but also characteristics that appear insensitive to particle composition. Loading of silica sand to 60 MPa results in a bilinear stress strain curve suggestive of elastic-plastic behavior. On unloading, the modulus is close to the loading "elastic" value. Coral sand is softer than silica sand on loading, and the modulus is almost constant and much less than for silica sand. However, the unloading modulus is very similar to silica sand. This may indicate that the particle rearrangement that occurs on unloading is relatively independent of the grain properties. All materials end up at a higher density than can be obtained with the starting particle mix. This suggests particles have crushed and filled some of the voids. Indeed, reduction of mean particle size is verified from post-test analysis. Coral sand, which has the greater reduction in void content, also exhibits increased particle breakup.

Acknowledgment The authors gratefully acknowledge the support of the Defense Threat Reduction Agency Grant No: HDTRA1-10-1-0049 and The United States National Science Foundation Grant No: DGE 0741714.

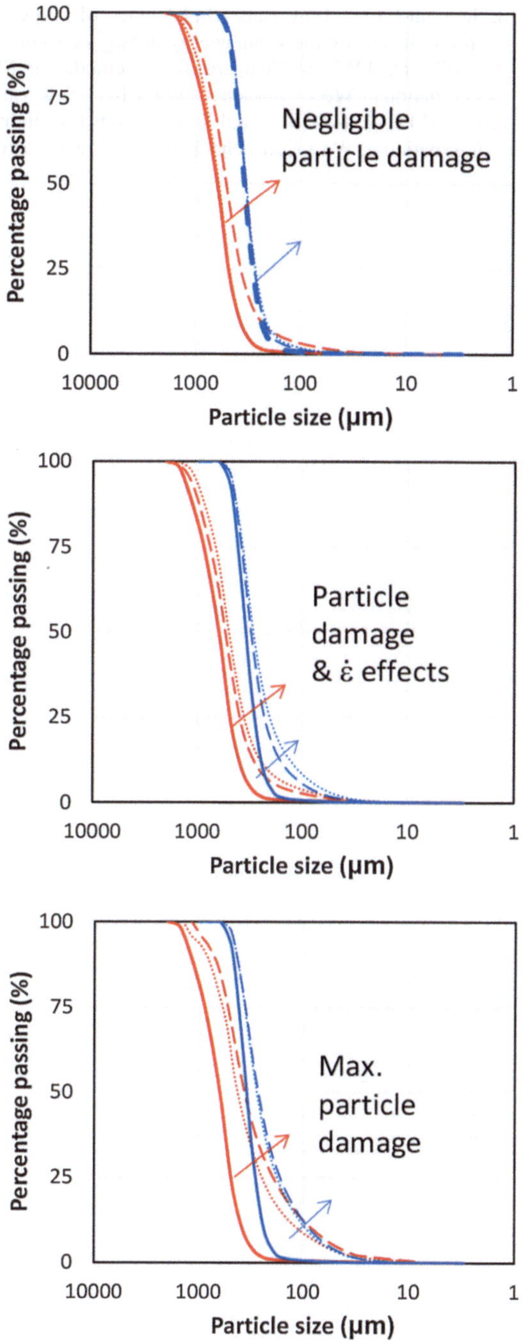

Fig. 12.4 Particle size distributions from original material and arrested tests. *Upper*: 5 % strain. *Middle*: 10 % strain. *Lower*: 25 or 30 % strain. *Blue curves*: silica. *Red curves*: coral

References

1. Russell, A.R., Einav, I.: Energy dissipation from particulate systems undergoing a single particle crushing event. Granul. Matter **15**(3), 299–314 (2013)
2. Lade, P.V., Yamamuro, J.A., Bopp, P.A.: Significance of particle crushing in granular materials. J. Geotech. Eng. **122**(4), 309–316 (1996)
3. Shahnazari, H., Rezvani, R.: Effective parameters for the particle breakage of calcareous sands: An experimental study. Eng. Geol. **159**, 98–105 (2013). doi:10.1016/j.enggeo.2013.03.005
4. Hardin, B.O.: Crushing of soil particles. J. Geotech. Eng. ASCE **111**, 1177–1192 (2007)
5. Lee, K.L., Farhoomand, I.: Compressibility and crushing of granular soil in anisotropic triaxial compression. Can. Geotech. J. **4**(1), 68–86 (1967)
6. McDowell, G.R., Bolton, M.D.: On the micromechanics of crushable aggregates. Geotechnique **48**(5), 667–679 (1998)
7. Coop, M.R., Lee, I.K.: The behavior of granular soils at elevated stresses. In: Houlsby, G.T., Schofield, A.N. (eds.) Predictive Soil Mechanics, pp. 186–199. Thomas Telford, London (1993)

8. Suescun-Florez, E., Kashuk, S., Iskander, M., Bless, S.: Predicting the uniaxial compressive response of granular media over a wide range of strain rates using the strain energy density concept. J. Dyn. Behav. Mater. **1–17** (2015)
9. ASTM D. 4253-14: Standard Test Methods for Maximum Index Density and Unit Weight of Soils Using a Vibratory Table. ASTM International, West Conshohocken (2014). www.astm.org
10. ASTM D. 4254-14: Standard Test Methods for Minimum Index Density and Unit Weight of Soils and Calculation of Relative Density. ASTM International, West Conshohocken (2014). www.astm.org

Chapter 13
Composite Plate Response to Shock Wave Loading

Douglas Jahnke, Vahid Azadeh Ranjbar, and Yiannis Andreopoulos

Abstract The mechanical response and characterization of composite materials under transient dynamic loading caused by shock or blast wave impingement is not well understood. Air blast is associated with a fast-traveling, high-pressure shock front followed by a lower pressure expansion wave. The timescales associated with the shock front are typically 10^3 faster than those associated with the expansion waves which follow. A new split-view Time-Resolved Stereo Digital Image Correlation system has been developed capable of measuring time dependent information of three component displacement vectors on two dimensional surfaces in a shock tube facility where transient aerodynamic loads on material specimens develop over the short time associated with the shock wave reflection time scales. To validate the techniques we embedded strain gauges in a S-2 glass epoxy 12.7 mm thick composite test specimen during in-house using vacuum assisted resin transfer molding fabrication. High-frequency-response, semiconductor, strain gauges were used in various combinations and locations in order to measure the transient strain rate during the impingement of the shock wave. In addition to strain gauges PZT transducers were also embedded which helped measure the frequency response of the test plate. In-house fabricated composite plates were tested under three different boundary conditions: clamped, quasi-simply supported, and bolted. Different displacements and deformation patterns were observed in each of these cases.

Keywords Composites • Shock wave • Boundary conditions

13.1 Introduction

Composites, heterogeneous material systems, are believed to exploit synergistic effects from the combination of dissimilar materials thereby developing unique property combinations not possible using other means. One class of composites is high-performance fiber-reinforced polymer (FRP). These have been increasingly used in structural applications over the last 30 years. Relative to traditional materials they have high specific-stiffness, high specific-strength, low specific-weight and are able to absorb significant energy from impacts. There are a many types of fibers and matrices available to use in FRPs. Each fiber and matrix combination can have different properties. Besides the variations in material properties in the fibers or matrices, the strength of the interface bond between the fiber and the matrix, and between the laminae as well as the lay-up sequences, ply orientations, and different forms of fiber arrangements within the matrix (particulate, planar, 2D/3D woven) all influence the final properties of the composite system. This allows composite properties such as modulus of elasticity, density, wave propagation speeds, and thermal conductivity to be adjusted. All these features have made FRP composites attractive as replacements for traditional materials.

These features also make them difficult to characterize in a systematic way. This vast number of variables makes it challenging to predict the overall structural and acoustic properties of a FRP composite from the details of the materials and construction. Thus, knowledge of the response of composite materials under dynamic loading is essential to develop a better understanding of its performance as a structural material. While such information starts to be slowly available very little is known about the behavior of such materials under the effects of dynamic mechanical extremes. Understanding such material systems requires performance testing data of the effects of dynamic events in order to fully document the behavior of this material.

D. Jahnke • V.A. Ranjbar • Y. Andreopoulos (✉)
Department of Mechanical Engineering, The City College of New York, 160 Convent Avenue, New York, NY 10031, USA
e-mail: andre@ccny.cuny.edu

D. Casem et al. (eds.), *Dynamic Behavior of Materials, Volume 1*, Conference Proceedings of the Society for Experimental Mechanics Series, DOI 10.1007/978-3-319-41132-3_13

13.2 Thin Composite Plate Response to Pressure Loading

The mechanical response and characterization of composite materials under transient dynamic loading caused by shock or blast wave impingement is not well understood. Air blast is associated with a fast-traveling, high-pressure shock front followed by a lower pressure expansion wave. The timescales associated with the shock front are typically 10^3 faster than those associated with the expansion waves [1]. Impingement of blast waves on structures result in a reflection of the wave off the surface followed by a substantial transient aerodynamic load which can cause significant deformation and damage to the structure. In addition, a complex aeroelastic interaction between the blast wave and the structure develops that can induce reverberation within an enclosure and cause substantial overpressure through multiple reflections of the wave. This reverberation can cause permanent hearing loss and both physiological and psychological damage. The response of advanced composite materials to dynamic loading caused by fast moving shock waves is associated with such thermomechanical extremes. This is relevant to a broad range of applications environments such as multifunctional survivability requirements of civilian or military vehicles. In addition, establishing their structural integrity under dynamic conditions at high strain rates and their post-impact behavior become pre-requisite in such crucial applications.

The assessment of the effects of blasts on structural systems was studied systematically after World War I. An early published work was Hopkinson's study of pressure from high explosives or bullet impacts [2]. One of the earliest analyses related to structures subjected to air-blast loading in the literature was performed by Taylor [3], where he developed a solution for a one-dimensional wave impinging on a free-standing plate to compute the momentum transmitted to the plate by the shock pulse. The prediction and measurement of the structural response of ship panels to free-field air-blast explosions were first investigated by Houlston et al. [4]. In this research square plates and stiffened panels subjected to air-blast and underwater shock loads were investigated. Later, assuming a form of displacement function and a modified Friedlander equation reflected blast overpressure loading that exponentially decays with time.

Gupta et al. [5] conducted a single-degree-of-freedom elastodynamic analysis of the response of a rectangular plate subjected to an explosive blast. Beshara [6] investigated the prediction of dynamic effects of unconfined explosions needed for the structural analysis of blast-loaded above-ground structures. Türkmen and Mecitoğlu [7, 8] carried out a study on the dynamic behavior of laminated composite plates subjected to blast loading. In this study, theoretical analysis and experiment results of the strain time histories were obtained.

Recently Xue and Hutchinson [9] studied the quasi-static dynamic responses of clamped circular sandwich plates subject to uniformly distributed impulsive loads. Numerical analysis on the plate neglecting the effects of fluid-structure interaction was carried out and a new constitutive relation for sandwich plates was introduced as well [10]. In similar research, the dynamic response of a clamped sandwich beam/plate was assessed. The fluid-structure interaction effect was evaluated utilizing Taylor's approximation by including a segregated fluid-structure interaction phase in research by Deshpande, Fleck, and Qui [11, 12]. Tekalur et al. [13] experimentally studied the effect of blast loading on both an E-glass fiber composite and carbon fiber composite material, by simulating a blast in a shock tube. Results suggested that the E-glass fiber composite experienced progressive damage during high-rate loading, whereas the carbon fiber composite experienced sudden failure. Similar work was conducted by Pankow, et al. [14].

The recent work by Gong and Andreopoulos [15] investigating the impact of shock waves on monolithic (aluminum and steel) and composite material plates has shown that, the interaction between the blast and the structure is mutual. Due to the elastic deformation of the plates and their reverberation, strong acoustic waves were generated on the external side of the impact which carry a significant signature of the plates' properties. Composite plates were found to suppress several of the modes of the wave patterns while metallic ones demonstrate a rich variety of interacting modes. The amplitude of the excited acoustic waves, however, was higher in the case of composite plates than in the case of steel plates. The frequency content of the strain signals on the surface of composite plates was not always the same with the content of the surface acceleration measured in free vibration experiments. Calculations using a coupled system of equations between the fluid and solid phases of monolithic materials provided predictions in good agreement with the measured values of modal frequencies. These theoretical results were also in agreement with the classical modal analysis results by using the Poisson-Kirchhoff theory for thin plates under axisymmetric or non-axisymmetric conditions. In addition to their analytical and theoretical work, the experimental results were predicted by a fully coupled computational model developed by Gong and Andreopoulos [16, 17] in which an unstructured adaptive mesh flow solver of second order accuracy in finite volume formulation, a finite element structure solver and a moving mesh algorithm were implemented in the numerical simulation of the interaction between a shock wave and a structure. This fully coupled approach accounted for the effects associated with the mutual interaction for the first time. The results illuminated complicated flow phenomena and structure vibration patterns, which in order to be detected experimentally require capabilities beyond those available by current experimental techniques. The numerical simulations also successfully modeled the aero-acoustic damping effects on the structure, which do not exist in previous numerical models. Further analysis of the results showed that the mutual interaction is not linear and that the non-linearity arises because the wave propagation in the fluid is not linear and it cascades a non-linear and non-uniform loading on the plate. Non-linearity intensifies when the plate is vibrating at high frequency while the wave propagation speed is low.

A blast produces transient loading of very short duration on a structural element. This loading is associated with the timescale of the shock reflection over the surface of the structure is (on the order of 1 μs). Because of the highly transient nature of the loading, the response of the structural element evolves over different timescales. It is well known that the material response in part determines the structural response. High shearing rates occur during the impact of the shock wave front on the structural element because of the extremely short timescales of the flow. Relaxation of these high shearing rates takes place immediately after the shock wave impact during the passage of the expansion wave which is associated with timescales that are 10^3 times or more slower than those of the shock wave.

In the timescale on the order of duration of the shock wave reflection, the transient response is localized and is governed by the wave propagation in the material. This response evokes a material response at the speed of stress development due to wave propagation in the material. Therefore, the structural response at the very short timescales produces strain rates higher than 10 s^{-1}. This initial response is significant as glass-polymer composites have shown strain rate effects resulting in strength increases of up to 20 % in this range in work by Khan et al. [18]. With time, the response of the structure evolves to produce vibration of the entire structural element at frequencies which are determined by the stiffness of the element and the boundary conditions. The vibrational response of the structural elements usually occurs on the millisecond timescale and produces strain rates in the range of 0.1–1.0 s^{-1}. Full-scale experiments involving actual geometries and charges are quite involved and costly, both in terms of preparation and measurements.

Moreover, full-scale testing of realistic blast scenarios is often prohibitively expensive and time consuming. Even the smallest of these full-scale tests can require explosive charges in the medium to extreme categories and outdoor laboratory experimentation [19]. At this scale, instrumentation becomes difficult and expensive, often yielding only point-wise piezoelectric pressure profiles at limited locations and a qualitative rather than quantitative evaluation of material deformation [20]. For these reasons, scaled-down experiments are highly desirable. However, the validity of such experiments has to be proven since scaling is not straightforward in all blast events.

The major advantage of a laboratory simulation as opposed to field measurements is its controllable environment which can provide a platform for systematic investigation of complex phenomena with well-established initial and boundary conditions. This is critically important in basic research where a better understanding of the underlying physics is sought or high fidelity experimental data are needed to test and verify advanced computational methods with specific predictive capabilities. In the case of monolithic materials, Justusson, et al. used a laboratory shock tube to examine aluminum deformation at high rates [21]. In the particular case of testing composite materials under blast loading existing methodologies are several and diverse since there is not an accepted standard. A tensile test, for instance, is a classic example of a widely used testing methodology (for example see ASTM D638-10) to establish the bulk properties of materials. An additional advantage to laboratory testing is the ability to vary boundary conditions.

Tekalur et al. [13] used a simply supported bending configuration to study the mechanical behavior and damage of composite specimens by static and blast loading . A shock tube apparatus was utilized to generate a planar shockwave with a controlled overpressure level. A flat rectangular panel was held in front of the end of the muzzle section. Thus the blast wave load is initially supplied over a circular area of the rectangular specimen which is simply supported. However, as the specimen is deformed the contact surface between the specimen and the shock tube wall decreases and gas pressure is applied beyond the original opening of the tube. Strain gauges are often used to measure strain rates as a function of time at specified locations on the surface of the material [15]. While a clamped boundary condition is a basic calculation analytically, it can be hard to achieve in the lab. Nurick et al. explored the effect of a clamped boundary condition on the onset of necking and tearing for circular plates [22]. Variation of the edge of the clamped boundary between a sharp edge and radii of 1.5 and 3.2 mm led to changes in deflection and the onset of thinning and tearing.

Most laboratory-scale research focuses on the deformation of plates subjected to blast loads. Nurick et al. [23] performed most of the initial work and established standard terms for the qualitative classification of steel plate failure modes and plate deformation measurement [22]. These tests include post-test measurements of maximum plate deflection and qualitative plate shape [23]. According to [23–25] the modes of failure are classified as:

- *Mode I:* Large plastic deformation without tearing of the specimen
- *Mode II:* Tensile tearing partial or complete at supports
- *Mode III:* Plastic shear sliding at supports.

Approaches using a shock tube platform appear to have certain distinct advantages over approaches using detonations. A major one is that the shock front in the shock tube is planar, thus yielding uniform pressure behind it which loads the structure. This simplifies the geometry and boundary conditions and therefore allows for a better evaluation of modeling issues associated with numerical simulations of the phenomena involved. In contrast, shock wave fronts generated by

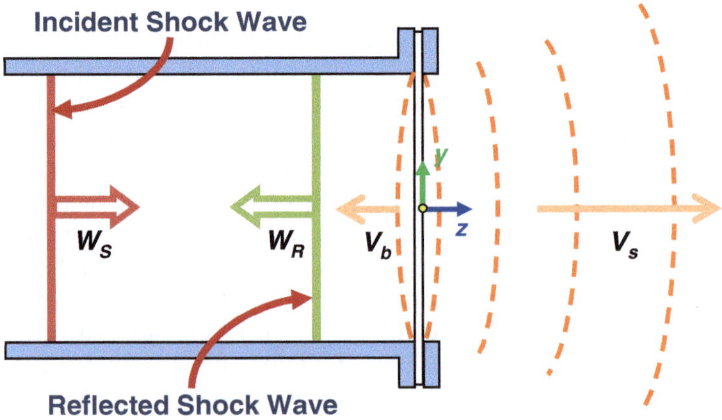

Fig. 13.1 Sound wave excitation through shock impingement

detonations are three dimensional in space and lead to complex pressure loadings which can mask the underlying basic physics of the impact. Once the physics of the impact is better understood in simple geometries then the next level of complication can be introduced.

The complicated interactions under investigation need to be set up by using a simple, yet relevant geometry of the composite material structure and simplified flow fields which have the potential to reveal basic features of the associated phenomena.

The pressure behind the incident shock wave, p_2, remains constant across the flow field before the shock reflection. As the shock wave contacts the surface of the end wall where the composite plate is installed, the process of reflection starts, the plate is loaded and deformation begins. If the plate were a rigid body the pressure behind the reflected shock, p_5 and would be given by the relation:

$$\frac{p_5}{p_2} = \frac{\left[\frac{\gamma+1}{\gamma-1} + 2\right]\frac{p_2}{p_1} - 1}{\frac{p_5}{p_2} + \left[\frac{\gamma+1}{\gamma-1}\right]}$$

When the plate is elastic, its deformation starts with the initial contact between the shock wave and the surface; however, it is not instantaneous. In tests with carbon and glass fiber composites, Arora et al. saw initial displacements lag between 1 and 2 ms behind shock wave impingement [26]. According to [27] during the initial displacement of the accelerating plate towards the $z > 0$ direction, expansion waves will be shed off in the $z < 0$ direction, the direction of the reflected shock, and compression waves will be shed off in the $z > 0$ direction on the external side of the plate. During the decelerating displacement of the plate in the other direction compression waves will be shed in the $z < 0$ direction behind the reflected shock wave and acoustic waves in the exterior side of the plate (see Fig. 13.1).

Another characteristic shared by composites is the ability to attenuate shock through resonance behavior. Work by Oved et al. showed resonance behavior due to wave reflections in the layers of the composite [28]. In particular this behavior, unexpectedly proved to be important at higher stresses (outside the linear elastic region) where elasto-plastic behavior was observed.

The fracture, damping, and stiffness properties of a composite also affect its ability to absorb energy and are dependent on the make-up and geometry of individual composite materials. In one case, Gibson et al. showed that adding a polymeric interleave in composite laminates can improve both damping and fracture toughness [29]. Other researchers such as Koratkar et al. [30], Zhou et al. [31], and Rajoria and Jalili [32] have shown that tremendous increases in the loss modulus of polycarbonates and epoxies have resulted from the addition of nanoparticles such as carbon nanotubes.

Previous work by Gupta and Nagesh [25] has shown that there is some sensitivity in the results, particularly on the onset of thinning and tearing at the boundary, to boundary fixation conditions.

13.3 Experimental Facilities

The experiments were carried out in the large-scale Shock Tube Research Facility of the Department of Mechanical Engineering at CCNY. The facility is partially shown schematically in Fig. 13.2 and is described in detail by Briassulis et al. [33] It has been used extensively for research on shock wave interactions with turbulence and vortices [1, 34, 35]. In the

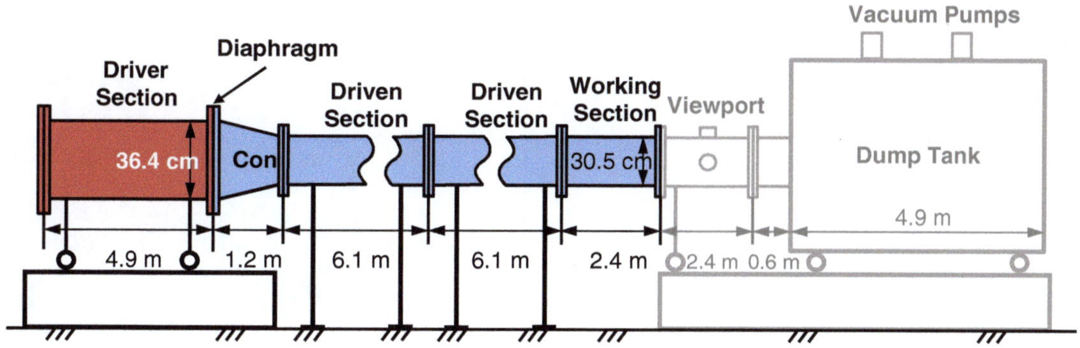

Fig. 13.2 Schematic of the CCNY high-resolution shock tube facility, with the unused features shown in gray

preset case it was used to study the impingement of shock waves on thin composite plates. The specimen was mounted at the end flange of the working section.

The rupture of the diaphragm between driver and driven sections released a shock wave which traveled downstream at pressure p_2 to the end of the working section where the test plates were mounted. The shock tube as configured for these experiments was 18.5 m (61′) in length with a working section 30.5 cm (12″) in diameter. The pressure in the shock tube was measured by transducers in the working section.

Aluminum diaphragms were used for all early experiments. These plates were 380 mm (15″) in diameter and 3.2 mm (1/8″) thick. A cross was machined to a specific depth to control the pressure at rapture (p_4). At lower pressures, up to around 690 kPa (100 psig), the aluminum diaphragms performed well. At higher pressures, the depth of the machined cut led to unreliable performance. Another material frequently used for diaphragms in shock tubes is polyethylene terephthalate (PET) commonly known by the DuPont trademark name Mylar. Later experiments were performed using 0.254 mm (10 mil) PET as the diaphragm material. A single layer of PET ruptured around 275 kPa (40 psig). Multiple layers have been used to reach higher pressures although the relationship between layers and pressure was found to be nonlinear and not entirely predictable.

13.4 Boundary Conditions

As mentioned earlier the work there is some sensitivity of the results to boundary fixation conditions [25]. Our initial results also provided indications that this is the case. In our work we configured three different boundary conditions, the original boundary condition which is referred to as clamped and two additional, quasi-simply supported and bolted. Descriptions and illustrations of the three boundary conditions are provided below.

13.4.1 Clamped

The clamped boundary condition was the as-built configuration of the shock tube for holding a circular test specimen. Figure 13.3 shows an illustration with an exploded view. The test specimen and gasket are held to the end of working section by the retaining ring. The end of the working section has a boss around the shock tube opening that fits into a rabbet in the retaining ring creating a tight seal. In order to minimize sensitivity of the particular boundary conditions on the vibrational and deformational pattern of the plate and to avoid introducing artificial nodes, additional care was taken in the way the bolts were tightened. The bolts are tightened in stages following a standard star-pattern as often specified for large pipe installations. First, the bolts are tightened snugly, in a star pattern, using an air powered impact wrench. It is important at this stage not to over tighten bolts thereby preventing the retaining ring from seating properly. Next, a torque wrench is used to tighten the bolts to the specified torque, again using a star pattern. Two passes were made using the torque wrench to ensure proper seating. If excessive movement was observed in the bolts on the second pass, i.e., a turn greater than a few degrees was required, a third pass would be made.

Fig. 13.3 Experimental clamped boundary condition

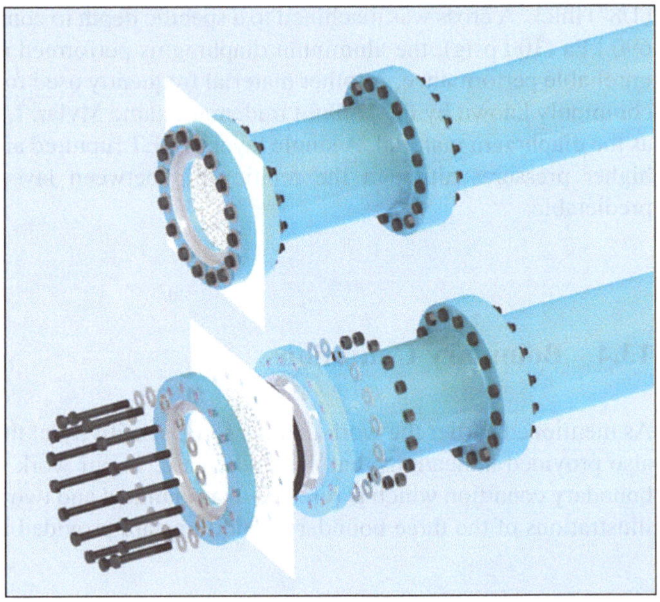

Fig. 13.4 Experimental bolted boundary condition

13.4.2 Bolted

The bolted mounting method was devised to minimize slipping or pulling out of the test specimen from the retaining ring. A larger rectangular test specimen was used and drilled with holes matching the bolt circle in the retaining ring. Figure 13.4 Experimental bolted boundary condition is an illustration of the bolted boundary condition with an exploded view. The gasket and test specimen were placed against the boss on the shock tube working end. The retaining ring was reversed so the rabbet was opposite test specimen. The clamping bolts pass through the test specimen and limit any radial movement. Figure 13.5 Clockwise from top left: composite plate drilled for bolt-through mounting, gasket in place on shock tube working section flange, plate placed on mounting bolts, retaining ring bolted in place shows a plate (in this case composite) drilled and painted for TRC-SDIC, the gasket attached to the plate mounting boss of the shock tube, the plate placed on the bolts, and, finally, the retaining ring bolted in place to complete the setup.

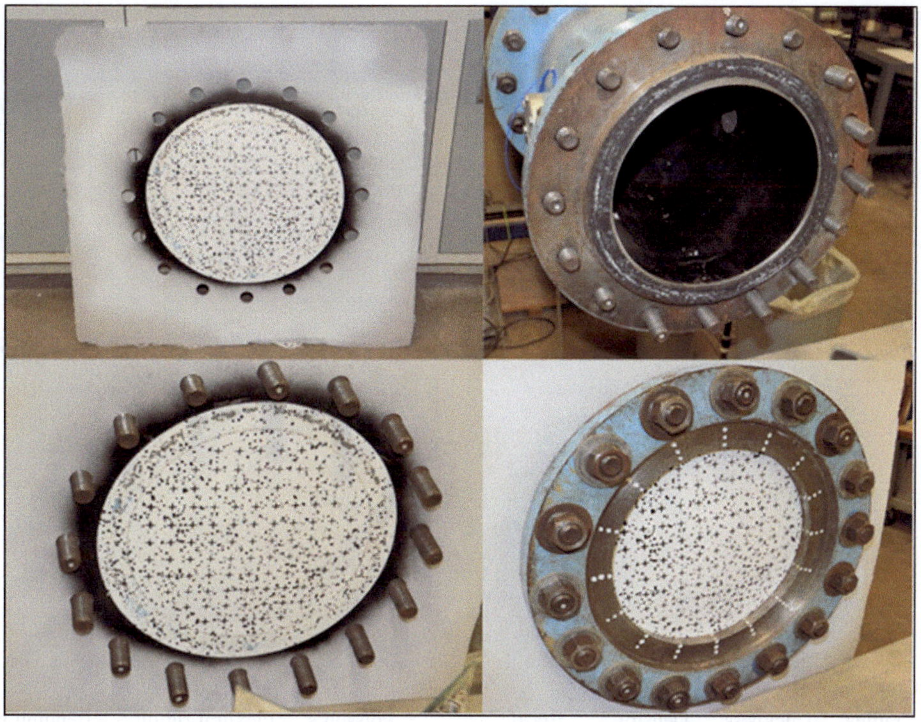

Fig. 13.5 Clockwise from top left: composite plate drilled for bolt-through mounting, gasket in place on shock tube working section flange, plate placed on mounting bolts, retaining ring bolted in place

Fig. 13.6 Detail of quasi-simple support gasket and cable

13.4.3 *Quasi-Simply Supported*

In this configuration the plate was supported on its back by a rubber gasket reinforced with a wire through its center (see Fig. 13.6). This arrangement can only be considered quasi-simple since in addition to the rubber gasket support, the retaining ring still had to be bolted in place, not tightly though, to hold the test specimen, so the free and unlimited movement in the plane of a true simple support was hindered, but still freer than the clamped mounting. Figure 13.7 is an illustration of the quasi-simply supported mounting with an exploded view. The gasket and the test specimen were placed against the boss on the working end of the shock tube. The quasi-simple support was then placed between the test specimen and the retaining ring. The retaining ring was again flipped so the rabbet faced away from the test specimen and then secured with the bolts. The bolts were made finger-tight to allow more freedom of movement and better mimic a simply supported boundary condition (Fig. 13.7).

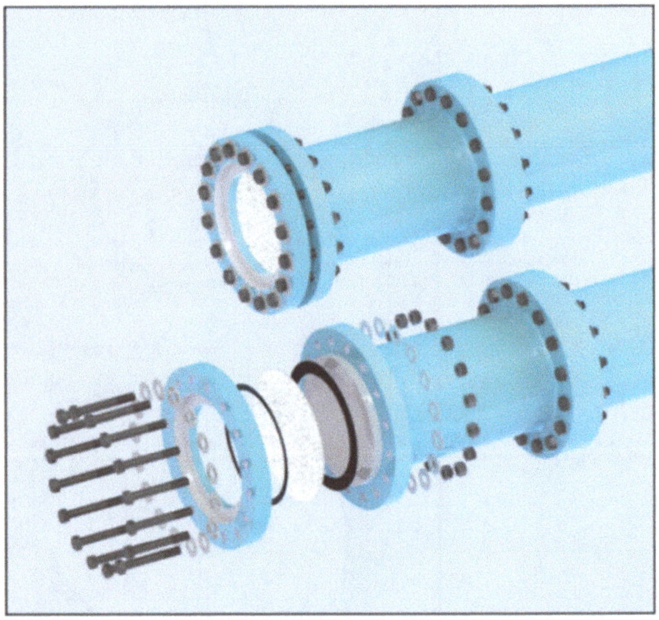

Fig. 13.7 Experimental quasi-simply supported boundary condition

Table 13.1 Parameters and properties of materials tested

Description	Free radius [mm]	Thickness [mm]	Density [kg/m^2]	Elastic modulus [GPa]	Poisson's ratio	Flexural rigidity [Pa m^3]	Bending wave-speed [m/s]	Timescale [ms]
Aluminum	152	3.2	2700	69	0.33	207	4.91	31.05
MIL steel	152	3.2	7900	200	0.30	586	4.83	31.52
Mild steel	152	3.2	7800	200	0.30	586	4.87	31.52
S2-glass/epoxy	152	3.2	1689	22	0.10	58.1	3.29	46.29
S2-glass/epoxy (thin)	152	1.6	1689	22	0.10	7.27	1.65	92.57
S2-glass/phenolic HJ1	152	3.2	1960	27	0.26	77.2	3.52	43.26

13.5 Test Specimens

The test specimens for the clamped and quasi-simply supported boundary conditions were thin circular plates 380 mm (15″) in diameter. The test specimens for the bolted boundary conditions were thin square plates 610 mm (24″) in width and height. Except where noted, all test specimens were 3.2 mm (1/8″) thick.

The S-2 glass epoxy composite test specimens were manufactured in-house using vacuum assisted resin transfer molding (VARTM). The woven S-2 glass fiber was sealed between two plastic plies along with additional plies of material to assist resin transfer, debagging, etc. In addition to composite materials, military grade steel and mild steel as well as 6060 Aluminum plates were also used as specimens.

Table 13.1 shows the parameters and properties of the materials tested. The S2-glass epoxy plate properties were determined through standard laboratory tests.

In addition to the pressure sensors, high-frequency-response, semiconductor, strain gauges made by Kulite Semiconductor Products (UHP 5000-60); high-frequency-response, semiconductor, backed bar strain gauges made by Micron Instruments (SS-060-033-1000 PB); and general purpose strain gauges made by Micro-Measurements/Vishay Precision Group (CEA-13-125UW-350) were used in various combinations and locations in order to measure the transient strain rate during the impingement of the shock wave. In early tests the strain gauges were surface mounted on the external face of the test specimen (facing outside the tube), but the large accelerations and deformations caused the gauges to detach or otherwise fail well before maximum deformation of the test specimen. Very little usable data was obtained. However, as the fabrication process of the composite plates was advanced, it was possible to embed such sensors into the composite plates during the manufacturing process.

13.6 Optical System and Setup

The experimental technique has been developed based on digital image correlation (DIC). The setup consists of a camera acquiring a split-view (from the left and from the right) of the motion using two separate mirror pairs located at to the left and right of the specimen axis that are inclined and tilted so as to capture the time-dependent motion of the whole plate specimen under blast/shock loading. The optical setup consists of two large size planar primary mirrors A & B viewing the specimen from two opposite locations tilted to about 30° to the specimen as shown schematically in Fig. 13.8. An optical system consisting of a combination of light reflecting (mirrors) and refracting elements (lenses) is called Catadioptric. The setup also includes a single high-frame-rate camera which can accommodate two simultaneous stereo images (the split-view) of the deforming structure on its CMOS chip and therefore it is a real Time-Resolved Catadioptric Stereo Digital Image Correlation (TR-SDIC) system. Figure 13.9 shows a photograph of the TR-SDIC setup. The typical camera setup for TRC-SDIC is shown in Fig. 13.8.

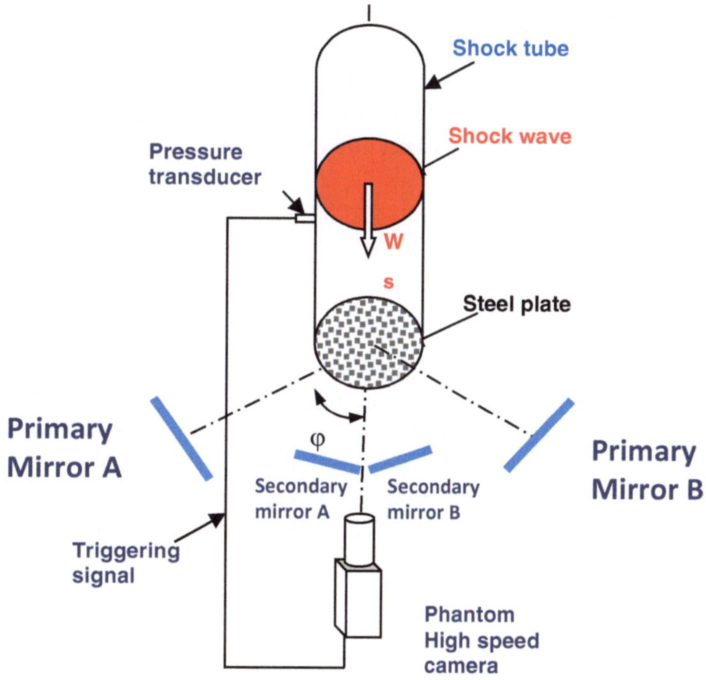

Fig. 13.8 TR-SDIC setup in the shock tube with two pair of mirrors

Fig. 13.9 Mirror setup for high-speed video capture of displacement for TRC-SDIC. The lens of the camera can be seen coming out of the tube section at the right

13.7 TRC-SDIC Validation

A major objective of this work is validation of the present optical techniques which will include static as well dynamic measurements of displacements and strain. In the beginning of our work we used a composite plate of 3.2 mm (1/8 in.) thickness with strain-gauges mounted on the external surface of the plate which was also viewed by the camera. However, the strain gauges pop off the plate during the dynamic impact of the shock wave on the internal surface of the plate because of large deformations and high accelerations developing during this dynamic incident. In order to overcome this problem, it was decided to embed the strain gauges in the S-2 glass epoxy 12.7 mm (1/2 in.) thick composite test specimens during in-house using vacuum assisted resin transfer molding (VARTM) fabrication. High-frequency-response, semiconductor, strain gauges made by Kulite Semiconductor Products (UHP 5000-60), high-frequency-response, semiconductor, backed bar strain gauges made by Micron Instruments (SS-060-033-1000 PB), and general purpose strain gauges made by Micro-Measurements/ Vishay Precision Group (CEA-13-125UW-350) were used in various combinations and locations in order to measure the transient strain rate during the impingement of the shock wave. In addition to strain gauges PZT transducers were also embedded which helped measure the frequency response of the test plate. Details of the setup can be found in [25].

Figure 13.10 shows a schematic of the composite plate with the locations of the embedded sensors marked. The strain gauges were mounted in radial/circumferential pairs to measure strain and were positioned carefully below the external surface of the plate under the first fiber layer. The 60 × 60 cm (2 × 2 ft) composite plate was mounted on the end flange of the 304 mm (1 ft) diameter shock tube by using bolts that passed through the specimen as well as the retaining ring to achieve a truer clamped boundary condition for static and dynamic testing.

13.7.1 Static Tests

The shock tube was pressurized slowly and the displacement of the plate was measured by our DIC techniques when the continuously monitored pressure had reached certain levels. The initial static test data from a 3.2 mm (1/8 in.) composite plate were obtained by TRC-SDIC and manually by traversing a digital caliper on a custom-made fixture and measuring at specific points along the specimen diameter. Unlike in the dynamic tests where the mirrors and camera are setup, a calibration shot taken, and the test was run in a short period of time with little chance of the setup moving, the static testing gave many opportunities for the setup to move. To eliminate this potential source of error, calibration shots were taken

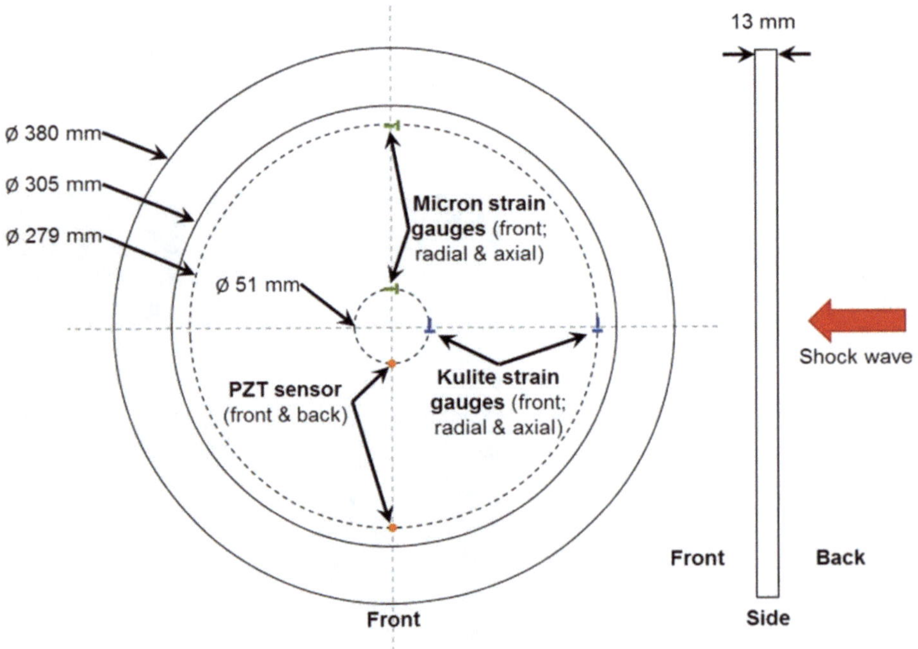

Fig. 13.10 Schematic of S-2 glass epoxy composite plate with locations of embedded sensors (Dimensions in inches)

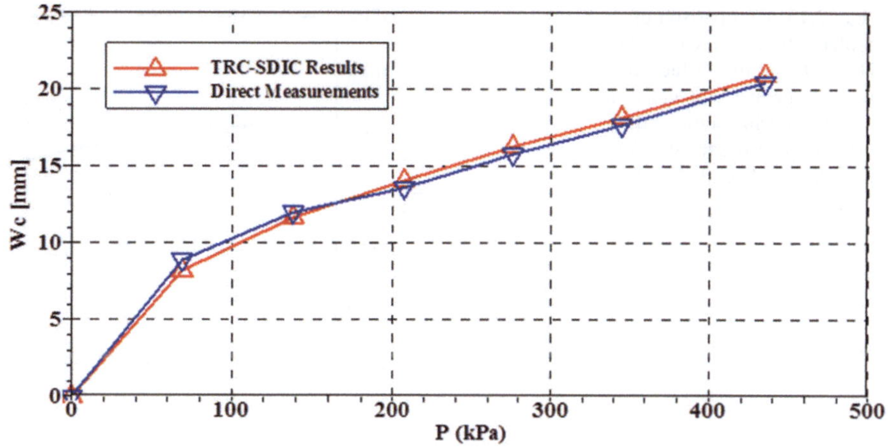

Fig. 13.11 Comparison of axial deformation at the center of the test plate caused by static pressure loading as measured experimentally with results from TRC-SDIC

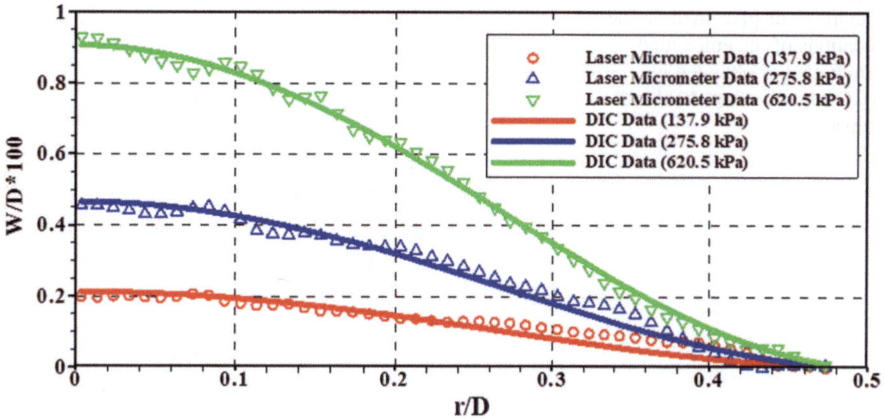

Fig. 13.12 Comparison of out of plane deformation W, during static tests between laser micrometer and DIC data at three different pressure levels. Free diameter of plate: 304 mm (12 in.)

before and after each frame. For these static measurements a laser position micrometer, model OPTO NCDT, was also used to determine the static displacement and deformation of the plate using a triangulation technique.

The results are shown in Fig. 13.11. The deformation at the center of the plate (maximum displacement) is plotted against the static pressure. The agreement between the two approaches is very good, with the largest variance around 8 %. This increases confidence in the TRC-SDIC results.

A second experiment under static conditions was carried out with the 12.7 mm (1/2 in.) composite plate with embedded strain gauges. In this case the displacement of the plate at the specified pressure was measured using the laser micrometer along one diameter. The measurements obtained by TRC-SDIC and the laser micrometer are compared in Fig. 13.12. The results are non-dimensionalized by the diameter of the plate, *D*. Maximum displacement has been observed at the center of plate at $r/D = 0$ at 620 kPa (90 psi) pressure and was found to be 2.7 mm. The displacement is substantially reduced towards the shock tube wall where the flange is mounted and it is zero at the location of the clamping area. As expected the displacement also depends on the applied static pressure. At the lower pressure levels of 275 kPa and 138 kPa (40 and 20 psi respectively) the measured displacements are also reduced. The laser micrometer data indicate some scatter which is well within the accuracy of the instrument and the uncertainties of measurements. It appears that the DIC data obtained by our optical setup and data processing algorithms based on fitting Bessel functions to the directly measured displacements, agree very well with the data obtained by the laser micrometer.

Comparison of the strains obtained by DIC and the embedded strain-gauges is shown in Fig. 13.13. The radial strain ε_{rr} in cylindrical coordinates is compressive (negative) on the external surface of the plate near the clamping location and changes sign to become tensile (positive) near the center of the plate. The compressive strain near the clamping is, in absolute terms, slightly higher than the tensile strain at the center of the plate. The location where the strain changes sign appears to be at $r/D = 0.283$ at all pressure levels. The strain gauges measured strain at locations below the external surface of the plate and if a linear distribution across the plate thickness, *h*, is assumed for the strain it will be proportional to $z/h/2$ where *z* is the location of the strain gauges from the middle plane. Thus $\varepsilon_{rr}(z) = \varepsilon_{rr}(h/2)2z/h$. The half-thickness is $h/2 = 6.4$ mm (1/4 in.). The actual location of the strain gauges, *z*, is rather difficult to determine accurately first because they can move slightly during the vacuum assisted infusion molding to not only different locations but also to different orientations and second because of their finite

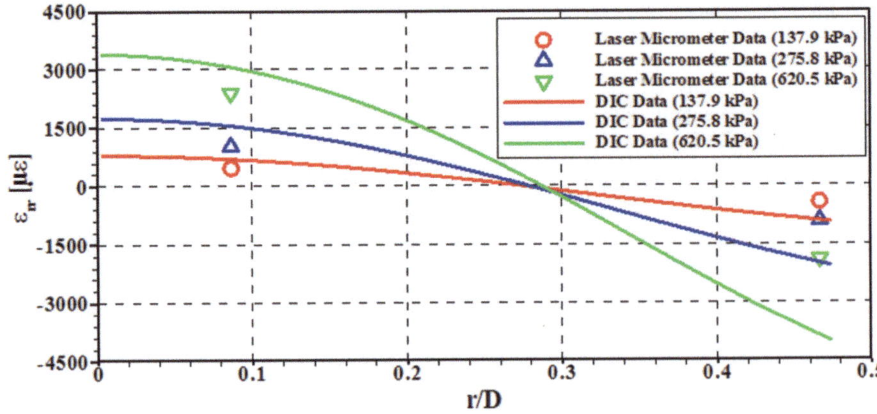

Fig. 13.13 Comparison of in-plane strain ε_{rr} measured by STC-SDIC at the surface and by strain-gauges embedded in the plate, during static at three different pressure levels. Free diameter of plate: 304 mm (12 in.)

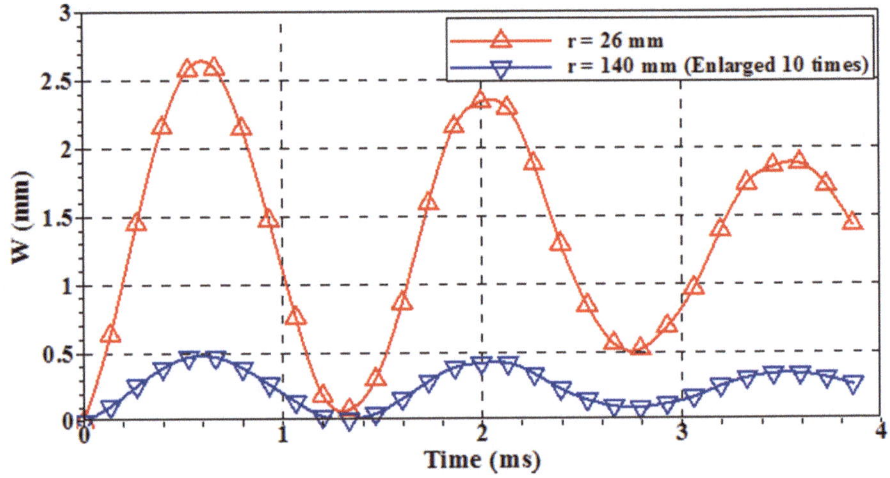

Fig. 13.14 Time-dependent signals of displacement at the locations of the embedded strain gauges obtained by TRC-SDIC during shock impact

thickness. Thus, the strain measurements obtained by the gauges are expected to be lower than the surface strains measured by DIC. This is the case in the data shown in Fig. 13.11b. The gauge located at $r = 0.08D$ consistently measured strains about 20 % less than the surface strains while the gauge at $r = 0.47D$ measured about 40 % of the corresponding surface strains $\varepsilon_{rr}(h/2)$. This relative difference is consistent at all pressures and this consistency provides confidence in the developed TRC-SDIC system.

13.7.2 *Dynamic Tests with Shock Impact*

A second experiment was carried out in which a shock of 620 kPa (90 psi) strength impinged on the plate and dynamic signals of strain were acquired from the gauges together with time-resolved DIC data. The time-dependent displacement of the plate at the locations of the strain gauges is shown in Fig. 13.14. The plate immediately after the shock impact starts to vibrate in the second mode with a maximum displacement of 2.6 mm at $r = 26$ mm $= 0.084D$ and 0.05 mm at $r = 140$ mm $= 0.46D$. A comparison of the strains measured by the gauges and those obtained by TRC-SDIC is shown in Fig. 13.12b. The two strain signals follow the same periodic variation with time which is due to the plate vibration. This indicates that the TRC-SDIC has the frequency resolution to capture the periodic variation in strain and displacement. At the location $r = 26$ mm the strain sensor measured about 10 % less than DIC's strain and at $r = 140$ mm the strain gauge measured 42 % less strain than DIC. This difference between the dynamic strains is very consistent with the static strains i.e. the strains measured by the embedded gauges are less than the strains at the surface measured by DIC by the same percentage (Fig. 13.15).

The agreement between the displacement data obtained by TRC-SDIC and by the laser micrometer or the caliper in static tests is very good, with the largest variance around 8 %. This increases confidence in the TRC-SDIC results not only in the static case, but also in dynamic cases where, from the TRC-SDIC process perspective there is no difference between static

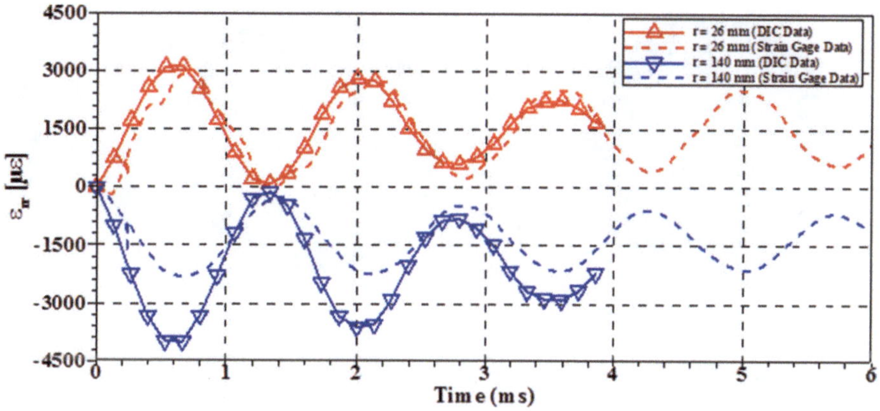

Fig. 13.15 Comparison of time-dependent strain signals obtained from the embedded gauges and from TRC-SDIC during shock impact

and dynamic images since in both cases sequential frames are compared to determine deformations without regard to time. Strain data are more difficult to obtain either by point sensors or global methods and are not very often provided in the literature. In the present work, we demonstrate for the first time a comparison between measurements obtained by embedded strain gauges and by our TRC-SDIC techniques. The agreement between these two methods is very good which provides validation of the approach and establishes confidence in the application of our TRC-SDIC techniques.

13.7.3 Results From the Time-Resolved Catadioptric Stereo Digital Image Correlation—Composite Specimen

The TRC-SDIC technique was used to measure out-of-plane displacement W (z-direction) of the external surface of an in-house manufactured composite specimen. Displacements were calculated from the high-speed camera images that were taken every 0.13 and are shown in Fig. 13.16 Out-of-plane total displacement (z-direction) at 0.13 ms intervals measured by the TRC-SDIC technique (composite specimen). As can be seen, the displacement at the center is increasing after the shock impact and then it decreases. The impact of the shock on the internal surface of the plate sets up a vibration pattern of the plate due to the strong coupling between the flow and the plate [15]. The sequence of the results shown in the frames in Fig. 13.16 Out-of-plane total displacement (z-direction) at 0.13 ms intervals measured by the TRC-SDIC technique (composite specimen) demonstrate an oscillatory behavior of the W-displacement which is mostly positive on the first frames with a maximum amplitude of about 24 mm at the center of plate and close to zero around its periphery nearby the flange. By the eighth frame the amplitude is reduced to 20–22 mm and another vibration cycle begins.

13.8 The Effects of Boundary Conditions

Despite what would appear to be a firmly clamped boundary condition, evidence of the test specimen slipping or pulling out slightly from the retaining ring was observed. For dynamic tests a pattern of radial scratches was often observed on the clamped part of the test specimen. These scratches became more prominent at higher pressures. Additionally some of the dynamic tests appeared to be more non-linear than expected. The results showed that the torque on the bolts applying the pressure to the retaining ring had a significant impact on the displacement of the test specimen. Using a steel plate allowed maximum torque (given the available torque wrench) to be applied without concern for crushing or initiating damage in the test specimen.

Since it was now clear that the regular boundary condition was not acting as a clamped condition additional methods for mounting the test specimens were developed as described above and additional tests were performed to see the effects of those conditions and to better classify the original clamped condition. Figure 13.17 shows the maximum center displacement for aluminum, steel, and composite plates under clamped, bolted, and quasi-simple boundary conditions. Exact repetition of the pressure level is difficult to achieve so some interpretation is required; however, the trends are as expected based on the relative holding power of the retaining ring normal to the shock tube axis. Figure 13.18 shows the elongation of the plate across the diameter. This elongation (or membrane stretching) data helps measure the slipping and contributes to calculating more accurate strain calculations.

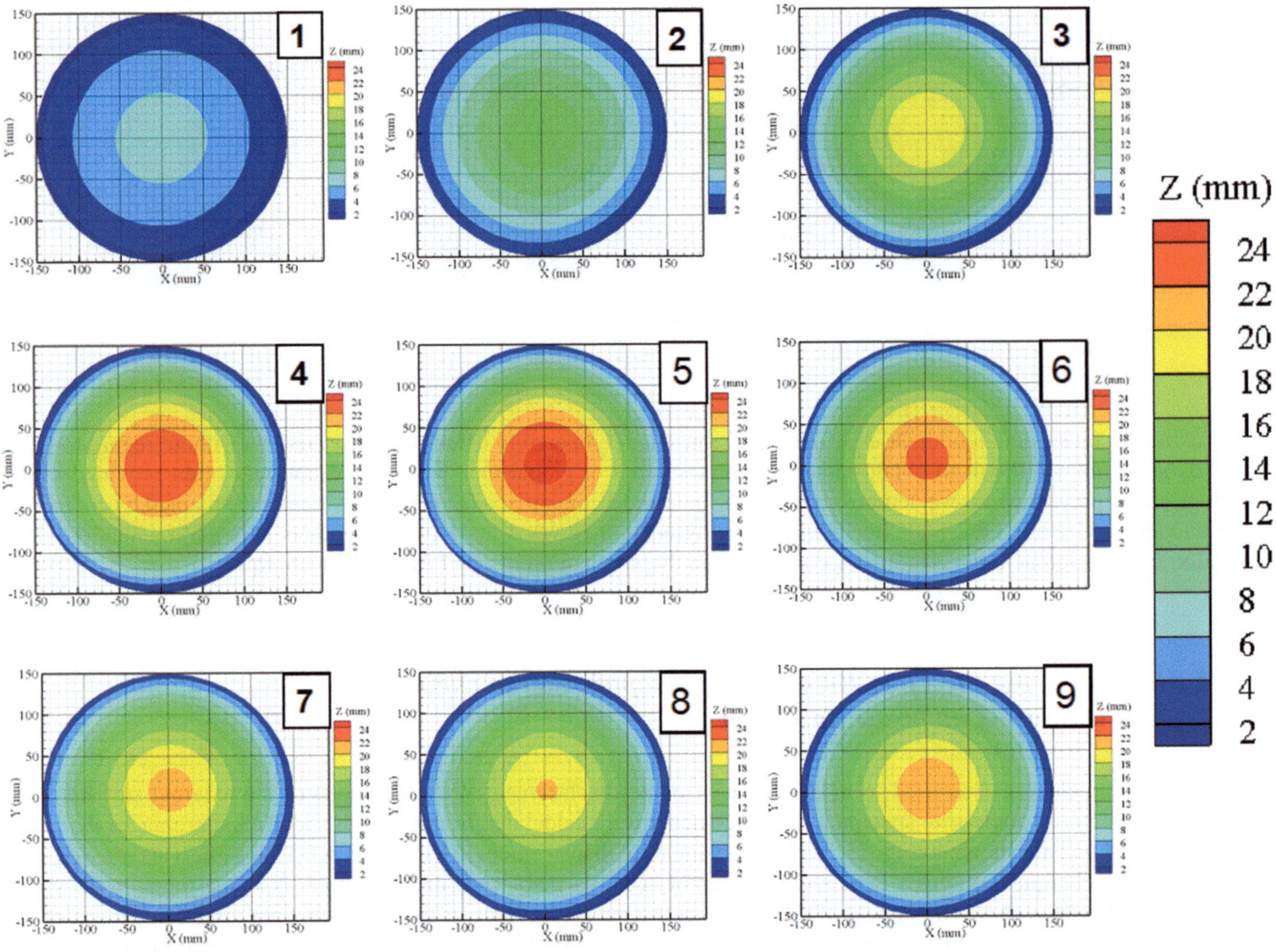

Fig. 13.16 Out-of-plane total displacement (z-direction) at 0.13 ms intervals measured by the TRC-SDIC technique (composite specimen)

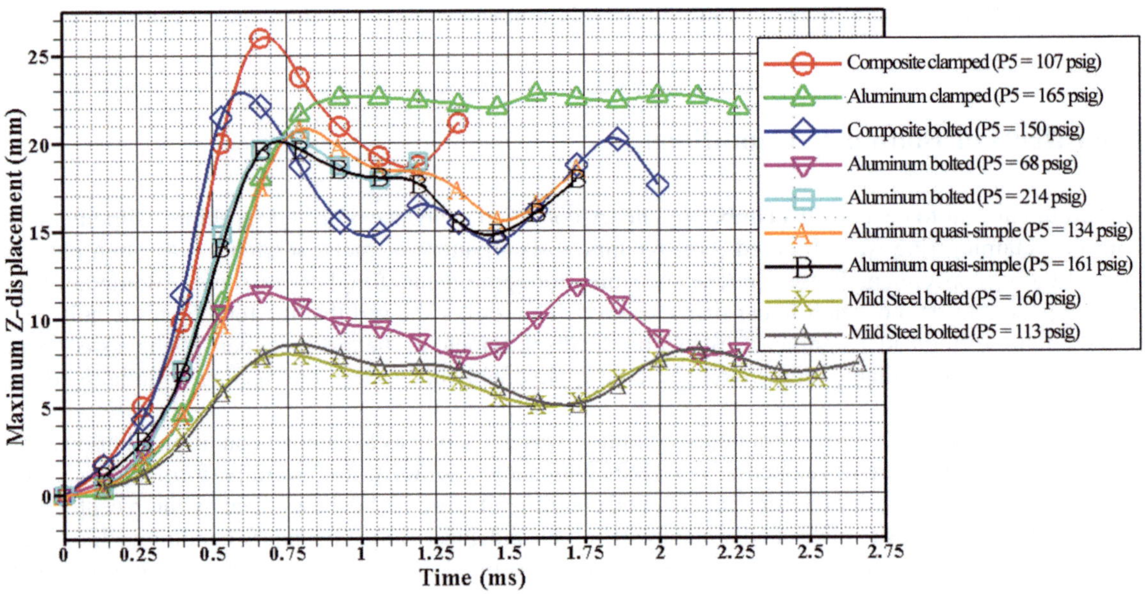

Fig. 13.17 Maximum center displacement as measured by DIC

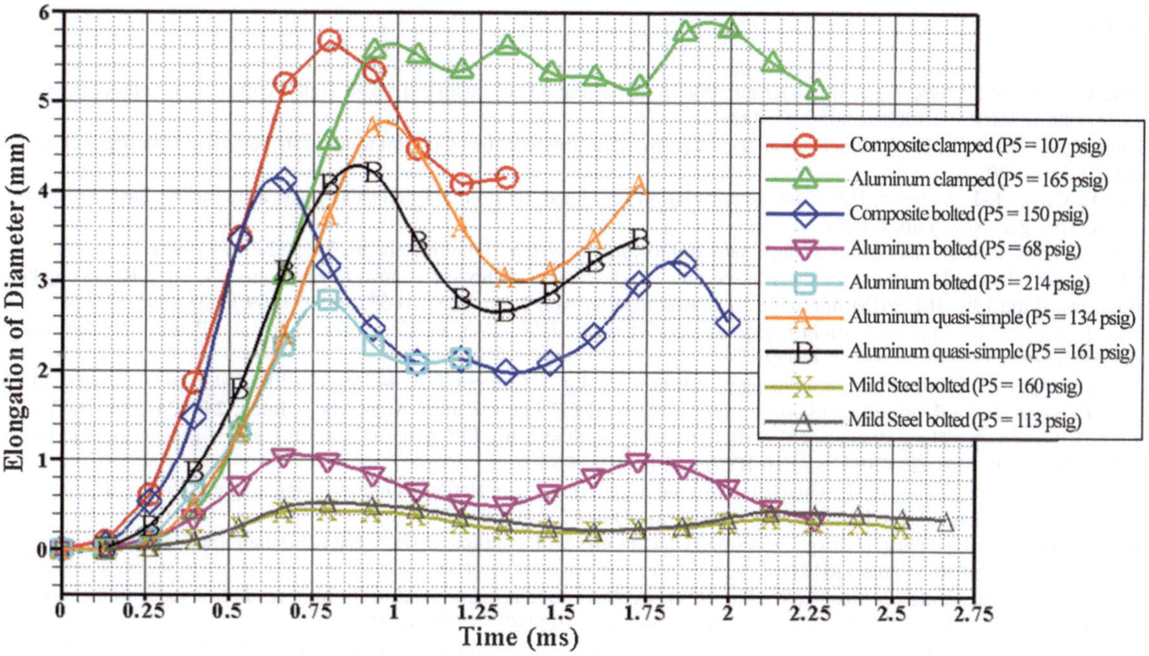

Fig 13.18 Elongation across the diameter as measured by DIC

13.9 Conclusions

In the present research we tested experimentally S2 HJ1 phenolic and S2 glass epoxy composites and provided details of their interaction with shock-waves For comparison of monolithic material response to similar loadings, tests were also conducted with aluminum specimens. A Time-Resolved Catadioptric Stereo Digital Image Correlation (TRC-SDIC) process was developed from traditional DIC techniques to determine the three-dimensional deformation of the specimens using a single camera. The TRC-SDIC process provided time-dependent full-field deformation measurements and allowed calculation of full-field strains. The system is capable of providing time dependent information of three component displacement vectors on two dimensional surfaces undergoing rapid deformation. The present layout offers several advantages over traditional systems with two different cameras:

1. It provides identical system parameters for the two views which minimizes their differences and thus facilitating robust stereo matching.
2. It reduces calibration time since only one camera is used.
3. Its cost is substantially lower than the cost of a system with two cameras.

In addition, in the present work we developed a calibration method to account for the motion of the mirrors during the experiments caused by acoustic waves emitted by the vibrating specimens and we experimented with the use of shape modes to filter/smooth/fit the experimental data with Bessel functions for elastic deformations.

At the end, we carried out extensive validation tests of our optical techniques which included comparisons with static and dynamic data obtained from embedded sensors in the specimens at the same time the images were acquired.

The effect boundary conditions i.e. the type of mounting the specimen to the end wall flange of the shock tube was investigated. Interactions with clamped, bolted or quasi-simply supported plates subjected to dynamic loading have been configured. Bolted specimens demonstrated the least maximum dynamic displacement and shortest elongation along one diameter

This material is based upon work supported by the US Department of Defense under Contract No. W56HZV-09-C-0569

References

1. Sutton, M., Wolters, W., Peters, W., et al.: Determination of displacements using an improved digital correlation method. Image Vis. Comput. **1**, 133–139 (1983). doi:10.1016/0262-8856(83)90064-1
2. Kahn-Jetter, Z.L., Chu, T.C.: Three-dimensional displacement measurements using digital image correlation and photogrammic analysis. Exp. Mech. **30**, 10–16 (1990)
3. Sutton, M.A., Orteu, J.-J., Schreier, H.W.: Image correlation for shape, motion and deformation measurements: basic concepts, theory and applications. Springer, New York (2009)
4. Goshtasby, A., Gruver, W.A.: Design of a single-lens stereo camera system. Pattern Recogn. **26**, 923–937 (1993)
5. Inaba, M., Hara, T., Inoue, H.: A stereo viewer based on a single camera with view-control mechanisms. In: Intell Robots Systems 93 IROS 93 Proceedings of the 1993 IEEE/RSJ International Conference on, vol. 3, pp. 1857–1865. doi:10.1109/IROS.1993.583888
6. Gluckman, J., Nayar, SK.: Planar catadioptric stereo: Geometry and calibration. In: Computer Vision and Pattern Recognition, 1999. IEEE Computer Society Conference on, vol. 1, (1999)
7. Gluckman, J., Nayar, S.: Catadioptric stereo using planar mirrors. Int. J. Comput. Vis. **44**, 65–79 (2001). doi:10.1023/A:1011172403203
8. Gluckman, J., Nayar, S.K.: Rectified catadioptric stereo sensors. IEEE Trans. Pattern Anal. Mach. Intell. **24**, 224–236 (2002)
9. Kreizer, M., Liberzon, A.: Three-dimensional particle tracking method using FPGA-based real-time image processing and four-view image splitter. Exp. Fluids **50**, 613–620 (2011). doi:10.1007/s00348-010-0964-3
10. Pankow, M., Justusson, B., Waas, A.M.: Three-dimensional digital image correlation technique using single high-speed camera for measuring large out-of-plane displacements at high framing rates. Appl. Opt. **49**, 3418–3427 (2010)
11. Elzawawy, A.: Time resolved particle image velocimetry techniques with continuous wave laser and their application to transient flows. Ph.D. Thesis, City University of New York (2012)
12. Briassulis, G., Agui, J.H., Andreopoulos, J., Watkins, C.B.: A shock tube research facility for high-resolution measurements of compressible turbulence. Exp. Thermal. Fluid Sci. **13**, 430–446 (1996). doi:10.1016/S0894-1777(96)00097-0
13. Subramaniam, K.V., Nian, W., Andreopoulos, Y.: Blast response simulation of an elastic structure: Evaluation of the fluid–structure interaction effect. Int. J. Impact Eng. **36**, 965–974 (2009). doi:10.1016/j.ijimpeng.2009.01.001
14. Pankow, M., Justusson, B., Salvi, A., Waas, A., Yen, C.-F., Ghiorse, S.: Shock response of 3d woven composites: an experimental investigation. Compos Struct **93**(5), 1337–1346 (2011). doi:10.1016/j.compstruct.2010.10.021
15. Gong, M., Andreopoulos, Y.: Coupled fluid-structure solver: The case of shock wave impact on monolithic and composite material plates. J. Comput. Phys. **228**, 4400–4434 (2009). doi:10.1016/j.jcp.2009.03.009
16. Kazemi-Kamyab, V., Subramaniam, K., Andreopoulos, Y.: Stress transmission in porous materials impacted by shock waves. J. Appl. Phys. **109**, 013523 (2011). doi:10.1063/1.3517791
17. Xia, S., Gdoutou, A., Ravichandran, G.: Diffraction assisted image correlation: A novel method for measuring three-dimensional deformation using Two-dimensional digital image correlation. Exp. Mech. **53**, 755–765 (2013). doi:10.1007/s11340-012-9687-0
18. Jahnke, D., Andreopoulos, Y.: Development of a time-resolved catadioptric stereo digital image correlation technique for the study of blast loading of composite material structures. ASME Conf. Proc. **2011**, 153–162 (2011)
19. Tiwari, V., Sutton, M.A., McNeill, S.R., et al.: Application of 3D image correlation for full-field transient plate deformation measurements during blast loading. International Journal of Impact Engineering 3**6**, 862–874 (2009). doi: 10.1016/j.ijimpeng.2008.09.010
20. Tsai, R.: An efficient and accurate camera calibration technique for 3D machine vision. In: Proceeding of the IEEE Conference on Computer Vision and Pattern Recognition, Miami Beach, FL, pp. 364–374, (1986)
21. Justusson, B., Pankow, M., Heinrich, C., Rudolph, M., Waas, A.M.: Use of a shock tube to determine the bi-axial yield of an aluminum alloy under high rates. Int. J. Impact Eng. **58**, 55–65 (2013)
22. Lowe, D.G.: Object recognition from local scale-invariant features. In: Computer Vision 1999 Proceedings of the Seventh IEEE International Conference on, vol. 2, pp. 1150–1157 (1999)
23. Wieneke, B.: Stereo-PIV using self-calibration on particle images. Exp. Fluids **39**, 267–280 (2005)
24. Jahnke, D., Azadeh-Ranjbar, V., Andreopoulos, Y.: Composite plates' response to shock wave impact. Compos. B, Submitted (2015)
25. Jahnke, D.: The effect of shockwave impingement on thin, woven glass fiber reinforced, polymer composite plates. Ph.D. Thesis, The City College of New York—Grove School of Engineering (2014)
26. Wang, E., Shukla, A.: Analytical and experimental evaluation of energies during shock wave loading. Int. J. Impact Eng. **37**(12), 1188–1196 (2010). doi:10.1016/j.ijimpeng.2010.07.003
27. Gong, M. Mutual interactions between shock waves and structures. PhD Thesis, City University of New York (2006)
28. Pankow, M., Justusson, B., Waas, A.M.: Three-dimensional digital image correlation technique using single high-speed camera for measuring large out of-plane displacements at high framing rates. Appl. Optics **17**, 3418–3427 (2010)
29. Gibson, R. F., Chen, Y., and Zhao, H.: Improvement of Vibration Damping Capacity and Fracture Toughness in Composite Laminates by the Use of Polymeric Interleaves, J. Eng. Mater. Technol., **123**(3), pp. 309–314 (2001)
30. Koratkar, N. A., Suhr, J., Joshi, A., Kane, R. S., Schadler, L. S., Ajayan, P. M., and Bartolucci, S.: Characterizing energy dissipation in single-walled carbon nanotube polycarbonate composites, Appl. Phys. Lett., **87**(6), p. (2005)
31. Zhou, X., Shin, E., Wang, K. W., and Bakis, C. E.: Interfacial damping characteristics of carbon nanotube-based composites, Dev. Carbon Nanotube Nanofibre Reinf. Polym., **64**(15), pp. 2425–2437 (2004)
32. Rajoria, H., and Jalili, N.: Passive vibration damping enhancement using carbon nanotube-epoxy reinforced composites, Compos. Sci. Technol., **65**(14), pp. 2079–2093 (2005)
33. Briassulis, G., Agui, J. H., Andreopoulos, J., and Watkins, C. B.: A shock tube research facility for highresolution measurements of compressible turbulence, Exp. Therm. Fluid Sci., **13**(4), pp. 430–446 (1996)
34. Andreopoulos, Y., Agui, J. H., and Briassulis, G.: Shock Wave Turbulence Interactions, Annu Rev Fluid Mech, **32**(1), pp. 309–345 (2000)
35. Briassulis, G., Agui, J. H., and Andreopoulos, Y.: The structure of weakly compressible grid-generated turbulence, J. Fluid Mech., 432, pp. 219–283 (2001)

Chapter 14
Initial Experimental Validation of an Eulerian Method for Modeling Composites

Christopher S. Meyer, Christopher T. Key, Bazle Z. (Gama) Haque, and John W. Gillespie Jr.

Abstract Impact loading response of unidirectional and plain weave fiber reinforced polymer composite materials is typically modeled using a Lagrangian method such as the finite element method. However, these methods often lack a coupled equation of state. In anisotropic materials, the pressure (equation of state) and deviatoric (strength) portions of the stress tensor are coupled: a shear stress can produce a volumetric response and a volumetric strain can produce a shear stress. High-velocity impacts of composite materials instigate a coupled pressure and stress response, so an equation of state is important in determining the non-uniform stress response of the material. A new composite model, which couples the pressure response to the constitutive response of the material, has been implemented in an Eulerian large deformation, strong shock wave, solid mechanics code. Experiments of steel projectiles perforating composite targets were numerically simulated to begin to validate this new composite model. This paper will discuss the coupled equation of state and strength response, and compare the results of these experiments with the results predicted by the model.

Keywords Numerical simulation • Glass fiber reinforced composites • Equation of state • Multi-constituent model

14.1 Introduction

Impact of fiber reinforced polymer composites normal to the fiber direction is dominated by the fiber material properties, and impact along the fiber direction is dominated by the reinforcement material properties. For sufficiently high velocity impacts, shock waves will propagate within a composite material. Fiber materials typically have greater stiffnesses than matrix materials and so sound travels faster along fibers than in reinforcement. Likewise, impedance mismatch between constituent materials will result in reflections and refractions of these waves. Adhesive bond strength between constituent materials will also influence wave propagation and delamination. Tsai and Prakash studied weak shock wave propagation in layered media, and found that wave propagation across laminar material is strongly influenced by impedance mismatch of laminae, but also lamina thickness, wave propagation distance, and inelastic material response [1]. Behavior of laminated media may be homogenized for impacts that induce a stress wave that travels much faster than the time it takes to cross a lamina thickness [2]. Damage in fiber reinforced polymer composites consists of matrix cracking, fiber failure, and fiber/matrix debonding, but a homogenized approach will not capture these effects, though progressive damage and failure may be considered. Also shock wave propagation in a homogenized material cannot address the differences in sound speed of the constituent materials. Under low-rate compressive loading, the thermodynamic state in pressure-volume space is akin to a stepped-equilibrium along isothermal or isentropic pressure-volume paths, but shock loading instigates larger compressions in materials, and the shocked change in thermodynamic state is akin to an instantaneous jump from the unshocked state to the shocked state on the Hugoniot pressure-volume path, and this jump requires an equation of state (EOS) to calculate the shocked thermodynamic pressure-volume state [3]. As anisotropic materials, composites require an anisotropic EOS [4–6], however such models are not yet widely available in computational mechanics codes. Also important is anisotropic constitutive relations [4, 7–10]. Lukyanov and Segletes provide a review of the subject [11].

C.S. Meyer (✉)
U.S. Army Research Laboratory, ATTN: RDRL-WML-H, Proving Ground, Aberdeen, MD 21005-5066, USA
e-mail: christopher.s.meyer12.civ@mail.mil

C.T. Key
Applied Physical Sciences, Groton, CT 06340, USA

B.Z.G. Haque • J.W. Gillespie Jr.
Center for Composite Materials, University of Delaware, Newark, DE 19716, USA

D. Casem et al. (eds.), *Dynamic Behavior of Materials, Volume 1*, Conference Proceedings of the Society for Experimental Mechanics Series, DOI 10.1007/978-3-319-41132-3_14

The impact problem is solved using computational mechanics codes, which assume materials form a continuum within a volume and apply boundary and initial conditions to solve equations of state requiring conservation of mass, momentum, and energy and constitutive equations relating stress to strain or deformation by the assignment of material properties. In these equations, the stress tensor is decomposed into a symmetric part and an antisymmetric part. The antisymmetric stress tensor involves torques and couples. The symmetric stress tensor is further decomposed into a direction-independent hydrostatic pressure and a direction-dependent deviatoric stress. In isotropic materials, the strain tensor may be decomposed into volumetric and deviatoric parts, from which a volumetric strain will produce only a change in size and a deviatoric strain will produce only a change in shape. Similarly, the volumetric strain, or change in volume, is calculated from the EOS, which relates pressure to volume, and the deviatoric strain is calculated from the constitutive relation, which relates strain to stress state. However, in anisotropic materials, volumetric and deviatoric strains are coupled so that a shear stress can produce a volumetric response and a volumetric strain can produce a shear stress. Therefore continuum modeling of anisotropic materials under large impact loads causing shocks or large deformations requires that the EOS and constitutive equations be coupled.

Computational solid mechanics codes use different numerical methods to solve the equations of motion, balance equations, and constitutive equations. The numerical method used depends upon the problem. Infinitesimal strain theory does not involve changes in extensive material properties such as density, so these problems often do not require equations of state and are typically solved using Lagrangian methods in which continua are discretized and these elements follow the displacement of the material. When faced with large deformations, Lagrangian methods simply remove or erode the distorted element, which is not physical. Meshless Lagrangian methods such as smooth particle hydrodynamics (SPH) were introduced to address large deformations, and are subject to other shortcomings such as tensile instability. Finite strain theory involves large deformations, such as found in high-velocity impacts and shock loading, and large compressions may change the material's density so that these problems require equations of state, and these problems are often solved using Eulerian methods in which space is discretized and the material is displaced through these fixed elements. Eulerian methods of course have their own shortcomings when modeling solid continua, so arbitrary Lagrangian Eulerian methods were introduced to capitalize on the best of both methods.

Composite materials are typically modeled using Lagrangian methods without coupled equations of state, which is ideal for example for composite aerospace structures under aerodynamic loads, composite fixed structures such as windmills under transverse loading, and structural loading of composite plates and beams. However, to model the large deformations and shocks caused by high-velocity impact loading of composite materials, which instigate a coupled pressure and stress response, an Eulerian method with a coupled EOS and constitutive model is useful. This work explores a new anisotropic material constitutive model, the multiple constituent model (MCM) [12–15], which couples pressure and deviatoric response of the material, and which has been implemented in CTH [16, 17], an Eulerian large deformation, strong shock wave, solid mechanics code. Experiments of steel projectiles perforating composite targets were numerically simulated to begin to validate this new composite model. This model and its validation is the subject of the present work.

14.2 Multiple Constituent Model

The multicontinuum theory for two [18] and three [12] constituents is built upon Hill's decomposition [19] of the macroscale composite stress/strain fields down to the mesoscale constituent stress/strain fields. Hill's decomposition [19] was developed to evaluate composite material stiffness properties based on known properties of the constituents, but multicontinuum theory works in the opposite direction, the theory evaluates constituent material properties based on known properties of the composite [17]. In the present work, unidirectional composites make use of two constituent multicontinuum theory [18] and plain weave composites make use of three constituent multicontinuum theory [12], which is simply twice-decomposed. The following is an overview, but detailed discussions and mathematical developments of the multicontinuum theory may be found in these references [12, 14, 17, 18].

MCM is a continuum model, so composite material stress/strain fields are homogenized, but composite stress/strain fields are also decomposed into stress/strain fields for fiber and matrix constituents, which enables improved accuracy in modeling intermediate and ultimate failure modes [17]. MCM takes the continuum assumption, the averaging of properties such as stress over a finite volume, and applies it to a continuum point, which is defined as occupying no volume [17]. Multiple constituents (e.g., fiber and matrix) with different properties exist at each continuum point, and this is defined as a multicontinuum [17]. Multicontinuum theory allows failure of the continuum to be predicted based on constituent states of stress rather than the composite state of stress [18], and use Hashin failure criteria [19] applied to individual constituents rather than composite. The kinetic theory of fracture for polymers [20–22], which uses applied stress to predict molecular bond failure, is the basis for the continuum damage evolution within MCM [17]. Matrix or fiber material properties are degraded as submicrocracks accumulate, where submicrocracks are consecutive failures of interatomic bonds in the polymer matrix [17]. Material property degradation is applied to constituents and the continuum, and apparently leads to a macroscopic response that resembles experimentally observed inelastic material behavior [17].

Composite strains are determined using a given finite element formulation. Composite strains are decomposed into constituent strains. Constituent stresses are determined from constituent strains based on a constituent volume-fraction-weighted linear elasticity with anisotropic (transversely isotropic) stiffness material properties determined from composite and constituent material properties using micromechanics and including thermal effects [17]. The micromechanics model for plain weave architecture is shown in Fig. 14.1, which is adapted from work by Barbero et al. [24]. Failure is determined from modified Hashin criteria [20], which uses constituent stresses rather than homogenized composite stresses, allowing distinct failure modes to be predicted for constituents [18]. Assuming transverse isotropy, the failure state of a given constituent is expressed in terms of transversely isotropic stress invariants [25]. Simplifying assumptions are made to ensure matrix dominated failure in the transverse direction and fiber dominated failure in the longitudinal direction of a unidirectional composite. The failure criterion expression coefficients, which are functions of known ultimate tensile, compressive, and shear strengths of constituents, are determined from pure tension, compression, and in-plane shear load cases for each constituent [17]. For plain weave architecture, two constituent theory is extended to three constituents: fiber bundles are themselves unidirectional composites, as shown in Fig. 14.2, but the warp and fill bundles are treated as individual

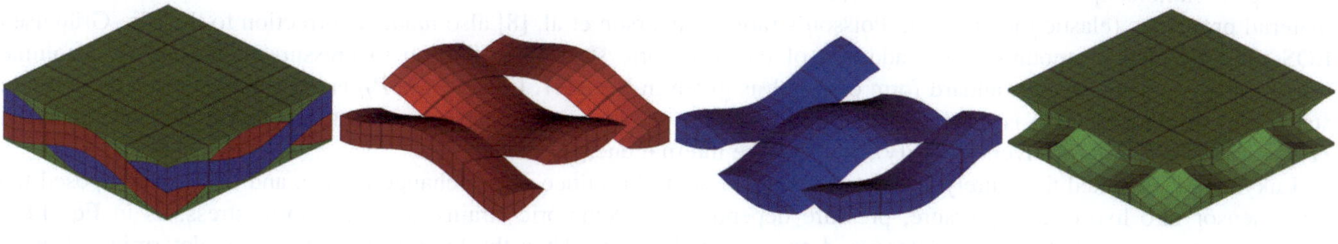

Fig. 14.1 3D model of plain weave architecture upon which micromechanics calculation of constituent material properties are calculated. From *left* to *right*: composite, warp fiber bundle constituent, fill fiber bundle constituent, and matrix constituent. Geometry adapted from Barbero et al. [24]

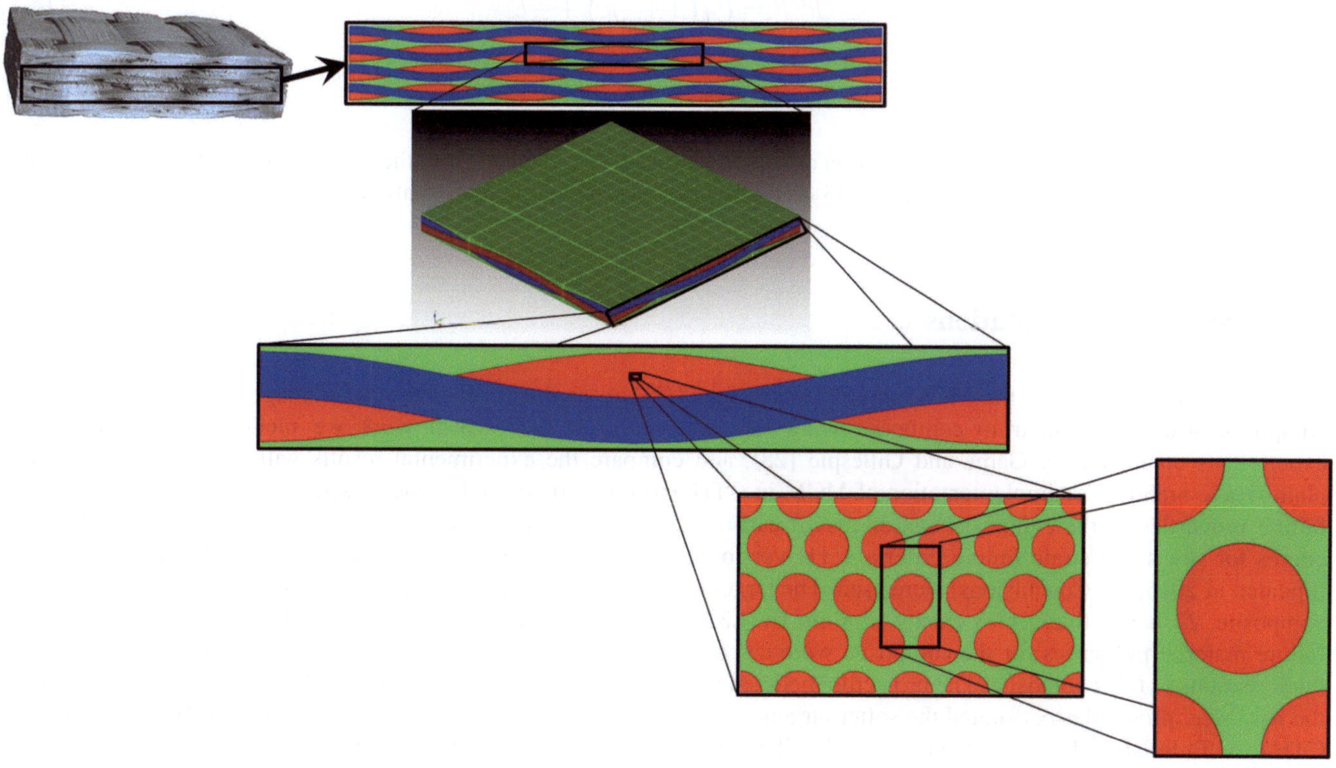

Fig. 14.2 View of macro to meso length scales of plain weave composite architecture

constituents, and despite the nonlinear shape of the fiber bundles, which means they are not truly transversely isotropic, the failure criteria for these constituents is assumed to be transversely isotropic [17]. MCM allows for a stress or strain based failure criterion [15, 17].

14.3 Equation of State Coupling

As discussed earlier, in large deformation hydrocodes (e.g., CTH [16]), the stress tensor is decomposed into pressure and deviatoric stress, and such codes require an equation of state (EOS) to determine the pressure (change in size or volume) and a constitutive model, sometimes called a strength model, to determine the deviatoric stress (change in shape). In anisotropic materials, pressure and deviatoric strain components are coupled. An applied hydrostatic pressure not only causes a change in size, but also induces strains that may be different in different directions. Likewise, an applied shear stress not only induces a change in shape, but may also lead to a change in size in different directions (volumetric strain). Therefore, the EOS and constitutive equations must be coupled.

Segletes [4, 11], O'Donoghue and Anderson et al. [7, 8], and Lukyanov [10, 11] have developed anisotropic constitutive relationships. Segletes decomposed hydrostatic and deviatoric terms [4]. O'Donoghue et al. [7] and Anderson et al. [8] developed relationships for pressure and deviatoric stress as functions of volumetric and deviatoric strains and anisotropic material properties (elastic modulus and Poisson's ratio). Anderson et al. [8] also made a correction to the Mie-Grüneisen EOS [26, 27], which amounts to the addition of the deviatoric strain contribution to pressure to the pressure-volume relationship of the EOS, the standard form of which is given in Eq. 14.1. In Eq. 14.1, P_H is the shocked material pressure state from the Hugoniot, μ is the bulk modulus, Γ is the well-known Gruneisen parameter, a thermodynamic material property, V is volume (or inverse density), and E is the internal energy density.

Lukyanov [10] stated that purely hydrostatic pressure should produce only a change in size, and so he decomposed the stress tensor into hydrostatic pressure, pressure dependent on deviatoric strains, and deviatoric stress, as in Eq. 14.2. Deviatoric stresses and strains are determined as described earlier. Then the hydrostatic pressure is determined from a given EOS, and the coupled pressure term is found from deviatoric strains and anisotropic stiffness and isotropic bulk modulus material properties as described by Lukyanov [10].

$$P^{EOS} = P_H\left(1 + \frac{\Gamma}{2}\mu\right) + \frac{\Gamma}{V}E \tag{14.1}$$

$$\sigma_{ij} = P^{EOS}\delta_{ij} + P^{dev}\left(S_{ij}\right) + S_{ij} \tag{14.2}$$

As currently implemented in CTH, the equation of state pressure, P^{EOS}, is found from the Mie-Grüneisen EOS [26, 27] given in Eq. 14.1, but theoretically any relevant EOS could be used to provide the hydrostatic pressure.

14.4 Numerical Simulations

Gorfain and Key [14] used the MCM implementation in CTH to evaluate prediction of damage extent in rib-stiffened composite structures similar to composite wing ribs and spars. In the present work we model ballistic perforation experiments conducted by Gama and Gillespie [28], and compare the experimental results with the unidirectional and plain weave architecture implementation of MCM in CTH and a micromechanics code, used to calculate the strength and failure properties of the composites as discussed earlier, and we evaluate impact velocity vs. residual velocity ($V_i - V_r$) curves for stress and strain failure criteria. CTH version 11.0 and the micromechanics code version as of October 2015 (updated in 2016, see Conclusions) were used. The target modeled was plain weave (5×5 tows/in.) S-2 Glass / SC15 epoxy composite, 22 layers thick, 17.8 cm by 17.8 cm by 1.32 cm with clamped boundary conditions. Stiffness, strength, and failure material properties for the composite were calculated as described above based on known material properties for thick sections of this composite. The projectile modeled was 4340 steel 0.22 caliber right circular cylinders. In experiments, the hard steel projectiles perforated the softer targets while remaining rigid, undeformed, and uneroded. However, because CTH is an Eulerian code, it is subject to mixed-cell erosion (that is, strength properties can be averaged in mixed cells such that a cell containing both steel and composite material could be effectively softened and thereby eroded, which is

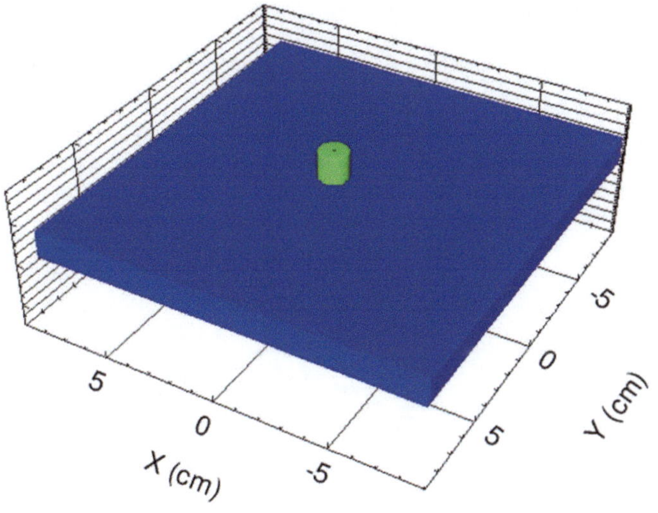

Fig. 14.3 Computational model used in all simulations. Strength and failure properties calculated using a micromechanics code for plain weave and unidirectional architectures were applied to the same volume for respective simulations

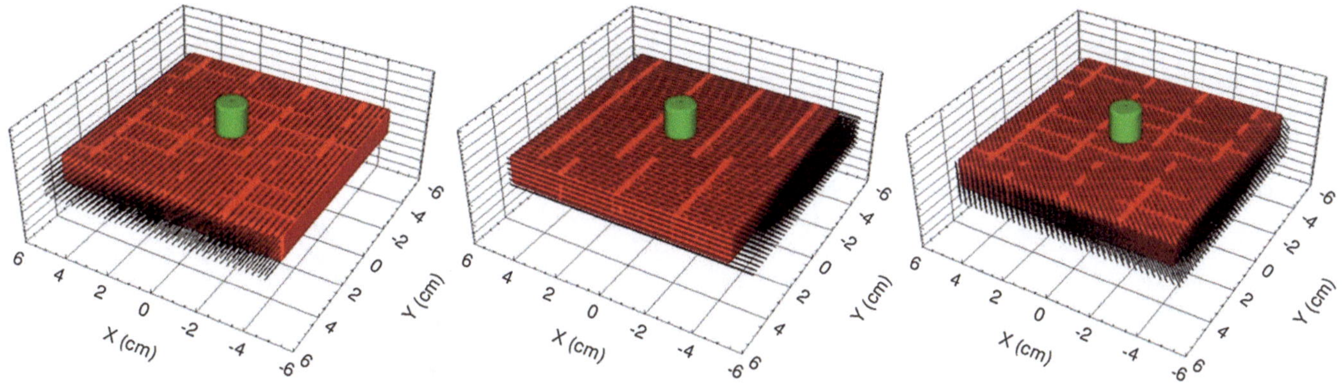

Fig. 14.4 Material orientation vector plots for a unidirectional composite. Layers shown are (*left* to *right*) 0, 90, and +45°

nonphysical), so to minimize this effect, projectiles were assigned numerically "infinite" material strength and failure properties. Also, as an Eulerian code, clamped boundary conditions cannot be applied, but instead semi-infinite boundary conditions were used and boundaries allowed stress waves to transmit out of the computational domain without reflection. Figure 14.3 shows the computational model. Figure 14.4 provides a cut-view of the computational model with material orientation vectors plotted for some of the layers of a unidirectional composite. Since the experimental results were for plain weave architecture, in unidirectional composite simulations, the material was assumed to be 1.32-cm-thickness 8-layer quasi-isotropic with orientations [0°/90°/+45°/−45°/−45°/+45°/90°/0°].

CTH flat mesh was used. A convergence study was conducted for cubic cell sizes with side lengths 0.25, 0.125, 0.0625, and 0.03125 cm, which produced models with 100 K, 1 M, 9 M, and 69 M elements respectively. A grant of Department of Defense High Performance Computing Modernization Program supercomputing time was used for the present work, and the above cell sizes ran to a computation time of 100 μs respectively as follows: 1.5 h on 512 cores, 1 h on 1024 cores, 8 h on 1024 cores, and 30 h on 4096 cores. Experimental data for the largest impact velocity, 1106 m/s, was used in the convergence study since CTH is known to produce the most accurate results for high velocity impacts of thick targets for eroding penetration of well characterized materials. The solution converged to a repeatable residual velocity for the smaller cell sizes (0.0625 and 0.03125 cm) as shown in Fig. 14.5. However, the error of the converged solution was about the same as the error for the mid-sized cells (0.125 cm), but the computational expense is significantly less for the mid-sized cells, therefore a cell size of 0.125 cm was used in the present work. Unfortunately, this level of grid refinement produced only 2 elements across the penetrator diameter, which likely contributes some error to the results, so future work will use adaptive mesh refinement to provide greater refinement across the penetrator diameter and less refinement away from the penetration action while maintaining optimum computational expense.

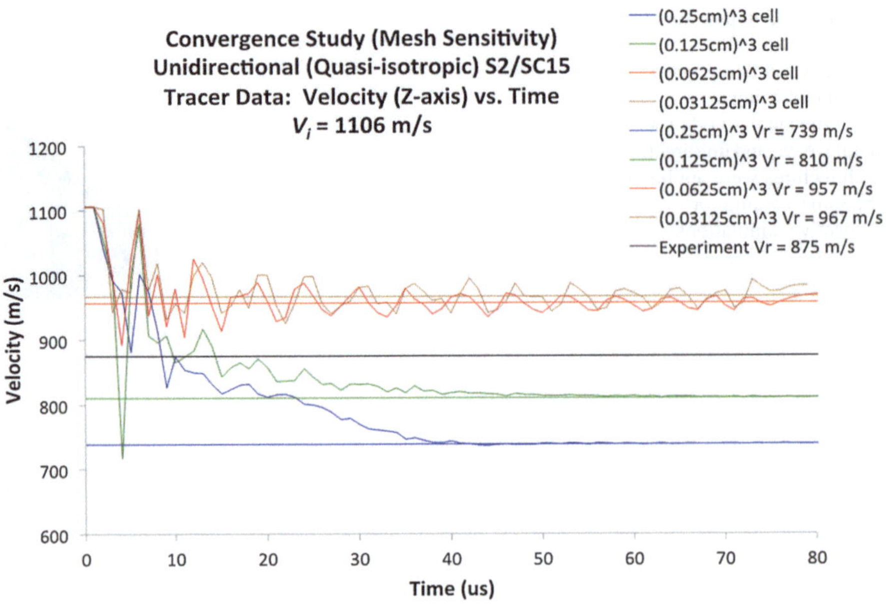

Fig. 14.5 Convergence study results

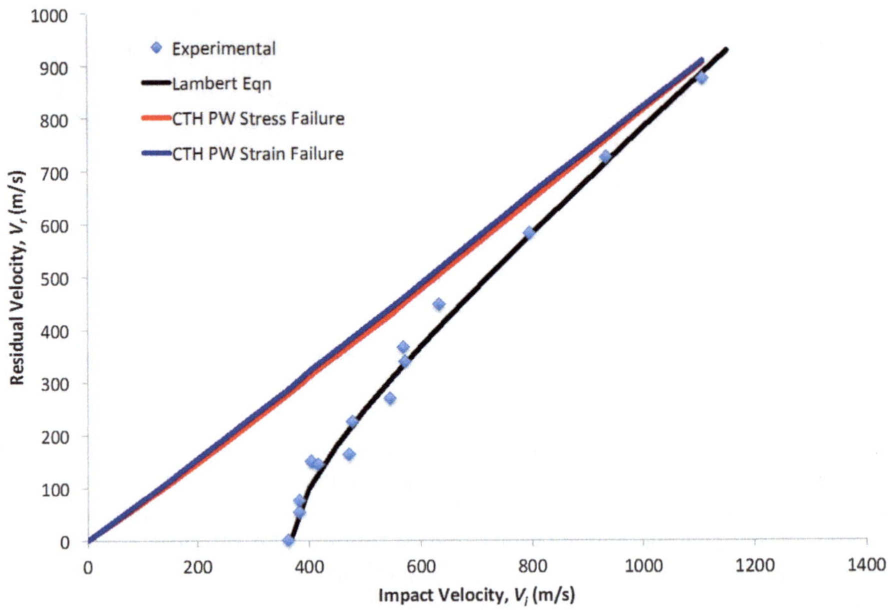

Fig. 14.6 Results for plain weave architecture strength and failure properties for stress-based and strain-based failure criteria

14.5 Results and Discussion

Residual velocity as a function of impact velocity ($V_i{-}V_r$) curves are plotted and compared with the experimental data, which is fit by the Lambert-Jonas equation [29] given by $V_r = 0.9039(V_i^{1.5834} - 367^{1.5834})^{1/1.5834}$ [28]. Figure 14.6 illustrates results for plain weave architecture material strength and failure properties for stress-based and strain-based failure criteria. Figure 14.7 illustrates results for unidirectional architecture material strength and failure properties for stress-based and strain-based failure criteria. Figure 14.8 provides a qualitative comparison of a non-penetrating impact velocity, 364 m/s, and a perforating impact velocity, 795 m/s on a unidirectional target. Results for the plain weave model indicate little difference between stress- and strain-based failure criteria, which is not expected. Plain weave results match experiment well at higher impact velocities, but deviate unphysically for lower impact velocities. Results for the unidirectional model for stress-based failure match experimental results well at higher impact velocities, but deviate unphysically for lower impact velocities. Results for the unidirectional model for strain-based failure criteria better follow the expected shape of the Vi−Vr curve, but underpredict the experimental results. An arbitrary reduction in the micromechanics calculated failure strength brought the predicted results closer to experiment, but this is an unphysical modification of material parameters.

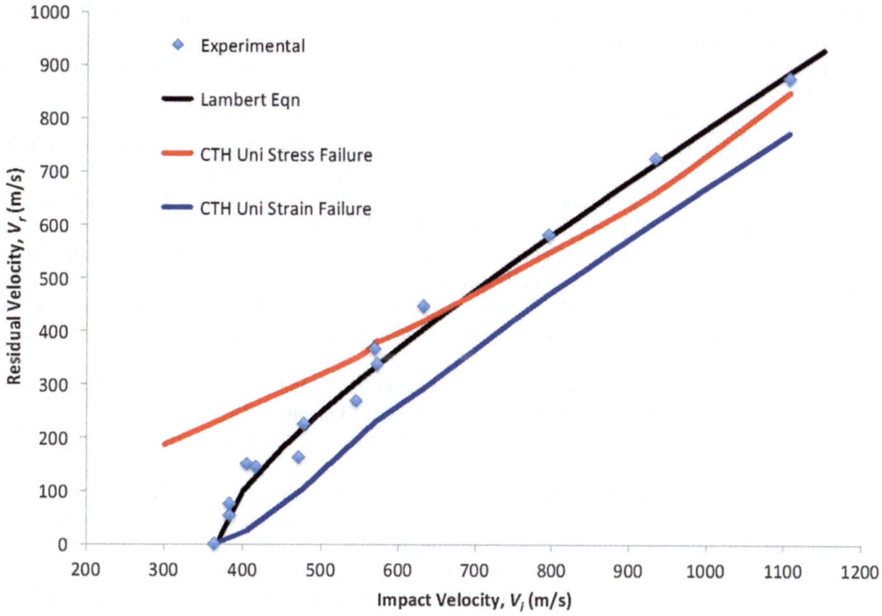

Fig. 14.7 Results for unidirectional architecture strength and failure properties for stress- and strain-based failure criteria

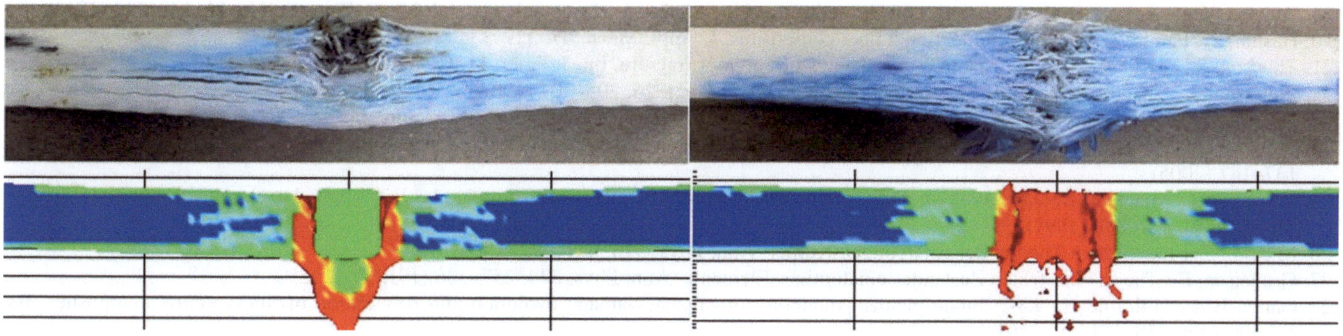

Fig. 14.8 Qualitative comparison of a non-penetrating impact velocity, on *left* 364 m/s, and a perforating impact velocity, on *right* 795 m/s and a unidirectional composite target

14.6 Conclusions

Results suggest there is an error in the implementation of the micromechanics calculation of material properties for the plain weave architecture or an error in model use. As of this writing, work is underway to determine repeatability of these results and revise the micromechanics code. Validation work of the MCM model remains ongoing. This model is expected to provide a new capability not available in other codes to model composite materials, for example, in armor applications under insults inducing large deformations and strong shocks.

References

1. Tsai, L., Prakash, V.: Structure of weak shock waves in 2-D layered material systems. Intl. J. Solids Struct. **42**, 727–750 (2005)
2. Zaretsky, E., deBotton, G., Perl, M.: The response of a glass fibers reinforced epoxy composite to an impact loading. Int. J. Solids Struct. **41**, 569–584 (2004)
3. Ahrens, T.J.: Equation of state. In: Asay, J.R., Shahinpoor, M. (eds.) High Pressure Shock Compression of Solids, pp. 75–114. Springer, New York (1993)

4. Segletes, S.: Deviatoric Constitutive Relationship for Anisotropic Materials, BRL-TR-2825. US Army Ballistic Research Laboratory, Aberdeen Proving Ground (1987)
5. Anderson, C.E., O'Donoghue Jr., P.E., Kerhut, D.: A mixture theory approach for the shock response of composite materials. J. Compos. Mater. **24**, 1159–1187 (1990)
6. Lukyanov, A.A.: An equation of state of a carbon-fibre epoxy composite under shock loading. Eur. Phys. J. B. **74**, 35–45 (2010)
7. O'Donoghue, P.E., Anderson Jr., C.E., Friesenhahn, G.J., Parr, C.H.: A constitutive formulation for anisotropic materials suitable for wave propagation computer programs. J. Compos. Mater. **26**(13), 1860–1884 (1992)
8. Anderson, C.E., Cox, P.A., Johnson, G.R., Maudlin, P.J.: A constitutive formulation for anisotropic materials suitable for wave propagation computer programs-II. Comput. Mech. **15**, 201–223 (1994)
9. Lukyanov, A.A.: Constitutive behavior of anisotropic materials under shock loading. Int. J. Plast. **24**, 140–176 (2008)
10. Lukyanov, A.A.: Thermodynamically consistent anisotropic plasticity model. J. Press. Vessel. Technol. **130**(2), 021701 (2008)
11. Lukyanov, A.A., Segletes, S.B.: Frontiers in the constitutive modeling of anisotropic shock waves. Appl. Mech. Rev. **64**, 1–6 (2012)
12. Key, C.T., Six, R.W., Hansen, A.C.: A three-constituent multicontinuum theory for Woven fabric composite materials. Compos. Sci. Technol. **63**(13), 1857–1864 (2003)
13. Key, C.T., Gorfain, J.: Part I: overview and application of the CTH Multiconstituent Composite Model (MCM) for ballistic impact simulations, Proceedings, 82nd Shock and Vibration Symposium, Baltimore, MD, 11 2011
14. Gorfain, J.E., Key, C.T.: Damage prediction of rib-stiffened composite structures subjected to ballistic impact. Int. J. Impact Eng. **57**, 159–172 (2013)
15. Key, C.T., Schumacher, S.C., Scott, A.C.: Evaluation of a strain based failure criterion for the multi-constituent composite model under shock loading. Proceedings, DYMAT 2015—11th International Conference on the Mechanical and Physical Behavior of Materials under Dynamic Loading, Ed. E. Cadoni, Lugano, Switzerland, 9 2015
16. McGlaun, J.M., Thompson, S.L., Elrick, M.G.: CTH: A three-dimensional shock wave physics code. Int. J. Impact Eng. **10**, 351–360 (1990)
17. Schumacher, S.C., Key, C.T.: CTH Reference Manual: Composite Capability and Technologies, SAND2009-0403. Sandia National Laboratories, Albuquerque (2009)
18. Mayes, J.S., Hansen, A.C.: Multicontinuum failure analysis of composite structural laminates. Mech. Compos. Mater. Struct. **8**(4), 249–262 (2001)
19. Hill, R.: Theory of mechanical properties of fibre reinforced materials—I. elastic behaviour. J. Mech. Phys. Solids. **12**, 199–212 (1964)
20. Hashin, Z.: Failure criteria for unidirectional fiber composites. J. Appl. Mech. **47**, 329–334 (1980)
21. Zhurkov, S.N., Kuksenko, V.S.: The micromechanics of polymer fracture. Int. J. Fract. **11**, 629–639 (1975)
22. Zhurkov, S.N.: Kinetic concept of the strength of solids. Int. J. Fract. **26**, 295–307 (1984)
23. Hansen, A.C., Baker-Jarvis, J.: A rate dependent kinetic theory of fracture for polymers. J. Fracture **44**, 221–231 (1990)
24. Barbero, E.J., Lonetti, P., Sikkil, K.K.: Finite element continuum damage modeling of plain weave reinforced composites. Compos. Part. B **37**, 137–147 (2005)
25. Hansen, A.C., Blackketter, D.M., Walrath, D.E.: An invariant-based flow rule for anisotropic plasticity applied to composite materials. J. Appl. Mech. **58**(4), 881–888 (1991)
26. Mie, G.: Zur kinetischen Theorie der einatomigen Körper. Ann. Phys. **316**, 657–697 (1903)
27. Grüneisen, E.: Theorie des festen Zustandes einatomiger Elemente. Ann. Phys. **344**, 257–306 (1912)
28. Gama, B.A., Gillespie Jr., J.W.: Finite element modeling of impact, damage evolution and penetration of thick-section composites. Int. J. Impact Eng. **38**, 181–197 (2011)
29. Lambert, J.P., Jonas, G.H., Ballistic Research Laboratory Report BRL-R-1852 (ADA021389) (1976)

Chapter 15
Characterization of High Strain Rate Dependency of 3D CFRP Materials

N. Tran, J. Berthe, M. Brieu, G. Portemont, and J. Schneider

Abstract New composite materials are increasingly used in aviation to reduce the mass of structures. Aeronautic structures have to be designed with respect to a broad range of mechanical loadings during their operational life. These loadings are considered in the design by numerous cases, from low up to high speeds. The motivation of the presented work is to establish and characterize the high strain rate dependency of the linear behavior of composites materials. More specifically, new generations of 3D carbon/epoxy composite materials are of interest because of their high mechanical performances, which require specific experimental developments to be done. Due to the large size of their textile Unit Cell and carbon fiber high strength and stiffness, unusual dynamic test capabilities are required, which leads to revisit the test protocols, specimens definition, instrumentation and exploitation techniques. The experimental method described in this work is applied to analyze the strain rate sensitivity of the mechanical behavior of such a 3D woven composite material. The experiments are done with a servo-hydraulic testing machine (ONERA) in a strain rates range varying between 10^{-4} and $10\ s^{-1}$. The linear mechanical behavior of the material in the warp, weft and 45° orientations is characterized. These tests, together with the new experimental protocol, permit to accurately reveal and measure the material behavior strain rate sensitivity, which proved to be large in the 45° direction.

Keywords Composite • 3D woven • Strain rate • Experimental • Characterization

15.1 Introduction

Currently, different woven composite material configurations exist. Woven architecture can be made in different textile types (taffeta, twill, 3D, 4D, 5D, ...). This study focuses on an epoxy reinforced by layer/layer unbalanced carbon fibers woven developed by SAFRAN Snecma. This Unit Cell, described in Fig. 15.1, as a significant size compared to classic 2D woven composites. The composite volumic warp/weft fibers ratio studied is low unbalanced.

This composite is made of two weaving directions, called warp (blue) and weft (red). Warp yarns weaving with different layers interlocks each layer together. This woven composite particularity prevents the fragile delamination phenomenon to appear. As Schneider [1] shows in quasi-static tensile tests ($\dot{\varepsilon} \simeq 10^{-4}\ s^{-1}$), 3D carbon/epoxy woven composite mechanical behaviour is orthotropic. This woven composite mechanical behaviour presents two domains, firstly a linear one in which the damage is negligible at the macroscopic scale. The second one is a non linear domain due to the development of damage (matrix cracks, fiber yarn cracks, decohesion fiber yarn/matrix).

The in-plane mechanical behaviour of this specific 3D woven composite has been mainly studied under quasi-static or fatigue loadings [1, 2, 3, 4]. In structure lifecycle, particulary for aeronautic applications, a wide loading speed range could be experimented for many reasons (bird strike, ice strike, ...). It is thus important to study the loading speed influence on materials mechanical behavior in order to design the structure. The present study focuses on the first part of the mechanical behaviour, assumed as linear, in various directions and for different strain rates.

N. Tran (✉) • J. Schneider
SAFRAN Snecma, Moissy-Cramayel 77550, France
e-mail: nicolas.tran@snecma.fr; julien.schneider@snecma.fr

J. Berthe • G. Portemont
Onera—The French Aerospace Lab, Lille F-59014, France
e-mail: julien.berthe@onera.fr; gerald.portemont@onera.fr

M. Brieu
Centrale Lille/LML, Villeneuve d'Ascq 59655, France
e-mail: mathias.brieu@ec-lille.fr

D. Casem et al. (eds.), *Dynamic Behavior of Materials, Volume 1*, Conference Proceedings of the Society for Experimental Mechanics Series, DOI 10.1007/978-3-319-41132-3_15

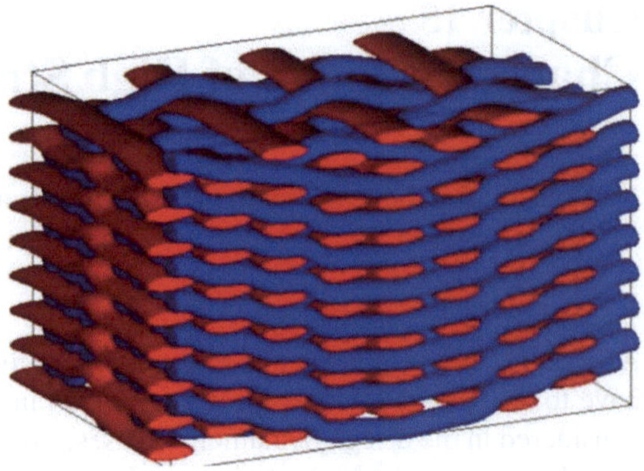

Fig. 15.1 Representation of a complex 3D woven fabric composite Unit Cell

15.1.1 Dynamical Testing Machine

A servo-hydraulic testing machine was chosen for the dynamical experimentations. This device is particulary used for materials characterization from quasi-static to dynamic loadings up to several 10 m s^{-1}.

In order to characterize the 3D woven composite quasi-static behavior, Schneider [1] proposed specific dog-bone tensile specimen geometry in warp and weft directions. The large size Unit Cell and the high strength and stiffness of the carbon fibers would lead to apply high load levels. The current dynamic experimental device can not reach such load levels. Tensile specimen geometry have to be minimized for dynamic tests. This dimensional change must not impact the intrinsic material properties characterization.

15.2 Dynamic Tensile Specimen

Rather than performing a costly and time consuming experimental campaign for designing the tensile specimen geometry, a numerical approach has been preferred. The numerical approach is based on numerical Finite Element (FE) tensile test simulations in warp and weft directions, directions in which the load levels are critical. The specimen dimensions influence was studied with respect to the elastic modulus.

15.2.1 Minimization of the Tensile Specimen Geometry

The mesh was realized with a modeling chain tool developed by SAFRAN Snecma [5]. From the reference specimen Zone Of Interest (ZOI) μ-tomography, a textile architectural mesh was made. This specific mesh takes into account possible heterogeneities generated by the manufacturing processes. In this case, carbon fiber yarns are considered elastic orthotropic, and epoxy matrix is considered elastic isotropic.

The ZOI width (W) and length (L) influences on the elastic modulus characterization were studied. The virtual specimens thickness is constant because of the interlocking pattern between each layer in this direction. In each case, displacement boundary conditions representative of a tensile test are applied to the faces orthogonal to the loading direction. The other faces are free in displacement and loading.

Elastic modulus obtained with warp tensile tests for different ZOI dimensions are described in Fig. 15.2a. Error bars correspond to the dispersion linked to the sample positions. Elastic moduli are normalized by the value obtained with the quasi-static tensile specimen test result (L = 1.75; W = 1). Volume dimensions are normalized by the Unit Cell ones.

For W = 1 (filled blue circle), elastic moduli converge quickly to the value obtained with the tested specimen assumed as the converged one. By decreasing the specimen width, W = 0.75 (filled red square) and W = 0.5 (filled brown circle), convergence to the reference solution were observed. A decrease in virtual specimen length by 43 % (L = 1) does not significantly alter the elastic modulus identification (error < 1.5 %). If the considered ZOI is too small (L = 0.25 and W = 0.5 as example) the elastic modulus identification varies a lot (± 4.4 % different from the mean value). In this case, a significant position sensitivity was noticed.

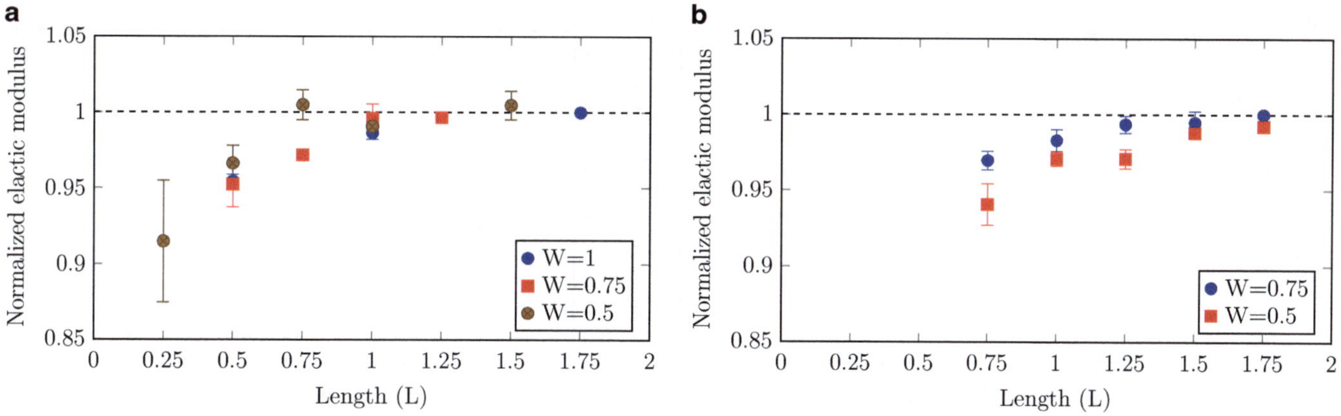

Fig. 15.2 Evolution of the normalized warp elastic modulus with the size and position of the specimen. (**a**) Warp tensile tests. (**b**) Weft tensile tests

Table 15.1 ZOI dimensions of the warp and weft dynamic tensile specimens

Direction	Length of the ZOI (L)	Width of the ZOI (W)
Warp	1.4 Unit Cells	0.75 Unit Cell
Weft	2.3 Unit Cells	0.75 Unit Cell

Elastic modulus obtained with weft tensile tests for different ZOI dimensions are described in Fig. 15.2b. In this case, virtual tests corresponding to the reference specimen width (W = 1) were not studied. Elastic moduli are normalized by the value obtained with the largest virtual specimen (L = 1.75; W = 0.75).

For W = 0.75 (filled blue circle), elastic moduli converge quickly to this reference value. As observed in warp direction, by decreasing the specimen width, for W = 0.5 (filled red square), the convergence to the reference solution was observed. Virtual specimen length decrease by 14 % (L = 1.5) does not significantly alter the elastic modulus, with 1.1 % different compared to the current reference value.

In both cases, it was observed that elastic modulus obtained with smaller virtual specimens width and length than the references may be admissible (error < 1.5 %). The tensile specimen width reduction limits load levels to apply. In the present study, a 0.75 Unit Cell specimen width in warp and weft directions allows to perform tensile tests with the dynamic servo-hydraulic machine. Dynamic tensile specimen dimensions are summarized in Table 15.1.

15.2.2 Experimental Validation

Dynamic tensile test specimens, designed by FE simulations, had to be experimentally validated before characterizing the 3D woven composite mechanical behavior strain rate sensitivity.

Quasi-static tensile tests ($10^{-4}\,s^{-1}$) of the dynamic specimens were realized with the servo-hydraulic machine. Longitudinal strain was evaluated by strain gage transversal strain by Digital Image Correlation (DIC). ZOI virtual gage dimension was used to identify the elastic modulus. As defined by the Snecma protocol, warp and weft elastic moduli are obtained by linear regression between 0.1 and 0.4 % longitudinal strain. Warp and weft elastic moduli obtained with the dynamic specimens are described in Fig. 15.3a, b (respectively open blue circle and open red circle).

Quasi-static tensile tests ($10^{-4}\,s^{-1}$) of the quasi-static specimens proposed by Schneider [1] were realized with electro-mechanical machine. Warp and weft elastic moduli obtained with the quasi-static tensile specimens are compared to the dynamic ones in Fig. 15.3a, b (respectively filled blue star and filled red star). The tensile specimen dimensions decrease has no impact on the 3D woven composite elastic modulus characterization. This experimental comparison shows a good correlation with the numerical approach.

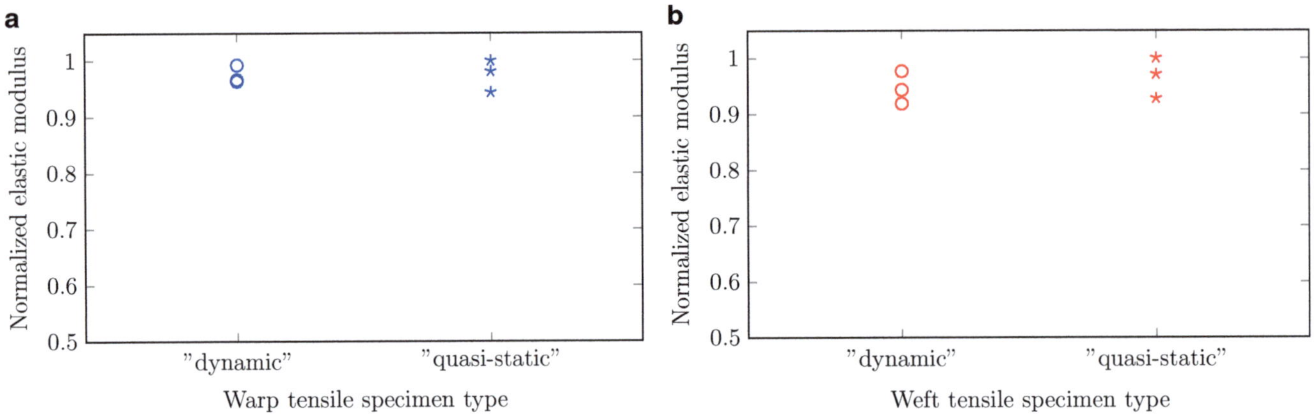

Fig. 15.3 Comparison of elastic moduli obtained with the reference specimens and the dynamic ones. (**a**) Warp tensile tests. (**b**) Weft tensile tests

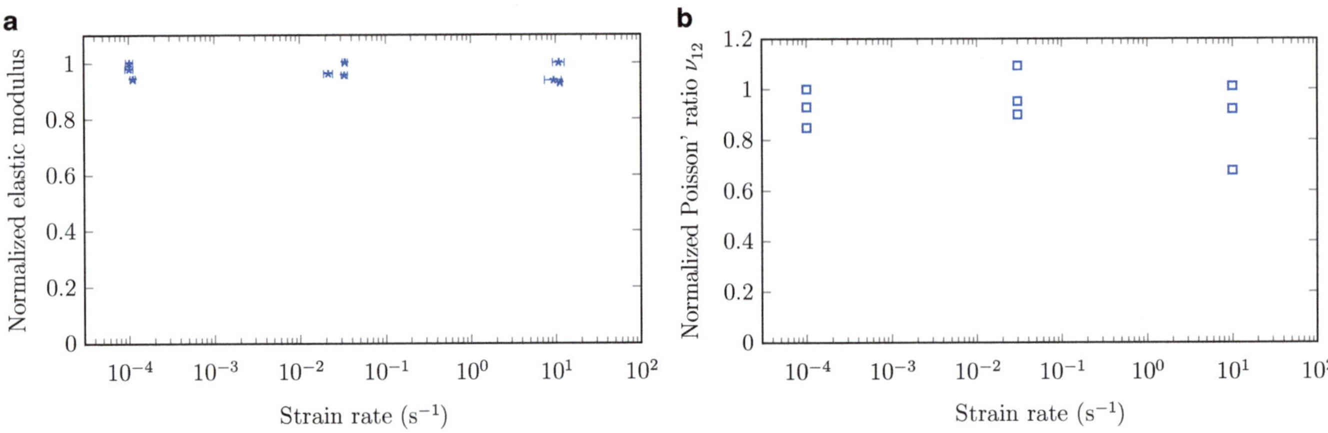

Fig. 15.4 Dynamic warp tensile tests. (**a**) Warp elastic modulus. (**b**) Poisson' ratio ν_{12}

15.3 Experimental Dynamic Tensile Tests

15.3.1 Warp Tensile Tests

Tensile tests on the dynamic warp and weft specimens were realized with the servo-hydraulic machine from quasi-static to dynamic strain rates. Tensile tests were made for 3 speeds: $1.7 \cdot 10^{-3}$ m s^{-1}, 0.01 ms^{-1} and 1 ms^{-1}. 3 tensile tests for each speed were carried out to evaluate the dispersion.

Elastic modulus obtained for different speeds in warp direction are described in Fig. 15.4a with respect to the mean strain rate, calculated in the elastic modulus measurement gap. Horizontal error bar in Fig. 15.4a represents the dispersion in strain rate in the elastic modulus measurement gap. Error bars are not significant for all warp tensile tests. Warp elastic modulus is normalized by the highest one at 10^{-4} s^{-1}. In warp direction, elastic modulus is independent of the strain rate between 10^{-4} and 10 s^{-1} and the dispersion is quite constant.

Poisson' ratio ν_{12} obtained in each warp tensile tests is described in Fig. 15.4b with respect to the strain rate. Poisson' ratio was evaluated by DIC in the same measurement gap as the elastic modulus. Poisson' ratio ν_{12} dispersion is higher than the one of elastic modulus. Low transversal strain levels in the Poisson' ratio measurement gap are close to the measurement uncertainly. Given these low measurement levels, the Poisson' ratio obtained is qualitative. An insensitivity of Poisson' ratio ν_{12} relative to the strain rate was observed.

In this case, warp mechanical behaviour under $\varepsilon_l < 0.4$ % of the 3D carbon/epoxy woven composite could be considered linear elastic.

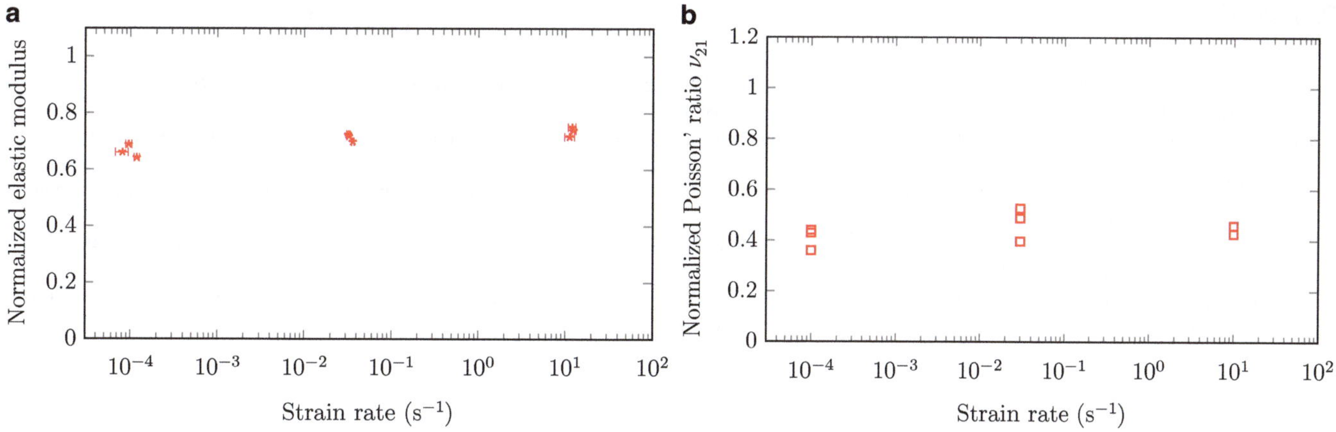

Fig. 15.5 Dynamic weft tensile tests. (**a**) Weft elastic modulus. (**b**) Poisson' ratio ν_{21}

15.3.2 Weft Tensile Tests

Tensile tests on the dynamic weft specimens were realized with servo-hydraulic machine from quasi-static to dynamic strain rates. 3 tensile tests per speed were ran for $1.7 \cdot 10^{-3}$ ms^{-1}, 0.01 ms^{-1} and 1 ms^{-1}.

Weft elastic moduli, normalized by quasi-static warp elastic modulus, are described in Fig. 15.5a. As shown in warp case, dispersion in weft elastic modulus is negligible. Weft elastic modulus strain rate sensitivity is not significant between 10^{-4} and 10 s^{-1}. Poisson' ratio also evaluated by DIC allows to make the same observation. In warp and weft directions, 3D carbon/epoxy woven composite mechanical behaviour under $\varepsilon_l < 0.4$ % can be considered linear elastic.

15.4 Conclusions

Quasi-static warp and weft 3D carbon/epoxy woven composite tensile specimens require to apply load levels that could not be reached with dynamic servo-hydraulic testing machine. Specific dynamic specimens were then developed in order to adapt the specimens strength to the dynamic testing machine capacity.

Reference tensile specimens geometry are one Unit Cell width. Dimensional and positional sensitivities of the elastic modulus characterization were studied by Finite Element tensile test simulations. In warp and weft cases it was observed that the elastic modulus quantification could be properly performed by reducing the reference width (W) and length (L). Elastic modulus obtained in both cases, with a 0.75 Unit cell width (L = 1 in warp direction and L = 1.5 in weft direction) is less than 1.5 % different from the reference, and relatively independent of the virtual specimen position. A length reduction specimen permits to reach upper strain rate levels than the reference one.

Elastic modulus obtained with the quasi-static and the dynamic tensile specimens at 10^{-4} s^{-1} prove a good correlation with respect to the virtual testing approach.

From specific dynamic warp and weft tensile specimens, quasi-static to dynamic tests were run in order to characterize the stain rate sensitivity of the 3D carbon/epoxy woven composite linear mechanical behaviour. Elastic modulus and the Poisson' ratio obtained in both cases show a independence to the strain rate between 10^{-4} and 10 s^{-1}.

In conclusion, warp and weft mechanical behaviour under $\varepsilon_l < 0.4$ % of the 3D carbon/epoxy woven composite can be considered linear elastic.

Acknowledgements This research was supported by SAFRAN Snecma. We thank our collegues from SAFRAN (Y. WIELHORSKI, T. FIXY, . . .), ONERA (M. RAGONET, A. DEUDON, . . .), Centrale Lille and LML who provided expertise and technical support that greatly assisted the research.

We thank E. Deletombe (ONERA) for assistance with particular scientific methodology and for comments that greatly improved the manuscript.

References

1. Schneider, J.: Caractérisation du comportement mécanique endommageable des matériaux composites tissés 3d interlock sous sollicitations quasi-statiques. Ph.D Thesis, Université de Technologie de Compiègne (2009)
2. Hurmane, A.: Analyse par un dialogue essais/calculs de la tenue en compression de structures composites tissées 3d. Ph.D Thesis, Université de Technologie de Compiègne (2015)
3. Henry, J.: étude et analyse des mécanismes d'endommagements en fatigue des composites à renfort tissus interlock. Ph.D Thesis, Université de Technologie de Compiègne (2013)
4. Rakotoarisoa, C.: Prévision de la durée de vie en fatigue des composites à matrice organique tissés interlock. Ph.D Thesis, Université de Technologie de Compiègne (2014)
5. Schneider, J., Hello, G., Aboura, Z., Benzeggagh, M., Marsal, D.: A meso-fe voxel model of an interlock woven composite. In: Proceeding of the International Conference in Composite Materials (ICCM) (2009)

Chapter 16
High-Strain Rate Compressive Behavior of a Clay Under Uniaxial Strain State

Huiyang Luo, Zhenxing Hu, Tingge Xu, and Hongbing Lu

Abstract Dynamic compressive behavior of a clay obtained from Boulder, Colorado was investigated on a 24.4 m split Hopkinson pressure bar. The as-received clay was first pulverized into powders, which were subsequently dried in oven at 105 °C for 2 weeks, and then sieved to collect powers with dimensions smaller than #50 (0.3 mm) using ASTM E-11 standard sieves. The sorted clay was compressed inside a hardened steel tube, by a tungsten carbide rod at one end in contact with the incident bar, and a brass rod at the other end in contact with the transmission bar. This assembly was subjected to repetitive shaking to consolidate the soil to attain a desired bulk mass density; it is then placed between the incident steel bar and the transmission aluminum bar for dynamic compression. Through measurements of both axial and transverse responses of the cylindrical clay specimens under confinement, both volumetric and deviatoric responses were determined under the 3D stress state involved. Both dry and moist clay specimens were characterized at high strain rates. The effect of moisture content on the compressive behavior was investigated. The compressibility as a function of axial stress was determined.

Keywords Modified split Hopkinson pressure bar (SHPB) • Mass density • Mason sand • Confinement • Nanoindentaion

16.1 Introduction

The mechanical behavior of a soil/clay under relatively high stress and high strain rates was not been well characterized. In order to simulate blast and ejecta more efficiently in soils, the current computational modeling relies heavily on the continuum approaches. With new development of multiscale multiphase hybrid Lagrangian particle-continuum computational approach, it requires the modelling parameters to be calibrated necessarily, and the coordinated laboratory experiments to validate the appropriate model [1, 2]. Recently, a modified split Hopkinson pressure bar (SHPB) with the use of pulse shaping has been used to characterize the mechanical behaviour of sand/soil under high strain rate, confined by a steel sleeve at high strain rates [3–6]. Their sand/soil specimen was directly sandwiched between incident and transmission bar ends, and confined by a cylindrical tube with the same inside diameter as the bar diameter. This arrangement caused the sand/soil with uncertain initial mass density. Recently, we developed successfully a technique to resolve this problem, to prepare a sand specimen with consistent initial mass density for dynamic compression [7], and investigate the compressive behavior of dense Eglin sand [8], Mason sand [9], and glass beads [10] at high strain rates. The constitutive behavior of sand at high strain rates were reviewed [11], and the modeling was possibly used to simulate the dynamic problem in soil mechanics under blast [12].

In this study, a clay obtained from the region around Boulder, Colorado was chosen as the materials for testing [1, 13, 14]. The clay was obtained from a stockpile of soil from a construction site on the University of Colorado Boulder campus, and is referred to as Boulder clay. The clay was processed after collection to remove all particles greater than the #50 sieve, and thus provided a more homogeneous and consistent material for experimental testing. Our developed technique continued to be used to prepare the clay specimens with consistent mass densities for experiments on a modified long SHPB. This technique was combined with some other techniques developed in the recent years, including strain gage attached on the outer surface of a confining cylinder tube [15, 16], to investigate the compressive behavior of clay under high strain rates.

H. Luo • Z. Hu • T. Xu • H. Lu (✉)
Department of Mechanical Engineering, The University of Texas at Dallas, Richardson, TX 75080, USA
e-mail: hongbing.lu@utdallas.edu

D. Casem et al. (eds.), *Dynamic Behavior of Materials, Volume 1*, Conference Proceedings of the Society for Experimental Mechanics Series, DOI 10.1007/978-3-319-41132-3_16

16.2 Experimental

In this investigation, a 24.4 m long modified SHPB was used to apply dynamic compression on clay with relatively high compressive strain with a pulse shaping technique, which can reach certain long loading times (up to ~3.4 ms). The SHPB, as shown schematically in Fig. 16.1a, consists of a solid 304 L stainless steel incident bars (Young's modulus 196 GPa, density 8.0 g/cm^3, and bar wave speed 4950 m/s), a solid 7075-T6 aluminium bar (Young's modulus 71.7 GPa, density 2.81 g/cm^3, and bar wave speed 5051 m/s) and a strain data acquisition system. The length of the incident bar and the transmission bar are 8.7 m and 3.66 m long, respectively, they have the same outer diameter of 19 mm.

To obtain clay samples with consistent mass densities, we developed a similar assembly to prepare clay specimen as that for our former sand specimen. This assembly was then inserted directly between two bar ends of the SHPB apparatus for dynamic compression. Unlike two cemented tungsten carbide (WC) rods for sand assembly, the clay assembly consists of one WC rod, one brass rod with both diameters of 12.7 mm, and a high carbon steel hollow cylinder with outside diameter 19.0 mm. for confinement. It is noted that the WC rod has a mechanical impedance value (the product of mass density, bar wave speed, and cross sectional area) that is 97 % of the corresponding values for the incident stainless steel bar with a diameter of 19 mm. The brass rod has mechanical impedance in 98 % of the values for the transmission aluminium bar with a diameter of 19 mm. Therefore, the dispersion due to the mismatch of the mechanical impedance between each bar end to insert rod can be ignored.

The as-received silty clay was moist, in form of non-regular clump. The clay clump was put inside of temperature chamber for drying at 105 °C for 2 weeks. Steel hamper was used to crush the clay clumps into smaller chunks/powders. Subsequently the crushed clay was sorted by #50 sieve (0.30 mm), to obtain dry clay samples, which are as shown in Fig. 16.2a. The sorted dry clay was first poured into the hollow steel cylinder with one end closed by the WC rod. Then the dry clay filled in the hollow cylinder assembly was shaken to a mass density of 1.30 g/cm^3 with length/diameter ratio of 0.716. Different weight levels of water were then added accurately into the chamber by using digital balance (Denver Instrument APX-300, 1 mg resolution). The open assembly was put inside a vacuum chamber for 30 min, to remove air and make the water evenly distributed inside clay. Then another brass rod was capped the testing chamber. A schematic diagram is shown in Fig. 16.1b. The total mass of the assembly less than 300 g can be shaken manually to consolidate the clay sample to attain different initial mass densities. The net mass of the clay (1.5 g) was determined by weighing the overall assembly and then subtracting the tare (steel sleeve and inserted rods). The length of the cavity occupied by clay was measured and controlled by change in distance of the end surfaces of the two WC rods. These values were then converted into bulk mass density of the clay specimen.

A strain gage was attached onto the external surface of the confining steel sleeve on the circumferential direction. The lateral deformation of the clay sample was measured by the strain gage, and the radial/circumferential stress was calculated from the strain considering a linear elastic response of the confining cylinder under lateral deformation below 0.2 %. This arrangement allows characterization of dynamic volumetric and deviatoric behaviours of clay. Further analysis yields other relationships, such as the void ratio—pressure relationship at high strain rates. In each experiment, the clay assembly was sandwiched between incident and transmission bars on SHPB, supported and aligned by two aluminum fixtures attached to the lateral surface of the ends of the incident and transmission bars. The WC rod side of the assembly is contacted on the incident steel bar end, while the brass rod side is contacted on the transmission aluminium bar end. Compact loading was applied through WC rods directly contact with the clay ends, on the clay specimen, then another end on brass rod, finally transited to the transmission bar.

Brass 260 disks of 7.6 mm in diameter with thicknesses varying from 1.6 to 3.1 mm were used as pulse shapers in the SHPB experiments by placing them on the impact surface of the incident bar. A thin layer of Dow Corning High-vacuum

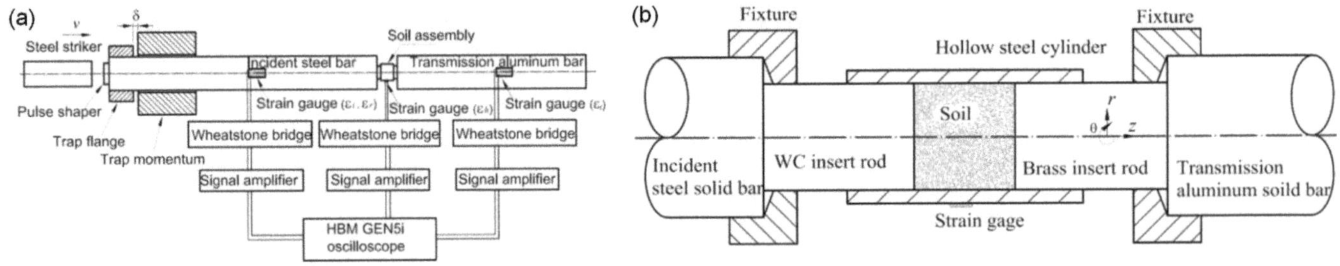

Fig. 16.1 Modified SHPB setup for dynamic compaction Boulder clay. (**a**). Schematic of SHPB apparatus; (**b**). Schematic of test section of the clay assembly (polar coordinates defined)

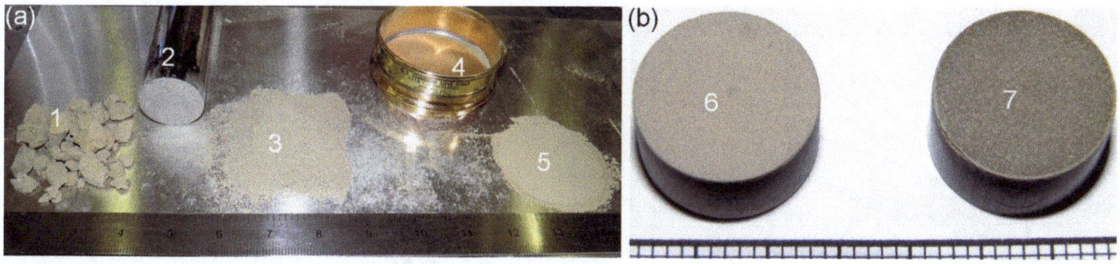

Fig. 16.2 Clay sample preparation and SHPB tested clay. (**a**). Preparation. 1: as-received clay; 2: steel hamper; 3: crushed clay; 4: sieve; 5: sorted clay; (**b**). Impacted specimen of dry clay (6) and partially saturated as-received clay (7)

grease (McMaster #2966 k52) was used at the interface between the pulse shaper and the incident end of the bar. A 2.13 m long steel striker, with the same diameter as that of both the incident and transmission bars, launched by compressed air in a gas gun, impacts the pulse shaper and generates a desired compressive wave traveling in the incident bar to induce a nearly constant strain rate in the clay specimen. Trap momentum was used to apply a single loading. A HBM Genesis digital oscilloscope (15-bit resolution, 25 Ms/s sampling rate) was used to acquire strain signals through Wheatstone bridges and a Vishay 2310B signal conditioning amplifier (Gain 1–11,000 continuously varying; 180 kHz frequency response, 180 kHz bandwidth and ±0.02 % linearity) connected to the strain gages.

A hollow cylinder was used to confine clay specimen passively. For a specimen confined by a hollow cylinder within the elastic range, the confining radial and circumferential pressures, as well as circumferential strain were well documented [15, 16]. The polar coordinates with r and θ was defined in Fig. 16.1b. In this study, we used the confinement method for dynamic compression of sand on SHPB to characterize the hydrostatic and deviatoric behaviors of clay at high strain rates. Since the incident bar is different from the transmission bar, the stress calculation is kept the unchanged while the axial strain and axial strain rate were a little different, well documented [17–20]. With the known axial, confining stress and strain histories, the mean stress (e.g., hydrostatic pressure, isotropic stress) component σ_m and the volumetric dilatation ε_m are well documented, as well as the octahedral shear stress τ_{oct} [15, 16]. With the known values for hydrostatic and shear stresses, the hydrostatic and deviatoric behaviors of clay under multi-axial stress state can be determined at high strain rates. Thus, the known 3-D stresses can be used as input for the modeling works in the future. In this investigation all tests were conducted at room temperature (~23 °C) and near 26 % relative humidity environment. Under each testing condition, eight or more specimens were tested to evaluate reproducibility of experimental results, and the average values are reported.

16.3 Results and Discussion

The physical properties of Boulder clay have been measured [13, 14]. The relevant grain size distribution parameters are determined as $D_{10} < 1.7 \times 10^{-4}$ mm, $D_{30} < 0.001$ mm, $D_{50} = 0.001$ mm, and soil index properties are liquid limit LL = 43 and plasticity index PI = 22. Boulder clay is classified as a low plasticity clay (CL) according to the Unified Soil Classification System (USCS). The specific gravity G_s was measured to be 2.70. From the standard Proctor compaction test, an optimal water content of 18 % corresponds to the maximum dry unit weight of 17.5 kN/m^3. The clay has an apparent preconsolidation stress of 0.30 MPa, a compression index c_c of 0.23, and a recompression index c_r of 0.041. In clay sample preparation, the first 2 days of drying of as-received big chucks cause 11 % wt (weight) loss. Steel hamper was used to crush the clay clumps into smaller fragments and chucks. Subsequently the clumps were crushed and allowed to dry for another 12 days, the final weight loss was 12 % wt. So the as-received clay was determined to contain 12 % wt moisture. The as-received clay was also crushed and sieved without drying. Then #50 sieve (0.30 mm) was used to sieve the crushed clay to obtain clay samples for high strain rate experiments. Another two different moisture levels (22 and 40 % wt) was achieved through adding extra water except the dry (0 % moisture) and as-received (12 % moisture). After SHPB impact, clay specimens (0.3 mm size) were compacted firmly into solid disks (Fig. 16.2b), even for dry clay. The compacted clay disk was not easily disintegrated. It indicated that the impacted clay was held together well by its own cohesion.

Typical stress-strain curves determined from this SHPB were shown in Fig. 16.3a for dry clay with density 1.30 g/cm^3 under strain rate 502 s^{-1}. They are axial stress-axial strain curve, mean stress-volumetric strain curve and shear-stress-shear strain curve. The stress components are plotted in logarithmic scale while the strains are plotted in linear scale. In these semi-logarithmic plots, the stress follows approximately a two-segment linear relationship with the corresponding strains. The first linear segment occurs within the initial 2 % compressive strain, most likely due to elastic deformation of clay

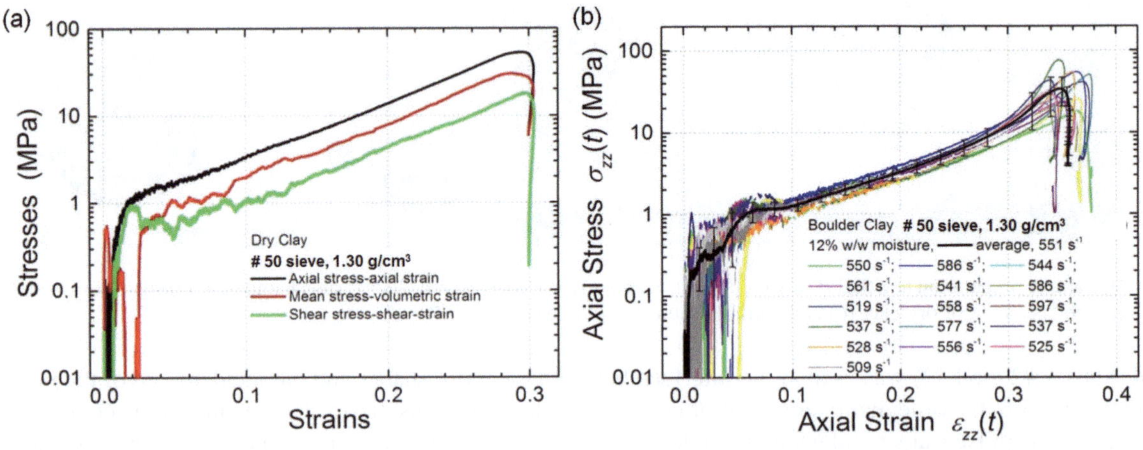

Fig. 16.3 SHPB results of dry clay and as-received clay. (**a**). Typical stress-strain curves; (**b**). Repeated axial stress-axial strain curves

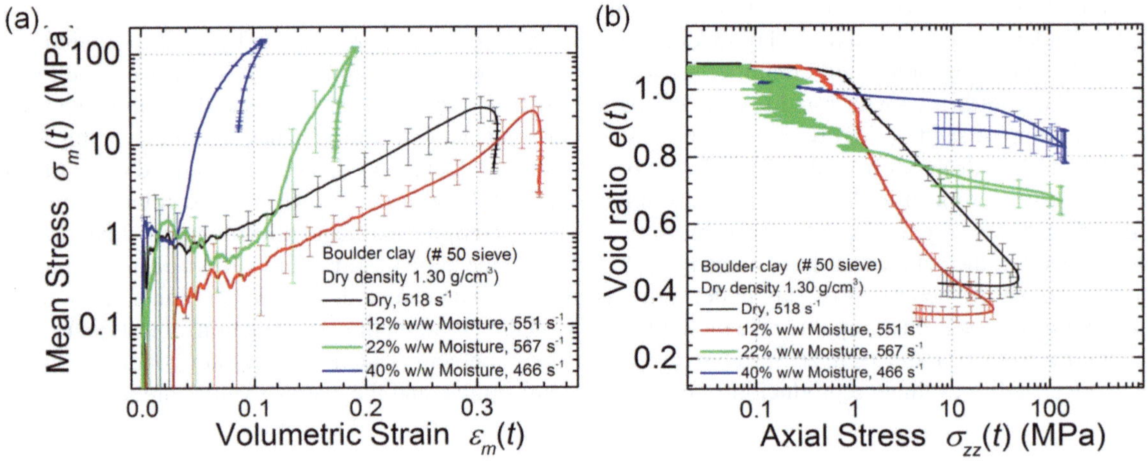

Fig. 16.4 SHPB results of clays with different moisture levels. (**a**). Mean stress-volumetric strain curves; (**b**). Compressibility *e*-log *p* curves

grains. It can be referred to as clay grain-level elastic behaviour, similar to the situation under quasi-static compression [21]. The second linear segment on a semi-logarithmic scale occurs at compressive strains larger than 2 %. It is perhaps associated with clay grain compaction through grain crushing, rearrangement of crushed fragments and intact grains under higher pressure. This stage can be referred to as grain-level plastic behaviour. At compressive strains between 2 and 4 %, there is a gradual transition region. It is noted that axial stress-axial strain curves are smooth; while mean stress- volumetric strain curve and shear stress-shear strain curve of the uncompact initial clay have a small oscillated yield-like stage in the transition region.

As shown in Fig. 16.3, all the stress-strain curves show clearly the relationship of exponential function [2] as $\varepsilon = A\log_{10}\sigma + B$ (*A*, *B* are constants) at strain larger than 3 %, indicating the compaction behaviour. Figure 16.3b shows the repeatability of SHPB experiments for clay from 16 test under strain rate of 400 ~ 600 s^{-1}. Finally, we report the average strain rate and average stress-strain curves with error bar representing the standard deviation. A good reproducible experiment after 4 % strain was achieved for clay specimen although there exists some scattering in boulder clay with #50 sieve size and low density 1.30 g/cm^3.

Figure 16.4a show mean stress-volumetric strain curves of clays with different moisture levels. It is noted that the 12 % w/w moisture level is the as-received sample in wet condition, the 40 % w/w water content is in saturated condition. In lower moisture level (12 % w/w as received clay), the load-bearing capability is evenly decreased to 1/3. When the moisture level increased to 22 % w/w, the load-bear capability is nearly unchanged below 13 % strain. After that, it increases significantly with higher slope and higher stress. This trend also happens for the full moisture-saturated clay, while the critical strain was decreased to 3 %. The effect of moisture levels on clay is not similar as sand, but more significant for load-bearing capability. The lower water content has weakened the mechanical properties of clay. But the higher moisture level has strengthened the clay properties. Figure 16.4b show the compressibility of clay as a function of pressure. The relationship is given in terms of the void ratio, *e*, and the applied axial stress, σ_z, often termed as *p*, defined in the soil mechanics. The measurement of both radial and circumferential stresses and strains in the clay sample in this work made it possible to report the relationship under high strain rate. Two linear segments clearly appear in the *e*-log *p* curve only for dry clay and lower moisture level of clay, representing an exponent relationship between *e* and *p*. For the higher partially saturated clay, the two stages were not clearly linear, but curved. The *e*-log *p* curve during unloading also follows nearly a linear relationship. At these high strain rates, with the increase of the moisture levels, the pre-consolidation pressure is slightly decreased and the compressibility is significantly dropped due that water begun to bear the load.

16.4 Conclusions

The dynamic compressive behaviour of Boulder clay under stiff confinement using steel tubes was investigated using a modified long split Hopkinson pressure bar (SHPB) at high strain rates (~500 s^{-1}) up to 150 MPa compressive stress. A clay specimen assembly using one WC insert and one brass insert inside a steel tube was employed, which allows to prepare clay samples with consistent initial mass densities. Clay samples of four different moisture levels: dry (0 %), as-received (12 %), partially-saturated (22 %) and saturated (40 %) at the same dry initial mass densities 1.30 g/cm^3 were compressed at high strain rates. With a strain gage attached on the external surface of the confining tubes, both the bulk and the deviatoric responses were determined. The curves of axial stress-axial strain, hydrostatic pressure-volumetric strain, shear stress and shear strain under 3D stress state of clay are determined up to 36 % compressive strain and up to 150 MPa axial compressive stress at high strain rates. The *e*-log *p* curves were used to determine the clay compressibility as a function of initial mass density. More sensitive moisture effects have been identified for clay than sand. The stress-strain relationships follow an exponential power relationship. The lower water content has weakened the mechanical properties of clay. But the higher moisture level has strengthened the clay properties. These initial results can be analysed further for constitutive modelling and for mesoscale simulations to understand the dynamics of clay under high pressure and high rate deformation.

Acknowledgment We acknowledge the support of ONR MURI grant N00014-11-1-0691 and DOE NEUP grant 09-818, NSF ECCS-1307997. Lu Also acknowledges the Louis A. Beecherl Jr. Chair for additional support.

References

1. Regueiro, R., Pak, R., McCartney, J., Sture, S., Yan, B., Duan, Z., Svoboda, J., Mun, W., Vasilyev, O., Kasimov, N., Brown-Dymkoski, E., Hansen, C., Li, S., Ren, B., Alshibli, K., Druckrey, A., Lu, H., Luo, H., Brannon, R., Bonifasi-Lista, C., Yarahmadi, A., Ghodrati, E., Colovos, J.: ONR MURI project on soil blast modeling and simulation. In: Dynamic Behavior of Materials: Conference Proceedings Society for Experimental Mechanics Series, vol. 1, pp. 341–353 (2014)
2. Luo, H., Xu, T., Wang, X., Lu, H.: Interaction of shock wave with granular materials. In: Antoun B. (ed.) Challenges in Mechanics of Time Dependent Materials. Proceedings of the 2015 Annual Conference on Experimental and Application Mechanics, vol. 2, pp. 35–43. Springer
3. Martin, B.E., Chen, W., Song, B., Akers, S.A.: Moisture effects on the high strain-rate behavior of sand. Mech. Mater. **41**(6), 786–798 (2009)
4. Song, B., Chen, W., Luk, V.: Impact compressive response of dry sand. Mech. Mater. **41**(6), 777–785 (2009)
5. Huang, J., Xu, S., Hu, S.: Effects of grain size and gradation on the dynamic responses of quartz sands. Int. J. Impact Eng. **59**, 1–10 (2013)
6. Martin, B.E., Kabir, E., Chen, W.: Undrained high-pressure and high strain-rate response of dry sand under triaxial loading. Int. J. Impact Eng. **54**, 51–63 (2013)
7. Luo, H., Lu, H., Cooper, W.L., Komanduri, R.: Effect of mass density on the compressive behavior of dry sand under confinement at high strain rates. Exp. Mech. **51**(9), 1499–1510 (2011)
8. Luo, H., Cooper, W.L., Lu, H.: Effects of particle size and moisture on the compressive behavior of dense Eglin sand under confinement at high strain rates. Int. J. Impact Eng. **65**, 40–55 (2014)
9. Luo, H., Du, Y., Hu, Z., Lu, H.: High-strain rate compressive behavior of dry Mason sand under confinement. In: Song, B., et al. (eds.) Dynamic Behavior of Materials, Proceedings of the 2014 Annual Conference on Experimental and Applied Mechanics, vol. 1, pp. 325–333. Springer
10. Luo, H., Du, Y., Hu, Z., Cooper, W.L., Lu, H.: High-strain rate compressive behavior of glass beads under confinement. Exp. Mech. **55**(5), 935–950 (2015)
11. Omidvar, M., Iskander, M., Bless, S.: Stress-strain behavior of sand at high strain rates. Int. J. Impact Eng. **49**, 192–213 (2012)
12. Higgins, W., Chakraborty, T., Basu, D.: A high strain-rate constitutive model for sand and its application in finite-element analysis of tunnels subjected to blast. Int. J. Numer. Anal. Methods Geomech. **37**(15), 2590–2610 (2013)
13. Svoboda, J.S., McCartney, J.S.: Shearing rate effects on dense sand and compacted clay. In: Dynamic Behavior of Materials. Conference Proceedings of the Society for Experimental Mechanics Series, vol. 1, pp. 389–395 (2014)

14. Mun, W., McCartney, J.S.: Compression mechanisms of unsaturated clay under high stresses. Can. Geotech. J. **52**(12), 2099–2112 (2015)
15. Ravi-Chandar, K., Ma, Z.: Inelastic deformation in polymers under multiaxial compression. Mech. Time-Depend Mater. **4**, 333–357 (2000)
16. Knauss, W.G., Emri, I., Lu, H.: Mechanics of Polymers: Viscoelasticity. In: Sharpe, W.N. (ed.) Handbook of Experimental Solid Mechanics, pp. 49–95. Springer, New York (2008)
17. Gray, GT III.: Classic split-Hopkinson pressure bar testing. In: Mechanical Testing and Evaluation. Metals Handbook. ASM, Materials Park, vol. 8, pp. 462–476 (2000)
18. Luo, H., Lu, H., Leventis, N.: The compressive behavior of isocyanate-crosslinked silica aerogel at high strain rates. Mech. Time-Depend Mater. **10**(2), 83–111 (2006)
19. Luo, H., Roy, S., Lu, H.: Dynamic compressive behavior of unidirectional IM7/5250-4 laminate after thermal oxidation. Compos. Sci. Technol. **72**(2), 159–166 (2012)
20. Luo, H., Lu, G., Roy, S., Lu, H.: Characterization of the visceoelastic behavior of bismaleimide resin before and after exposure to high temperatures. Mech. Time-Depend Mater. **17**(3), 369–399 (2013)
21. Hagerty, M.M., Hite, D.R., Ullrich, C.R., Hagerty, D.J.: One-dimensional high-pressure compression of granular material. J. Geotech. Eng. **119**(1), 1–18 (1993)

Chapter 17
Mesoscopic Modelling of Ultra-High Performance Fiber Reinforced Concrete Under Dynamic Loading

P. Forquin, J.L. Zinszner, and B. Lukic

Abstract For the two last decades numerous research works have been developed on the mechanical behavior of UHPFRC (Ultra-High Performance Fiber Reinforced Concrete) under quasi-static and dynamic loadings. However fibers orientation remains a major problem as during the implementation of fibers in concrete structure a random distribution and orientation of fibers is difficult to achieve. In this study four UHPFRC have been tested in dynamic tension and numerically simulated. The first concrete is made without any fibers. A random orientation of fibers is considered in the second concrete. Fibers oriented parallel or orthogonally to the loading direction (tensile loading) are considered in the two last cases. The four sets of concrete have been subjected to dynamic tensile loading by means of spalling tests. Next, a mesoscopic numerical simulation has been developed by considering a biphasic model: the concrete matrix is modelled by applying the DFH (Denoual-Forquin-Hild) anisotropic damage model to 3D finite-elements. In addition two nodes-finite elements are introduced in the 3D mesh to simulate numerically the presence of fibers by considering three orientations: fibers randomly distributed, parallel or orthogonal to the loading direction. As observed in the experiments, a small influence of fibers is observed regarding the peak-stress whereas a strong influence of fibers orientation is noted regarding the post-peak tensile response of UHPFRC.

Keywords Concrete • Spalling tests • Damage • Fibers • mesoscopic modeling

17.1 Introduction

For several years, the dynamic behavior of Ultra-High Performance Concretes (UHPC) is investigated due to their exquisite material properties. The quasi-static compressive strength of concretes has increased considerably over the last 20 years [1]. In the middle of the 1990s, the use of superplastifiers, silica fumes and quartz grains led to production of UHPC concrete materials with a compressive strength of about 200 MPa [2], that were outstanding compared with that of ordinary concrete. Research programs that aim at evaluating ballistic performance of ultra-high strength concrete are currently under way. One of the goals is to improve the understanding and modeling of damage mechanisms under impact [3]. In the present study, a UHPC similar to that of Ductal® concrete produced by Lafarge is investigated under different loading conditions. Besides its compressive strength (200 MPa), this UHPC is of interest to be produced on an industrial scale. Significant increase of flexural strength and ductility is obtained by adding fibers during the mixing process [4]. However, the potential effects of fiber content under extreme loading conditions such as impact of rigid projectile needs to be thoroughly investigated. The present work aims at evaluating the influence of fibers under three different types of loadings: dynamic tensile loading, confined compression loading and dynamic shear loading.

Indeed, during the impact of a rigid projectile against a concrete target severe damage modes are observed such as spalling on the front face, radial cracking in the whole target and, in case of the perforation, scabbing on the rear face [5]. Moreover, high confining pressures are observed in front of the projectile involving mechanisms such as micro-cracking and collapse of porosity [6]. Both confined and tensile behaviors are influencing the penetration resistance of the target [7]. Therefore, the constitutive model used in numerical simulations of a penetrating rigid projectile in a concrete target, should take into account the concrete confined behavior and the tensile resistance of concrete at high strain rates.

In the present work, dynamic tensile tests (spalling tests) are performed with a Hopkinson Pressure Bar. Furthermore, the tensile behavior of UHPC is investigated through mesoscopic modeling. Finally, the numerical predictions are compared to experimental data to assess the predictive capabilities of the mesoscopic model.

P. Forquin (✉) • J.L. Zinszner • B. Lukic
Soils Solids Structures Risks (3SR) Laboratory, Grenoble-Alps University, 38041, Grenoble Cedex 9, France
e-mail: pascal.forquin@3sr-grenoble.fr

D. Casem et al. (eds.), *Dynamic Behavior of Materials, Volume 1*, Conference Proceedings of the Society for Experimental Mechanics Series, DOI 10.1007/978-3-319-41132-3_17

Table 17.1 Material composition and mechanical properties of an Ultra-High Performance Concrete [3]

Composition		UHPC
Cement	[kg/m^3]	*730*
Silica fume	[kg/m^3]	*235*
Crushed quartz grains	[kg/m^3]	*220*
Sand	[kg/m^3]	*885*
Superplasticizer	[kg/m^3]	*10*
Water	[l/m^3]	*160*
Steel fibers	[vol. %]	*2 %*
W/(C + SF)		*0.17*
Mechanical Properties		
Density	[kg/m^3]	*2396*
Young's modulus	[GPa]	55
Quasi-static compressive strength	[MPa]	200

17.2 Tested Concrete

17.2.1 Composition of UHPC

The material investigated is an Ultra High Performance Concrete (UHPC) whose mechanical properties and composition are gathered in Table 17.1. In contrast to conventional concrete, apart from cement, sand and water, it contains finely crushed quartz grains and silica fume. The optimized granular mixture of UHPC, where the maximal grain size is approximately 0.6 mm (sand), and extremely low water to cementious material (cement and silica fumes) ratio allows casting a concrete with outstanding mechanical properties. The use of superplasticizer in order to increase the workability is of great importance. In the present work, four sets of specimens have been considered for dynamic spalling testing: one which contains 2 % volumetric ratio of short steel fibers (0.2 mm diameter and 7 mm in length) with 3 types of orientation (random orientation, well or badly oriented fibers) and the other without additional fiber reinforcement.

17.2.2 Mixing and Casting of UHPC and Concrete Samples

The specimens used in spalling testing are cylinders of diameter around 46 and 140 mm in length. The specimens were cored out from large blocks (270 $\times$ 270 $\times$ 170 mm^3) more than 40 days after the mix was poured into watertight plywood moulds. All the samples had been stored in water saturated by lime in order to avoid the dissolution of portlandite into water.

17.3 Experimental Results

The device used to perform the Spalling experiments is described on the Fig. 17.1. The gas gun, of caliber 50 mm, is used to launch small cylindrical projectile at impact velocity ranging from 1 to 20 m/s. The spalling test set-up used in this study is composed of a 50 mm-long spherical-cap-ended projectile and a Hopkinson bar both with a diameter of 45 mm. An illustration of the experimental set-up is given on Fig. 17.1. A short incident compressive pulse is created and propagates through the Hopkinson bar. Then, the pulse is transmitted to the sample and reflects at the free-end of the specimen as a tensile pulse leading to a dynamic tensile loading within the sample. The particle velocity at the rear-face is recorded with a laser interferometer, enabling velocity measurements up to 20 m/s with 1.5 MHz bandwidth [8].

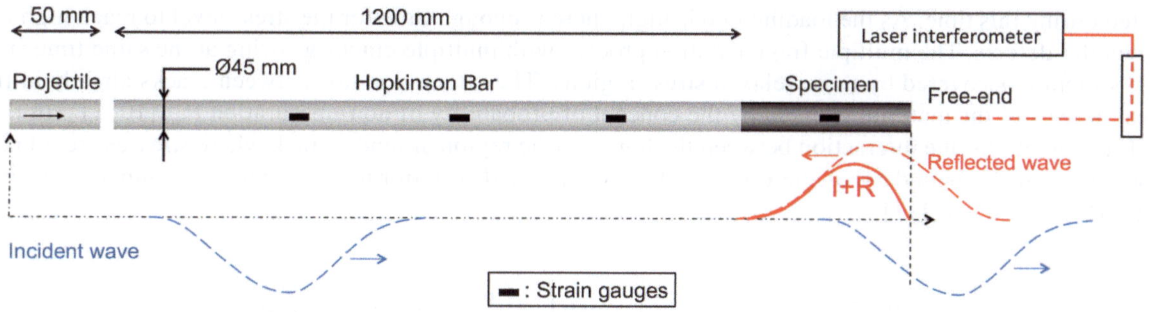

Fig. 17.1 Experimental device for spalling tests

Fig. 17.2 Particle velocity profiles recorded on the rear free surface of samples subjected to spalling test

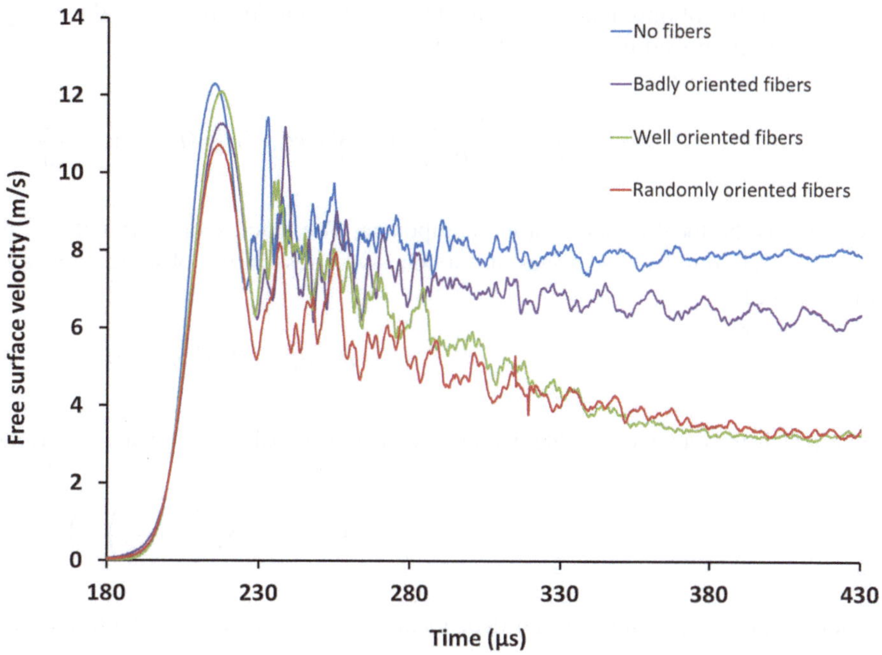

The particle velocity profile recorded on the rear free surface of four different samples is plotted on the Fig. 17.2. Based on the Novikov et al. [9] approach, one can estimate the dynamic tensile strength of the each sample from the pullback velocity (the difference between the maximum speed and the speed corresponding to rebound) assuming a linear elastic behavior of the material before to reaching the tensile ultimate strength:

$$\sigma_{dyn} = \frac{1}{2}\rho C_0 \Delta V_{pb} \tag{17.1}$$

where ρ is the density, C_0 is the one dimensional wave velocity (about 4790 m/s) in the material and ΔV_{pb} is the pullback velocity. A dynamic tensile strength of 30 MPa is obtained whatever the tested sample.

17.4 DFH Anisotropic Damage Model

In the present work, the Denoual-Forquin-Hild modeling [10, 11] is employed to model the tensile behavior of UHPC without fibers. Under high strain rate conditions, several cracks are initiated and propagate from the initial defects leading to multiple fragmentation. In that case, while the weakest defect is activated and propagates, several other cracks are initiated during this time. As the loading rate is high, there is enough time for the stress level to reach high levels and activate smaller defects. The multiple fragmentation process with multiple cracks growing at the same time stops when the whole structure is covered by these relaxed stress regions. The interaction law between cracks already initiated and the critical defects of the material is given by the concept of probability of non-obscuration P_{no} [10, 11]. In the case of multiple fragmentations, the interaction between the horizon (the region around a crack where stresses are relaxed due to crack opening) and the boundary of the domain Ω is small and if a uniform stress field is assumed, the obscuration probability P_o is written as [11]

$$P_o(T) = 1 - P_{no}(T) = 1 - \exp\left(-\int_0^T \frac{d\lambda_t}{dt}[\sigma(t)]\ Z_o(T-t)dt\right) \quad (17.2)$$

In Eq. 17.2, Z_o is the obscured zone, σ the local principal stress component, T the current time and t the crack initiation time. The probability of obscuration is defined for each eigen direction i and the change of P_i is expressed in differential form, in order to be employed in an FE code using Eq. 17.3, as

$$\frac{d^2}{dt^2}\left(\frac{1}{1-P_i}\frac{dP_i}{dt}\right) = 3!S(kC_0)^3\lambda_t[\sigma_i(t)] \text{ when } \frac{d\sigma_i}{dt} > 0 \text{ and } \sigma_i > 0 \quad (17.3)$$

where σ_i is the local principal stress component. The macroscopic strength of the material is described as the microscopic stress in the non-obscured zones and a cohesive stress in the obscured zones that controls the softening behavior of the material [12].

$$\Sigma_i = (1 - P_{oi})\sigma_i + (P_{oi})^{\alpha_D}\sigma_{coh}(\varepsilon) = (1 - D_i)\sigma_i, \quad (17.4)$$

where D_i is the damage variable defined for each principal direction and σ_{coh} is the residual strength in the obscuration zone:

$$\sigma_{coh} = \sigma_o^d \exp\left(-\left(\frac{\varepsilon}{\varepsilon_0^d}\right)^{n_d}\right) \quad (17.5)$$

where α_D, σ_o^d, ε_o^d, n_d are material-dependent parameters that were identified for the non-fibered concrete.

17.5 Numerical Simulations

A series of numerical simulation were conducted with Abaqus Explicit Finite-Element code. The concrete sample is meshed with C3D8R elements and 2-nodes T3D2 elements are used for the fibers. The fibers are embedded in the concrete mesh. An elasto-plastic behavior (yield strength: 1800 MPa) is assumed for the fibers whereas the DFH model is used to model the concrete matrix. The experimental compressive pulse is applied to the "contact" sample end. Finally, the particle velocity measured on the rear free-end is compared to experimental data on the next figures considering randomly oriented fibers (Fig. 17.3), badly oriented fibers (Fig. 17.4) and well oriented fibers (Fig. 17.5).

Finally, the mesoscopic calculations enable to predict the rebound of particle velocity that corresponds to the peak strength of the concrete. In addition, the continuous decrease of particle velocity noted in the cases of randomly-oriented and well-oriented fibers is well captured by the mesoscopic modeling whereas particle velocity plateau noted in the case of badly oriented fibers is well predicted.

Fig. 17.3 Spalling test performed with randomly oriented fibers. Comparison of experimental and numerical data

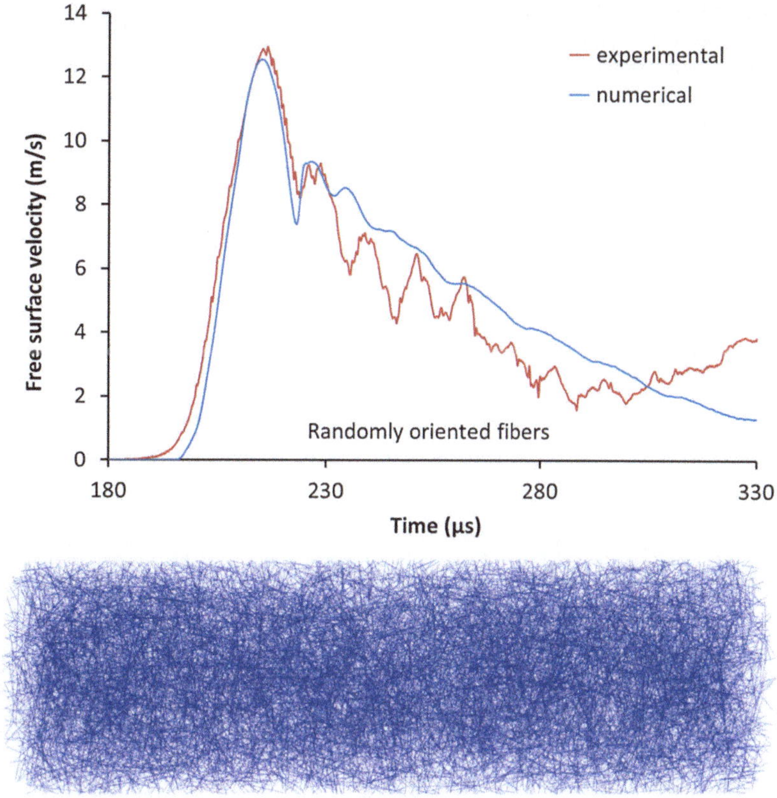

Fig. 17.4 Spalling test performed with badly oriented fibers. Comparison of experimental and numerical data

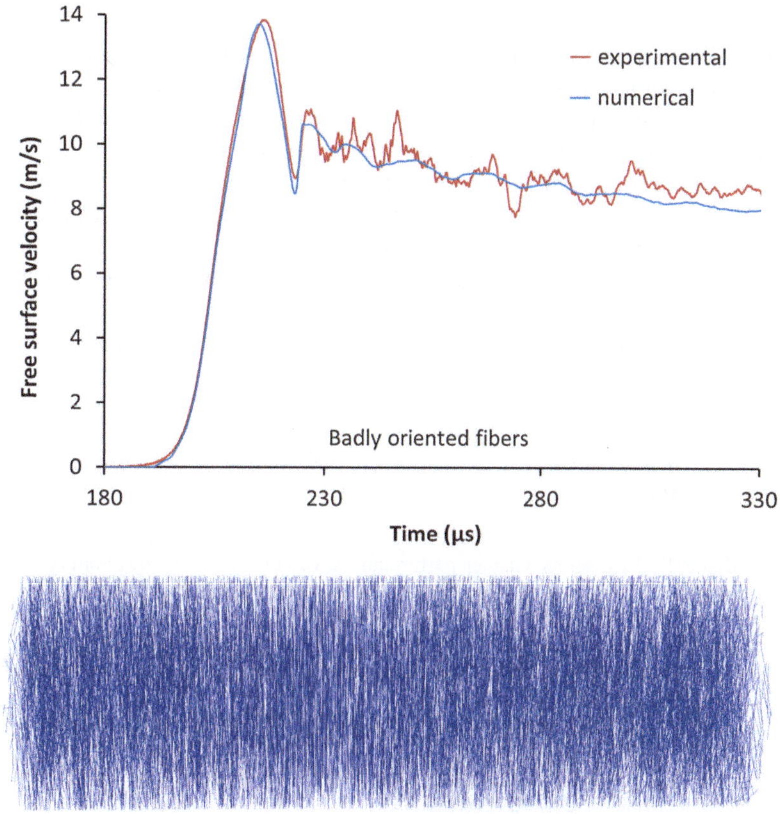

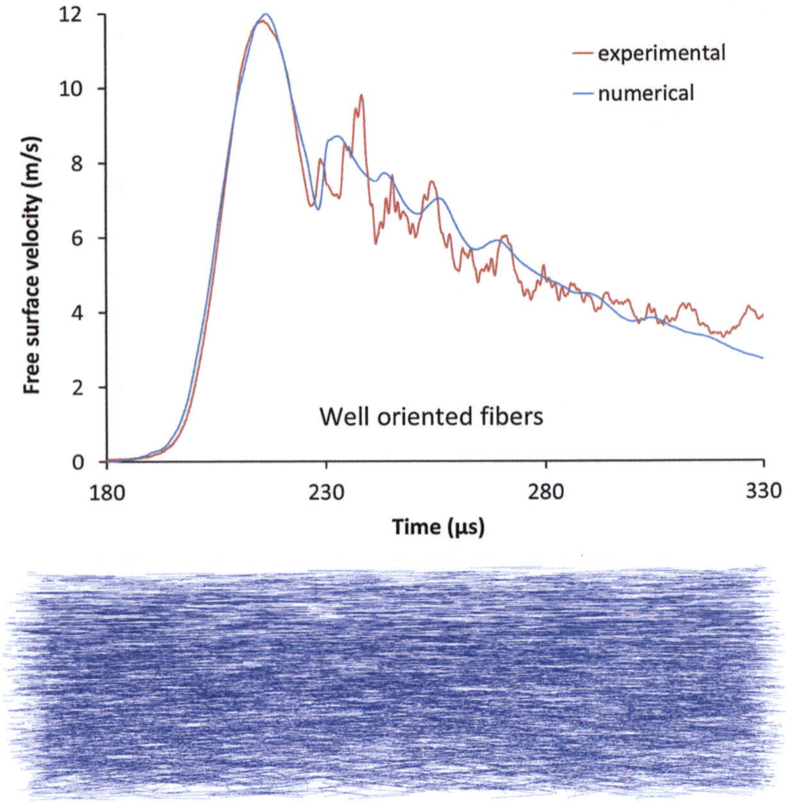

Fig. 17.5 Spalling test performed with well oriented fibers. Comparison of experimental and numerical data

17.6 Conclusion

In conclusion, the dynamic tensile behavior of an UHPC (Ultra-High Performance Concrete) with and without steel fiber reinforcement has been investigated through spalling tests. A Hopkinson Pressure Bar (HPB) device is used and a laser interferometer is pointed out toward the rear free-end of the samples allows deducing the particle velocity. Whereas the peak stress obtained through the Novikov formula is weakly influenced by fibers and fibers orientation, a clear influence of fibers is noted regarding the tail of velocity profiles. In addition, a mesoscopic modeling was used to simulate numerically the experiments by differentiating the behavior of the concrete matrix (use of DFH Denoual-Forquin-Hild anisotropic damage model) and the behavior of the fibers (elasto-plastic model). Finally the mesoscopic calculations are able to reproduce the peak strength and the post-peak responses of the fibered concretes whatever the orientation of fibers.

Aknowledgment This work was supported by the CEA (French Alternative Energies and Atomic Energy Commission, Centre de Gramat). This support is gratefully acknowledged.

References

1. Malier Y.: Les bétons à hautes performances, l'Ecole Nationale des Ponts et Chaussées, France, 1992
2. Richard P., Cheyrezy M.: Composition of reactive powder concretes, vol. 25, no. 7, pp. 1501–1511, 1995
3. Forquin, P., Hild, F.: Dynamic fragmentation of an ultra-high strength concrete during edge-on impact tests. J. Eng. Mech. **4**(134), 302–315 (2008)
4. Bayard O.: Approche multi-échelles du comportement mécanique des bétons à ultra hautes performances renforcés par des fibres courtes, Ecole Normale Supérieure de Cachan, 2003
5. Li, Q.M., Reid, S.R., Wen, H.M., Telford, A.R.: Local impact effects of hard missiles on concrete targets. Int. J. Impact Eng. **32**(1–4), 224–284 (2005)
6. Forquin, P., Arias, A., Zaera, R.: Role of porosity in controlling the mechanical and impact behaviours of cement-based materials. Int. J. Impact Eng. **35**(3), 133–146 (2008)
7. Forquin, P., Sallier, L., Pontiroli, C.: A numerical study on the influence of free water content on the ballistic performances of plain concrete targets. Mech. Mater. **89**, 176–189 (2015)

8. Erzar, B., Forquin, P.: An experimental method to determine the tensile strength of concrete at high rates of strain. Exp. Mech. **50**(7), 941–955 (2010)
9. Novikov, S.A., Divnov, I.I., Ivanov, A.G.: The study of fracture of steel, aluminium and copper under explosive loading. Fiz. Met. Metalloved **21**, 608–615 (1966)
10. Denoual, C., Hild, F.: A damage model for the dynamic fragmentation of brittle solids. Comput. Methods Appl. Mech. Eng. **183**(3), 247–258 (2000)
11. Forquin, P., Hild, F.: A probabilistic damage model of the dynamic fragmentation process in brittle materials. Adv. Appl. Mech. **44**, 1–72 (2010)
12. Erzar, B., Forquin, P.: Analysis and modelling of the cohesion strength of concrete at high strain-rates. Int. J. Solids Struct. **51**(14), 2559–2574 (2014)

Chapter 18
Comparison of Failure Mechanisms Due to Shock Propagation in Forged, Layered, and Additive Manufactured Titanium Alloy

Melissa Matthes, Brendan O'Toole, Mohamed Trabia, Shawoon Roy, Richard Jennings, Eric Bodenchak, Matthew Boswell, Thomas Graves, Robert Hixson, Edward Daykin, Cameron Hawkins, Zach Fussell, Austin Daykin, and Michael Heika

Abstract The objective of this paper is to propose experimental techniques for studying the behavior of titanium alloy, Ti-6Al-4 V (Grade 5), under shock loading. Single-layer and multi-layered stacks of forged titanium, and additive manufactured (AM) titanium plates were considered. In these experiments, target materials were subjected to ballistic impact using a two-stage light gas gun. A Photonic Doppler Velocimetry (PDV) diagnostics system was used to measure free-surface velocity on the back of each target. The experimental measurements were used to describe the behavior of these materials under shock loading. In addition to velocity measurements, physical damage and spall crack formation were monitored.

Keywords Titanium • Hyper-velocity impact • Photonic Doppler Velocimetry (PDV) • Additive Manufacturing

18.1 Introduction

Ti-6Al-4 V is a commonly used titanium alloy that accounts for 45 % of total titanium usage in the world [1]. Due to its exceptionally high strength-to-weight ratio, Ti6Al4V is extensively used in aerospace and nuclear industries. Recently, Additive Manufacturing (AM) techniques were introduced, bringing the potential of creating complex parts quickly and on demand. AM techniques have the potential for tailoring geometry and optimizing structural performance, in addition to minimizing the amount of material wasted for odd shaped parts and thereby reducing cost. However, there are many unanswered questions related to the effect of using AM techniques with titanium alloys, especially under shock conditions. A possible way of understanding the behavior of an AM plate is to compare its behavior with that of a forged single plate and stacked plates, with the same thickness, under similar loading conditions.

The following is a summary of some of the recent research in this area. Quasi-static analysis of AM titanium shows only 3–5 % lower mechanical properties than that of a forged counterpart [2, 3]. Researchers studied the mechanical characteristics of AM metallic components under static loading conditions, which revealed that there was scatter in mechanical properties for wire-feed deposits, even some heat treatments were not useful for enhancing the strength and ductility [4]. Some dynamic loading conditions have been studied [5] showing more dense specimens exhibit superior strength, but high-velocity impact conditions have not been studied. Studies were conducted on the ballistic performance of single-layer and multi-layered aluminum plates that were impacted by different end types of projectiles [6]. It was found that the ballistic resistance of in-contact multi-layered plates was weaker compared to that of single-layer plates of equal thickness. A similar study conducted impact experiments on steel plates struck by ordinary bullets to explore the outcome of target arrangement on the ballistic performance [7]. They determined that single-layer plates were more effective than multi-layered plates of corresponding thickness.

The focus of this research is the Electron Beam Additive Manufactured (EBAM) Titanium Alloy. This AM process builds up a part using an electron beam gun that deposits metal from a wire feedstock, layer by layer, through a repeated process of melting and solidifying material. The final part can be machined to obtain its final shape. In this work, we conduct a series of

M. Matthes (✉) • B. O'Toole • M. Trabia • S. Roy • R. Jennings • E. Bodenchak • M. Boswell
University of Nevada, Las Vegas, 4505 S. Maryland Parkway, Las Vegas, NV 89154-4027, USA
e-mail: matthes2@unlv.nevada.edu

T. Graves • R. Hixson
National Security Technologies, LLC, New Mexico Operations, Los Alamos Office, 182 Eastgate Dr, Los Alamos, NM 87544, USA

E. Daykin • C. Hawkins • Z. Fussell • A. Daykin • M. Heika
National Security Technologies, LLC, North Las Vegas, 2621 Losee Road, North Las Vegas, NV 89030, USA

D. Casem et al. (eds.), *Dynamic Behavior of Materials, Volume 1*, Conference Proceedings of the Society for Experimental Mechanics Series, DOI 10.1007/978-3-319-41132-3_18

ballistic impact experiments using a two-stage light gas gun to impact Ti-6Al-4 V target plates. Three cases are considered: single forged plate, multi-layered stack, and EBAM single plate. A Photonic Doppler Velocimetry (PDV) diagnostics system was used to measure free-surface velocity of several points near the center of impact on the back of each target. PDV is a heterodyne laser interferometer system that measures the beat frequency of coherent laser light to determine measuring position using fiber optic probes [8]. Physical damage and spall crack formation are also measured. These experiments are used to assess the effect of AM on Ti6Al4V behavior under shock conditions.

18.2 Experimental Setup

18.2.1 Two-Stage Light Gas Gun

A series of hyper-velocity impact experiments were conducted using a two-stage light gas gun. It uses a powder breech to propel a plastic piston into a pump tube that is filled with hydrogen. The gas is compressed as the piston travels through the pump tube and a petal valve, that separates the pressurized gas from the launch tube, ruptures. The projectile is then accelerated down the launch tube and into the target chamber. The projectile is a 5.6 mm diameter Lexan cylinder (8.6 mm long) with a mass of 0.25 g. The projectile impacts the target plate, which is bolted to a mounting plate, inside the target chamber at velocities between 4.8 and 6.9 km/s (Fig. 18.1).

18.2.2 Materials

Forged Titanium 6AL-4 V (annealed) was chosen for single and multiple layered stacks. The thickness of the targets varied from 3.2 mm to 12.7 mm. The AM plates were fabricated using EBAM techniques. Six AM preforms of an extra low interstitial (ELI) grade of Titanium 6, AMS-4956 were deposited onto a substrate of Titanium 6, Grade 4 AMS 4911, Fig. 18.2. The six preforms were heat treated and produced using varying deposition parameters, producing three coupons with two deposits each:

- 'A' (high speed, raster on); layer height 3.2–3.4 mm
- 'B' (high speed, raster off); layer height 3.8–3.4 mm
- 'C' (low speed, raster on); layer height 3.6–3.7 mm

The density of the forged targets is compared to the density of each type of AM target, with results shown in Table 18.1. The AM targets exhibit lower density than the forged targets, with AM 'C' having the lowest density. As mentioned, the AM material may have voids that are 0.5 mm (0.020″) in diameter. These results indicate some level of porosity, which leads to reduced strength and different dynamic behavior.

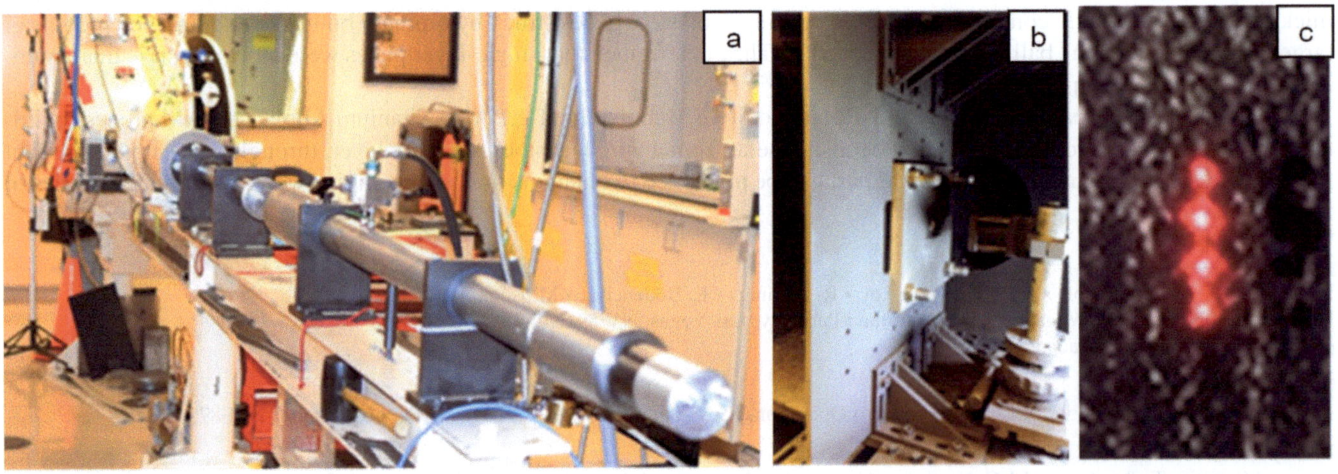

Fig. 18.1 (**a**) Two-stage light gas gun at UNLV, (**b**) Target chamber assembly, (**c**) Typical 4-probe PDV arrangement in the target chamber

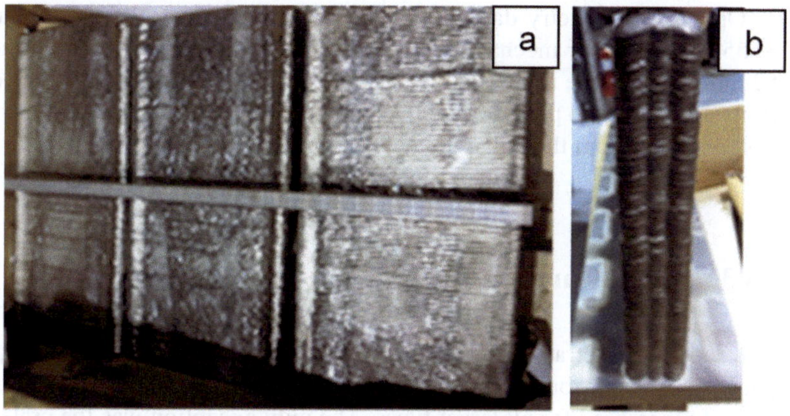

Fig. 18.2 Electron beam additive manufactured (EBAM) Titanium alloy test material, (**a**) Front view, (**b**) Side view

Table 18.1 Density of forged vs. AM targets

	Forged Ti	AM 'A'	AM 'B'	AM 'C'
Density (g/cm^3)	4.368	4.363	4.318	4.178
Percentage difference (%)	–	0.124	1.16	4.36

Table 18.2 Target descriptions and velocity ranges

Test #	Target plates	Impact velocity (km/s)	Probe locations
1	Forged Ti, single plate (12.7 mm)	4.838	1 mm–1 mm–1 mm
2	Forged Ti, single plate (12.7 mm)	5.202	1 mm–1 mm–1 mm
3	Forged Ti, single plate (12.7 mm)	N/A	1 mm–1 mm–1 mm
4	Forged Ti, single plate (12.7 mm)	5.655	1 mm–1 mm–1 mm
5	AM Ti 'A' (12.7 mm)	5.175	1 mm–1 mm–1 mm
6	Forged Ti, four plates (3.2 mm, each)	5.395	1 mm–1 mm–1 mm
7	Forged Ti, four plates (3.2 mm, each)	5.603	1 mm–1 mm–1 mm
8	AM Ti 'A' (12.7 mm)	5.552	1 mm–1 mm–1 mm
9	Am Ti 'B' (12.7 mm)	5.613	1 mm–1 mm–1 mm
10	AM Ti 'C' (12.7 mm)	5.634	1 mm–1 mm–1 mm
11	Forged Ti, single plate (12.7 mm)	6.145	2 mm–1 mm–3 mm
12	AM Ti 'A' (12.7 mm)	5.976	2 mm–2 mm–4 mm
13	AM Ti 'B' (12.7 mm)	6.08	2 mm–2 mm–4 mm
14	AM Ti 'C' (12.7 mm)	5.907	2 mm–2 mm–4 mm
15	Forged Ti, four plates (3.2 mm)	6.773	2 mm–2 mm–4 mm
16	Forged Ti, single plate (12.7 mm)	6.612	2 mm–2 mm–4 mm
17	Forged Ti, four plates (3.2 mm)	6.158	2 mm–2 mm–4 mm
18	AM Ti 'B' (12.7 mm)	5.255	2 mm–2 mm–4 mm
19	AM Ti 'C' (12.7 mm)	5.14	2 mm–2 mm–4 mm
20	Forged Ti, single plate (12.7 mm)	6.597	2 mm–2 mm–4 mm

18.2.3 Testing

Targets were cut from the AM performs using water jet techniques and then machined to final dimensions. The experiments were designed to study the plastic deformation during the impact process. Table 18.2 shows the test matrix. A range of projectile velocities were chosen so that deformation varied from a barely visible bulge on the back surface to a nearly complete penetration. A wide range of velocities were used since initial experiments were performed, starting with relatively low projectile velocities and gradually increasing the velocity until inchoate penetration was achieved.

Optical velocimetry data was taken from the back of the target plates with a four-channel PDV system. A 50.8 mm × 558.8 mm linear mechanical transfer (MT) array was mounted in the target chamber with a distance of approximately 52 mm between the array and the back of each test plate. The spacing between data collection points could be manipulated from shot to shot to provide better coverage of the impact zone of the target plates. The spacing was modified to achieve the best readings from the bulge on the back of the target plate.

18.3 Results and Discussion

The depth of penetration and the shape of the plastically deformed regions were quantified by measuring the crater diameter, penetration depth bulge, and back surface (Table 18.3). These physical measurements followed an increasing trend as velocity increased for all target types. The only exception was the crater diameter for the four plates of 3.2 mm, which did not follow any specific trend among the various velocities. This was perhaps due to the first plate undergoing complete penetration and vibration between the plates.

Figures 18.3 and 18.4 show the cross-sections of typical forged 12.7 mm Ti and AM 'A' targets, respectively, to illustrate the differences in response of these two target types. In both cases, as impact velocity increases, internal fracturing occurs, leading to spalling at the higher impact velocities. It is noted that spalling became evident at Fig. 18.3d with a velocity of

Table 18.3 Deformation results single-layer plates

Test #	Crater diameter (mm)	Penetration depth (mm)	Back surface bulge (mm)
1	17 ± 1	5.8 ± 0.1	N/A
2	18 ± 1	6.3 ± 0.1	0.3 ± 0.1
3	19 ± 1	7.2 ± 0.1	1.0 ± 0.1
4	20 ± 1	N/A	1.0 ± 0.1
5	20 × 21 ± 1	6.1 ± 0.1	0.9 ± 0.1
6,a	22 ± 1	N/A	N/A
6,b	26 ± 1	4.1 ± 0.2	3.0 ± 0.1
6,c	27 ± 1	3.3 ± 0.2	2.4 ± 0.1
6,d	27 ± 1	2.2 ± 0.2	1.7 ± 0.2
7,a	25 ± 1	N/A	N/A
7,b	19 ± 1	4.3 ± 0.2	3.2 ± 0.1
7,c	17 ± 1	2.7 ± 0.2	7.0 ± 0.1
7,d	27 ± 1	2.4 ± 0.1	2.2 ± 0.1
8	20 × 21 ± 1	8.6 ± 0 0.1	1.0 ± 0.1
9	20 × 21 ± 1	5.9 ± 0.2	2.1 ± 0.2
10	20 ± 1	6.3 ± 0.3	3.2 ± 0.1
11	21 ± 1	6.4 ± 0.1	1.2 ± 0.1
12	21 × 23 ± 1	8.1 ± 0.2	4.5 ± 0.1
13	21 × 22 ± 1	8.2 ± 0.1	5.2 ± 0.1
14	20 × 23 ± 1	8.0 ± 0.1	4.9 ± 0.1
15,a	22 ± 1	N/A	N/A
15,b	22 ± 1	6.8 ± 0.3	5.6 ± 0.2
15,c	25 ± 1	5.5 ± 0.1	4.4 ± 0.1
15,d	23 ± 1	4.9 ± 0.1	3.4 ± 0.2
16	23 ± 1	7.4 ± 0.1	4.2 ± 0.1
17,a	22 ± 1	N/A	N/A
17,b	23 ± 1	5.7 ± 0.1	4.3 ± 0.1
17,c	23 ± 1	4.4 ± 0.1	3.2 ± 0.1
17,d	25 ± 1	3.7 ± 0.4	3.1 ± 0.1
18	19 × 20 ± 1	5.9 ± 0.1	1.2 ± 0.1
19	18 × 20 ± 1	6.0 ± 0.1	1.0 ± 0.2
20	22 ± 0.2	7.8 ± 0.2	N/A

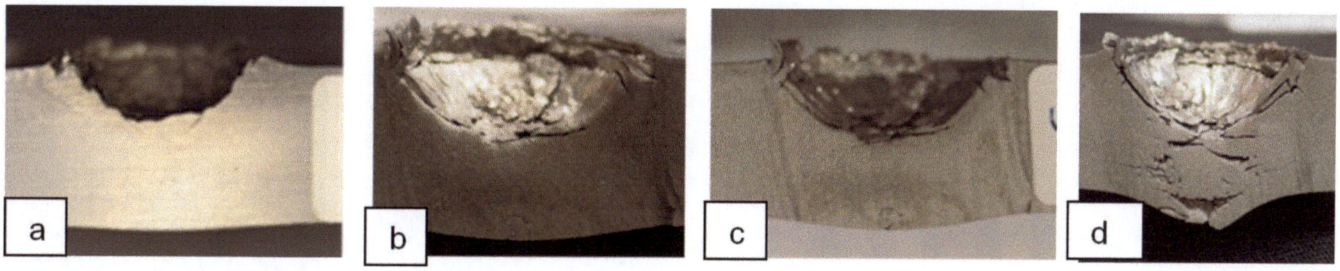

Fig. 18.3 Deformation vs. impact velocity for forged 12.7 mm Ti plate at velocities of (**a**) 4.838 km/s, (**b**) 5.655 km/s, (**c**) 6.145 km/s, and (**d**) 6.612 km/s

Fig. 18.4 Deformation vs. impact velocity for AM 'A' 12.7 mm Ti plate at velocities of (**a**) 5.175 km/s, (**b**) 5.552 km/s, and (**c**) 5.976 km/s

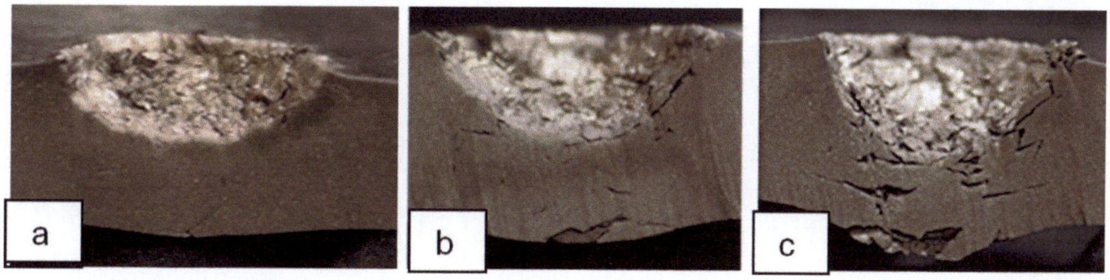

Fig. 18.5 Deformation vs. impact velocity for AM 'B' 12.7 mm Ti plate at velocities of (**a**) 5.255 km/s, (**b**) 5.613 km/s, and (**c**) 6.08 km/s

Fig. 18.6 Deformation vs. impact velocity for AM 'B' 12.7 mm Ti plate at velocities of (**a**) 5.14 km/s, (**b**) 5.634 km/s, and (**c**) 5.907 km/s

6.612 km/s in the case of the forged material. On the other hand, AM material exhibits spalling at lower velocities as shown in Fig. 18.4c at 5.976 km/s, Fig. 18.5c at 6.08 km/s, and Fig. 18.6c at 5.907 km/s. This difference in behavior may be explained by the lower density of the AM material (Table 18.1).

Additionally, the discontinuities in crystal structure due to the continuous process of melting and solidification may affect the performance of the AM materials, even after undergoing heat treatment. It is also observed that the AM plates under higher velocity impacts exhibit non-uniform crack propagation throughout the material along with brittle and fragmentation failure mechanisms that are occurring simultaneously, Fig. 18.4c. The same figure shows that the material is exhibiting shear cracking. Figure 18.7 reveals a plugging failure mode which did not occur in any other specimen at equal velocities.

Fig. 18.7 AM 'A' shows severe shear failure at 5.976 km/s

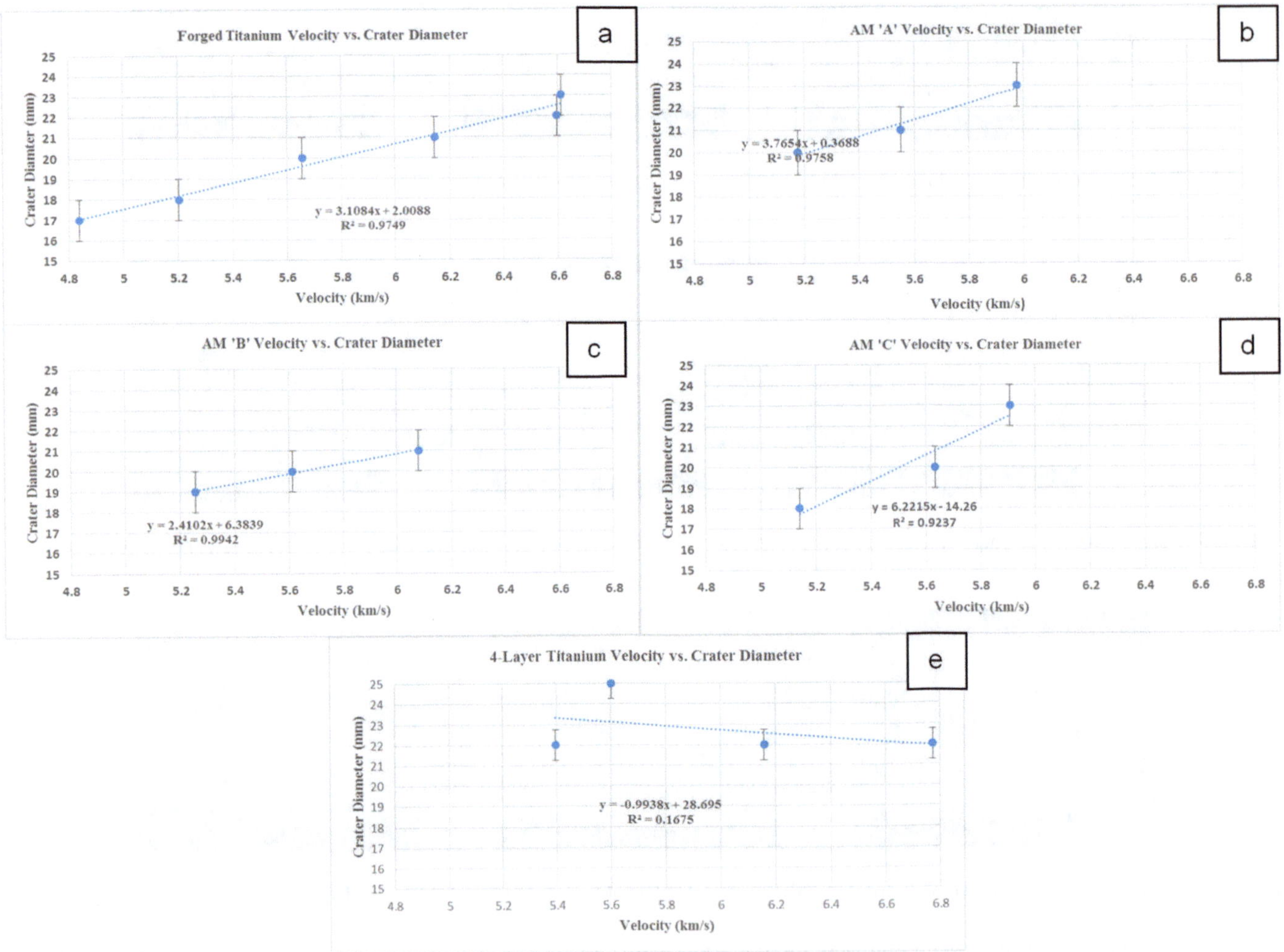

Fig. 18.8 Velocity vs. crater diameter trends (**a**) Forged Ti, (**b**) AM 'A' Ti, (**c**) AM 'B' Ti, (**d**) AM 'C' Ti, (**e**) 4-layer

Figure 18.8 shows the regression for the single forged and AM plates of velocity versus crater diameter. The results show a linear relationship between permanent deformation and impact velocity. There was not a similar trend in the 4-layer plate configuration.

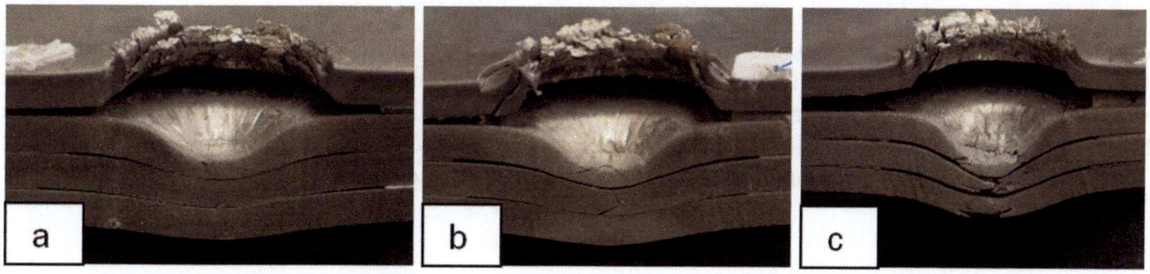

Fig. 18.9 Deformation vs. impact velocity for 4-layer plates at velocities of (**a**) 5.1 km/s, (**b**) 5.603 km/s, (**c**) 6.773 km/s

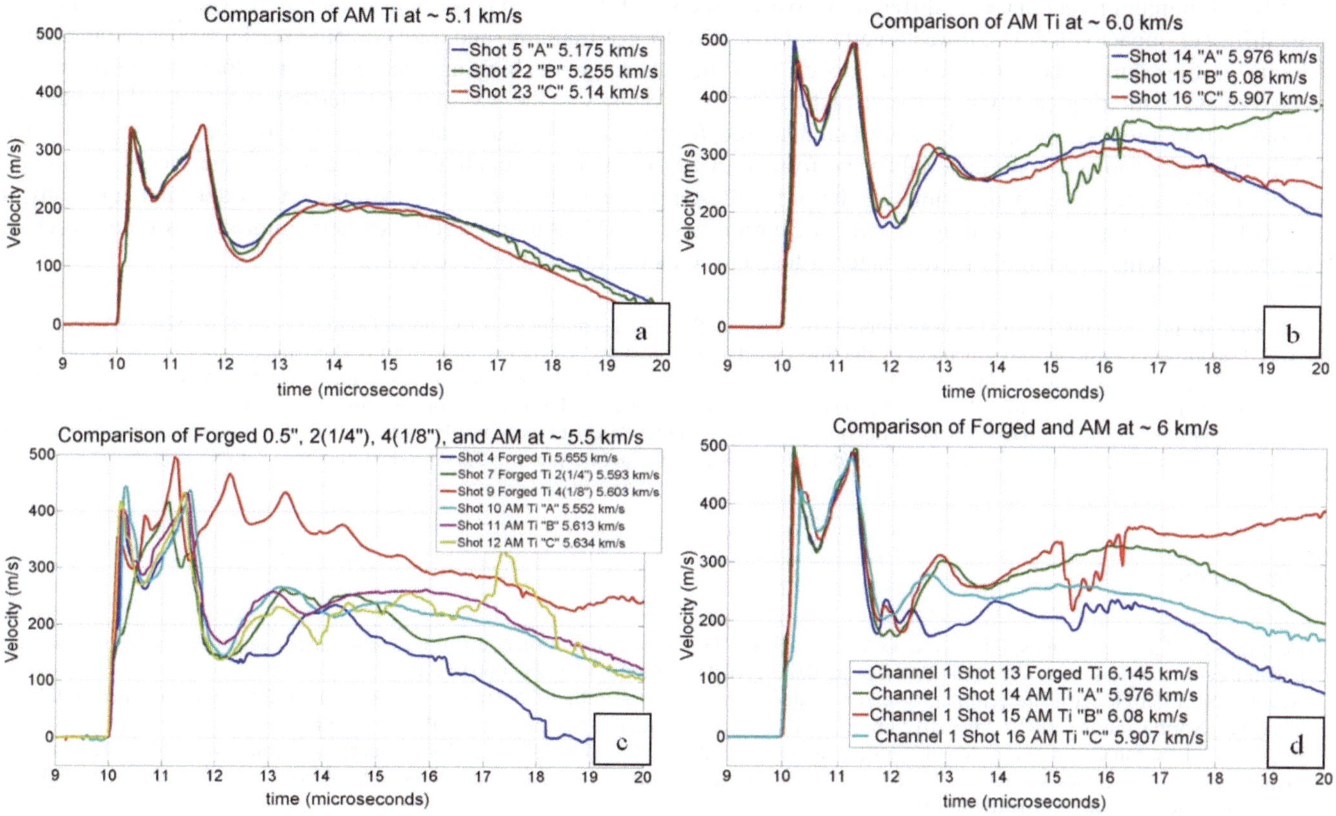

Fig. 18.10 Velocity traces of (**a**) AM Titanium at v = ~5.1 km/s, and (**b**) AM Titanium at v = ~6.0 km/s (**c**) forged and stacked Titanium, (**d**) forged and AM Titanium

It is noted that the 4-layer forged titanium alloy experienced first-layer penetration at all tested velocities. The first-layer also experiences material pull back causing the first-layer and second-layer to deform in different directions, which creates a gap between the first- and second-layer, as shown in Fig. 18.9. It was also found that the performance of the in-contact multi-layered plates experienced more damage, than that of a single-layer plate of equal thickness. Additionally, the AM material showed more signs of failure at slower velocities than that of the multi-layered plates.

Figure 18.10 shows velocity traces of some selected tests. Typically, the probe closest to the impact center has the most displacement and will show the maximum velocity profile. All velocity traces shown are from channel one, representative of the maximum velocity profile of that specific target. Shown in Fig 18.10a, b are the velocity profiles of the three different types of AM materials at two different velocities, 5.190 ± 0.059 and 5.990 ± 0.087 km/s. The figures show that the velocity profiles are similar for all types of AM materials at these velocities. Spalling occurred for all three AM types at the 5.990 km/s velocity, with Type 'C' spalling more than the others, possibly due to its lower density.

Figure 18.10c shows a velocity profile for each of the six types of targets tested. At a velocity of 5.608 ± 0.035 km/s the 12.7 mm forged and AM target profiles are similar and both show no or little spall. The stacked targets vibrate more, especially the four stacked plates, which exhibit a significantly different profile, as shown by the red line. Figure 18.10d compares the 12.7 mm forged and AM targets at a velocity of 6.027 ± 0.106 km/s. All four profiles are similar, especially during the early times of the impact. As the impact progresses, the AM targets begin to spall, however the forged targets do not. No similarities between AM plates and multi-layer stacked plates could be verified.

18.4 Conclusion

The experiments were successfully completed to explore the dynamic properties of forged titanium, stacked forged titanium, and EBAM titanium target plates of different deposition rates. Differences in deformation and failure were documented for three different impact velocities. It was ascertained that AM, forged titanium, and multi-layered stacks produce comparable velocity profiles during the early stage of impact, with the AM targets exhibiting spall at lower velocities and the multi-layered stacks exhibiting vibrations between plates. The lower density of the AM material, due to porosity developed during the fabrication process, most likely contributed to the formation of fractures at lower velocities. It is suggested that additional microscopy of target materials be performed to assist with understanding differences in failure modes. Computational models can be developed to simulate the projectile-target interaction to verify the material models and equations of state (EOS). A better understanding of dynamic properties of AM titanium alloy can help improve different types of engineering systems by minimizing the material lost and reducing the cost of production.

Acknowledgments This manuscript has been authored by National Security Technologies, LLC, under Contract No. DE AC52-06NA25946 with the U.S. Department of Energy and supported by the Site-Directed Research and Development Program. The United States Government retains and the publisher, by accepting the article for publication, acknowledges that the United States Government retains a non-exclusive, paid-up, irrevocable, world-wide license to publish or reproduce the published form of this manuscript, or allow others to do so, for United States Government purposes. The U.S. Department of Energy will provide public access to these results of federally sponsored research in accordance with the DOE Public Access Plan (http://energy.gov/downloads/doe-public-access-plan). NSTec product number DOE/NV/25946-2751.

References

1. Henriques, V.A.R., de Campos, P.P., Cairo, C.A.A., Bressiani, J.C.: Production of titanium alloys for advanced aerospace systems by powder metallurgy. Mater. Res. **8**(4), 443–446 (2005). http://doi.org/10.1590/S1516-14392005000400015
2. Needler, S.: F-35 Direct manufacturing: Material qualification results, AeroMat Conference and Exposition 2012, presentation (WDJ1.1), Charlotte, North Carolina, 18–21 June 2012 https://asm.confex.com/asm/aero12/webprogram/Paper30786.html
3. Phelps, H.: Electron Beam Direct Manufacturing (EBDM) on the F-35 Lightning II, Lockheed martin presentation at NC state advanced manufacturing & logistics symposium, Raleigh NC, 17 Oct 2013 http://camal.ncsu.edu/wp-content/uploads/2013/10/Hank-Phelps-EBDM-Presentation-to-NC-State-Final-RevA.pdf
4. Brandl, E., Palm, F., Michailov, V., Viehweger, B., Leyens, C.: Mechanical properties of additive manufactured titanium (Ti–6Al–4V) blocks deposited by a solid-state laser and wire. Mater. Des. **32**(10), 4665–4675 (2011)
5. Refael, F., Daniel, R., Amnon, S.: Dynamic mechanical behavior of additively manufactured Ti6Al4V with controlled voids. J. Appl. Mech. **82** (4), 041004 (2015). doi:10.1115/1.4029745. Paper No: JAM-14-1539
6. Radin, J., Goldsmith, W.: Normal projectile penetration and perforation of layered targets. Int. J. Impact Eng. **7**(2), 229–259 (1988)
7. Almohandes, A.A., Abdel-Kader, M.S., Eleiche, A.M.: Experimental investigation of the ballistic of steel-fiberglass reinforced polyester laminated plates. Compos. Part B Eng. **27**, 447–458 (1996)
8. Strand, O.T., Goosman, D.R., Martinez, C., Whitworth, T.L., Kuhlow, W.W.: Compact system for high-speed velocimetry using heterodyne techniques. Rev. Scic. Instrum. **77**(8), 083108 (2006). doi:10.1063/1.2336749

Chapter 19
Instrumented Penetration of Metal Alloys During High-Velocity Impacts

P. Jannotti, B. Schuster, R. Doney, T. Walter, and D. Andrews

Abstract A methodology is presented for characterizing the failure behavior of metallic targets due to high-velocity and hypervelocity impacts. Time-resolved sub-scale terminal ballistic experiments were performed at approximately 1.2 km/s to assess the feasibility of using high-speed optical imaging, photon Doppler velocimetry, and high-speed 3D digital image correlation for measuring back face deformation. Spherical copper impactors were fired into aluminum alloy targets with thickness equal to one half the impactor diameter. The approach has implications for determining the susceptibility of metallic targets to different failure modes including bulk plastic deformation resulting in tensile failure, cratering, plugging, spallation and adiabatic shear band formation. Results will be used to assist in validation of large-scale computational models used to model ballistic impact.

Keywords Impact • Deformation • High-speed imaging • DIC • PDV

19.1 Introduction

At present, problems exist both in the efficacy of full-scale experimentation as well as the ability of computational models to accurately capture the complex physics during impact events. This emphasizes the need for high-fidelity, data rich sub-scale experimentation which aims to not only identify and characterize the time-dependent failure of materials, but also to verify and validate the relevant physics for computational models. To this end, real-time experimental analysis is essential to the study of high rate ballistic phenomena.

In recent years, popular experimental techniques to capture real-time deformation behavior of ballistic events include high-speed imaging, digital image correlation (DIC) and photon Doppler velocimetry (PDV). Digital image correlation, specifically 3D DIC, enables the user to obtain full-field out-of-plane displacement and velocity fields through pattern-based optical tracking algorithms [1]. Sample preparation consists merely of apply a high contrast pattern to the sample surface. Thus, with the proliferation of ultrahigh speed cameras offering high temporal and spatial resolution as well as relatively long record times (tens to hundreds of microseconds), it is a natural extension for 3D DIC to be used in ballistic testing. Photon Doppler velocimetry, on the other hand, operates on the principle of Doppler shifts, measuring the velocity of a surface with reflected laser light [2]. PDV can sample at the nanosecond level over very long record length (hundreds of microseconds), and is able to measure velocities extending into the tens of km/s range. The use of PDV for rear surface velocity measurements, therefore, offers a wealth of supplementary information to high-speed imaging-based techniques such as accurate time of impact and breakout, evidence of fracture, plugging, or shear banding, and residual projectile velocities.

The current study presents time-resolved sub-scale terminal ballistic experiments that will be used to assess the susceptibility of metal alloys to different failure modes. A methodology is presented for analyzing the damage evolution *in situ* using a combination of high-speed optical imaging, high-speed 3-D digital image correlation (DIC), and photon Doppler velocimetry (PDV). The results of this approach will provide an in-depth examination of operative failure mechanisms and provide model validation for hydrocodes used to simulate the impact events. The intent of this work is to conduct such experiments at ordinance and hypervelocities on a range of metal alloys, but preliminary feasibility studies were performed at 1.2 km/s on aluminum alloy targets.

P. Jannotti (✉) • B. Schuster • R. Doney • T. Walter • D. Andrews
U.S. Army Research Laboratory, Aberdeen Proving Ground, Aberdeen, MD 21005, USA
e-mail: Phillip.A.Jannotti.ctr@mail.mil

D. Casem et al. (eds.), *Dynamic Behavior of Materials, Volume 1*, Conference Proceedings of the Society for Experimental Mechanics Series, DOI 10.1007/978-3-319-41132-3_19

19.2 Experimental

The impact experiments were conducted using spherical copper (C101; Grade 200) projectiles of 0.5″ diameter at 1.2 km/s. Copper was chosen due to its widely characterized and well-known properties and deformation behavior. The projectiles were fired from a 0.50-caliber smooth-bore powder gun using a two-piece discarding sabot with obturator. Projectile striking velocity (V_S) was measured using two make-screens set at a predetermined distance. The impact occurred at normal incidence on metal target plates (6″ × 6″) of aluminum alloy of thickness 0.25″. Relevant physical, mechanical, and shock properties for the projectile and target material are listed in Table 19.1.

In situ diagnostic techniques employed to characterize the time-dependent failure behavior included high-speed imaging, DIC, and PDV. A make screen placed directly in front of the impact surface of the target provided a common trigger signal for the high-speed cameras, illumination, and PDV. The make screen was connected directly into the camera(s), and contact of the metallic (i.e. conductive) penetrator on the make screen triggered the camera. The cameras then generated a 5 V TLL pulse out of the auxiliary ports to the light strobes and PDV system.

For high-speed imaging and DIC, high-speed cameras (Specialized Imaging Kirana) were used to capture the damage sequence at 2 million frames per second (fps) with an interframe time of 0.5 microseconds. A series of tests were performed using only one camera oriented edge-on with the target to capture the back face damage evolution by obtaining high-speed shadowgraphs (see Fig. 19.1a). Here, 2D image analysis was employed to track the back face deflection profile and velocity histories using edge detection and curve fitting algorithms. Another series of tests employed two cameras in a stereographic configuration to acquire images of "speckled" targets for 3D DIC analysis. For 3D DIC, the two cameras were positioned 1275 mm from the target and spaced 725 mm from one another in order to view the target at a relative (perspective) angle of 30° (see Fig. 19.1b). A first surface mirror was placed behind the target at 45° to achieve the desired view of the rear face of the target. The images were then processed using VIC-3D™ (Correlated Solutions) to obtain a 3D measurement and visualization of the time-resolved deformation process. Illumination for the high-speed imaging and DIC was provided by two light strobes (Photogenic 2500DR-UV Powerlight).

In each of the tests where edge-on high speed imaging was employed, PDV was used concurrently to provide time-resolved rear surface velocity measurements. The PDV system used in the current study was described by Ostrand et al.[2]. Each PDV probe consisted of a GRIN lens attached to a fiber pigtail (AC Photonics 1CL15A070LSD01); collimated light over a 300 mm working distance to a spot size of ~0.8 mm. Probes were aligned parallel to the shot line using a polycarbonate fixture. This facilitated measurement of velocities at several positions on the rear surface of the target sample, as shown in Fig 19.1a. Velocity records were measured directly on the shot line and at several distances away from the shot line. The PDV data was recorded using an effective bandwidth of 16 GHz, a sampling rate of up to 80 Gigasamples/s, and a maximum of 2 Gigapoints (Keysight DSOV174A Infiniium V-Series oscilloscope). The data was then reduced using proprietary PDV post-processing software developed by W.K. Bowman and D.S. Lowry to compute displacement- and velocity-time pairs for each channel.

Table 19.1 Properties of the projectile and target materials (Matweb.com)

Material	Density (g/cm^3)	Elastic modulus, E (GPa)	Yield strength, σ_y (MPa)	Hardness, H (GPa)	Poisson's ratio, ν	Shock impedance, Z	Zplate/ Zprojectile
Copper (C101)	8.89–8.94	115	69–365	0.75–0.9	0.31	32.0	–
Aluminum (6061-T6)	2.70	69	276	1.07	0.33	13.6	0.43

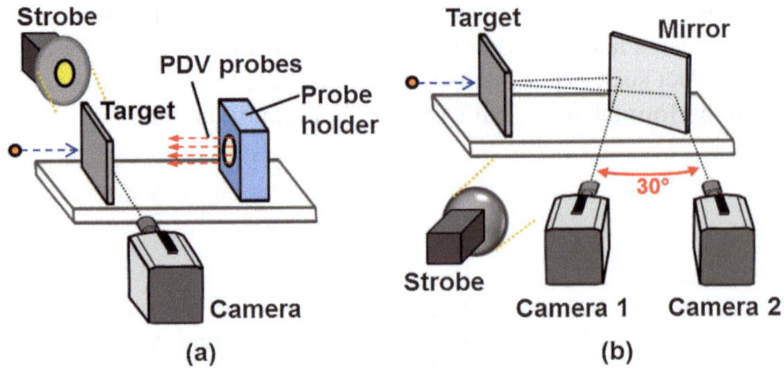

Fig. 19.1 Schematic of experimental test setup: (**a**) Edge-on (2D) image analysis and PDV, (**b**) 3D DIC

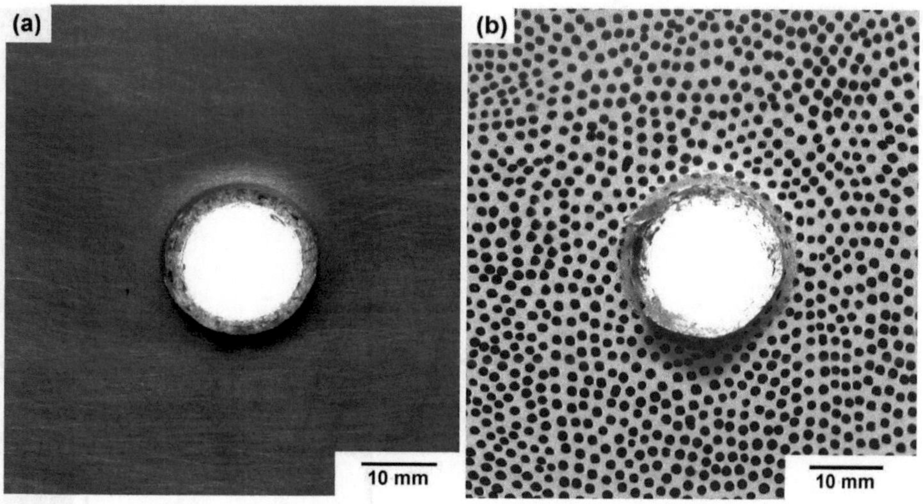

Fig. 19.2 Post-mortem back face images of (**a**) bare and (**b**) "speckled" targets which have been impacted by 0.5″ copper spheres to demonstrate the affected area

19.3 Results and Discussion

To validate the technique, aluminum alloy targets were impacted at ~1.2 km/s and a comparison of the resultant data obtained from high-speed image analysis, 3D DIC, PDV was made. Figure 19.2 illustrates the damage that occurred due to copper impacts at more than 1 km/s. Note that the through hole generated (0.60–0.75″) was slightly larger than the impactor diameter of 0.5″, and takes place in approximately 10–15 microseconds. Thus, the evolved deformation represents a highly localized damage region.

19.3.1 High-Speed Imaging

Figure 19.3 shows high-speed shadowgraph images that were captured during the impact as well as images converted into outlines using edge-detection algorithms. The images show the respective times after damage begins to evolve on the back surface. The images reveal that a crater is formed in the target plate due to plastic flow (to accommodate the volume of the striking projectile), creating a bulge on the rear face of the plate. The high shear stresses imposed on the target material just ahead of the projectile generate a plug. Failure in the metal target occurs within ~ 15 microseconds, and a plug is ejected from the rear of the plate. The images also show the deformed shape of the projectile long after impact, which becomes semi-circular and has a residual velocity of more than 800–900 m/s.

19.3.2 Edge-On (2D) Image Analysis

Using an edge-on shadowgraph imaging technique, the images were processed using MATLAB to compute the back face deflection and velocity fields (see Fig. 19.4a,b). Note the bulge symmetry, which is expected for a "reasonably" isotropic material. Also, notice that the velocity of the bulge is approximately constant at all times during the damage process. From the data, the maximum values of deflection and velocity during the deformation process were determined and tracked to evaluate the displacement- and velocity-time histories at the shot line. Here, it was observed that the displacement history is approximately linear to almost 8 mm over 13 microseconds prior to failure and spalling. Similarly, a steady rise in the back face velocity is seen until a near constant velocity of 850–900 m/s is reached. Beyond 13 microseconds, the back face ruptures and a spall is ejected.

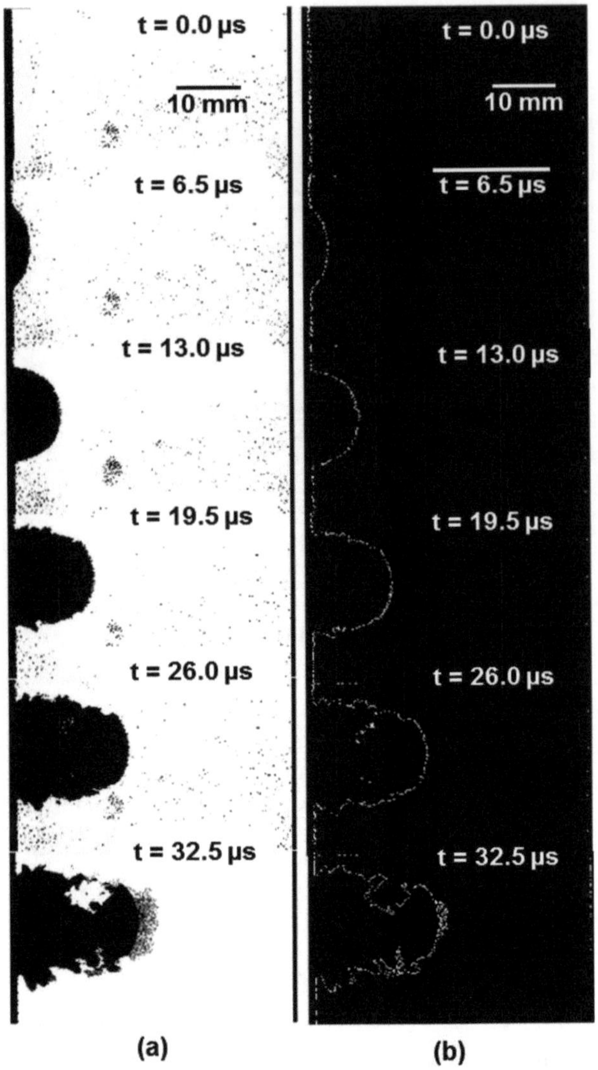

Fig. 19.3 High-speed images of the time-resolved damage evolution in 0.25″ aluminum alloy due to a 0.5″ copper sphere impact at ~1200 m/s: (**a**) shadowgraph images, (**b**) images processed using edge-detection

19.3.3 Digital Image Correlation (DIC)

Using two high-speed cameras in a stereographic arrangement, 3D digital image correlation was performed and compared to data determined using high-speed image analysis. The maximum displacement determine using DIC was ~9 mm (at around 15 microseconds), as seen in Fig. 19.5a,b. At 9 mm, the paint was no longer adhered to the metal plate back face and no correlations could be made. This represents a significant plastic deformation of the metal plate that can be captured using DIC prior to rupture. By tracking the center of the rear surface bulge (at the shot line), the deflection- and velocity-time histories were extracted. Similar to what was seen using edge-on image analysis, the back face deflection rose steadily to nearly 9 mm within 15 microseconds before back face rupture took place. Beyond this point, DIC correlations could not be made indicating that breakout of the back face had occurred. From the velocity-time history, it was discerned that the velocity of the bulge increased rapidly to 800 m/s before tapering off to ~900 m/s, within 15 microseconds. Interesting, after examining the time-resolved strain evolution computed from VIC-3D (not shown for brevity), it was found that the material fails where the tensile stress gradient is maximum, rather than at the point of maximum tensile stress.

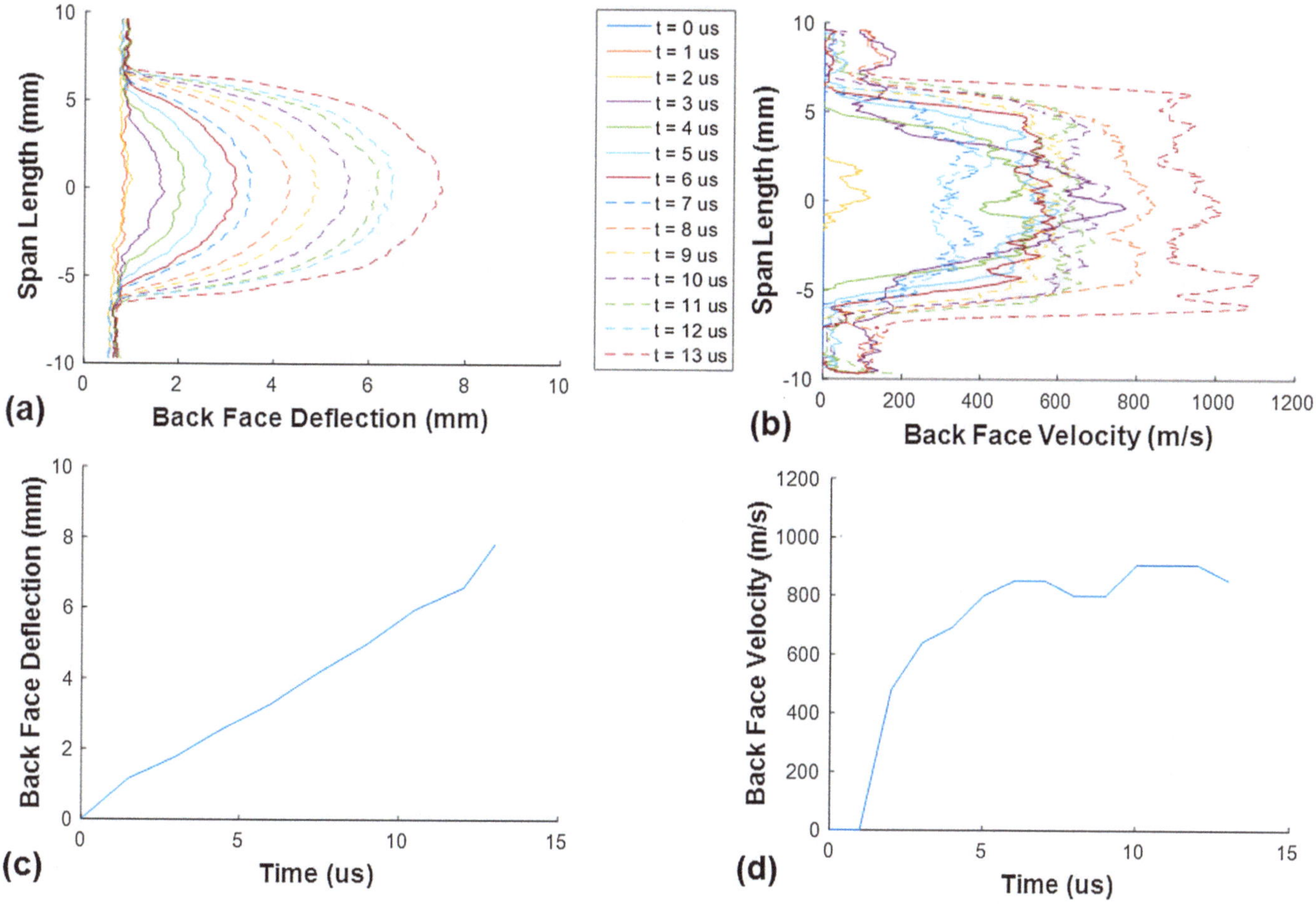

Fig. 19.4 Back face deformation data determined by 2D image analysis: Back face (**a**) deflection and (**b**) velocity profiles, and (**c**) deflection- and (**d**) velocity-time histories at the shot line (i.e., center of the bulge)

19.3.4 Phonon Doppler Velocimetry

PDV has recently become a mainstay of time-resolved data acquisition in shock physics and terminal ballistics. It is only fitting that it is used as a means of validation for the other two methodologies. Figure 19.6 shows the displacement- and velocity-time pairs. Note that one PDV probe was aligned along the shot line while another probe was aligned far away from the shot line to provide a baseline (control) for the other probe. The probe away from the shot line denoted a velocity of approximately zero as one would expect, and is not shown. The arrow in each plot designates the approximate time of rear face failure (as determined from the high-speed images), where a plug is ejected from the back face of the metal target plate. In Fig. 19.6a, the plate back face reaches nearly 850 m/s prior to failure. Beyond that point, the plug continues to accelerate to 1050 m/s (equivalent to 77 % of the striking energy). Recall that the striking velocity of the copper projectile was ~1200 m/s. Thus, the metal target dissipates 23 % of the initial impact energy.

19.4 Discussion

The results indicate an agreement between the various techniques for measuring back face deflection-and velocity-time histories. Of the techniques, PDV is currently the most widely accepted method for obtaining back face displacement and velocity data. However, both edge-on shadowgraph imaging offer an ability to discern deflection and velocity profiles (2D) in addition to provide time histories at a given point (1D). Moreover, DIC offers 3D analysis opportunities, while edge-on imaging is limited to 2D data analysis. For the current case of aluminum alloy, a "reasonably" isotropic material, both methods seem aptly suited, but 3D DIC provides greater flexibility in examining a wider range of materials such as

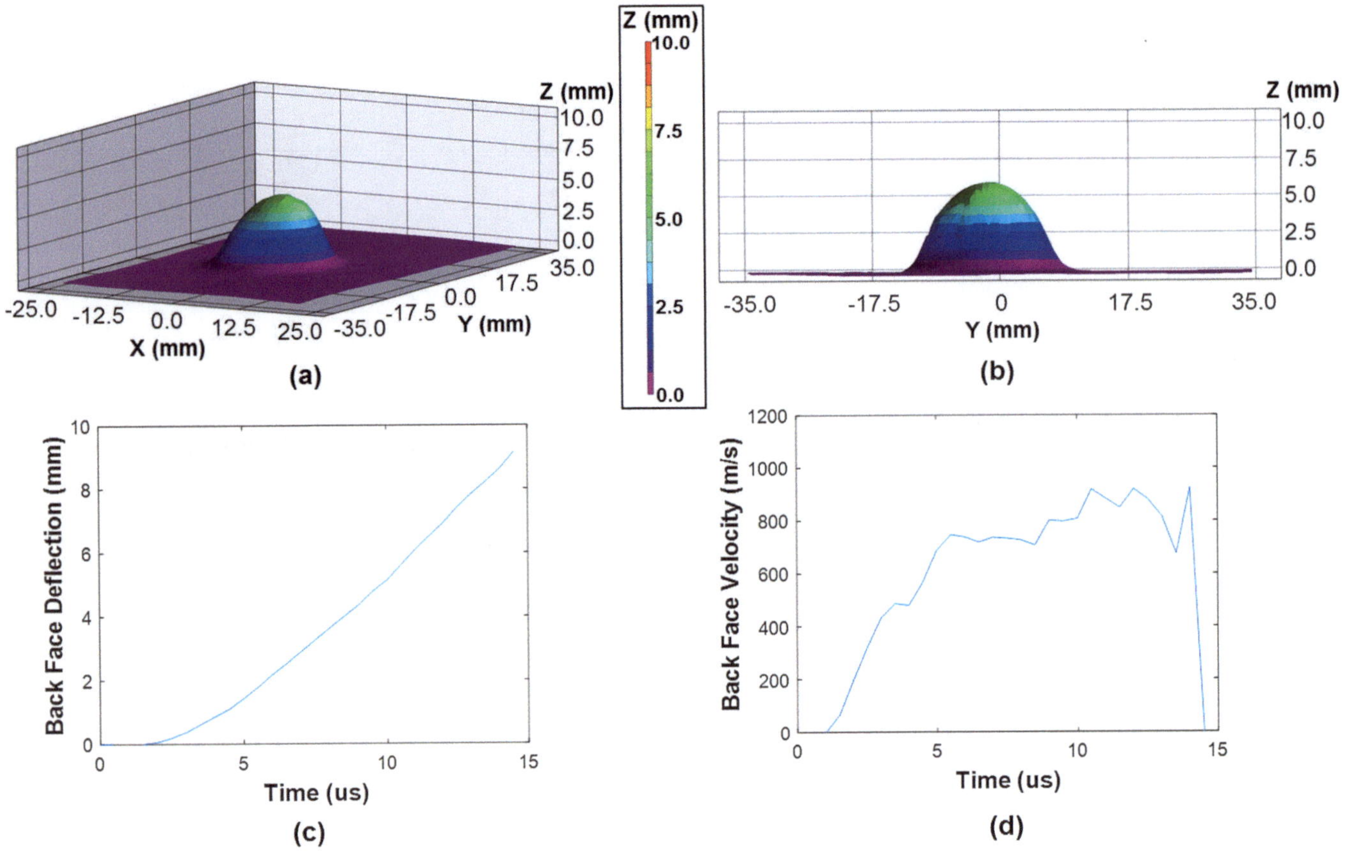

Fig. 19.5 Back face deformation data computed from 3D DIC: Deflection profiles as viewed from (**a**) an oblique angle and (**b**) edge-on as well as (**c**) deflection- and (**d**) velocity-time histories

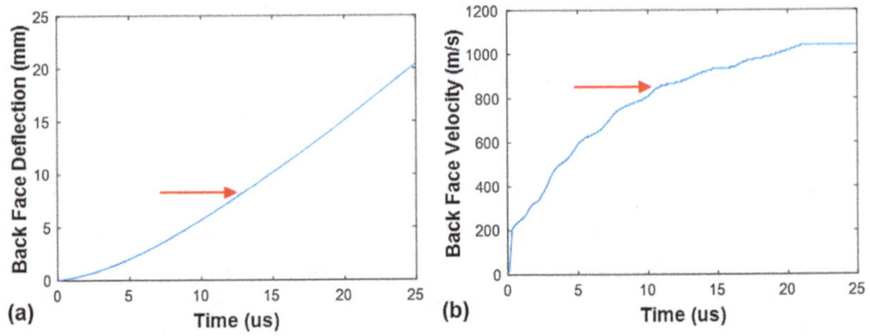

Fig. 19.6 Back face deformation data computed from PDV: (**a**) Deflection- and (**b**) Velocity-time pairs. The *red arrow* denotes the approximate time of failure of the metal plate

magnesium alloys which exhibit pronounced anisotropy and localization. The downside of DIC in general is that the quality of the results is firmly reliant on the quality of the calibration and speckle pattern. The primary issue presented by the calibration process is the time required to properly calibrate prior to each test, which is not a problem for edge-on imaging. Also, given the highly localized nature of the damage process the ability to accurately determine out-of-plane displacements becomes difficult, especially using cameras in a stereographic arrangement. For example, once the back face begins to bulge and the pattern becomes highly inclined to the original back face plane, it is difficult to maintain proper visual and contrast on the pattern as it tilts severely towards/away from the cameras. This can lead to difficulties with the cameras being able to clearly capture patterns on either side of the bulge. Another limitation includes decohesion of the paint from the back surface due to shock stresses. These issues, however, do not manifest until the metal plate bulges ~10 mm. Depending on the requirements of the materials being tested and the data required for model validation, DIC remains a prime candidate for tracking and characterizing back face deformation due to high-velocity ballistic impacts. If 3D analysis is not necessitated, edge-on (2D) image analysis technique captures the same basic information as DIC in addition to impact and residual velocities.

19.5 Conclusions

A methodology is presented for high-speed shadowgraph image analysis, 3D DIC, and PDV during terminal ballistic impacts. The approach shows great promise for being applied to hypervelocity ballistic impact problems in the 1.5–2.5 km/s regime. The general principles of the technique are easily scalable into the hypervelocity range, especially given that high-speed cameras are easily capable of attaining frame rates of 5–10 Mfps (0.1–0.2 microseconds interframe times). The experimental framework presented may be utilized to provide time-resolved measurements of both the displacement- and velocity-fields (2D and 3D), as well as displacement- and velocity-time histories at any point of interest. In future investigations, the approach will be refined to use PDV to also measure projectile deceleration on the impact side of the metal target, include a wider range of striking velocities from 1 to more than 2 km/s, investigate target thickness to projectile diameter ratios of 0.5–2, and investigate a wide range of plate materials.

References

1. Tiwari, V., Sutton, M.A., McNeill, S.R., Xu, S., Deng, X., Fourney, W.L., Bretall, D.: Application of 3D image correlation for full-field transient plate deformation measurements during blast loading. International Journal of Impact Engineering **36**(6), 862–874 (2009)
2. Strand, O.T., Goosman, D.R., Martinez, C., Whitworth, T.L., Kuhlow, W.W.: Compact system for high-speed velocimetry using heterodyne techniques. Review of Scientific Instruments **77**, 083108 (2006)

Chapter 20
Confined Underwater Implosions Using 3D Digital Image Correlation

Helio Matos, Sachin Gupta, James M. LeBlanc, and Arun Shukla

Abstract This study experimentally investigates fluid structure interactions occurring during confined implosions using high-speed digital image correlation (DIC). Aluminum tubular specimens are placed inside a confining cylindrical structure with one end open to a pressurized environment. These specimens are exposed to hydrostatic pressure, which is slowly increased until they collapse onto themselves. The implosion event is viewed through an acrylic window on the confining structure. Full field deformation and velocities are captured with DIC and are synchronized with the pressure history. Experiments show that implosion inside a confining structure leads to extremely high oscillating water hammer effects. Both peak structural velocities and hammer impulses increase linearly with increasing collapse pressure.

Keywords Fluid structure interaction • Implosion • Digital image correlation • Water hammer • Confined environment

20.1 Introduction

Structures will undergo external hydrostatic pressures when used in underwater applications. These external hydrostatic pressures can lead to implosions, which are highly violent in nature with resulting high-velocity water motions and strong shock waves [1]. Several investigations have been reported by researchers in naval and marine communities on the mechanics and fluid-structure interaction during a free-field implosion process [2–8].

Although the mechanics of implosion is well established for free-field implosions, the studies reporting implosions in confining environments are limited. Author's recent study on sympathetic implosion inside an open-ended confining tube shows that hammer waves can potentially damage even relatively stronger implodable volumes inside the confining tube. This study measured the dynamic pressure history inside the open ended confining tube to understand the evolution of water hammer waves at the onset of implosion [9]. Any changes in the design of the structure will alter the fluid structure interaction process and thus the strength of hammer waves.

There is a need to understand the evolution of these waves in order to predict the peak strength and total impulse of these detrimental water waves. No studies currently reports on relating the generation of water hammer waves with the measured structural deformations of the implodable volume. Hence, this paper illustrates this research gap by investigating the underwater implosion mechanics in the open-ended confining tube using 3-D DIC.

20.2 Experimental Setup

20.2.1 Implodable Volume and Open-Ended Confining Tube

The implosion experiments are conducted inside a pressure vessel facility. The implodable volumes are made out of 6061-T6 seamless extruded tubing with a nominal outer diameter of 38.1 mm and an unsupported length of 254 mm. The wall

H. Matos (✉) • S. Gupta • A. Shukla
Dynamic Photo Mechanics Laboratory, Department of Mechanical, Industrial and Systems Engineering,
University of Rhode Island, Kingston, RI 02881, USA
e-mail: hmatos@uri.edu

J.M. LeBlanc
Naval Undersea Warfare Center (Division Newport), Newport, RI 02841, USA

D. Casem et al. (eds.), *Dynamic Behavior of Materials, Volume 1*, Conference Proceedings
of the Society for Experimental Mechanics Series, DOI 10.1007/978-3-319-41132-3_20

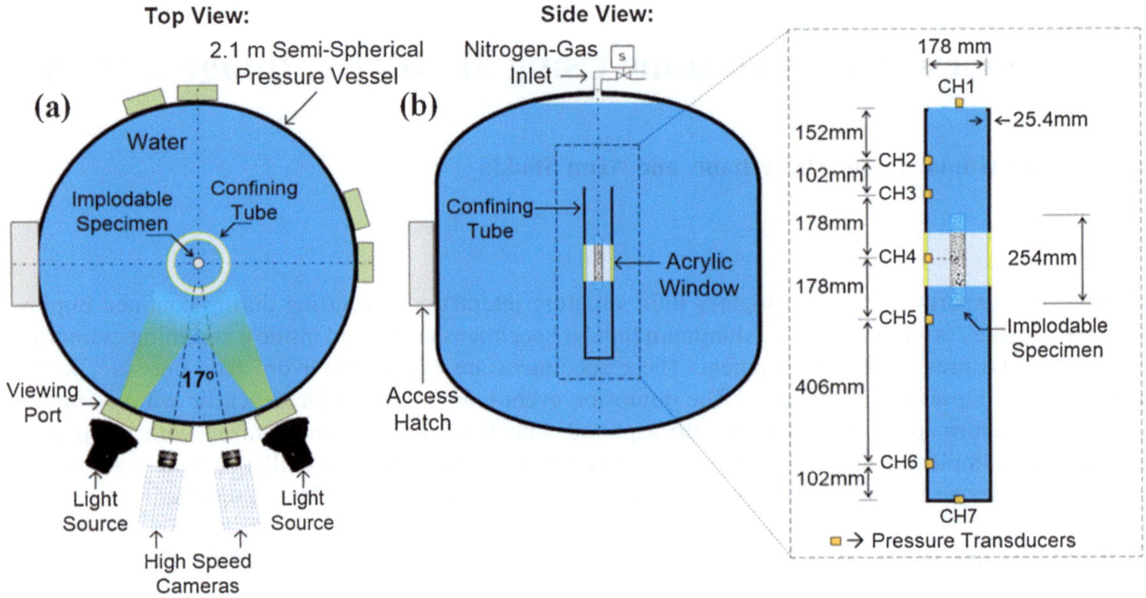

Fig. 20.1 Experimental setup viewed from the (**a**) top, (**b**) side and detailed side section

thickness of the implodable volume is 0.73 mm yielding a collapse pressure of 1.50 MPa. The implodable volumes are sealed on both ends using aluminum caps and O-rings for seals.

The implodable specimens are placed concentrically inside a confining tube with one open end which sits inside the underwater pressure vessel facility [5]. The schematic of pressure vessel facility with confining tube and implodable volume is shown in Fig. 20.1. The pressure vessel is pressurized using compressed nitrogen gas from the top of the chamber using a solenoid inlet valve.

The confining tube has an inner diameter of 178 mm, wall thickness of 25.4 mm, and length of 1.12 m. It is equipped with a 152 mm long cylindrical transparent acrylic optical window. The inner diameter and the wall thickness of cylindrical window section are identical as aluminum section to maintain constant cross-sectional area. The bottom end of the confining tube is closed with an end-plate and the top end is open to the high-pressure water environment of the pressure vessel, hence it is a semi-confining nature. Both high-frequency dynamic face pressure sensors (PCB-113B22) and tourmaline blast pressure sensors (PCB 138A05) are installed in the inner wall of the confining tube as shown in Fig. 20.1. The signal from the pressure sensors is recorded using a 200 kHz bandwidth recorder at a sampling rate of 2 MHz. The real time deformation of the implosion event is captured using a pair of Photron SA-1 high-speed cameras at 30,000 fps. A random intensity pattern is applied on the surface of the implodable volume using flat paint and the pattern is illuminated using a pair of high-intensity arc lamps [5].

20.2.2 Calibration of 3-D Digital Image Correlation Technique for Cylindrical Window

The calibration of extrinsic and intrinsic parameters by using a submerged calibration target can result in high accuracy in both in-plane and out-of-plane displacement measurements using 3-D DIC [5]. As the experimental setup used in this article contains an additional medium in the optical path of the cameras (i.e. cylindrical acrylic window as shown in Fig. 20.1), the technique proposed in [5] requires recalibration in order to estimate the accuracy of measured DIC in-plane/out-of-plane displacements for objects placed inside a cylindrical window. Therefore, two sets of calibration are conducted in this study. The first is an accuracy-calibration which is performed in a custom designed tank (Fig. 20.2) to evaluate the DIC accuracy. The second is an experimental-calibration which is performed in the experimental setup (Fig. 20.1) in order to obtain the relative camera parameters needed to run experiments.

The accuracy-calibration is performed, and the accuracy of the DIC technique can be evaluated as shown in Fig. 20.3. Then the experimental-calibration is performed, and the precision of measured DIC displacement is further estimated quantifying the radius of the implodable using the DIC technique inside the experimental setup (Fig. 20.1). The radius is

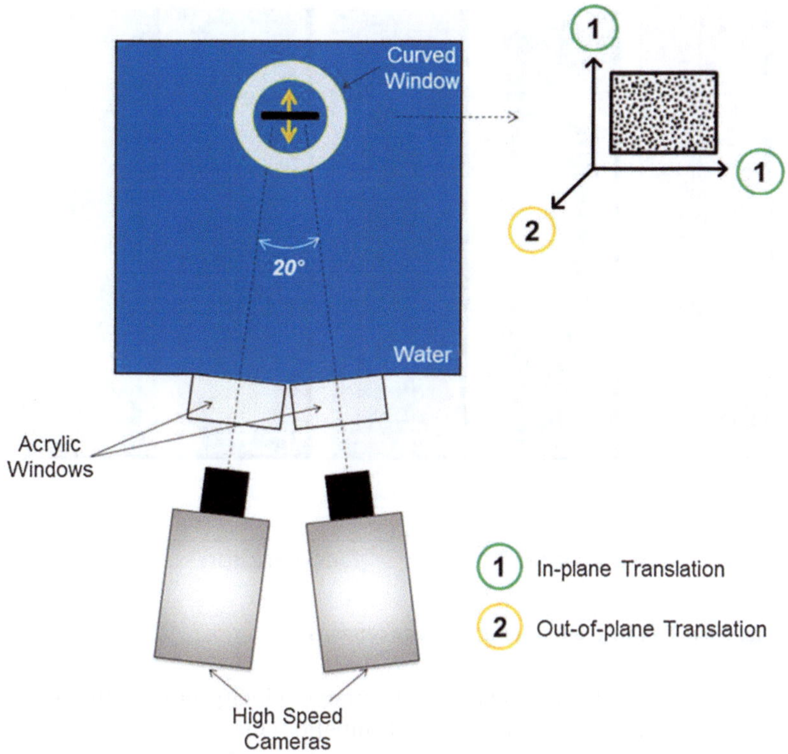

Fig. 20.2 Custom designed tank setup for underwater DIC calibration experiments

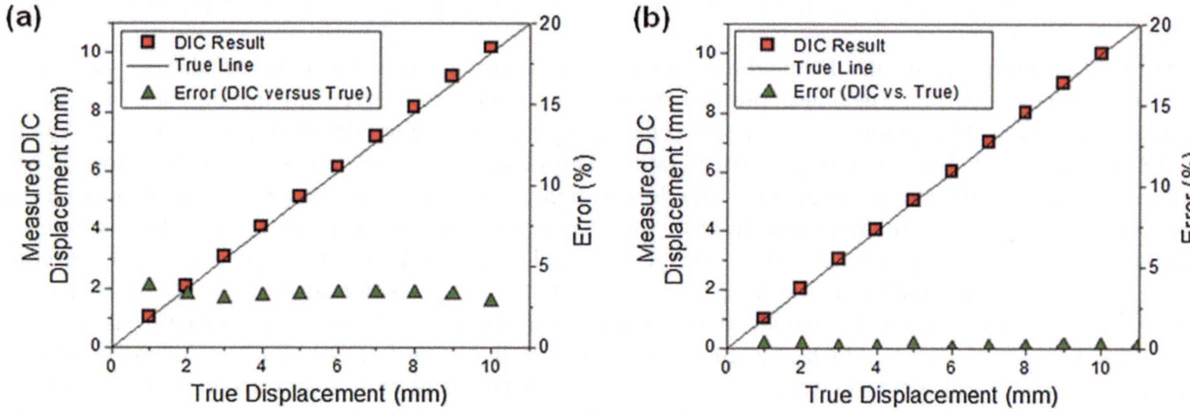

Fig. 20.3 Underwater DIC experimental-calibration (**a**) out-of-plane and (**b**) in-plane displacements

found to 19.07 mm ± 0.22 mm (with 95 % confidence interval). As the true radius of the specimen is 19.05 mm, the maximum deviation from true radius is found to be 2.25 %. Thus, it can be established that both the shape and deformation of submerged objects behind a cylindrical window can be measured accurately.

20.3 Experimental Results and Discussion

The DIC velocity contours are shown in Fig. 20.4. Time t = 0 ms indicates the time when the dynamic pressure at the nearest sensor drops to 99 % of the collapse pressure. The velocity contours in Fig. 20.4 show that the evolution of velocity is significantly different from a free-field implosion experiments. The implodable accelerates up to time t = 1.3 ms and reaches a peak center point velocity of 12 m/s. During this period, the velocity variation along the length is almost linear similar to earlier studies [10]. The deformation process after this instance differs from free-field situations. In a free field

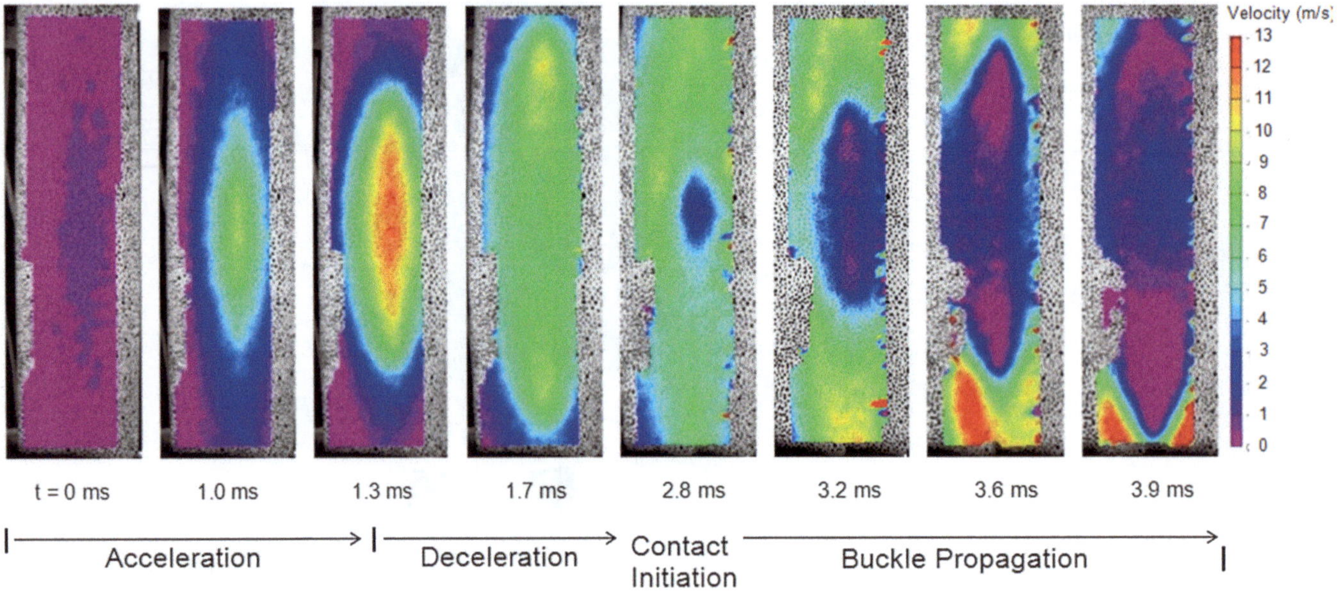

Fig. 20.4 Full-field out-of-plane velocity contours

case, the implodable reaches its maximum velocity just before contact initiation, thus, the kinetic energy of implodable reaches its maxima prior to contact initiation. On the contrary in the open tube confined implosion case, the implodable undergoes deceleration at the center for about 1.5 ms prior to contact. This phenomenon suggests that the resistance to deformation for implodable exceeds beyond the driving force applied by surrounding fluid's pressure in open tube confined implosion. In turn, the kinetic energy gained during acceleration phase is consumed in further plastically deforming the implodable.

The deformation profile along the longitudinal direction has been shown to be a linear/half-sine wave for mode-2 cylindrical geometry [10, 11]. The linear profiled deformation mode is observed till the instance of peak velocity. Subsequently, it is seen that the points away from the center along the longitudinal direction gain velocity as seen between t = 1.5–2.8 ms, while central region undergoes deceleration. This deformation behavior appears to be unique to open tube confined implosion. This indicates the transition of the longitudinal deformation mode from a lower order mode (half-sine wave) to a higher order mode (multiple sine waves). The primary reason for such transition is that the strain energy associated with higher-order longitudinal modes at this instance is lower than the strain energy of lower order mode.

The deceleration in the implodable is followed by contact initiation at the center point at t = 2.8 ms. Later, the point contact grows in both diametrical and longitudinal direction as seen in time t = 3.2 ms. This contact front is seen to grow with an average velocity of 70 m/s, which is smaller as compared to that from free-field implosion experiments (between 150 and 200 m/s [5]). This lower contact front propagation velocity and the deceleration of walls prior to wall contact are the effects of the open-tube confined environment. A typical implosion is a relatively longer event (of the order of ~1 ms) as compared to the radial reflection-free time period inside the confining tube (~120 μs). Thus, the confining tube inhibits the free-propagation of low-pressure implosion waves during initial collapse period of implodable. As a result, these waves reflect back from the inner walls of the confining tube and superimpose leading to large dynamic pressure drop. Therefore, the pressure in near-field of implodable is always smaller as compared to an equivalent free-field implosion experiment. With significant pressure drop, the driving forces to continue the implosion process decreases over time and hence the overall structural velocities are smaller.

20.4 Conclusions

An experimental investigation is conducted to understand the evolution of water hammer type waves resulting from an implosion occurring inside an open-ended confining tube. Both pressure measurements along with high-speed DIC measurements are conducted. The findings of this study are as follows:

- The calibration using a submerged calibration grid can successfully account for the refractive index mismatch between the water/cylindrical acrylic window/flat acrylic window/air. The calibration experiments reveal that the both the in-plane and out-of-plane measurements can be measured using this modified 3-D DIC calibration procedure within 5 % error.
- The presence of open-ended confining tube around an implodable significantly affects the implosion process. The velocity during the collapse is highly reduced due to the large pressure drop from the superposition of low-pressure implosion waves inside the confining tube. The peak velocity and the average velocity prior to wall contact are found to be linearly varying with the collapse pressure.
- The confining environment of open-ended confining tube causes the implodable volume to decelerate well before the initiation of wall contact. This duration is also found to increase with decreasing collapse pressure and it reaches a significant duration of 1.5 ms for the lowest collapse pressure experiments conducted in this study.

Acknowledgment The authors kindly acknowledge the financial support provided by Dr. Yapa D.S. Rajapakse and Dr. Stephen E. Turner of the Office of Naval Research under Grant Nos. N00014-10-1-0662 and N00014-12-1-0382 respectively.

References

1. LeBlanc, J.M., Ambrico, J.M., Turner, S.E.: Underwater implosion mechanics: experimental and computational overview. In: Blast Mitigation—Experimental and Numerical Studies. Springer, New York (2014)
2. Urick, R.J.: Implosions as sources of underwater sound. J. Acoust. Soc. Am. **35**, 2026–2027 (1963). doi:10.1121/1.1918898
3. Turner, S.E., Ambrico, J.M.: Underwater implosion of cylindrical metal tubes. J. Appl. Mech. **80**, 011013 (2012). doi:10.1115/1.4006944
4. Farhat, C., Wang, K.G., Main, A., et al.: Dynamic implosion of underwater cylindrical shells: Experiments and Computations. Int. J. Solids Struct. **50**, 2943–2961 (2013). doi:10.1016/j.ijsolstr.2013.05.006
5. Gupta, S., Parameswaran, V., Sutton, M.A., Shukla, A.: Study of dynamic underwater implosion mechanics using digital image correlation. Proc. R Soc. Math. Phys. Eng. Sci. **470**, 20140576 (2014). doi:10.1098/rspa.2014.0576
6. Pinto, M., Gupta, S., Shukla, A.: Hydrostatic implosion of GFRP composite tubes studied by digital image correlation. J. Press. Vessel. Technol. **137**, 051302 (2015). doi:10.1115/1.4029657
7. Pinto, M., Gupta, S., Shukla, A.: Study of implosion of carbon/epoxy composite hollow cylinders using 3-D Digital Image Correlation. Compos. Struct. **119**, 272–286 (2015). doi:10.1016/j.compstruct.2014.08.040
8. Chamberlin, R.E., Guzas, E.L., Ambrico, J.M.: Energy balance during underwater implosion of ductile metallic cylinders. J. Acoust. Soc. Am. **136**, 2489–2496 (2014). doi:10.1121/1.4896744
9. Gupta, S., LeBlanc, J.M., Shukla, A.: Sympathetic underwater implosion in a confining environment. Extreme. Mech. Lett. **3**, 123–129 (2015). doi:10.1016/j.eml.2015.03.007
10. Gish, L.A., Wierzbicki, T.: Estimation of the underwater implosion pulse from cylindrical metal shells. Int. J. Impact Eng. **77**, 166–175 (2015). doi:10.1016/j.ijimpeng.2014.11.018
11. Timoshenko, S.P., Gere, J.M.: Theory of Elastic Stability. Dover, Mineola (1963)

Chapter 21
Response of Composite Cylinders Subjected to Near Field Underwater Explosions

E. Gauch, J. LeBlanc, C. Shillings, and A. Shukla

Abstract Experiments were conducted on woven E-glass/epoxy roll wrapped cylinders in three configurations; base composite, and base composite with a thin (100 % composite thickness) and thick (200 % composite thickness) polyurea coating. Each cylinder configuration was subjected to near-field UNDEX loading at charge standoff distances of 2.5 cm and 5.1 cm inside of a large diameter test tank. Results show that the application of a polyurea coating is effective for reducing damage in the cylinders.

Keywords Composite • UNDEX • Near Field • Polyurea • Shock

21.1 Introduction

Composite materials have several characteristics which make them particularly appealing in marine environments such as high strength to weight ratios and superior resistance to corrosion. When structures composed of composite materials are fielded in a marine environment they may be subjected to harsh loading conditions such as UNDEX loading, both near and far field in addition to operational loads. Maximizing the benefit of these materials, particularly for minimum weight, requires a full understanding of the response of these materials to such loadings in order to avoid overly conservative designs.

Studies on the response of composites subjected to UNDEX have generally focused on far field loading in which the encroaching shock front is nearly planar and there is no interaction between the UNDEX bubble and the structure. LeBlanc and Shukla [1, 2] have studied the response of both flat and curved E-glass/epoxy composite plates to far field loading both experimentally and computationally. In [3] Avachat and Zhou investigated the response of monolithic as well as sandwich structure composite cylinders to underwater impulsive loading. Mouritz, et al., [4], conducted a study of the development of damage in a glass reinforced composite subjected to underwater explosive loading at increasing pressures.

Near-field loading is generally characterized by a spherical shock front impinging upon the structure as well as interaction of the UNDEX bubble and the target structure. This can lead to highly localized damage and response in the structure rather than the more global character of the far field loading. In LeBlanc, et al., [5], coated and non-coated flat E-glass/epoxy plates were subjected to near field UNDEX loading. Deflections and damage extents were compared across the plate configurations. Brett, et al., [6, 7], presented a study of steel cylinders subjected to near field UNDEX. They observed that at standoff distances less than the UNDEX bubble radius the bubble was attracted to the cylinder and collapsed upon it resulting in a significant structural response.

Recently polyurea has found interest as a potential blast mitigating coating. It is an easy to apply polymer that becomes quite stiff at high rates of loading. Several studies have been conducted to determine polyurea's ability to reduce structural response to blast loading as well as reduce damage in materials. LeBlanc, et al., [5, 8, 9] studied the response of composite plates coated with polyurea to UNDEX loading. It was determined that both location and thickness of the coating were important considerations in efforts to reduce damage and deflection. Tekalur, et al., [10] and Gardner, et al., [11] studied monolithic and sandwich composites, respectively, subjected to air blast loading. It was found that polyurea was able to mitigate damage and deflection in the monolithic plates. For the sandwich composites blast resistance was improved by placing the polyurea between the back face sheet and the foam core; performance was degraded when the polyurea was applied between the front face sheet and the foam core.

E. Gauch • J. LeBlanc (✉)
Naval Undersea Warfare Center, Division Newport, 1179 Howell St., Newport, RI 02841, USA
e-mail: James.M.LeBlanc@Navy.Mil

C. Shillings • A. Shukla
Univeristy of Rhode Island, Dept. of Mechanical Engineering, 92 Upper Colege Road, Kingston, RI 02881

D. Casem et al. (eds.), *Dynamic Behavior of Materials, Volume 1*, Conference Proceedings of the Society for Experimental Mechanics Series, DOI 10.1007/978-3-319-41132-3_21

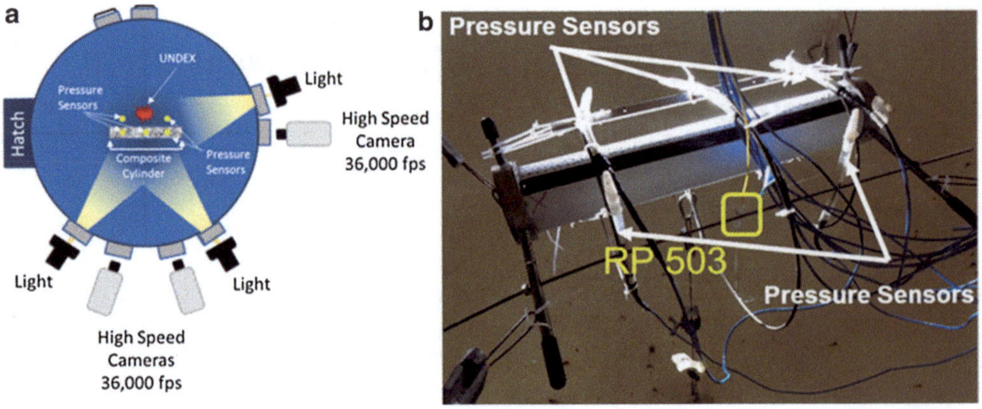

Fig. 21.1 Test configuration, (a) Tank schematic, (b) Cylinder in support cage

21.2 Materials

This investigation tested woven, roll wrapped E-glass/epoxy composite cylinders (APC Composites, Inc.) in a base configuration comprised solely of the composite material as well as the base composite with an applied polymeric coating in two different thicknesses. The layup schedule of the cylinders is [0/45/45/0] with a cured ply thickness of approximately 0.25 mm. The composite cylinders had an outside diameter of 7.44 cm, thickness of 1.14 mm and unsupported length of 38.1 cm. Thin (100 % of composite thickness) and thick (200 % of composite thickness) Polyurea (Dragonshield-BC) coatings were applied via spray cast to the outer surface of the cylinders.

21.3 Experimental Set-up

All experiments were conducted in the large diameter (2.1 m) water filled spherical test tanked located in the University of Rhode Island Dynamic Photomechanics Laboratory (DPML). The test tank is rated to withstand pressures up to 6.89 MPa in addition to 4 g of RDX explosive. An array of windows along the horizontal axis of the test tank allow for full viewing and recording of experiments as well as illumination of the test article. RP-503 explosive charges were used in this study. A schematic of the test set-up can be seen in Fig. 21.1.

The cylinders were mounted and held in the center of the tank via cables suspended from pad eyes located along the tank walls above and below the specimen. Pressures in the vicinity of the cylinder were recorded using PCB 138A05 tourmaline dynamic pressure sensors and data was monitored and recorded using a Dash 8HF-HS data recorder.

Three high speed video cameras (FastCam SA1, Photron USA, Inc., San Diego, CA) were used to capture video during experiments for assessment of bubble interaction and digital image correlation (DIC) analysis of the structural response. One camera was mounted to align with the longitudinal axis of the cylinder, providing a side view of the UNDEX event. The remaining two cameras were arranged to provide a stereoscopic view of the cylinder on the opposite side of the explosive (front view). High intensity lights were used to provide the necessary light for high speed video capture. Frame rates of 36,000 fps were used.

21.4 Experimental Methodology

Each cylinder configuration (base composite, thick coating and thin coating) was tested at least twice at each of two charge stand-offs, 2.54 and 5.08 cm. End caps, with rubber o-ring seals, were fixed to the ends of the cylinders to ensure they remained air filled during filling of the test tank. The charge distance to the cylinder surface was maintained by fixing the charge within a support cage with monofilament line; see Fig. 21.1b. The explosive lead wires were passed from the interior of the test tank to the exterior via a high pressure pass-through in the tank wall. Following placement of the set-up within the tank the hatch was secured and the tank flooded with water, leaving a small air pocket at the very top. Although the tank is capable of being pressurized via the introduction of compressed nitrogen into the air pocket these experiments were conducted at ambient pressure.

Once filling was complete the explosive lead wires were connected to a detonation box which supplied the amperage required to ignite the explosive.

21.5 Results and Discussion

At a charge standoff of 5.1 cm from the surface of the cylinder the damage in the uncoated cylinder is primarily characterized by two longitudinal cracks through the thickness of the cylinder as well as three circumferential through-thickness cracks at the termination of the longitudinal cracks. In addition there are small scale longitudinal and circumferential cracks along the interior of the cylinder which do not extend through the thickness of the composite.

At the same charge standoff of 5.1 cm the cylinders coated with the thin polyurea coating had much less extensive damage than the non-coated cylinders. The damage in the thinly coated cylinders consists of longitudinal cracks extending 8.9 cm on either side of the point nearest the explosive. Additionally, there is a circumferential crack originating at the same point which extends 1.9 cm on either side. While the primary cracks do not extend through the coating they can be felt beneath the surface. Additionally, circumferential cracks of approximately 1.3 cm can be seen at ±70°.

The test specimens with the thick polyurea coating displayed a dramatically reduced level of damage, particularly as compared with the uncoated cylinders. At locations ±60° from the point closest to the charge there are areas of damage which include circumferential cracks of approximately 2.5 cm in length as well as longitudinal cracks of approximately 1.3 cm in length. These cracks can be felt below the surface of the coating, but do not extend through it. Figure 21.2 depicts the post-test damage states of the cylinders test with a 5.1 cm charge standoff.

As would be expected, the damage in the cylinders tested with a 2.5 cm charge standoff was far more severe than for the 5.1 cm standoff, Fig. 21.3. In the case of the uncoated cylinder the damage was dominated by a large hole to the left of the charge location as well as a longitudinal crack extending 23 cm along the length of the cylinder and a circumferential crack originating near the center of the cylinder, the point closest to the charge detonation. There is another large crack which can be observed extending at an angle of 50° from the longitudinal axis of the cylinder.

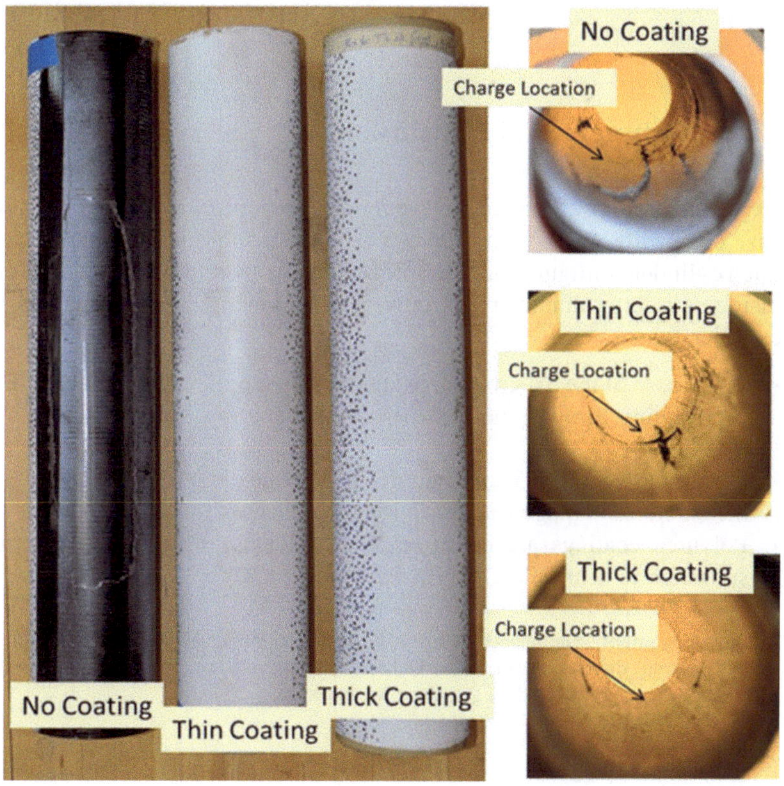

Fig. 21.2 Observed damage for 5.1 cm charge standoff

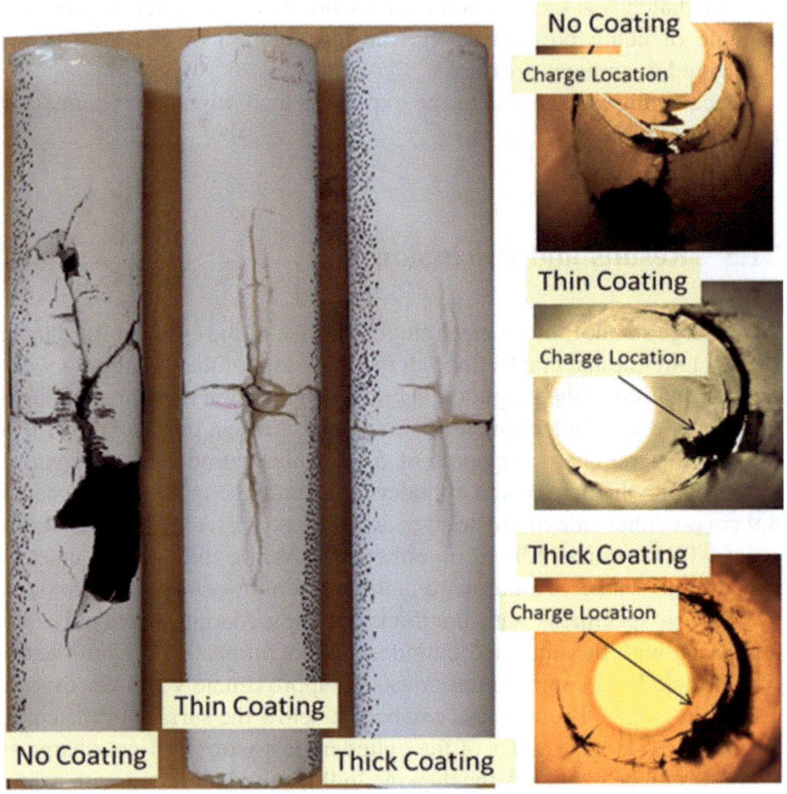

Fig. 21.3 Observed damage for 2.5 cm charge standoff

For the thinly coated cylinder circumferential and longitudinal cracks, through the thickness of the composite and the coating, emanated from the point closest to the explosive. The damage in the thickly coated cylinder was similar to the thinly coated cylinder although the longitudinal crack did not extend through the coating and was not as extensive.

21.6 Summary and Conclusions

Experiments were conducted on woven E-glass/epoxy roll wrapped cylinders in three configurations; base composite, and base composite with a thin (100 % base composite thickness) and thick (200 % composite thickness) polyurea coating. Each cylinder configuration was subjected to near-field UNDEX loading at charge standoff distances of 2.5 cm and 5.1 cm inside of a large diameter test tank. The following conclusions can be drawn regarding the damage sustained by the cylinders:

- In the case of the 5.1 cm charge standoff the application of even a thin coating of polyurea prevented any damage from propagating through the thickness of the coating and significantly reduced the amount of damage to the composite itself.
- For the case of the 2.5 cm standoff the damage did propagate through the thickness of the cylinder and coating, however, the extent of damage was much reduced for both the thin and thick coatings.
- Polyurea coatings are effective at significantly reducing the amount of damage sustained by composite cylinders subjected to near field UNDEX loading.

Acknowledgement The financial support of the Naval Undersea Warfare Center, Division Newport In-house Laboratory Independent Research (ILIR) program directed by Mr. Neil Dubois is greatly acknowledged as well as the financial support of the Office of Naval Research Solid Mechanics Program directed by Dr. Yapa Rajapakse.

References

1. LeBlanc, J., Shukla, A.: Dynamic response and damage evolution in composite materials subjected to underwater explosive loading: an experimental and computational study. Compos. Struct. **92**(10), 2421–2430 (2010)
2. LeBlanc, J., Shukla, A.: Dynamic response of curved composite plates to underwater explosive loading: Experimental and computational comparisons. Compos. Struct. **93**(11), 3072–3081 (2011)
3. Avachat, S., Zhou, M.: Response of cylindrical composite structures to underwater impulsive loading. Procedia Eng. **88**, 69–76 (2014)
4. Mouritz, A., Saunders, D., Buckley, S.: The damage and failure of GRP laminates by underwater explosion shock loading. Compos. **25**(6), 431–437 (1994)
5. LeBlanc, J., Shillings, C., Gauch, E., Livolsi, F., Shukla, A.: Near field underwater explosion response of polyurea coated composite plates. Exp. Mech. **56**(4), 569–581 (2016). doi:10.1007/s11340-015-0071-8
6. Brett, J., Yiannakopolus, G.: A study of explosive effects in close proximity to a submerged cylinder. Int. J. Impact Eng. **35**(4), 206–225 (2008)
7. Brett, J., Yiannakopolus, G., van der Schaaf, P.: Time-resolved measurement of the deformation of submerged cylinders subjected to loading from a nearby explosion. Int. J. Impact Mech. **24**(9), 875–890 (2000)
8. LeBlanc, J., Shukla, A.: Response of polyurea coated flat composite plates to underwater explosive loading. J. Compos. Mater. **49**(8), 965–980 (2015)
9. LeBlanc, J., Gardner, N., Shukla, A.: Effect of polyurea coatings on the response of curved e-glass / vinyl ester composite plates to underwater explosive loading. Compos. B **44**(1), 565–574 (2013)
10. Tekalur, S.A., Shukla, A., Shivakumar, K.: Blast resistance of polyurea based layered composite materials. Compos. Struct. **84**(3), 271–281 (2008)
11. Gardner, N., Wang, E., Kumar, P., Shukla, A.: Blast mitigation in a sandwich composite using graded core and polyurea interlayer. Exp. Mech. **52**(2), 119–133 (2012)

Chapter 22
Microstructural Effects on the Spall Properties of 5083 Aluminum: Equal-Channel Angular Extrusion (ECAE) Plus Cold Rolling

C.L. Williams, T. Sano, T.R. Walter, and L.J. Kecskes

Abstract 5083 Aluminum alloy is a light weight and strain-hardened material used in high strain-rate applications such as those experienced under shock loading. Symmetric real-time (in-situ) and end-state (recovery) plate impact experiments were conducted to study the spall response and the effects of microstructure on the spall properties of both 5083-H321 and 5083-ECAE + 30 % cold-rolled (CR) aluminum alloys shock loaded to approximately 1.46 GPa (0.2 km/s) and 2.96 GPa (0.4 km/s). The results show that mechanically processing the 5083-H321 aluminum by ECAE, followed by subsequent rolling significantly increases the Hugoniot Elastic Limit (HEL) by 78 %. However, this significant increase in HEL was at the expense of spall strength. The spall strength of the 5083-ECAE + 30 % CR aluminum dropped by 37 % and 23 % when compared to their 5083-H321 aluminum counterpart at shock stresses of approximately 1.46 GPa (0.2 km/s) and 2.96 GPa (0.4 km/s) respectively. This reduction in spall strength is attributed to the re-alignment of the manganese (Mn)-rich intermetallic second phase particles during mechanical processing (i.e., ECAE and subsequent cold rolling) which are consequently more conducive to spallation.

Keywords Shock • Spall • Inclusions • Microstructure • Failure

22.1 Introduction

5083 aluminum is a light weight, strain hardened, corrosion resistant, and high strength alloy commonly used in high strain-rate applications such as those experienced under shock loading. It is not a heat treatable alloy but significant strengthening effects can be achieved through alloying with magnesium and manganese (Mn) followed by mechanical processing such as cold working. Magnesium is added for solid solution strengthening and manganese for refining the grain structure through the formation of dispersoid particles which pin grain boundaries.

The spall response of mechanically processed 5083 aluminum (i.e., cold and hot rolled, extruded, etc.) has been previously studied by several researchers [1–4]. Results from the work of Boteler and Dandekar [1] show no spall strength dependency on peak shock stress ranging from 1.58 to 2.78 GPa for 5083-H131 aluminum. Appleby-Thomas and Hazell [2] were able to show that 5083-H32 armor grade aluminum exhibited no strengthening effects as a function of peak shock stress. From microstructural analyses conducted on shock recovered samples, they determined that spall failure in 5083-H32 aluminum initiated and propagated from one inclusion to another [2]. Whelchel et al. [3] studied the spall behavior of 5083-H116 aluminum and found that the Hugoniot Elastic Limit (HEL) in the transverse direction exhibited the highest value and the lowest HEL was observed along the rolling direction. In addition, they determined that the spall strength in the rolling direction was higher than that in the transverse direction and from microstructural studies, they determined that spall damage propagated along grain boundaries.

In addition to studying the spall behavior of 5083-H116 aluminum, Whelchel et al. [4] also studied the spall behavior of 5083-H321 aluminum which was mechanically processed using Equal-Channel Angular Extrusion (ECAE) and then rolled to yield a uniform grain structure throughout the plate. This processing technique produced an Ultra Fine Grain (UFG) structure with an average grain size of approximately 400 nm. From this research work, they showed that samples mechanically processed using ECAE alone did not produce alignment of the second phase particles favorable to spall failure and therefore, increases in spall strength and HEL were observed with minor orientation dependence. However, when the ECAE material was further rolled, the second phase particles were aligned preferentially along grain boundaries and this alignment of particles is favorable to spall failure. Subsequently rolling the material results in significant increases in HEL but considerable decreases in the spall strength [4]. The objective of this research is to further develop a better understanding

C.L. Williams (✉) • T. Sano • T.R. Walter • L.J. Kecskes
U. S. Army Research Laboratory, Aberdeen Proving Ground, Aberdeen, MD 21005-5066, USA
e-mail: cyril.l.williams.civ@mail.mil

D. Casem et al. (eds.), *Dynamic Behavior of Materials, Volume 1*, Conference Proceedings of the Society for Experimental Mechanics Series, DOI 10.1007/978-3-319-41132-3_22

of the microstructural aspects in particular the role of second phase particles in 5083-H321 aluminum which has been mechanically processed using ECAE then cold rolled and shock loaded. This objective is achieved by studying the spall response using real-time (in-situ) shock experiments and the acquired results are augmented by end-state (recovery) shock experiments.

22.2 Materials and Experimental Methods

22.2.1 Materials

The as-received rolled 5083-H321 aluminum plate was mechanically processed using ECAE at 250 °C. After each pass through the die, the 5083-H321 aluminum plate was rotated 90° about the through-thickness normal of the plate for a total of four passes. The resulting ECAE plate was further cold rolled at room temperature to 30 % reduction in thickness. By employing ECAE processing and subsequent 30 % cold rolling, the microstructure throughout the 5083-H321 aluminum plate was transformed to a uniform highly refined grain structure with an average grain size of approximately 400 nm [5]. Whelchel et al. [4] have previously determined that the microstructure developed after four passes reveal large inclusions and after cold rolling, the inclusions were broken into small particles, which eventually aligned themselves along grain boundaries. It is noteworthy to point out that the 5083 aluminum plate used for this research is identical to that used by Whelchel et al. [4]. All plate impact samples were cut through the thickness of the 5083 ECAE + 30 % cold-rolled (CR) aluminum plate using wire Electro-Discharge Machining (EDM). The average specimen dimensions were 42.950 ± 0.044 mm in diameter and 6.044 ± 0.047 mm thick. The measured density (ρ) was determined to be 2691.453 ± 57.245 kg/m^3. The longitudinal (C_L) and shear wave (C_S) speeds were measured to be 6.394 ± 0.040 km/s and 3.162 ± 0.022 km/s respectively, and the calculated elastic properties were; elastic modulus (E): 72.023 ± 1.229 GPa, shear modulus (G): 26.913 ± 0.469 GPa, bulk modulus (K): 74.141 ± 1.592 GPa, Lame's constant (λ): 56.199 ± 1.423 GPa, and Poisson's ratio (ν): 0.338 ± 0.003. The bulk sound speed (C_O) was computed using the bulk modulus to be 5.249 ± 40.581 km/s. These properties are consistent with those reported in the open literature.

22.2.2 Plate Impact Experiments

A total of eight shock experiments were conducted for this investigation using a single stage 102 mm (slotted bore) diameter gas gun at the shock physics laboratory, US Army Research Laboratory (ARL), Aberdeen Proving Ground (APG). Four real-time (in-situ) shock experiments were conducted to study the spall responses of both 5083-H321 and 5083-ECAE + 30 % CR aluminum at velocities of approximately 0.2 km/s and 0.4 km/s. The free surface velocity-time histories for all real-time experiments were acquired using Velocity Interferometry System for Any Reflector (VISAR). In addition, four end-state (shock recovery) experiments were conducted to augment all four real-time shock experiments in order to study the effects of microstructure on the spall properties of both materials. All eight experiments were symmetric implying that both the flyer and sample materials were identical (same impedance). The basic loading conditions and procedure for both real-time and end-state shock experiments have been previously described by Williams et al. [6–8].

22.3 Results and Discussion

Figure 22.1 shows the free surface velocity-time histories of 5083-H321 aluminum shocked at 0.203 km/s, corresponding to 1.45 GPa peak shock stress and 0.394 km/s, corresponding to 2.88 GPa peak shock stress. Both velocity profiles are characteristic of the spall response of metals and metallic alloys. Similarly, the velocity-time histories of 5083-ECAE + 30 % CR aluminum shocked at 0.196 km/s, corresponding to 1.40 GPa peak shock stress and 0.395 km/s, corresponding to 2.89 GPa peak shock stress are shown in Fig. 22.2. The velocity profile for the 5083-ECAE + 30 % CR aluminum shocked at 1.40 GPa is not a characteristic spall signal of metals and metallic alloys. The velocity profile beyond the elastic-plastic release phase (beyond ~1.5 μs) is distinctly different from the other profiles and it is characteristic of incipient spall (i.e., spallation was arrested due to insufficient stress required to drive the spall process to completion). A comparison between the HEL and

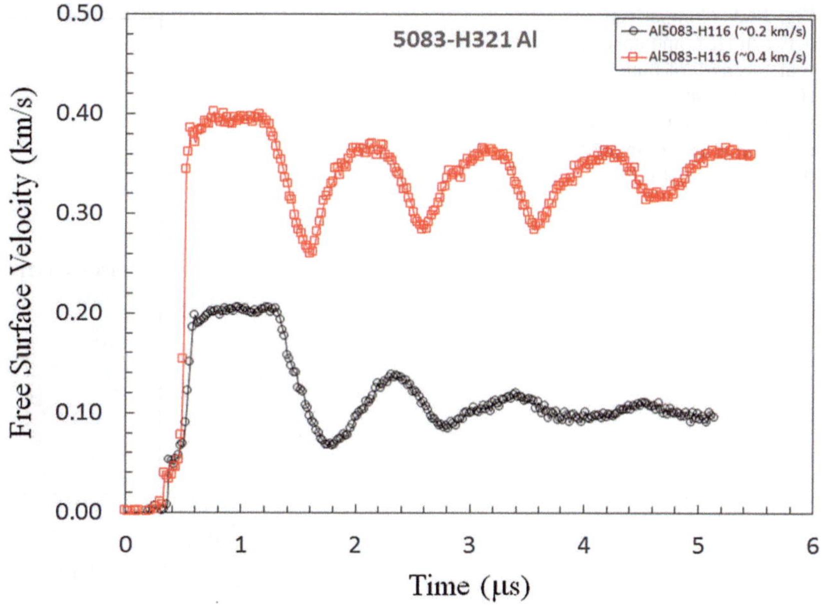

Fig. 22.1 Free surface velocity profiles for 5083-H321 aluminum at approximately 0.2 km/s and 0.4 km/s, respectively

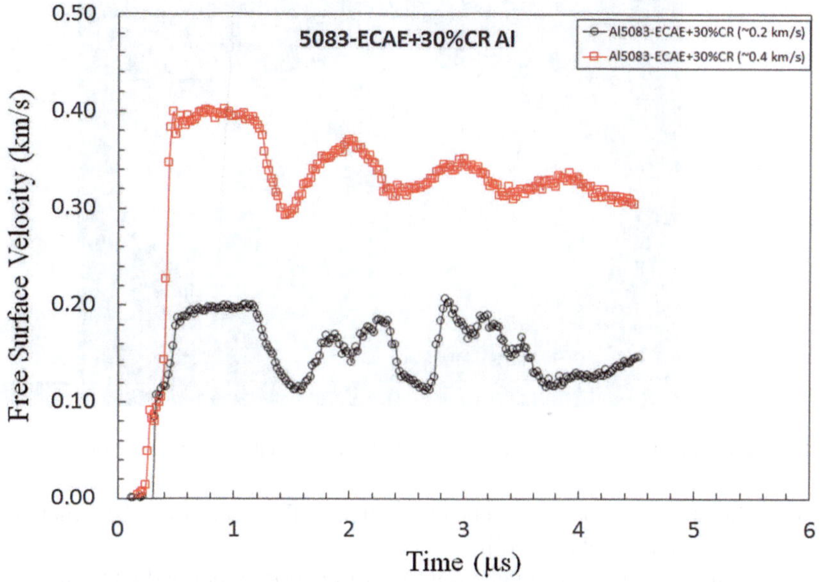

Fig. 22.2 Free surface velocity profiles for 5083-ECAE + 30 % CR aluminum at approximately 0.2 km/s and 0.4 km/s, respectively

uncorrected spall strength (not corrected for elastic-plastic effects) of both 5083-H321 and 5083-ECAE + 30 % CR aluminum acquired at approximately 0.2 and 0.4 km/s are shown in Fig. 22.3. Note that uncorrected spall strength is reported in this paper for convenience. On one hand, the HELs of the 5083-H321 aluminum were determined to be 0.435 GPa and 0.409 GPa at peak shock stresses of 1.45 GPa and 2.88 GPa, respectively. The uncorrected spall strengths of the same material were determined to be 0.947 and 0.938 GPa at the same peak shock stresses, respectively. This represents a 1 % change (insignificant) in spall strength between 1.45 and 2.88 GPa peak shock stress. On the other hand, the HELs of the 5083-ECAE + 30 % CR aluminum were determined to be 0.774 GPa and 0.731 GPa at peak shock stresses of 1.40 GPa and 2.89 GPa, respectively. The corresponding uncorrected spall strengths were determined to be 0.601 GPa and 0.719 GPa, respectively, for the same peak shock stresses. Increasing the peak shock stress from 1.40 and 2.89 GPa produces a 20 % increase in spall strength (uncorrected) of the 5083-ECAE + 30 % CR aluminum. The results in Fig. 22.3 also show that by mechanically processing the 5083-H321 using ECAE and 30 % CR, the average HEL increases from 0.422 ± 0.018 GPa to 0.753 ± 0.030 GPa representing a 78 % increase. However, the same is not true for the spall strength; mechanically processing the 5083-H321 aluminum using ECAE and 30 % CR reduces the corresponding spall strengths acquired at approximately 0.2 km/s and 0.4 km/s by 37 % and 23 %, respectively. This is perhaps due to the migration and eventual alignment of these particles along grain boundaries during the ECAE and rolling processes. These particles consequently serve as potential void nucleation sites favoring the spall process as

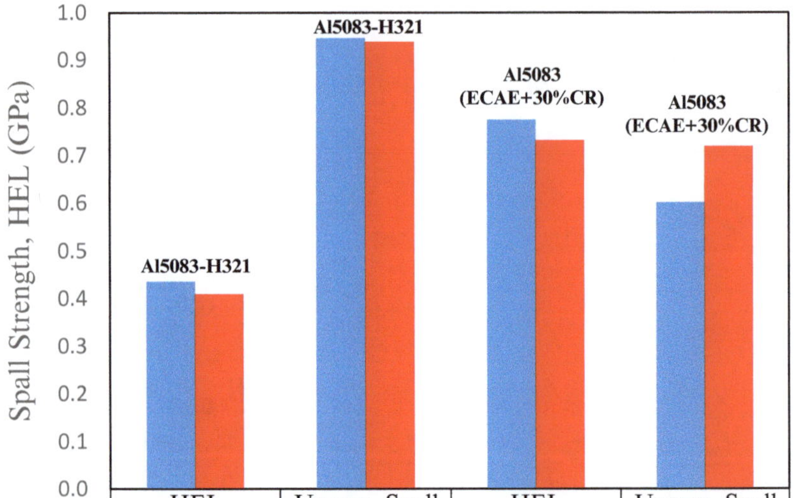

	HEL	Uncorr. Spall	HEL	Uncorr. Spall
~200 m/s	0.435	0.947	0.774	0.601
~400 m/s	0.409	0.938	0.731	0.719

Fig. 22.3 Comparison between HEL and uncorrected spall strength of both 5083-H321 and 5083-ECAE + 30 % CR aluminum acquired at approximately 0.2 km/s and 0.4 km/s

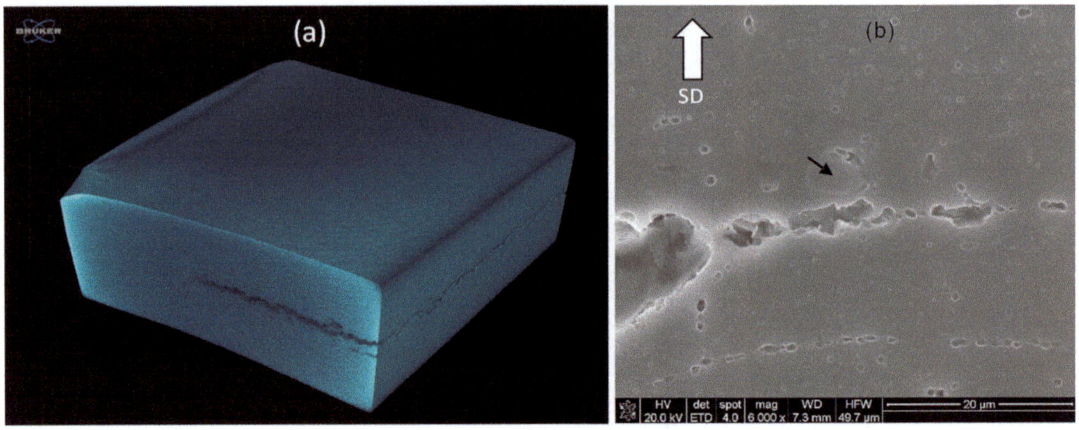

Fig. 22.4 The (**a**) X-ray microCT scan and (**b**) SEM residual microstructure of 5083-H321 aluminum sample shocked at approximately 1.46 GPa

determined from shock recovery experiments. The real-time experimental results discussed here are consistent with those reported by Whelchel et al. [4].

Results from spall recovery (end-state) experiments conducted at peak shock stresses of approximately 1.46 GPa for 5083-H321 aluminum and 5083-ECAE + 30 % CR aluminum are shown in Figs. 22.4 and 22.5, respectively. Although the velocity-time history shown in Fig. 22.1 for 5083-H321 aluminum shocked at approximately 1.46 GPa (0.2 km/s) is characteristic of spallation, the recovered sample did not reveal complete spallation; but X-ray micro-computed tomography (X-ray microCT) revealed spallation in the interior of the material as shown in Fig. 22.4a. Spallation was also revealed in the interior of the material after the sample was sectioned for microstructural analysis. The Scanning Electron Microscopy (SEM) micrograph shown in Fig. 22.4b represents the residual microstructure in the vicinity of the spall plane of the 5083-H321 aluminum shocked at approximately 1.46 GPa. The central region of the micrograph shows nucleated voids that appear to have grown and coalesced with each other along a direction normal to the shock stress (depicted by a white arrow marked SD in the figure). In addition, a second phase inter-metallic particle (see black arrow in the figure) which was determined to contain manganese (Mn) by Energy Dispersive X-ray Spectroscopy (EDS) appears to have been de-bonded from the matrix aluminum during the shock compression phase. This de-bonding between the second phase inter-metallic particle and matrix aluminum may serve as a potential void nucleation site if positioned within the spall plane. Similarly, the X-ray microCT scan and SEM micrograph of the 5083-ECAE + 30 % CR aluminum shock loaded to approximately 1.46 GPa are shown in Fig. 22.5. The 5083-ECAE + 30 % CR aluminum show similar results to that of 5083-H321 aluminum, in that, the material spalled internally but did not separate into two halves as shown in the X-ray microCT scan in

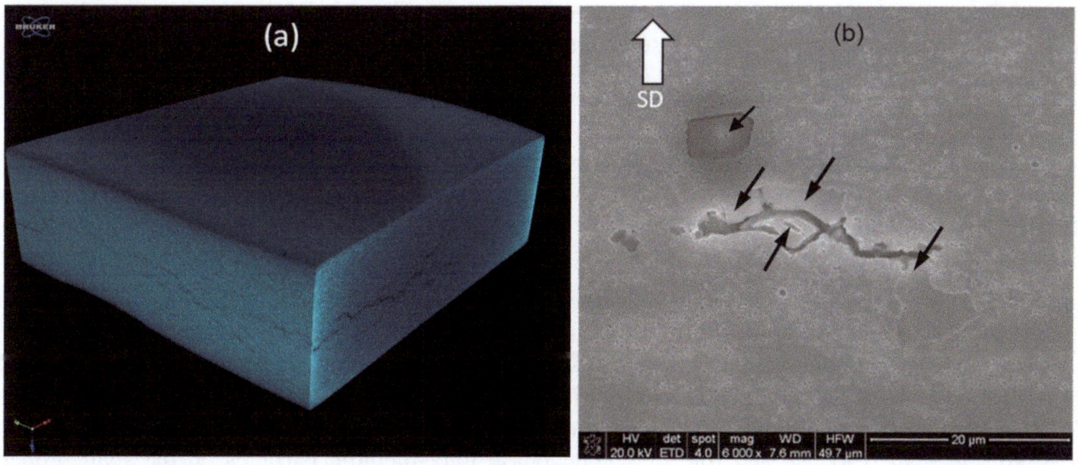

Fig. 22.5 The (**a**) X-ray microCT scan and (**b**) SEM residual microstructure of 5083-ECAE + 30 % CR aluminum sample shocked at approximately 1.46 GPa

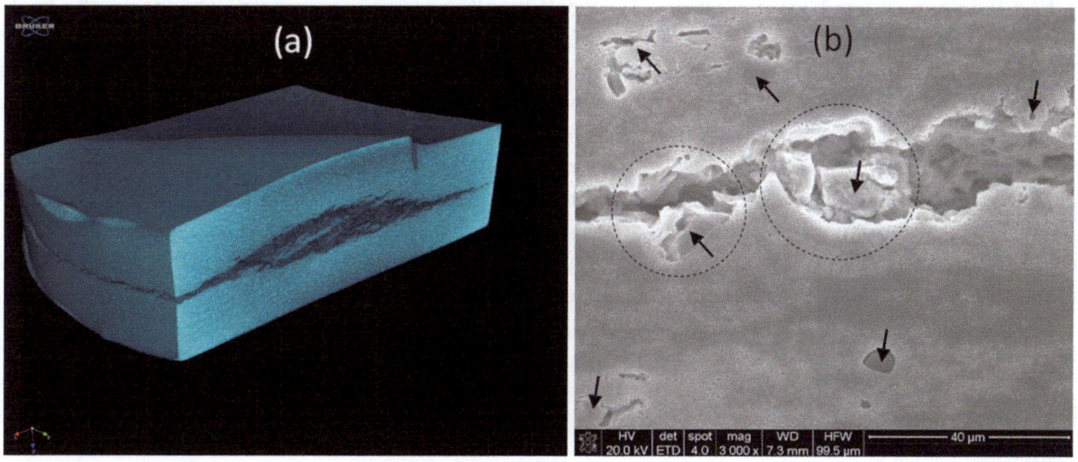

Fig. 22.6 The (**a**) X-ray microCT scan and (**b**) SEM residual microstructure of 5083-ECAE-30%CR aluminum sample shocked at approximately 2.96 GPa

Fig. 22.5a. Spallation in the interior of the 5083-H321 aluminum appears to be highly localized while that of the 5083-ECAE + 30 % CR aluminum appears to be less localized and the damage zone appears to be relatively large consisting of multiple cracks. Unlike the 5083-H321 aluminum, numerous second phase particles are visible within the crack opening suggesting that the particles were potential void nucleation sites as shown in Fig. 22.5b. It is noteworthy to point out that the numerous pits observed in the micrograph are artifacts of chemical polishing.

The X-ray microCT scan and SEM micrograph in Fig. 22.6 were obtained from the 5083-ECAE + 30 % CR aluminum sample shock loaded to 2.96 GPa. Although the 5083-ECAE-30 % CR aluminum sample appears to have totally spalled during the experiment, the spalled sample did not separate into two halves as shown in the X-ray microCT tomography scan in Fig. 22.6a. Far field damage from the center of the specimen was less severe than that observed at the center. The SEM micrograph shown in Fig. 22.6b reveals a large crack normal to the shock direction (SD), formed as a consequence of the nucleation, growth, and coalescence of voids. Second phase inter-metallic Mn-rich particles are visible along the crack length and are approximately 40 μm apart from each other. Two of the particles (shown within the circular dashed line) appear to be locations were voids nucleated, then grew and coalesced to form cracks because of the curvature of the aluminum matrix surrounding the particles. The width of the crack between both circles is narrow, suggesting a potential location where the two voids coalesced. Also, numerous second phase inter-metallic particles are revealed in the micrograph. However, these particles are farther away from the spall plane and therefore, were not able to grow and coalesce with neighboring particles. Almost all the inter-metallic particles outside the spall zone were de-bonded from the aluminum matrix. This observation is further evidence that particle-matrix de-bonding occurs during the shock compression phase and not during the tensile spall process. The micrograph also reveals ductile dimples within the interior of the crack resulting from void nucleation, growth, and coalescence. Results obtained from spall recovery experiments conducted on 5083-H321 aluminum at approximately 2.96 GPa shock stress are not shown here, however, the material totally spalled and separated into two halves.

22.4 Summary

As-received 5083-H321 aluminum plate was mechanically processed using ECAE and then cold rolled to 30 % reduction in thickness to achieve a highly refined ultra-fine grain structure through the thickness of the plate resulting in an average grain size of approximately 400 nm. Plate impact specimens were machined from the through-thickness direction of the mechanically processed plate using wire EDM. Symmetric plate impact experiments were conducted on both 5083-H321 and 5083-ECAE + 30 % CR aluminum alloys. Real-time (in-situ) and end-state (recovery) shock experiments were conducted to study the spall response and role of second phase inter-metallic particles on the spall properties of both 5083-H321 and 5083-ECAE + 30 % CR aluminum alloys. The results from real-time plate impact experiments show that, by mechanically processing 5083-H321 aluminum using ECAE and then cold rolled to 30 % reduction in height, the average HEL increases by 78 %. However, this significant improvement on the dynamic yield strength under uniaxial strain conditions (HEL) was not realized for spallation. The spall results show no significant change in the spall strength of 5083-H321 aluminum shock loaded at 1.45 GPa and 2.88 GPa shock stress, respectively. However, the spall strength of the 5083-ECAE + 30 % CR aluminum shock loaded at the same stress levels decreased by 37 % and 23 %, respectively, when compared to their 5083-H321 aluminum counterpart. End-state recovery experimental results show that the second phase inter-metallic Mn-rich particles de-bonded from the aluminum matrix during shock compression and were potential void nucleation sites in both materials. Furthermore, recovery results also show that spallation occur in both materials by void nucleation, growth, and coalescence. However, spall failure occur along the re-aligned inter-metallic Mn-rich particles for the 5083-ECAE + 30 % CR aluminum and this perhaps is responsible for the reduction in spall strength. More experimental investigations are needed to develop a better understanding of the microstructural aspects in particular the role of second phase inter-metallic particles under shock compression and consequent failure of 5083 aluminum alloys.

References

1. Boteler, J.M., Dandekar, D.P.: Dynamic response of 5083-H131 aluminum alloy. J. Appl. Phys. **100**, 054902 (2006)
2. Appleby-Thomas, G.J., Hazell, P.J.: A study on the strength of an armour-grade aluminum under high strain-rate loading. J. Appl. Phys. **107**, 123508 (2010)
3. Whelchel, R.L., Kennedy, G.B., Dwivedi, S.K., Sanders Jr., T.H., Thadhani, N.N.: Spall behavior of rolled aluminum 5083-H116. J. Appl. Phys. **113**, 233506 (2013)
4. Whelchel, R.L., Thadhani, N.N., Sanders, T.H., Kecskes, L.J., Williams, C.L.: Spall properties of Al5083 plate fabricated using equi-channel angular pressing (ECAP) and rolling. J. Phys. Conf. Ser. **500**(11), 112066 (2014)
5. Jin, H., Gallerneault, M., Segal, V.M., Young, P.J., Lloyd, D.J.: Grain structure and texture in aluminum alloy AA5083 after equal angular channel extrusion, warm rolling and subsequent annealing. Mater. Sci. Technol. **27**, 789–792 (2011)
6. Williams, C.L., Ramesh, K.T., Dandekar, D.P.: The spall response of 1100-O aluminum. J. Appl. Phys. **111**(12), 123528 (2012)
7. Williams, C.L., Chen, C.Q., Ramesh, K.T., Dandekar, D.P.: On the shock stress, substructure evolution, and spall response of commercially Pure1100-O aluminum. Mater. Sci. Eng. A **618**, 596–604 (2014)
8. Williams, C.L., Chen, C.Q., Ramesh, K.T., Dandekar, D.P.: The effects of cold rolling on the microstructural and spall response of 1100 aluminum. J. Appl. Phys. **114**(9), 093502 (2013)

Chapter 23
Experimental Study of the Dynamic Fragmentation in Transparent Ceramic Subjected to Projectile Impact

P. Forquin and J.L. Zinszner

Abstract Transparent ceramics, such as monocrystalline sapphire, are very interesting materials for many applications including armor systems. Indeed, these technical ceramics present a very high compressive strength, a high Hugoniot Elastic Limit and a low density. However, due to their brittleness and low tensile strength, a fragmentation of the ceramic target occurs under ballistic impact. In the present study, the fragmentation process in a transparent ceramic subjected to projectile impact is investigated. To achieve this goal, edge-on impact tests have been performed at various impact speeds. The use of an ultra-high speed camera at a frame-rate set to one million frames per second allows visualizing "in real time" the whole fragmentation process resulting from the initiation and propagation of multiple cracks in the targets. Depending on the impact speed, cracks initiate from the impacted edge, the rear edge or from the two lateral surfaces of the targets. It is also observed that cracks propagate following specific directions related to crystallographic planes. Finally, the impact velocity and the orientation of the crystal are seen to play a major role on the final cracking pattern of the target.

Keywords Monocrystalline sapphire • Dynamic fragmentation • Edge-On Impact • High-speed imaging

23.1 Introduction

Due to their very high hardness and Hugoniot Elastic Limit [1, 2] and their low density, monocrystalline sapphires are known to be among the best ceramics to be used in transparent armor configurations. However, an intense fragmentation process composed of numerous oriented cracks develops in the impacted ceramic [3]. This fragmentation process has a direct influence on the ballistic efficiency of the target. That is why its understanding and modeling remains a major issue.

In the present study monocrystalline sapphire targets are subjected to edge-on impact (EOI) tests. Small cylindrical projectiles launched at impact speeds ranging from 190 to 230 m/s are impacting the edge of the ceramic tile, so a fragmentation process develops in about 20 μs. An ultra-high speed camera is used to visualize the fragmentation process in the ceramic. The preliminary experimental results are detailed in the present contribution.

23.2 Experimental Configuration

The device used to perform the EOI experiments is described on the Fig. 23.1. The gas gun, of caliber 20 mm, is used to launch small cylindrical projectiles at impact velocity ranging from 10 to 300 m/s. In the present work, the projectiles are made of high-strength steel. They are 10 mm in diameter and 11 mm in length and their mass is 6.6 g. In addition, their hemispherical end cap of large radius (about 100 mm) allows ensuring better contact conditions when impacting the target. The camera used to visualize the target is a Kirana ultra-high speed camera. A laser interferometer is used for triggering the camera. The resolution of the images is 924×768 pixels. The interframe time is set to 0.5 μs. The target is a monocrystalline sapphire ceramic of size $60 \times 60 \times 4\ mm^3$. The impacted edge is $60 \times 4\ mm^2$.

Two edge-on impact tests performed with the C axis parallel to the impact direction are presented first. Two edge-on impact tests performed with the C axis orthogonal to the impact direction are then presented.

P. Forquin (✉) • J.L. Zinszner
Soils, Solids, Structures, Risks (3SR) Laboratory, Grenoble-Alps University, BP53 38041, Grenoble Cedex 9, 1270 rue de la piscine, Saint martin d'Hères, 38400 France
e-mail: pascal.forquin@3sr-grenoble.fr

D. Casem et al. (eds.), *Dynamic Behavior of Materials, Volume 1*, Conference Proceedings of the Society for Experimental Mechanics Series, DOI 10.1007/978-3-319-41132-3_23

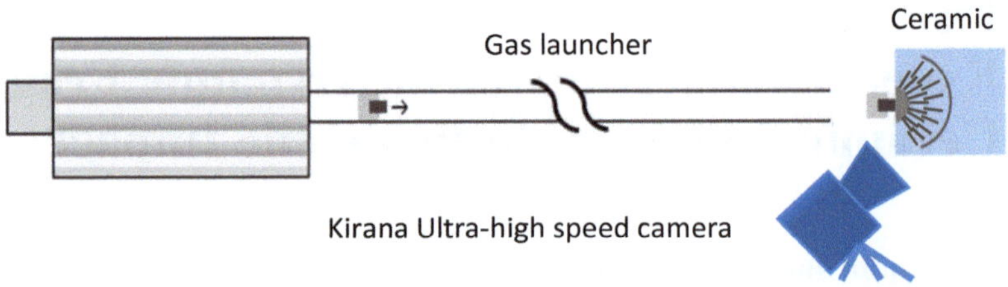

Fig. 23.1 Sketch of the experimental device (3SR lab.)

23.3 Edge-On Impact Tests Performed with the C Axis Parallel to the Impact Direction

Edge-on impact experiments in open configuration (visualization of fragmentation by means of an ultra-high speed camera) or in sarcophagus configuration (to investigate the post-mortem fragmentation pattern of the impacted ceramics) were previously performed on transparent [3] or non-transparent ceramics [4, 5, 6]. In the present work four EOI experiments were conducted.

Two edge-on impact tests are performed considering a C axis of monocrystalline sapphire parallel to the impact direction. The S1 impact test is performed with an impact velocity of 200 m/s. The Fig. 23.2 presents a sequence of 8 images obtained with the Kirana camera in the EOI test. All cracks are propagating following three main directions: the C axis direction and two "inclined" directions oriented at about 59° and 44° in relation to the C axis direction. Just after impact ($t = 2.5$ μs), one can see the initiation and the propagation of two inclined cracks initiated near the impact point. During the crack propagation, "secondary cracks" are also initiated on their lips and propagate following the C axis direction. In addition, a "circular crack" initiated near the impact point is observed at $t = 5, 8, 9.5$ μs. Height microseconds after impact, due to the reflection of the incident longitudinal wave on the rear free surface, cracks are initiated on the rear edge and propagate following the three previous directions.

The second EOI test (S3 test) is performed with an impact velocity of 230 m/s to study the influence of the impact velocity (Fig. 23.3). The impact axis is parallel to the C axis of the sapphire crystal. As observed in S1 test all cracks are propagating following three directions, the C axis direction and two "inclined" directions, oriented at about 54 and 43° in relation to the C axis direction. In particular, 5 μs after impact, one can observe one crack propagating following the C axis in addition to the cracks following the "inclined" directions (Fig. 23.3).

23.4 Edge-On Impact Tests Performed with the C Axis Orthogonal to the Impact Direction

Two more edge-on impact tests were performed by positioning the C axis of the monocrystalline sapphire orthogonally to the impact direction. The S2 impact test is performed with an impact velocity of 191 m/s. The Fig. 23.4 presents a sequence of 8 images obtained with the Kirana camera. The failure pattern in the sapphire target clearly differs from the ones observed in previous tests. In the first microseconds after impact, only cracks propagating following the two "inclined" directions are visible. The cracks are oriented at 42° and 58° in relation to the C axis direction. Moreover, the crack oriented at 42° is seen to change of orientation when approaching the free surface, propagating then with an angle estimated about 58°. Furthermore, very few cracks are noted to propagate on the C axis direction.

The S4 impact test (Fig. 23.5) was performed with an impact velocity of 227 m/s by positioning the impact axis orthogonal to the C axis of the sapphire crystal. In the first microseconds after impact, the failure pattern is consistent with the one observed in the S2 target.

23.5 Conclusion

In this work, 4 edge-on impact tests performed on monocrystalline sapphire are presented. The influences of the impact velocity and crystal orientation in relation to the impact direction have been studied. It is observed that due to monocrystalline nature of the sapphire, all the cracks are oriented following three specific directions regarding the crystal orientation. The impact velocity has also a clear influence. Indeed, above an impact speed of 220 m/s more cracks are observed in the target.

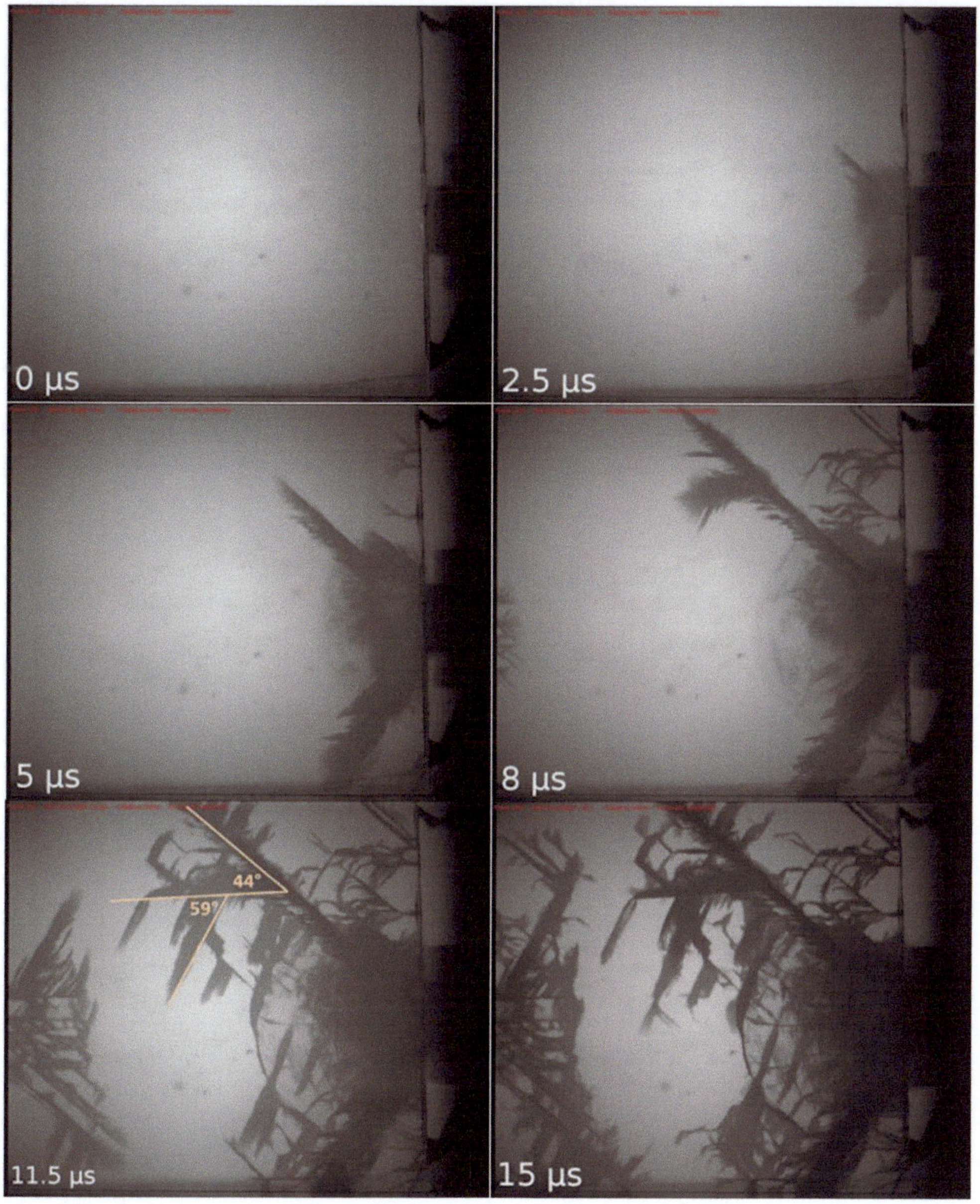

Fig. 23.2 "S1" edge-on impact test performed at 200 m/s. Impact direction parallel to the C axis

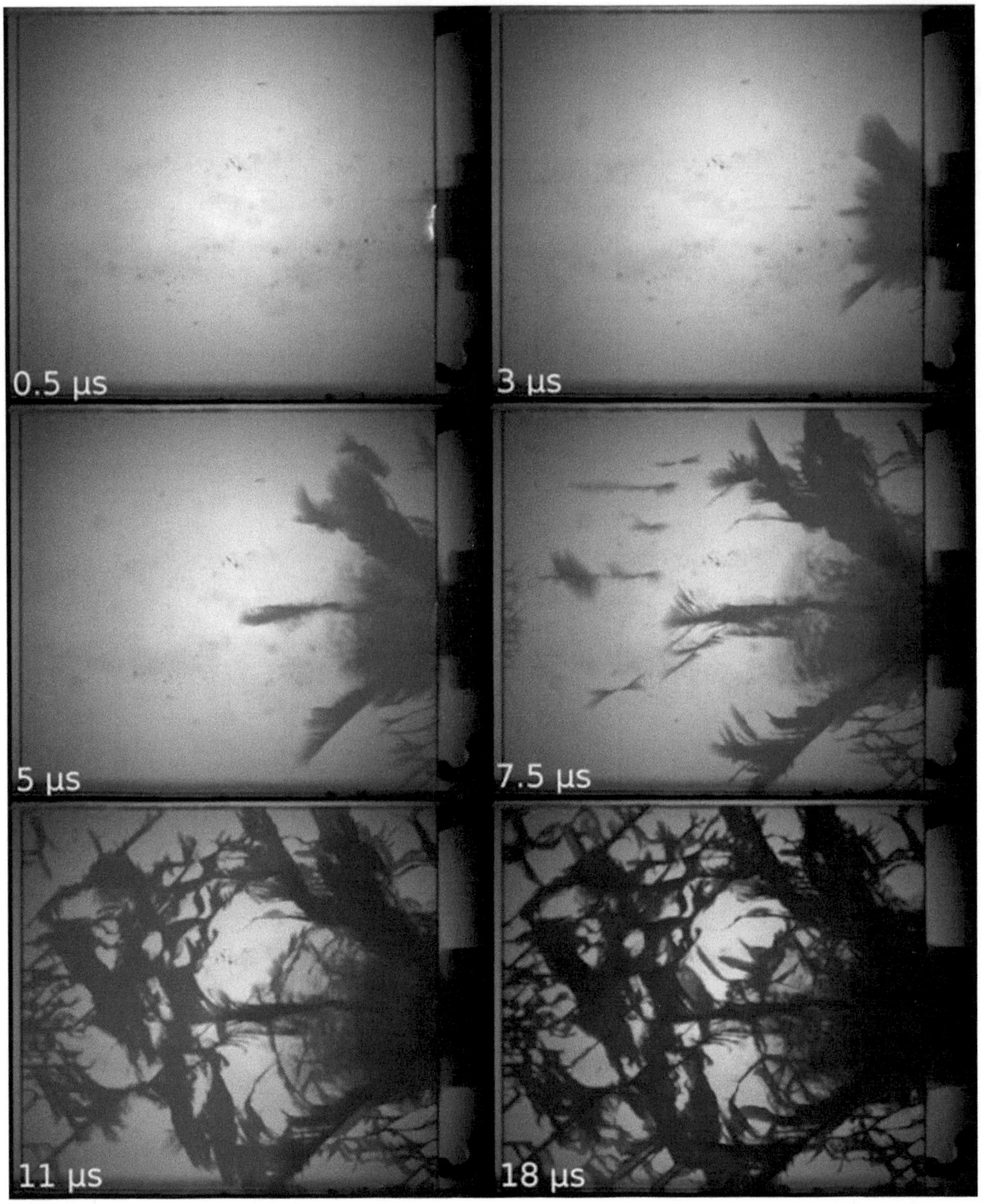

Fig. 23.3 "S3" edge-on impact test performed at 230 m/s. Impact direction parallel to the C axis

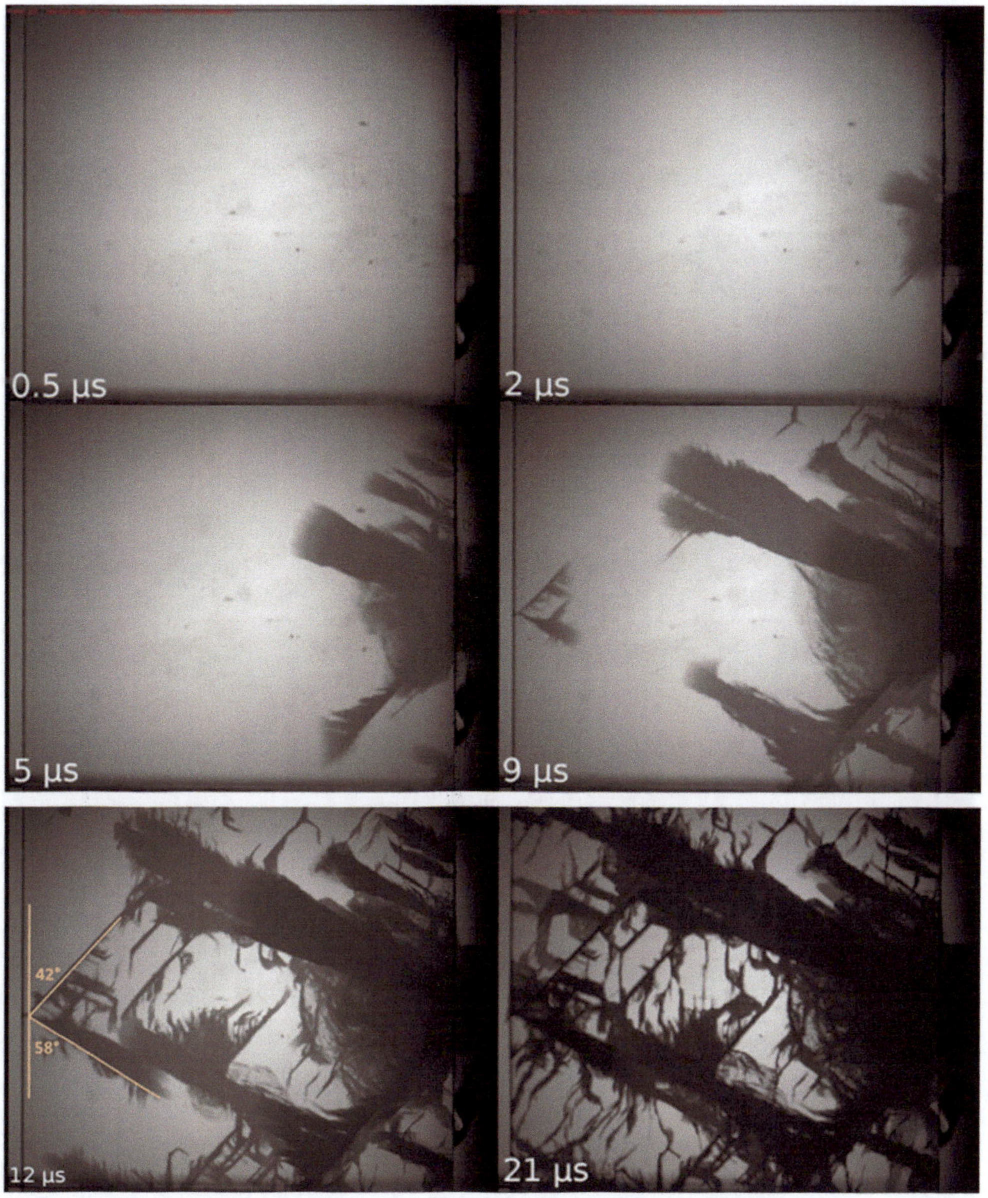

Fig. 23.4 "S2" edge-on impact test performed at 191 m/s. Impact direction orthogonal to the C axis

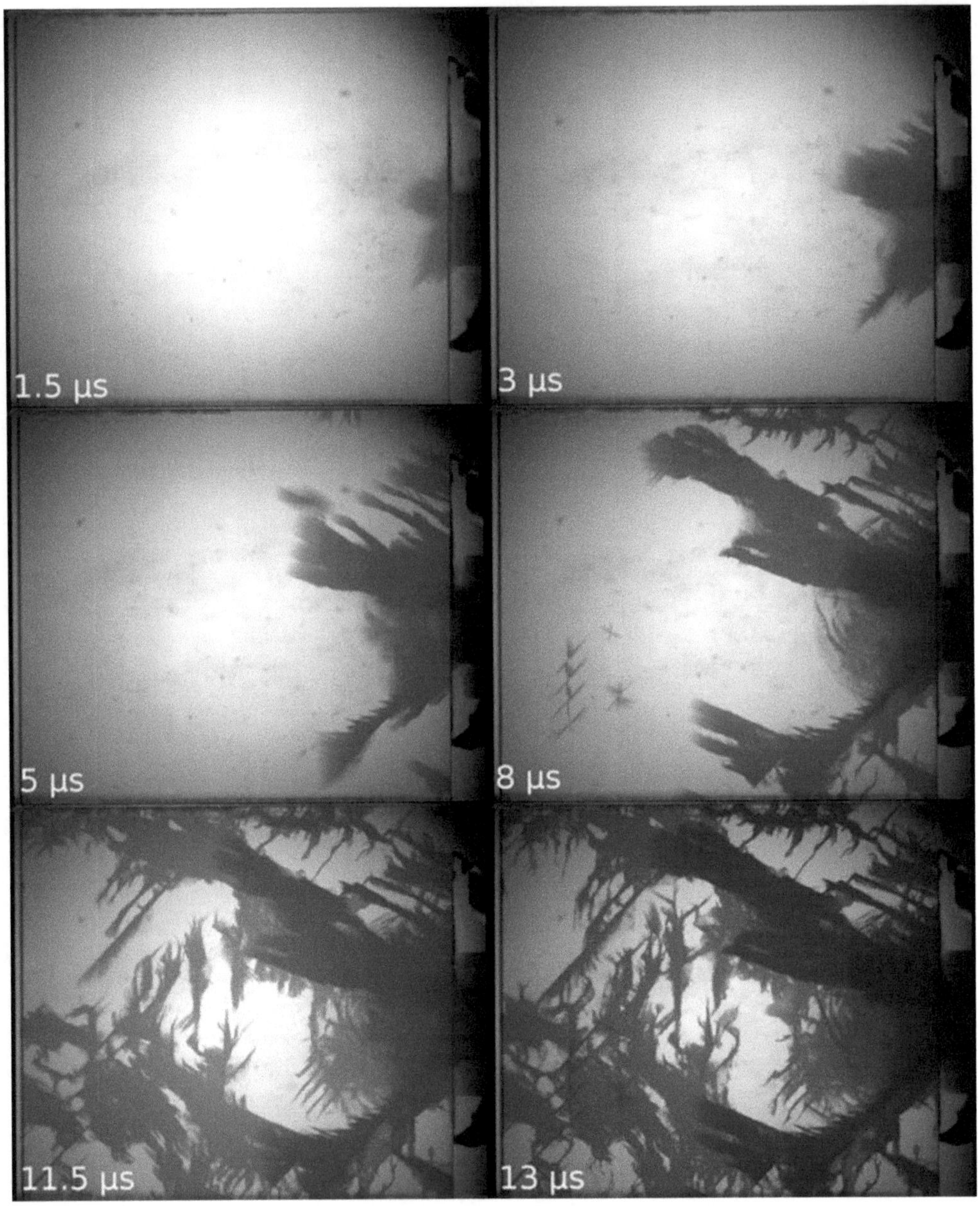

Fig. 23.5 S4 edge-on impact test performed at 227 m/s. Impact direction orthogonal to the C axis

Acknowledgement This work was supported by Saint-Gobain Company. This support is gratefully acknowledged.

References

1. Munsen, D.E., Lawrence, R.J.: Dynamic deformation of polycrystalline alumina. J. Appl. Phys. **50**(10), 6272–6282 (1979)
2. Kanel, G.I., Nellis, W.J., Savinykh, A.S., Razorenov, S.V., Rajendran, A.M.: Response of seven crystallographic orientations of sapphire crystals to shock stresses of 16–86 GPa. J. Appl. Phys. **106**, 043524 (2009)
3. Strassburger E., Patel P., McCauley J.W.: Visualization and analysis of impact damage in sapphire. Ballistics 2011: 26th International Symposium on Ballistics, Vol. 1 and 2, 1772–1782, 2011
4. Forquin, P., Tran, L., Louvigné, P.-F., Rota, L., Hild, F.: Effect of aluminum reinforcement on the dynamic fragmentation of SiC ceramics. Int. J. Impact Eng. **28**, 1061–1076 (2003)
5. Riou, P., Denoual, C., Cottenot, C.E.: Visualization of the damage evolution in impacted silicon carbide ceramics. Int. J. Impact Eng. **21**(4), 225–235 (1998)
6. Zinszner, J.L., Forquin, P., Rossiquet, G.: Experimental and numerical analysis of the dynamic fragmentation in a SiC ceramic under impact. Int. J. Impact Eng. **76**, 9–19 (2015)

Chapter 24
Instrumented Projectiles for Dynamic Testing

Guojing Li, Dahsin Liu, and Dan Schleh

Abstract Drop weight impact testers (DWIT) are commonly used in experiments for characterizing material properties and structural performance. With large mass, DWIT can provide large impact energy. However, the impact velocity involved in DWIT is usually low due to the fact that the impact velocity is proportional to the square-root of dropping height, i.e. $v = (2gh)^{1/2}$. Even if the impact velocity can be increased, the wave propagation involved in impact may become entangled, resulting in distorted wave for dynamic analysis. This study investigates a mechanical technique to reduce the effect due to wave propagation and its application to dynamic impact measurements.

Keywords Instrumentaion • Projectile • Testing • Dynamic • Strain gauges

24.1 Measurements of Dynamic Forces

Measuring force and deformation histories are perhaps the most fundamental activities in mechanical investigations. Many deformation-based measuring techniques, such as electrical resistance strain gages (RSG) and optical methods, like laser interferometry and digital image correlation, have been successfully used for measuring deformations caused by static loading. For measuring deformations associated with dynamic loading, optical methods in conjunction with a high-speed movie camera have been commonly used due to their capability of whole-field measurements. Once the deformation fields are obtained, they may be converted into associated force fields, with the use of Hooke's law and corresponding material properties, to complete the mechanical investigations.

Though being a point-by-point technique, RSG is much simpler in practice and more tolerable to environmental effects than optical methods. However, the strain wave due to dynamic loading can bounce back and forth between the ends of the structure on which the RSG is mounted. The overlap of the incoming and rebounding strain waves can then result in a completely distorted strain history. Consequently, the initial strain wave which bears the most important information of the dynamic loading can be lost.

Though highly demanded in many engineering applications, no reliable technique for measuring dynamic forces is available until the introduction of an instrumented projectile by the authors. It is the objective of this study to extend the recently developed technique to further improve their accuracy and applications.

24.2 The Innovation of an Instrumented Projectile (IP)

When the nose of a cylindrical rod impacts on a target, a strain wave will be generated in the rod. Subsequently, the strain wave will propagate to the tail of the rod. Once reaching there, the wave will rebound from it and propagate back to the nose. The wave will continue to bounce back and forth between the two ends of the rod until it is completely attenuated. If the rod

G. Li • D. Liu (✉)
Department of Mechanical Engineering, Michigan State University, East Lansing, MI 48864, USA
e-mail: liu@msu.edu

D. Schleh
Liuman Technologies, Lansing, MI 48910, USA

D. Casem et al. (eds.), *Dynamic Behavior of Materials, Volume 1*, Conference Proceedings of the Society for Experimental Mechanics Series, DOI 10.1007/978-3-319-41132-3_24

is short, numerous wave propagations will overlap one another and result in a largely distorted wave form which is very different from the initially impact-induced one.

With a special geometry design in the rod, it is possible to reduce the strain wave with less rebounding back to the nose. The initial strain wave caused by impact can then be largely isolated. The instrumented projectile (IP), based on geometrical manipulation, has been developed recently by the authors [1]. The short IP (about 6-in. long) consists of a solid neck and a hollow body. A pair of RSG's is mounted on the opposite sides of the neck. A data acquisition board and a battery have also been installed inside the IP's body. The IP has been used for impact testing. Experimental results have shown the high similarity of the wave propagation from the short IP and that from a 6-ft long cylindrical bar (which can separate the incoming wave from the rebounding one) when the IP and the long bar collide with each other.

The wave propagation in a short rod without a special geometry is completely different from that in a long cylindrical bar when the two collide with each other. However, the wave propagation in a short rod with a special geometry is almost identical to that in a long cylindrical bar when the two collide with each other. The ability to eliminate the wave rebounding based on the special geometry design is thus demonstrated. Accordingly, the initial strain wave which bears the most important information of impact-induce force history can be clearly identified by the short instrumented projectile.

24.3 Research for Improving Experimental Accuracy

Based on the innovation of the wave elimination technique and the success of the IP design, more applications based on the IP to improve engineering dynamic measurements have been explored. Some tasks are listed below and in this research investigation.

1. Drop-weight impact tester (DWIT) has been commonly used in both academia and industries due to its simplicity. However, the issue concerning wave propagation has been completely neglected in the design of the DWIT currently available in the market. As a task of this research study, it is to evaluate the effect of wave propagation involved in the DWIT by comparing the result from the DWIT with that from the IP (which preserves the initial impact-induced wave). The goal of this study is to demonstrate the importance of eliminating the wave propagation in correctly measuring dynamic forces.
2. Split Hopkinson's pressure bar (SHPB) – SHPB has been commonly used for characterizing dynamic constitutive relations of materials at high strain rates which are required for modeling the performance of materials and structures subjected to high-velocity loading. SHPB is usually organized by aligning two long bars (such as 1.5 m each and mounted with strain gages) with a gas gun while the specimen to be tested is situated between the two long bars. When an impactor is shot out of the gas gun and collides onto the first long bar, a strain wave will be generated. The wave will then propagate through the bar into the specimen before entering the second bar. Based on the wave propagations in the two long bars measured by electrical resistance strain gages mounted on them, the constitutive relation of the specimen material tested can be established for the high strain rate performed. Since the IP can record the incoming strain wave without the rebounding one, it can be used to replace the second long bar. The length of the SHPB can then be reduced to about half. The shorter SHPB should prove to be more convenient for laboratory operations.

Besides advancing the dynamic testing techniques given above, this research program can also be extended to other applications of the IP.

(a) A more accurate drop-weight impact tester – The ones currently used in engineering and research institutes neglect the wave propagation involved in the DWIT testing.
(b) An instrumented ballistic impactor – The ones currently used are non-instrumented and are not useful for scientific investigations.
(c) A shorter Hopkinson's pressure bar – The ones currently used are twice longer than the proposed one and are not convenient for labs with restricted space.
(d) An instrumented Taylor's impactor – The ones currently used are based on post-test measurements instead of in situ measurements.

It is the ultimate goal of this study is to eventually achieve significant advancement in dynamic material characterizations and structure tests in both academia and industries.

24.4 Horizontal Impact Testing Method for Comparing Tests between the Instrumented Projectile (IP) and a Gas Gun

For an IP validation test, the test window needs to be identified for the IP and the validation device. In this study, a long bar with a strain gage is used. The test window of interest is defined as the time duration between two points during which the event of interest occurs. The test window must be determined relative to the trigger signals of the data recording device used. A dummy IP may be used to find the test window for the bar. A triggering signal for the data acquisition device and an oscilloscope should be established. In the case of using a bar, an IR emitter and an collector are used. As the IP passes between them, a change in voltage in an associated circuit occurs. This voltage change can then be used as the trigger for the point in time onwards from which data should be collected. The IR sensors are placed at a point where the IP passes them prior to impact with the end of the bar to which the strain gages are attached.

Multiple tests with the dummy IP striking at the end of the bar should be perform to identify the appropriate test settings in the oscilloscope to maximize the number of data points to be collected for the event. This may be done by locating the test window and then reducing the 'Time/Div' setting on the oscilloscope so that it is small enough to allow sufficient data resolution but large enough to allow the entire desired event to be captured. It is good to begin with sampling a large time window and then reducing it once the event of interest is identified.

For the testing conducted, a total of 200,000 data points are usually taken per test with a sampling rate of 10 M/s. All passageways around the gas gun holding chamber should be open during testing. The loading and the pressure used to drive the IP have effects on the velocity of the IP and the subsequent force at impact. When needed, these parameters should be changed conservatively to minimize the potential damage to the IP and the bar. Once the data points are captured with a data acquisition card in conjunction with a scope program, the data must then be saved and processed. The channel producing the data from the bar should also be saved. Similarly, the time data should be saved along with the sampling frequency so the time data can be processed later (Fig. 24.1).

Once the experiment is completed, the data from the IP and that from the bar data may be plotted on a single graph. The data sets may be shifted along the time scale to have the onsets of initial loading peaks coincide with each other. The data set from the bar often contains several data points prior to the onset of loading at very high force values relative to the rest of the data sets. These points are deemed erroneous and thus are not considered as realistic representation of the loading. Figure 24.2 provides an example of experimental result.

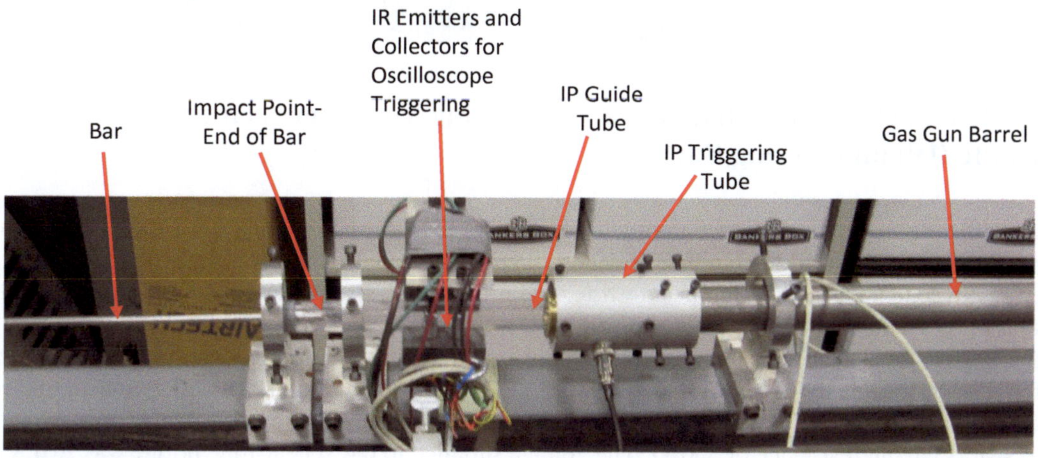

Fig. 24.1 Gas Gun—Bar interface

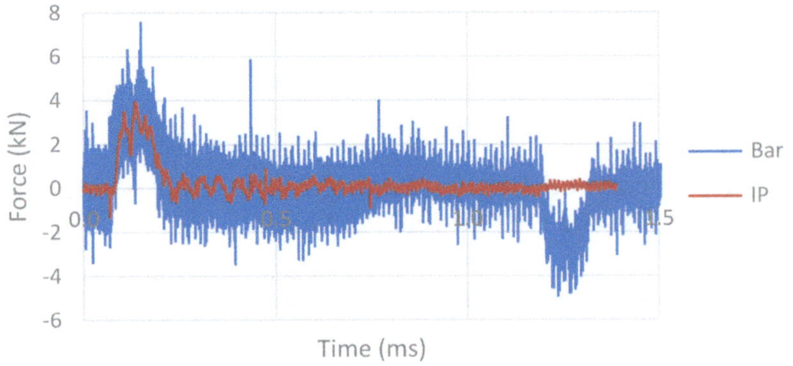

Fig. 24.2 Matched data between IP and gas gun

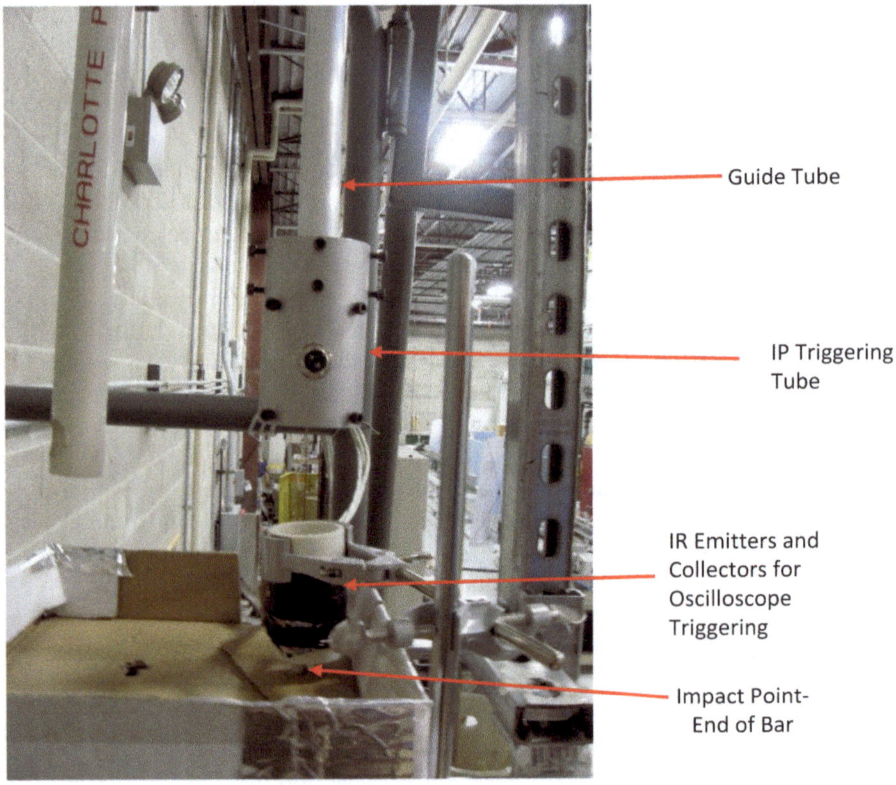

Fig. 24.3 Vertical drop bar interface

24.5 Vertical Impact Testing Method for Comparing Tests Between the Instrumented Projectile (IP) and a Long Bar

Testing procedures involving vertical impact are relatively simpler than those involving horizontal impact. Instead of using compressed gas to propel a projectile into a bar, gravity is involved. Similar basic procedures and details as the horizontal impact can be used to conduct the vertical impact testing. The projectile should be dropped from a consistent height through tube which is properly aligned with the impact rod (Fig. 24.3).

Different types of rods may be used in both vertical drop-weight impact and horizontal gas gun tests. In this study, the rod impacted on the end had a diameter of 0.375″ and a Young's modulus of 73.1 GPa. The traveling time of the IP's was around 26 ms and the sampling rate for the IP was 1 MHz. For the vertical drop bar, a Wheatstone bridge with two active arms was used with all resistors having resistance of 350 Ω. The voltage supplied to the Wheatstone bridge was 2.01 V. The gain was 100. The gage factor for the strain gages was 2.150. Figure 24.4 shows the experimental results. The result from the IP and that from the Bar seem to agree quite well.

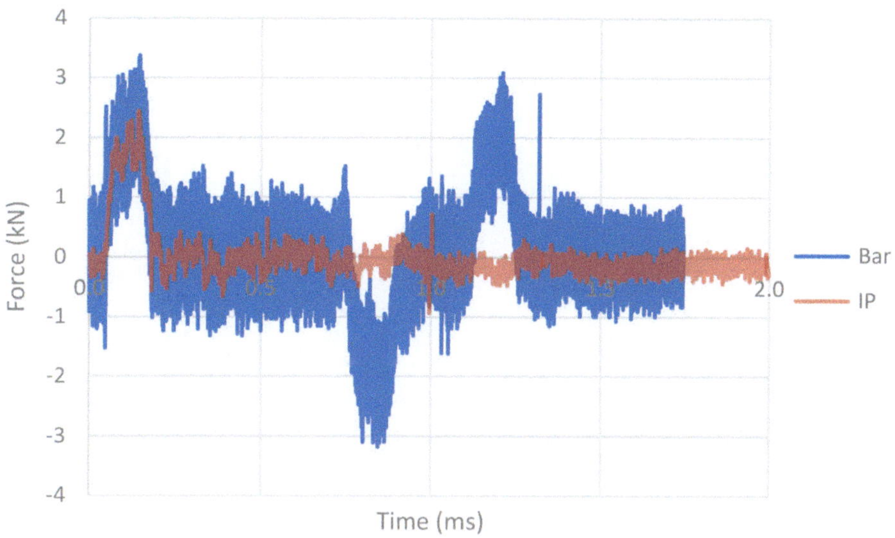

Fig. 24.4 Data comparison between IP and Bar data from a vertical drop bar test

24.6 Conclusions

This study presents both horizontal and vertical impact tests to validate the recently established instrumented projectile. Both testing results have confirmed the usefulness of the IP in providing the force histories during dynamic impact testing.

Acknowledgment The authors wish to express their sincere thanks of the financial support from Army Research Laboratory and the valuable comments from Dr. Chianfong Yen.

Reference

1. Li, G., Liu, D., Schleh, D.: An Instrumented projectile for dynamic material characterization. Proceedings of the 2015 SEM Annual Conference & Exposition on Experimental & Applied Mechanics, Costa Mesa, CA, 8–10 June 2015

Chapter 25
NIST Mini-Kolsky Bar: Historical Review

R.L. Rhorer, J.H. Kim, and S.P. Mates

Abstract The Society for Experimental Mechanics (SEM) has sponsored a series of technical paper sessions titled "Novel Testing Techniques" at their annual meetings. These sessions were organized by the Dynamic Behavior of Materials Technical Division of SEM and started in 2008. One of the novel techniques that we first learned about by attending SEM was the use of a small-size Kolsky bar system especially designed for the testing of polymer-single fibers. The Mini-Kolsky Bar was added to the National Institute of Standards and Technology (NIST) Kolsky Bar Laboratory based, in part, on the work presented at SEM conferences. A number of informal discussions at the annual SEM conferences added to the understanding and design details as we constructed our first small tension Kolsky bar. Subsequent developments of the NIST Mini-Kolsky bar, including improved gripping techniques were presented and discussed at SEM conferences. This paper reviews some of the work presented in the SEM's Novel Testing Techniques sessions and discusses the history of additional follow-on work precipitated by the original papers.

Keywords Kolsky bar • Hopkinson bar • Dynamic fiber testing • Kolsky bar history • Experimental mechanics history

25.1 Introduction

We have developed a tension Kolsky bar for the dynamic testing of single fibers in our laboratory at the National Institute of Standards and Technology (NIST). We refer to this apparatus as the "Mini-Kolsky Bar". Our approach, and the design of this system, did not originate at NIST—it is based on work by a number of researchers from several different labs. Professor Wayne Chen and his colleagues at Purdue University have been particularly helpful in our design effort [1]. This NIST Mini-Kolsky bar has been very successfully used in performing hundreds of tests on several different types of single polymer fibers [2, 3].

The purpose of this current paper is to review the history of developing the NIST Mini-Kolsky Bar. In particular, we highlight contributions to our work resulting from our involvement in the Society for Experimental Mechanics (SEM). The conference papers from the SEM annual meetings, in particular the Dynamic Behavior of Materials Technical Division and a series of sessions from 2008 to 2015 called "Novel Testing Techniques", have played a role in our application of the Mini-Kolsky bar. Also, the informal interactions facilitated by the SEM meetings have influenced our development and research approaches, including our development of a new fiber gripping technique. Our work at NIST has both benefitted from the SEM meetings and it has been a venue for us to report some of our contributions.

There are two reasons that reviewing the history of the NIST Mini-Kolsky bar are important. First, reviewing the ideas and designs can guide us in our future work: What approaches have worked and what questions remain? Where are the gaps in our understanding? We summarize some of these thoughts in Sect. 25.5.

Second, a historical review can contribute to our ongoing effort to quantify the uncertainties of Kolsky Bar measurements. For example, we know that alignment is important, but can we estimate an uncertainty in our results based on an assessment of our bar alignment? An approach to communicating uncertainties of the Kolsky Bar is presented in Sect. 25.6. The building of an uncertainty budget, filling in all the numbers in this outline, is beyond the scope of this paper; as a matter of fact, we suggest that this will be an ongoing cooperative effort with input from many different research

R.L. Rhorer (✉)
Guest Researcher, Engineering Laboratory, National Institute of Standards and Technology, 100 Bureau Drive, Gaithersburg, MD 20899, USA
e-mail: rrhorer@rcn.com

J.H. Kim • S.P. Mates
Materials Measurement Laboratory, National Institute of Standards and Technology, 100 Bureau Drive, Gaithersburg, MD 20899, USA

D. Casem et al. (eds.), *Dynamic Behavior of Materials, Volume 1*, Conference Proceedings of the Society for Experimental Mechanics Series, DOI 10.1007/978-3-319-41132-3_25

institutions. Improved uncertainty assessments will assist in establishing broadly accepted standards for Kolsky bar dynamic material testing.

In preparing this conference paper we have reviewed our Mini-Kolsky bar project from the standpoint of where the pertinent ideas came from—or at least where we learned of the basics for a specialized dynamic polymer fiber testing system. We do not claim any unique inventions on our part, and we apologize if we have not given proper credit to all those who have made special contributions to the techniques of using a small Kolsky bar for single-fiber dynamic testing. An evolutionary scenario seems best to describe the history of the Kolsky bar work—small, but important, developments. These developments have come over many years to bring the state-of-the-art to the point where we started building the Mini-Kolsky Bar at NIST.

25.2 Background: Dynamic Material Properties

There has always been a practical need to understand how materials behave under load: "How much does the object deform when you push on it?" or, "Will this beam break when you drop a rock on it?" Robert Hooke (1635–1703) is credited with the idea that there is a mathematical relationship between the load on a component and its deformation, now referred to as Hooke's law. Thomas Young (1773–1829) contributed the idea that there was a material property—the modulus of elasticity often referred to as Young's modulus—that relates the stress (load per area) in the component to its strain (change in length per unit length). (Young is given credit for the idea, although it can probably be traced to a paper by Leonard Euler in 1727.) In Young's own words [4]:

> [W]e may express the elasticity of any substance by the weight of a certain column of the same substance, which may be denominated the modulus of its elasticity, and of which the weight is such, that any addition to it would increase it in the same proportion, as the weight added would shorten, by its pressure, a portion of the substance of equal diameter.

Even though Young's wording is difficult to understand, the idea is correct. Applying new mathematical approaches and the principles of classical mechanics scientists developed the ideas of Hooke and Young, and others, into what we now call "Solid Mechanics". It seems a safe bet that many practical scientists understood that when applying loads rapidly that materials behave differently than when slowly loaded, and experiments to measure dynamic properties were needed. John Hopkinson, a professor at Cambridge University in the nineteenth century, is often cited for early scientific experiments to measure dynamic properties. In 1872 he published results documenting a considerably higher strength for an iron wire—up to a factor of two—when the load was applied by impact. John Hopkinson's son Bertram followed on with dynamic experiments in the early part of the twentieth century. This interesting history of the Hopkinson's work is summarized by both Kolsky [5] and Rhinehart [6] in books written in the early 1950s.

The description of the relationship between stress and strain, now referred to as the constitutive equation for a material, relies on experiments involving challenging measurements. An excellent and concise description of the measurement problem we are discussing here was presented by Professor James F. Bell in the 1960s [7]:

> On a cylindrical solid the simultaneous measurement of an axial force and the axial deformation it produces is conceptually one of the simplest experiments in experimental physics. That a vast literature has been written during the last two centuries to describe the functional relation between load and deformation in such experiments attests to the complexity of nature.

For slow rates—quasi-static type loading—the simultaneous measurement of load and displacement might appear as a straight forward experimental problem. This is the traditional material testing where a load frame applies a force to a specimen. The magnitude of the force is measured with a load cell. The extension or compression of the sample is measured by a transducer on the specimen or motion of the frame, and the results are a load-displacement curve. A great deal of research work has led to new techniques and specialized equipment for determining load-displacement curves. The load frames and associated instrumentation, such as modern servo-hydraulic testing machines, have become very sophisticated instruments. However, testing under high rates is still beyond the load-frame approach. Historically dynamic testing has developed from impact experiments with some now common approaches evolving from the Hopkinson work. In these approaches stress waves in elastic bars provide the dynamic loading. In recent decades there have been efforts to bridge, or combine, aspects of the traditional load frame testing approaches and the dynamic stress-wave techniques. The goal of the different methods is aimed at answering Bell's statement of the problem quoted above.

Much of Bell's research work in the 1950s and 1960s was directed at measuring the deformation—or strain—directly on the sample under dynamic loads. By producing a fine grating on a small cylindrical specimen and using an optical diffraction technique, an accurate time history of the specimen strain could be recorded. Bell's approach has high resolution, both in

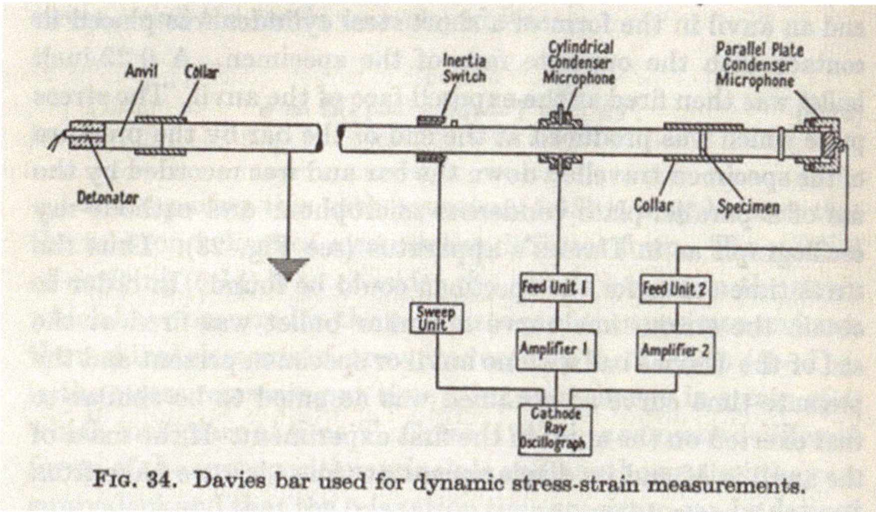

FIG. 34. Davies bar used for dynamic stress-strain measurements.

Fig. 25.1 Schematic drawing of the historic "Davies bar" used by H. Kolsky [8]

strain and time, and can measure large plastic strains, but it does appear that it would be difficult and time consuming to make the specimens.

In the same era as Bell's diffraction grating work, techniques related to Hopkinson's impact experiments were being developed. In particular, the approach of using two long bars with a small sample held between the bars became popular. Historically the approach appears to be two Hopkinson pressure bars sandwiching the short specimen, and hence the name "split Hopkinson pressure bar" is used. Also, the basic theory of the split bar was developed and clearly presented by H. Kolsky; hence, the name "Kolsky bar" has also become common. The basic concept of this now widely used apparatus is illustrated in Fig. 25.1 [8]. This apparatus looks much like our modern apparatus except the "condenser microphones" have been replaced with common metal foil strain gages, and the impact is usually applied by an air gun launching a projectile rather than an explosive charge at the end. This drawing is copied from Kolsky's book where he labeled it a "Davies bar". There has been some debate over the years which name should be used—"split Hopkinson pressure bar" or "Kolsky bar"—but maybe we should be calling it a "Davies bar." At NIST, in about 2000, we started developing our new dynamic materials measurement laboratory and decided to use the name "Kolsky bar" because it seemed to be a more general term (pressure bar might imply a compression bar).

The specimens are easy to make for the Kolsky bar compared to Bell's approach, but the strain is not measured directly in the specimen as stated as a requirement in his description of the problem. In the Kolsky bar the strain in the specimen is calculated from elastic strain measurements on the bar. The experimental problem is how to accurately measure the elastic strain as a function of time and relate this strain measurement to the motion of the end of the bar; and then, specify the resulting deformation of the specimen. Much of the practical work related to Kolsky bar research in the past 50 years has been in determining the plastic-strain time-history in the specimen as related to the strain-time recorded signal from a strain gage on the bar some distance from the specimen location.

Historically, metal resistance strain gages—originally thin wire gages that evolved into the now common metal foil gages—were used in impact experiments. In the early 1950s there was a considerable effort in developing the wire strain gages, including work at NIST (then called National Bureau of Standards (NBS)). A special conference was held as part of the 50th anniversary of NBS in 1951 highlighting progress in strain gages. The NBS Director, A.V. Astin, wrote in the introduction to the book publishing the papers of the conference: "Work is in progress on strain gages consisting of a conducting coating applied by an evaporation technique" [9]. The development of the Kolsky bar as a tool for dynamic material property measurements appears to be in parallel to the development of the metal foil strain gages.

However, there were concerns that the gages, calibrated in a quasi-static fixture, would not respond with the same sensitivity in dynamic applications and high strains. In 1970, Professor William Sharpe published a very important paper which we have referred to in answering questions related to strain gages: "Dynamic Plastic Response of Foil Gages" [10]. In this paper, Professor Sharpe compares the metal foil gages with output from an interferometric strain measuring technique in impact experiments. There was reasonable agreement between the metal foil gages and interferometric gage up to about 8 % strain.

Applying strain gages directly to metal and ceramic specimens have been used by some researchers, but it is impossible with the size of the single fiber . It is worth noting that the modern high-speed video recording capability coupled with advances in the relatively new field of digital image correlation (DIC), when applied to Kolsky bar experiments, may provide the direct strain measurements in the specimen as advocated by Professor Bell; however, the small size fibers prevent the use of this technique.

The adaption of the Kolsky bar approach to tension would appear to be straight forward—apply a tension impact pulse rather than a compression pulse. However, there are a number of practical problems associated with tension testing such as gripping the specimen. By 2008 the NIST Kolsky lab had established the capabilities and gained experience in high rate testing of metals, including a unique pulse-heating capability; however, we did not have the equipment or experience in tension testing of polymers. So, to answer a need for a capability to test single polymer fibers we turned to others for the necessary specialized techniques. About the same time, there was an increased interest at the Society for Experimental Mechanics (SEM) meetings regarding applying the traditional dynamic methods to novel material testing techniques. Therefore we have benefitted from participating in the SEM conferences.

25.3 Background: Novel Testing Techniques Sessions at SEM Annual Meetings

For more than half a century the Society for Experimental Mechanics (SEM) has been a technical community where researchers developing the Kolsky bar techniques have presented and discussed their work. (Note: we have chosen to use the name "Kolsky bar" in this paper to apply to a variety of dynamic test systems including several variations of the "Hopkinson bar", such as the split-Hopkinson pressure bar.) The society started in the 1940s and was originally called the Society for Experimental Stress Analysis (SESA) and evolved into our current technical society called SEM. SESA/SEM played an important role in the development, and applications, of variable resistance strain gages, such as the famous SR-4 gage (described in an interesting article in the historic Handbook of Experimental Stress Analysis [11].) Also, from the beginning the society has been a home for the development of dynamic measurement techniques, using strain gages and other transducers. So, with the advantage of hindsight, it seems natural that the SEM was the society to highlight Kolsky bar system design and development.

There are significant SEM papers related to Kolsky bar development from the 1960s through 2000, but from a quick review there appears to be an increase in the reports on measuring dynamic properties for materials other than metals, especially using Kolsky bars, in the past two decades. With the establishment of the Dynamic Behavior of Materials Technical Division, and in particular a series of sessions called "Novel Testing Techniques", the SEM annual meetings became an excellent forum for researchers interested in applying dynamic tests to new materials. Therefore, for those of us building new dynamic test apparatuses the Novel Testing Techniques sessions provided a place to learn about recent developments, interact with researchers currently active in the field, and to present our ideas for rapid feedback.

In reviewing the technical program for the SEM annual meetings from 2008 to 2015, there have been 23 sessions called "Novel Testing Techniques." (In 2013 three of these sessions were actually called "Novel Techniques" and in 2014 three sessions were called "Novel Testing.") In the 23 sessions there were 103 papers listed with authors from many different universities and institutions, international as well as many locations in the USA. About a third of the papers were from three institutions (Purdue University, Army Research Laboratory, and Sandia National Laboratory).

It appears that all the papers addressed dynamic measurements, and a majority of the reported research involved an apparatus of the Kolsky or split-Hopkinson type. Of the 103 papers, 15 titles included the word "Hopkinson" (or a related acronym) and 24 titles included the word "Kolsky".

In terms of our work at NIST, we were particularly interested in these sessions because of our effort to perform dynamic tests on polymer fibers. Dynamic tension testing seems like an obvious extension of traditional tension tests—just make the test machine move faster. The challenges of increasing the speed in conventional test machines were introduced in the first paper in 2008 Novel Testing Techniques I: "Analysis of Dynamic Tensile Testing" by X. Xiao of General Motors Corporation [12]. This paper discusses the dynamic stress equilibrium issue. It is interesting to note that this is a key issue for many dynamic tests and is often mentioned in related work. Several papers refer to pulse shaping techniques in regard to achieving force equilibrium in testing soft materials and other new applications. In our work two papers that provided a valuable background in pulse shaping and understanding the force equilibrium condition were published in *Experimental Mechanics* by Chen [13] and by Song [14].

Discussions of the uncertainty of the measurement results (addressing the question "How good are these measurements?") is an underlying theme of much of the work presented in the Novel Testing Techniques. All researchers have an interest in this question, but we are particularly interested at NIST because it is part of our mission to characterize uncertainty of measurement processes. In the last session of the 2015 SEM meeting (Novel Testing Techniques II) one of the papers addressed this directly: "Data Reduction Uncertainties in Kolsky Bar Experiments on Metals", E.E. Nishida, B. Song, E. Corona, Sandia National Laboratories [15]. This paper presentation was valuable. Also, the discussion following this paper was significant—there were a number of questions and comments concerning other uncertainties that have not yet

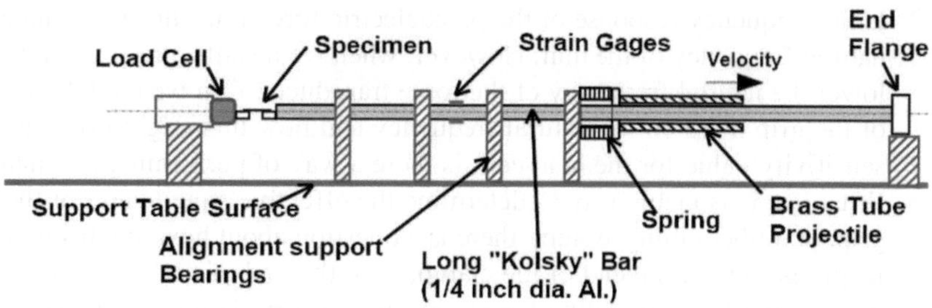

Fig. 25.2 Schematic drawing of NIST tension mini-Kolsky bar for fiber testing

been addressed such as the effect of bar straightness. The need for additional work in the ongoing efforts to understand the uncertainty of dynamic testing results may indicate a need for continuing the novel testing technique sessions in the future!

25.4 NIST's Mini-Kolsky Bar

The fiber-testing Kolsky bar that we built at NIST, starting in 2009, is shown schematically in Fig. 25.2. The drawing labels the major components. The idea and basic configuration of our system are based on the apparatus we observed in a visit to Dr. Wayne Chen's laboratory at Purdue University. The long bar with a strain gage is an obvious connection to the traditional Kolsky bar. The use of the load cell to replace the transmitted bar is a reminder of the traditional load frame approach: the mini-Kolsky bar shown could be thought of as a traditional testing machine with simply the moving frame being replaced with a high velocity ram. The end velocity of the bar, often referred to as the particle velocity, is caused by the impact of the hollow tube on the end flange producing a stress wave to propagate down the bar. The required rapid loading is achieved, but the extension of the specimen is limited by the length of the striker tube.

We were familiar with Kolsky bar testing of metals from several years of work with our larger (15 mm diameter) bar system; however, the mini-Kolsky bar presented some new challenges. We had the experience and necessary equipment to apply and record strain gage and load cell data at the high rates required. However, as we started our work on the Mini-Kolsky bar we had questions about the fiber attachment methods, the use of the piezoelectric load cell, details about the pulse shaping needed for testing single polymer fibers, and gripping techniques for the small fibers (approximately 10 μm diameter). Interactions with other researchers were valuable in addressing these concerns.

The pulse length in our Mini-Kolsky Bar design is limited by the length of bar that can be used unsupported by bearings. And, the velocity of impact is limited by the force of the spring and it was difficult to add a stiffer spring because of the limits of hand compression and release of the spring. To address a desire for a longer pulse and higher impact velocity, a project to update the Mini-Kolsky Bar was undertaken by one of the authors (Jae Hyun Kim). This system uses an air gun to fire a projectile down a long barrel that becomes part of the incident bar. This new system was modeled after the large tension Kolsky bar in the NIST laboratory. The design of this tension system is presented in a paper on the tension testing of a special steel alloy by Mates [16]. There will be future discussion papers about the updated Mini-Kolsky Bar.

25.5 Unanswered Questions

In reviewing our work with the Mini-Kolsky Bar, we have recognized that we still have some unanswered questions, or gaps in our understanding. These questions may provide topics for papers and discussions at future "Novel testing techniques" sessions. Some of our questions are summarized below:

1. Do we have an accurate measure of the change in the fiber length during a test? By using the strain gage signal of the "return pulse" we calculate the end motion of the bar. We have compared the end motion calculated from the strain gage signal with a DIC measurement of the end motion, and there is agreement within the measurement uncertainties. We have installed a laser displacement gage on the NIST Mini-Kolsky Bar (adapted from the system presented by Lim [17]). All three methods agree on the displacement time history of the end of the bar; however, we have yet to verify that there is no relative motion of the force transducer end. How can the stability of the force transducer be verified? Can a laser displacement gage be adapted to measure the actual motions at each end of the specimen?

2. Is the frequency response of the piezoelectric force transducer adequate? The manufacturer of the force gage provides a natural frequency of the unit. However, when we install a gripping system to hold on to the fiber end, the added mass will lower the natural frequency of the force transducer. Can we model, or perform a test experiment, to determine the effect of the grip mass on the natural frequency and how this might affect the force-time history? We use the manufacturer's sensitivity value for the load cell; is there a way of performing a dynamic calibration of the cell?
3. What is a reasonable way to determine the effective gage length of the specimen? With either the direct clamping type grip, or a fiber gluing system, there is a question about how much slip may occur inside the grip. Assuming that the gage length is right at the end of the clamp, or at the end of the glue blob, may produce some uncertainty of the gage length number used in calculating the strain-time history. Is there a modeling or experimental techniques that will provide a quantitative uncertainty estimate of the effective gage length?
4. What method should be used to calculate the stress in the specimen? To calculate stress from the force measurement we need the cross sectional area of the specimen. How should the area be determined? We have used scanning electron microscope measurements of the diameters of several fibers and then calculated an average area. Is there a better way of determining a cross sectional area for calculating stress?
5. What is the effect of pre-tension on the results from the Mini-Kolsky Bar? In a traditional load frame type test a fiber can be loaded with a slight amount of slack and the load—displacement data can be adjusted to take into effect the removal of any slack in the specimen. This does not work in a dynamic test. To ensure that there is no slack, it may be necessary to pre-load slightly. How does this preload affect the measurement results?

25.6 Outline for Uncertainty Budget

Before a technique like testing of single polymer fibers can become a standard test, there needs to be an assessment of the instrument's measurement results uncertainty. One method of summarizing and communicating the overall performance of a measuring technique is referred to as an "uncertainty budget". This approach involves analyzing each aspect of a technique including an estimate of the uncertainty of each error component. Then all of the individual uncertainties are combined into an overall uncertainty of the final measurement results.

Researchers usually look at their data with some sort of statistical approach. The "scatter" in the data can be expressed in terms of a standard deviation. The "expanded uncertainty" can be expressed in terms of the standard deviation times a coverage factor. The variation or scatter when testing different samples can be a result of material property variations; however, it can also be a result of inconsistencies, or unknowns, in the test apparatus or test technique. Our long-term goal is to be able to quantify variables in the test equipment and procedure; that is, to determine uncertainties in our results related to the test technique so we can evaluate variations in the material properties of the specimen. Test results variations cannot all be attributed to variations in material properties of the specimens.

A proposed approach for preparing an uncertainty budget for the Mini-Kolsky Bar apparatus is to analyze the causes of the uncertainties by separation into four components:

1. the uncertainty estimate for the measured strain gage signal as related to the time history of the end motion of the input bar
   (a) gage factor and sensitivity calibration uncertainties
   (b) gage bonding effects
   (c) dynamic response of metal foil gages
   (d) gage length effects on the strain gage signal (mechanical filtering)
   (e) effect of gage in the center of the bar related to end motion
      - straightness of bar and alignment effects
      - bearing drag effects
2. Uncertainty estimate for the force-time signal recorded from the piezoelectric load cell
   (a) sensitivity calibration uncertainty
   (b) frequency response effects
   (c) alignment of the load cell effect
3. Uncertainty of the force-displacement history of the specimen

(a) force non-equilibrium effects
(b) pulse shaping effects and repeatability
(c) slack or pre-tension effect
(d) effect of slipping of fiber in grip

4. Uncertainty of the stress-strain response of the specimen material

(a) uncertainty of sample size (diameter and area calculation) for calculating stress
(b) uncertainties in calculating strain from displacement

We have only begun the rather lengthy task of filling in numbers for this outline. This project will need the input of many researchers and the support of several different research organizations. That is why we see the value of discussing this effort as part of the SEM technical paper sessions such as the Novel Testing Techniques.

The approach of combining individual uncertainties into the four major groups; and then, combining the groups into an overall uncertainty is an effort beyond the scope of this paper. We propose that the approach used for estimating the strain gage measurement uncertainty in Rhorer [18] is a starting point for the more complete uncertainty analysis outlined here.

25.7 Conclusions and Future Work

The Mini-Kolsky Bar has been an effective research tool at NIST. The design and construction of the system has been impacted by our participation in SEM conferences. The background for the Kolsky bar has a long history, building on dynamic testing research conducted over the past century. This historical review has highlighted several areas where additional work could be beneficial to achieve the ultimate in "simultaneously measuring the axial force and the axial deformation" of the single polymer fibers.

References

1. Personal communications with Professor Weinong Wayne Chen of Purdue University over several years including informal discussions at SEM Annual Meetings
2. Kim, J.H., Heckert, N.A., Kang, K.-L., McDonough, W.G., Holmes, G.A., Rice, K.D.: Statistical characterization for tensile properties of co-polymer aramid fibers: Loading rate effects. Society for Experimental Mechanics, Annual Conference, Costa Mesa, California, 8–11 June 2015
3. Kim, J.H., Rhorer, R.L., Kobayashi, H., McDonough, W.G., Holmes, G.A.: Effects of fiber gripping methods on single fiber tensile test using Kolsky bar. Society for Experimental Mechanics, Annual Conference, Indianapolis, Indiana, USA, 7–10 June 2010
4. Robinson, A.: The last Man Who knew everything: Thomas young, the anonymous polymath Who proved Newton wrong, explained How We See, cured the sick, and deciphered the Rosetta stone, among other feats of genius, p. 126. Pi Press, New York (2006)
5. Kolsky, H.: Stress waves in solids. Dover Publications, New York (1963). An unabridged and corrected republication of the work first published by the Clarendon Press, Oxford, 1953
6. Rinehart, J.S., Pearson, J.: Behavior of metals under impulsive loads. Dover Publication, New York (1965). An unabridged and corrected republication of the work first published by the American Society of Metals in 1954
7. Bell, J.F.: The physics of large deformation of crystalline solids. Springer tracts in natural philosophy, vol. 14, p. 1. Springer, Berlin (1968)
8. Kolsky, Ibid., p. 154
9. Astin, A.V.: Characteristics and application for resistance strain gages, circular 528. National Bureau of Standards, Washington, DC (1954). Proceedings of the NBS Semi-centennial Symposium on Resistance Strain Gages held at the NBS on November 8 and 9, 1951
10. Sharpe Jr., W.N.: Dynamic plastic response of foil gages. Exp. Mech. **10**(10), 408–414 (1970)
11. Hetenyi, M. (ed.): Handbook of experimental stress analysis, pp. 170–180. John Wiley & Sons, New York (1950)
12. Xiao, X.: Analysis of dynamic tensile testing. Society for Experimental Mechanics, Annual Conference, Orlando, USA, 2–5 June 2008
13. Chen, W., Zhang, B., Forrestal, M.J.: A split Hopkinson bar technique for low-impedance materials. Exp. Mech. **39**(2), 81–85 (1999)
14. Song, B., Chen, W.: Dynamic stress equilibration in split Hopkinson pressure bar tests on soft materials. Exp. Mech. **44**(3), 300–312 (2004)
15. Nishida, E.E., Song, B., Corona, E.: Data reduction uncertainties in Kolsky bar experiments on metals", Society for Experimental Mechanics, Annual Conference, Costa Mesa, California, 8–11 June 2015
16. Mates, S., Abu-Farha, F.: Dynamic tensile behavior of a quenched and partitioned high strength steel using a Kolsky bar. Society for Experimental Mechanics, Annual Conference, Costa Mesa, California, 8–11, June 2015
17. Lim, J., Chen, W.W., Zheng, J.Q.: Dynamic small strain measurements of Kevlar 129 single fibers with a miniaturized tension Kolsky bar. Polym. Test. **29**, 701–705 (2010)
18. Rhorer, R.L.: Uncertainty of strain gage measurements on Kolsky bars. Society for Experimental Mechanics, Annual Conference, Greenville, South Carolina, 2–5 June 2014

Chapter 26
A General Approach to Evaluate the Dynamic Fracture Toughness of Materials

Ali Fahad Fahem and Addis Kidane

Abstract In this paper a combined experimental and numerical approach is proposed to evaluate the dynamic fracture toughness of materials. A circular tube specimen, made of Aluminum alloy, 7050-T7651, having a spiral crack on its outer surface is used to demonstrate the technique. A torsional Hopkinson bar is used to generate a dynamic torsion pulse, which in turn creates predominantly a tensile load along the crack line of the specimen. The torque applied on the sample is obtained from strain gages attached on the bars using one dimensional wave equation. Commercial FE package, ABAQUS, is used to simulate the dynamic fracture parameters. In this case the subspace projection method and standard implicit integration in ABAQUS with time increment are used, assuming the system is linear. The angle of twist associated to the torque measured on the specimen is used as input in the model. The dynamic stress intensity factor is determined using minimum strain energy theory. Digital image correlation is used to measure the deformation field near the crack tip. The measured strain/displacement fields are used to determine the exact time at which the crack initiated. The result show that, all the three classical modes of fracture are existing, but mode I is at least one order magnitude higher than the others. Also the dynamic SIF of the Aluminum alloy 7075-T6571 is higher than the quasi-static SIF (i.e $K_{Ic}{}^{d}=1.36K_{Ic}$), The value obtained in this experiment is in well agreement with the values documented in the literature.

Keywords Dynamics fracture toughness $K_{Ic}{}^{d}$ • Torsional stress wave • Mode I fracture • Spiral notch • Torsional fracture

26.1 Introduction

Most mechanical structural components are subjected to dynamic loading in service, and they must be designed to survive high strain rate loading during operation conditions. Dynamic fracture is the main concern especially in brittle materials, and understanding the dynamic fracture initiation and propagation behavior of materials has always been an interest topic in materials and mechanics community. Fracture mechanics in general has been a subject of interest for decades and different parameters such as, Stress Intensity Factor (SIF), elastic energy release rate, J-integral, crack-tip opining displacement, and crack-tip opining angle, have been develop and used to characterize the fracture properties of most engineering materials under quasi-static loading conditions [1]. To select and judge which material may be best fit for specific application, these parameters are important, and usually, they are obtained experimentally by using standers methods, such as ASTM E399 for the stress intensity factor K_{Ic}, E1820 for elastic-plastic toughness J_{Ic}, E1920 to evaluate CTOD, and so on [2, 3].

On the other hand, dynamic fracture is important in structures that are subjected to a sudden and high strain loading, such as pressurized water reactors (PWR) under thermal shock loading due to loss of coolant accident [4], oil and gas pipelines under high-pressure conditions, fuselages of airplanes at cruising altitude and more. Unfortunately, the quasi-static methods cannot be used to accurately determine the fracture parameters in materials under dynamic loading conditions [5]. For these reasons many researchers have been working to developing an experimental method that can be applied to evaluate the dynamic fracture toughness of materials [6–8]. In general, there are two different methods, with some modifications, that have been used to estimate the dynamic fracture toughness of materials, Charpy V-notch test and three-point bending test in a modified Hopkinson pressure bar apparatus.

A.F. Fahem (✉)
Department of Mechanical Engineering, University of South Carolina, Columbia, SC 29209, USA

Department of Mechanical Engineering, University of Al-Qadisiyah, Al-Diwaniyah, Qadisiyah Province, Iraq
e-mail: afahem@email.sc.edu

A. Kidane
Department of Mechanical Engineering, University of South Carolina, Columbia, SC 29209, USA
e-mail: kidanea@cec.sc.edu

D. Casem et al. (eds.), *Dynamic Behavior of Materials, Volume 1*, Conference Proceedings of the Society for Experimental Mechanics Series, DOI 10.1007/978-3-319-41132-3_26

Charpy V-notch test uses the Empirical correlation formulas to estimate fracture toughness from Charpy Fracture energy. One of the limitation of the Charpy test is, the fracture energy can only be measured on standardized specimen (geometry, size, notch sharpness and loading rate) and only at intermediate loading rate. Due to its simplicity, the standard Charpy test is still very popular to characterize fracture toughness, despite of its well-known theoretical weaknesses [9, 10].

In recent years, a modified split Hopkinson pressure bar (MSHPB) apparatus is used to investigate the dynamic initiation toughness of materials. The principle of modified MSHPB is similar to general SHPB experiment, except in the case of MSHPB only the incident and striker bar are required. In MSHPB experiment, a three-point bend or four-point single edge specimen will be sandwiched between the incident bar and a supporting frame. The striker bar, made of the same material as of the incident bar, will be projected towards the incident bar by the help of compressed air gun and will generate a well-defined loading pulse that propagates towards the specimen. Once the compressive stress wave reaches the specimen, some of the wave will propagate to the specimen and some of the waves reflected back to the incident bar. Using a strain gage bonded in the incident bar, both the incident and reflected waves can be recorded. Using one dimensional wave theory the force acting on the sample can be calculated as a function of time. Then the fracture intensity factor can be calculated using quasi-static equation [11, 12]. The weakness of this method is that this technique can be used to calculate the fracture intensity factor only if the time of fracture is sufficiently long enough to neglect the inertia effect. To satisfy this condition and to avoid the transient effect, in most cases the experiment has to be conducted at lower speed. This limits the application of this method, in the case of only lower strain rate loading [5]. Also, this technique has limitation associated with sample size and geometry.

A new approach to evaluate the quasi-static fracture toughness of materials is proposed recently using spiral notch specimen under pure torsion loading [13]. This technique has been demonstrated by using different materials, such as ceramics, metal, polymer and concerts under quasi-static loading conditions [13]. More recently, the potential of the technique to be used for studying the dynamic fracture properties of materials has been indicated [14, 15]. The main advantage of the spiral notch specimen under torsion loading over the MSHP fracture experiment is the possibility of eliminating the inertia effect on the fracture load.

In this work, a general experimental and numerical method to measure the dynamic fracture toughness $K_{IC}{}^{d}$, of materials is proposed. The experiments are conducted on a Torsional split Hopkinson specimen using spiral notch Aluminum alloy 7050-T7651 samples. FF simulation is conducting using ABAQUS and the data obtained from the experiment as input. Finally, the dynamic stress intensity factor is obtained.

26.2 Experimental Approach

The TSHB has 2 m long incident and transmitter bar made of 1 in. diameter Aluminum 7075 alloy.

26.2.1 Material and Specimen Geometry

An Aluminum alloy 7050-T7651 is studied. The mechanical properties of the material are given in Table 26.1. A Spiral Notch Torsion Specimen (SNTS) is used to investigate the dynamic fracture properties. A tubular cylinder specimen with wall thickness of 2 mm is machined from Aluminum alloy rod. The cylindrical specimen has a 45° spiral notch grove on the outer surface as shown Fig. 26.1. The outer diameter and gage length of the specimen are 19 and 60 mm respectively and the depth of the grove is 1 mm. The specimen is sandwiched between the incident and transmitter bar and glued with epoxy. The dynamic tension load perpendicular to the grooved line (opening mode) is generated by subjected the specimen under shear loading.

Table 26.1 Properties of material under static condition (Aluminum alloy 7050-T7651)

Density	Modulus of elasticity	Poisson's ratio	Yield stress	Shear modulus	Fracture toughness K_{Ic}
2.81 g/cm^3	71.7 GPa	0.33	530 MPa	26.9 GPa	28 MPa$\sqrt{m}$

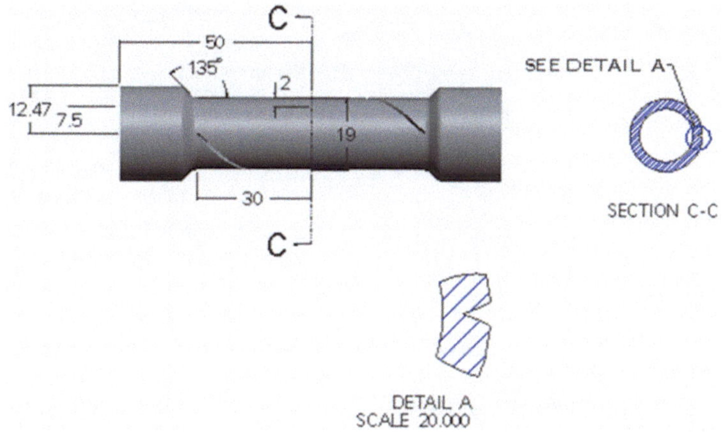

Fig. 26.1 Specimen Geometry

26.2.2 Experimental Setup

Torsional Hopkinson bar is used to generate the opening mode load on the SNET specimen. It is an experimental technique generally used to study properties of martial under high strain rate shear loading. Its principle is the same as that of usual Hopkinson pressure bar except the loading direction. The experimental analysis is based on one dimensional wave theory. A hydraulic driven torsional split Hopkinson bar device shown in Fig. 26.2, is used in this experiment. Detail descriptions of the torsional setup used in this work is available in the litterateur [16, 17] and only briefly described here. It has two, incident and transmitter, 2 m long, circular bars of 25.4 mm in diameter made from aluminum alloy 7075-T6. The incident bar has a clamp mechanism at 200 mm from one end of the bar where a rotary actuator is attached. The specimen is sandwiched between the incident and the transmitter bar and glued with J-B Weld epoxy. During testing, half of the incident bar is clamped from twisting and a torsional strain is stored in the portion of the bar between the clamp and the rotary actuator. By suddenly releasing the clamp, a torsional wave is allowed to propagate towards the specimen. When the wave reached the specimen, some of the wave transmitted through the specimen and some of the wave reflected back. The incident, transmitted and reflected strain signals are recorded using the strain gages attached at the middle of the incident and transmitter bars. The torque applied on the specimen are calculated from these signal using one dimensional wave equations.

The average strain rate that cause a fracture in a specimen can be calculated from Eq. (26.1), which is dependent on the parameter of specimen and bars [16].

$$\dot{\gamma}_s(t) = \frac{2CD_S}{L_S D}\dot{\gamma}(t) \tag{26.1}$$

The average shear strain in the specimen is given Eq. (26.2)

$$\gamma_s(t) = \frac{2CD_S}{L_S D}\int_0^t \dot{\gamma}_R(t)\cdot dt \tag{26.2}$$

26.2.3 High Speed Imaging and Digital Image Correlation

To capture the local full field deformation i.e. 3D surface displacements and strain, of the region around the crack tip, digital image correlation is used. The specimen is speckled with black and white spray paint ash shown in Fig. 26.2b. Two high speed cameras, SAX2 by Photon Inc., are used to capture set of images during loading at a framing rate of 50,000 frames/s. The images are post proceed using commercial software Vic 3D by correlated solution Inc. and the displacement and strain fields around the crack tip region is obtained.

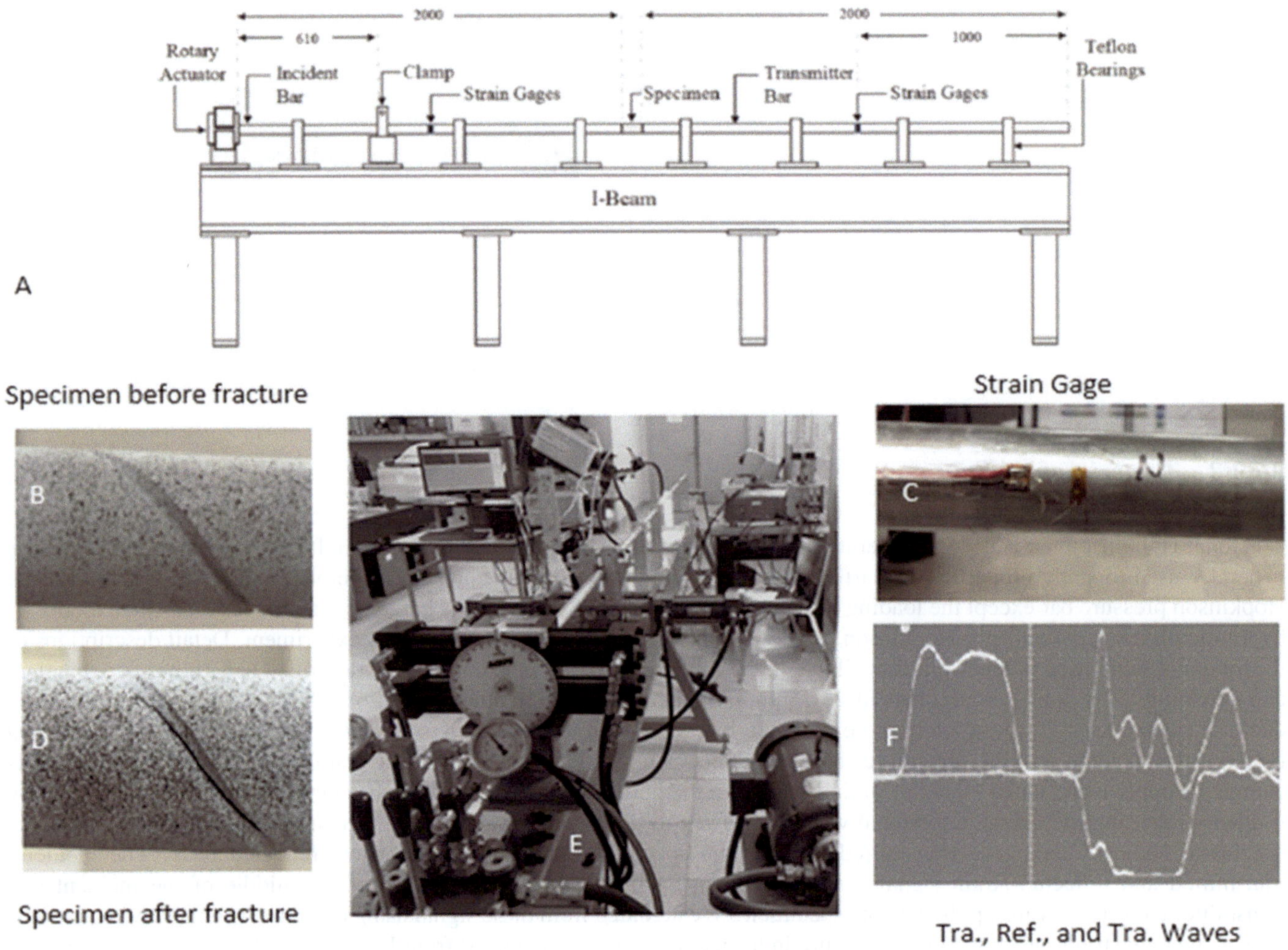

Fig. 26.2 Experimental setup used in this study (**a**) schematic diagram (**b**) specimen before loading (**c**), strain gage (**d**) fracture specimen (**e**) actual photo of the setup (**f**) typical signs recorded by the strain gages

26.3 Numerical Analysis

Since there is no analytical solution, for spiral notch sample subjected to dynamic loading, numerical method is needed to calculate the fracture parameters. Finite element analysis FEA is used to calculate the dynamics stress intensity factors of the spiral notch crack around a cylindrical tube Fig. 26.3c. Using Abaqus dynamic implicit method, a mode is developed for one quarter of the actual length of the sample as shown in Fig. 26.3. In the FE modeling, one end is fixed in three dimensions (*r*, *θ*, *and z*), and the impulse torsional load is applied on the other end as a displacement boundary condition as shown Fig. 26.3d.

The singular elements around the crack tip were built with a polar FE quadratic hexahedral elements. In order to obtain a refined wedge element around the crack tip, first a small diameter cylinder is created, then the cylinder is meshed by dividing it into 40 slices (9° for each from the center). This generates concentrated refined mesh around the crack tip, with the smallest element size of 20 μm as shown in Fig. 26.3b. The 3D solid model with a structure element quadratic hexahedral C3D20RH type is used with pitch angel of 15°. The total number of nodes are 288,130 and the total number of elements are 66,700 as shown in Fig. 26.3E.

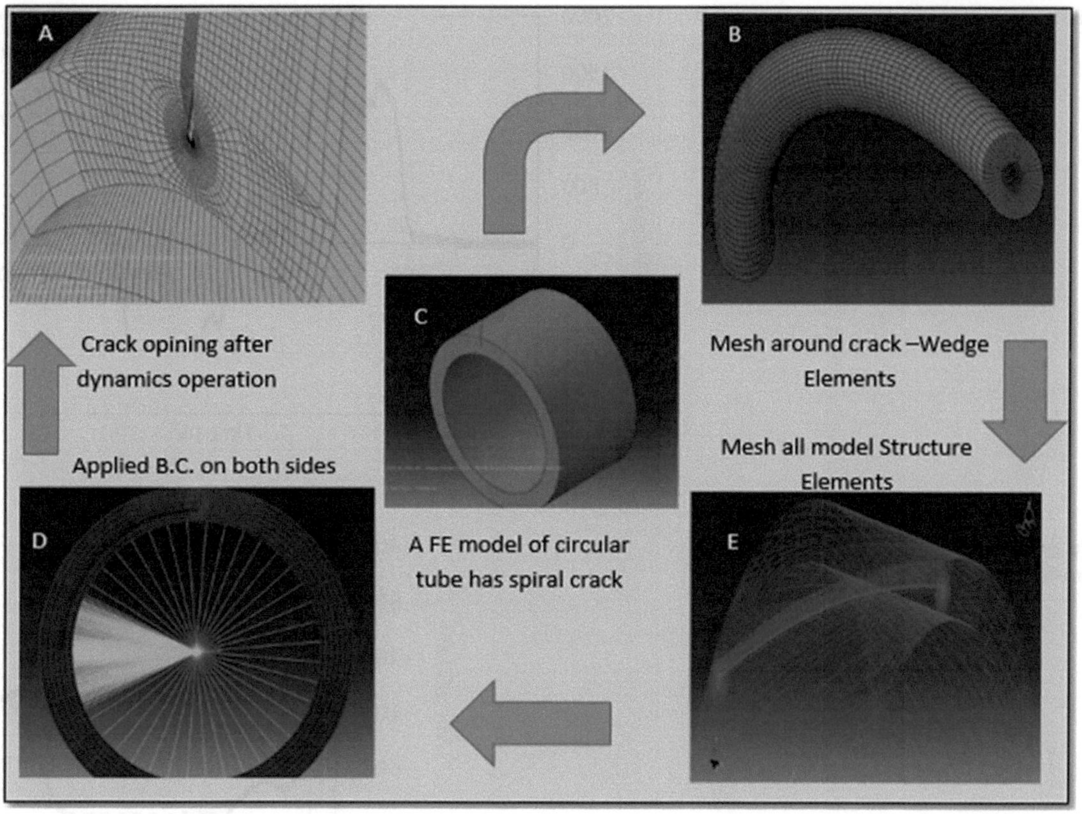

Fig. 26.3 FE Model (**a**) mesh (**b**) refined mesh around the crack tip (**c**) the one quarter spiral model (**d**) boundary condition (**e**) elements

26.4 Result and Discussion

Typical incident, transmitted and reflected signals measured in this experiment are shown in Figs. 26.4–26.5. The torque applied on the specimen as a function of time is calculated based on the strain gage signals and Eqs. (26.3 and 26.4).

$$T_1(t) = \frac{2GJ}{D}(\gamma_I(t) + \gamma_R(t)) \tag{26.3}$$

$$T_2(t) = \frac{2GJ}{D}(\gamma_T(t)) \tag{26.4}$$

Where $T_1(t)$ a torque calculated on the incident bar side and $T_2(t)$ a torque calculated on the transmitter bar. Ideally if the inertia is negligible the two value should be very close. In this experiment, the two torques are averaged and the associated twist angle as a function of time is calculated and plotted as shown in Fig. 26.6.

As shown in the figure, the maximum load is reached at about 83 μs. According, the angle of twist applied to the sample is calculated based on the strain signal shown in Fig. 26.5. The angle of twist in radian is plotted as a function of time shown in Fig. 26.6. The angle of twist as a function of time is given as input displacement to the FE model to estimate the dynamic stress intensity factors fracture.

In order to capture the exact time at which the fracture initiated, the full field displacement and full field strain obtained from the DIC are used. Displacements and strains at two points each across the crack front line are extracted and plotted in Figs. 26.7a–f. The deformation field (displacement and strain) of the material shown in Fig. 26.7, can be divided to three parts. First, a rigid body motion, from t = 0 to 40 μs, in this period the opening mode strain (e_{yy}) is almost zero. The second part is in the period between 40 and 80 μs, in this time range, the specimen deforms elastically and there is no plastic deformation or fracture. It is clear that the displacements have a distinct future at a time exactly 83 μs, indicating plastic

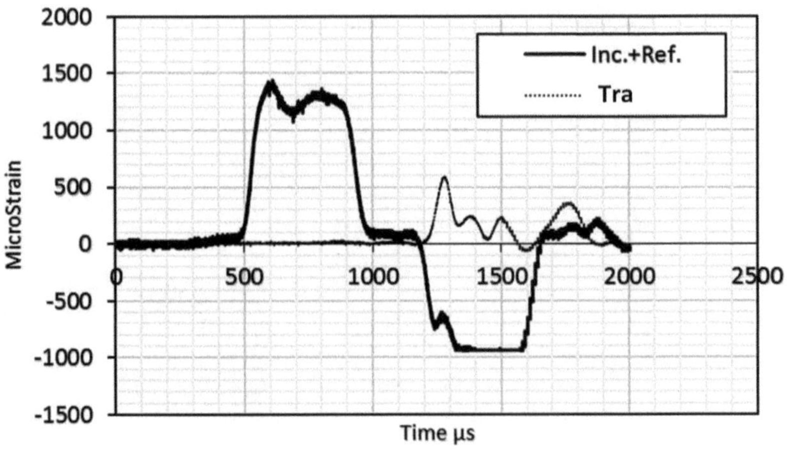

Fig. 26.4 Strain gages data

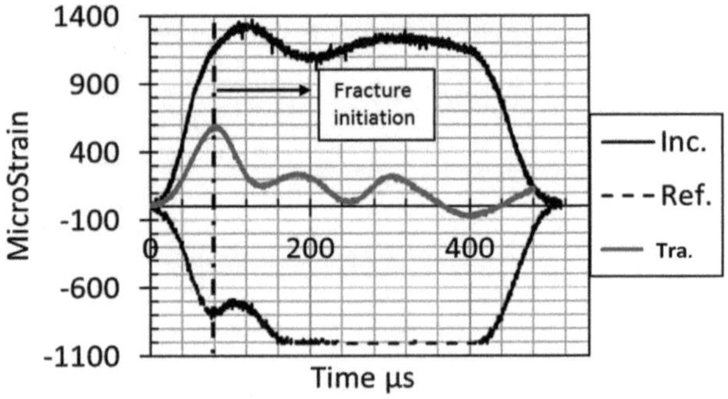

Fig. 26.5 Merge and center effected macrostrains

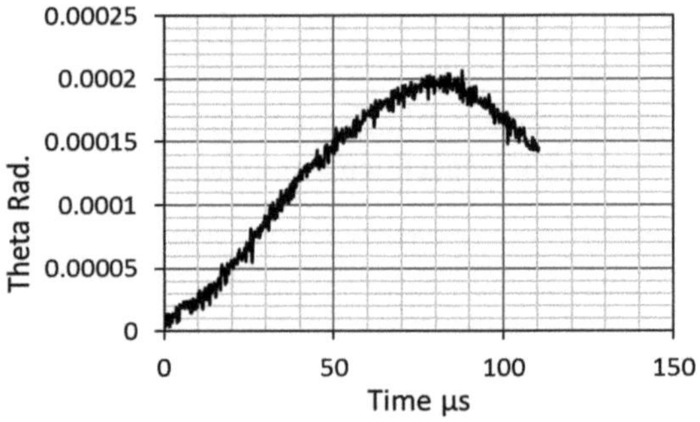

Fig. 26.6 Load profile

deformation is happing and the fracture is initiated at this time. For example, the gap between the opening strain (e_{yy}) of the two points across the crack front line is maximum at this point. In addition, the slop of the out of plane displacements of the two points changes at this time.

The mode I of DSIF as function of time across the crack front line is obtained and plotted as shown in Fig. 26.8. As shown in the figure, the DSIF is higher and almost constant at the middle region of the specimen. It is observed that, the crack is initiated at the middle of the crack front and propagated towards the end of the notch. To avoid the boundary effect and the associated nonlinearity, on the fracture parameters, the value in the middle is averaged and presented.

Figures 26.9, 26.10 and 26.11 shows the average DSIF for mode I, II, and III respectively, as function of time. As shown in the figures, the mode I DSIF is at least one order magnitude larger than the other two modes. This indicates that mode I is predominately the driving feature for the fracture. Knowing the exact time at which the crack initiated, the DSIF at initiation can be calculated. In this case, the mode I DSIF at initiation (at 83 µs) for Aluminum alloy 7075-T6571, is found to be $38.109 \mp 2\ MPa\sqrt{m}$. The value is higher than the quasi-static SIF value of $28\ MPa\sqrt{m}$ reported in the literature [18]. It is well documented that, the dynamics stress intensity factor in metal should be at least 40 % higher

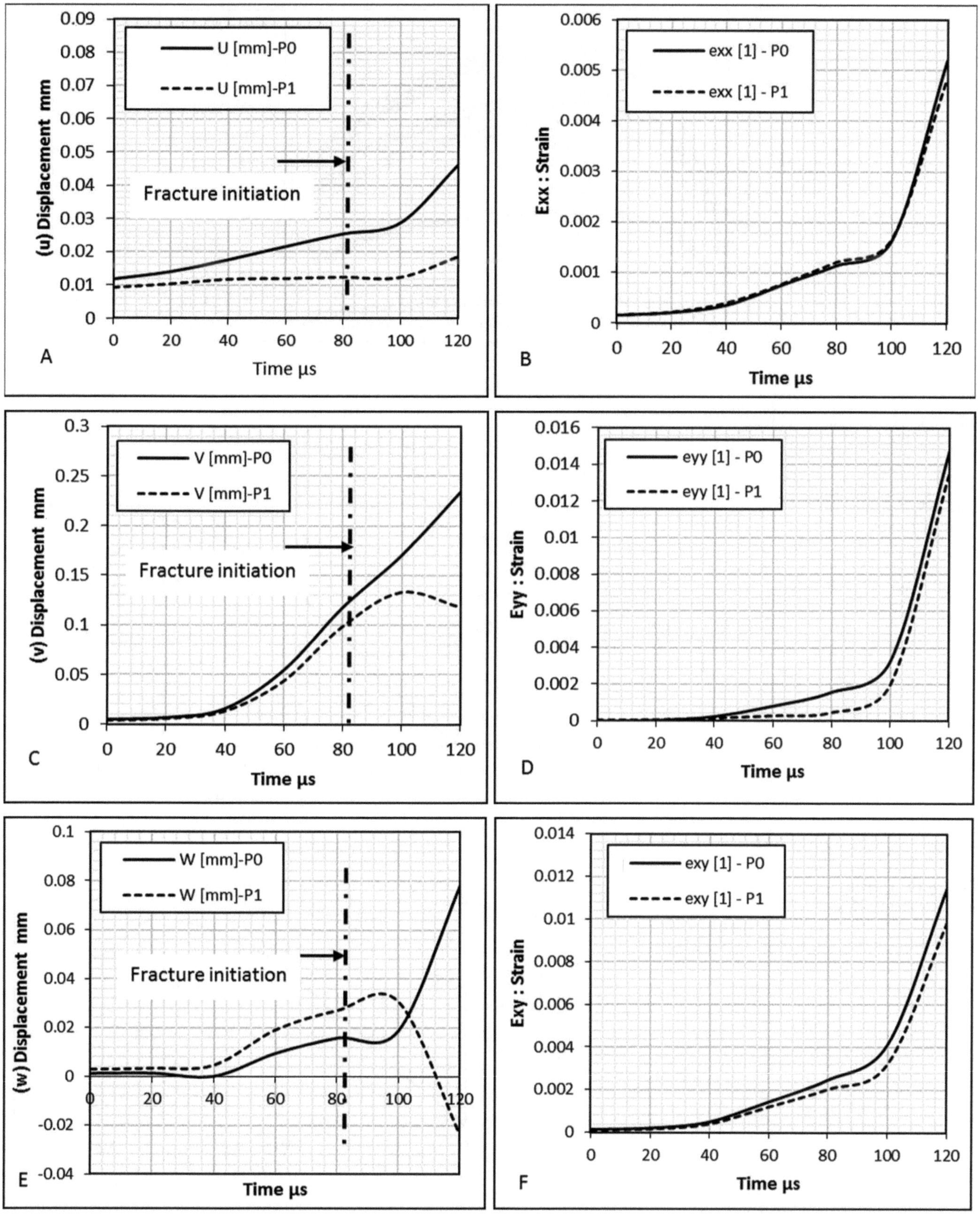

Fig. 26.7 Full field CTO displacement

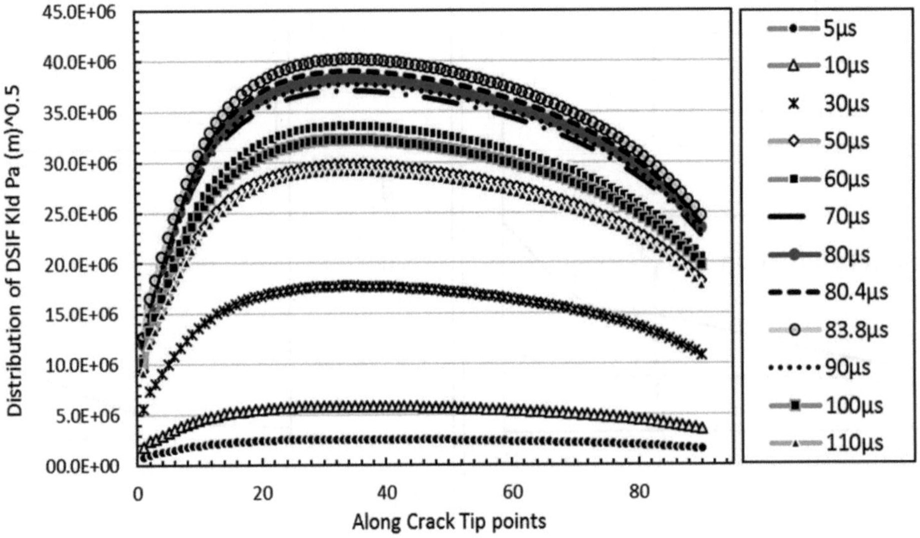

Fig. 26.8 Distribution of DSIF K_{Id}

Fig. 26.9 Dynamic stress intensity factors K_{Id}

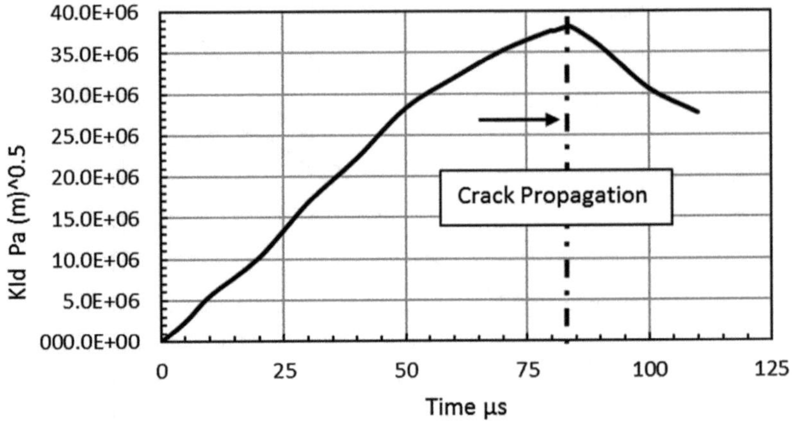

Fig. 26.10 Dynamic stress intensity factors K_{IId}

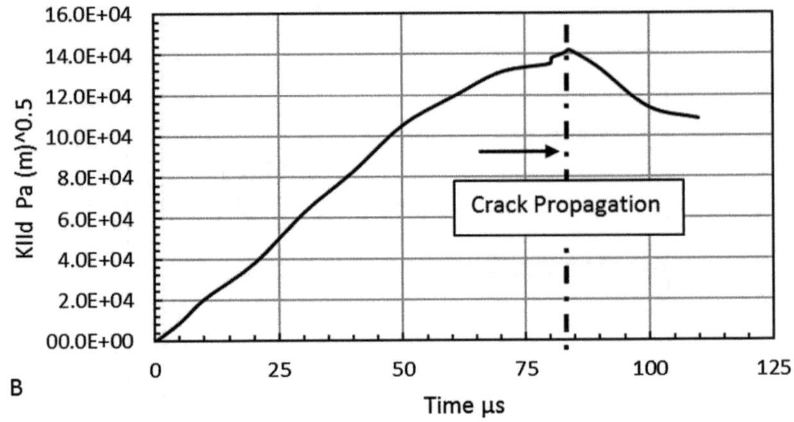

Fig. 26.11 Dynamic stress intensity factors K_{IIId}

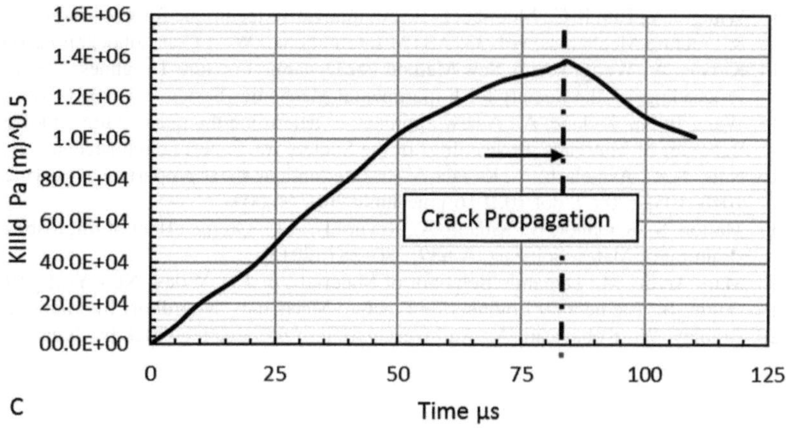

than the quasi-static static stress intensity factor [19, 20]. The value obtained for strain rate sensitive Aluminum alloy 7075-T6571, in this work is 35 % higher than the quasi-static value, which is close to what is reported in the literature [19, 20].

26.5 Conclusion

A new approach to estimate the dynamic fracture toughness $K_{Ic}{}^{d}$ of materials at high strain rate loading is proposed by extending a similar work at quasi-static loading. A spiral crack with 45° pitch angle around the circumference of a circular tube made of Aluminum alloy 7075-T6571 is used to study the dynamic mode I fracture properties of materials. The load is generated by subjecting the specimen to dynamic shear loading using torsional Hopkinson bar. The torque and the corresponding angle of twist applied on the specimen is obtained from the torsional Hopkinson experiment. Further the time at which the fracture initiated is obtained using 3D digital image correlation, by measuring and comparing displacements and strains at two points across the crack front line. To calculate the fracture parameters, a three-dimension FE model is develop using ABAQUS using the experimental data as input. Finally, the DSIF for all three modes are calculated. It if found that the mode I DSIF is at least one order magnitude higher that the other two modes. It is also found that the DISF of the Aluminum alloy 7075-T6571 used in this study is higher that the quasi-static SIF $(i.eK_{Ic}^{d} = 1.36K_{Ic})$. The value obtained in this experiment is in well agreement with the values documented in the literature. The method need to be verified using different engineering materials.

References

1. Howard, K., Dana, M.: Mechanical Testing and Evaluation, vol. 8, pp. 425–677. ASM Handbook, Material Park (2000)
2. Sih, G.: Mechanics of Fracture Initiation and Propagation. Kluwer Academic, Dordrecht (1991)
3. Zhu, X.-K., James, A.: Review of fracture toughness (G, K, J, CTOD, CTOA) testing and standardization. Eng. Fract. Mech. **85**, 1–46 (2012)
4. Griesbach T.J.: Fracture toughness characterization of Japanese reactor pressure vessel steels. Nuclear Power Division, EPRI, 2006
5. Ravi-Chandar, K.: Dynamic fracture. Elsevier, Amsterdam (2004)
6. Xu, Z., Li, Y.: A novel method in determination of dynamic fracture toughness under mixed mode I/II impact loading. Int. J. Solids Struct. **49**, 366–376 (2012). doi:10.1016/j.ijsolstr.2011.10.011
7. Loya, J.A., Fernández-Sáez, J.: Three-dimensional effects on the dynamic fracture determination of Al 7075-T651 using TPB specimens. Int. J. Solids Struct. **45**, 2203–2219 (2008). doi:10.1016/j.ijsolstr.2007.11.027
8. Jiang, F., Rohatgi, A., Vecchio, K., Cheney, J.: Analysis of the dynamic responses for a pre-cracked three-point bend specimen. Int. J. Fract. **127**(2), 147–165 (2004). doi:10.1023/B:FRAC.0000035058.03627.30
9. Schindler H.J.: The correlation between Charpy fracture energy and fracture toghness from a theortical point of view. Proceedings of the 12th European Conference on Fracture, Sheffield, 841–847, 1998
10. Bannister A.C.: Determination of fracture toughness from Charpy impact energy: Procedure and validation, British Steel plc., 1998
11. Rubio, L., Femandes, J., Navarro, C.: Determination of dynamic fracture-initiation toughness using three-point bendiang test in modified Hopkinson pressure Bar. Exp. Mech. **43**(4), 379–386 (2003)
12. Kidane, A., Shukla, A.: Quasi-static and dynamic fracture initiation toughness of Ti/TiB layered functionally graded material under thermo-mechanical loading. Eng. Frac. Mech. **77**, 479–491 (2010)

13. Wang, J.A., Liu, K.C., McCabe, D.E.: An innovative technique for measuring fracture toughness of metallic and ceramic materials. In: Fatigue & Fracture Mechanics, vol. 33. ASTM International, West Conshohocken (2003)
14. Kidane, A., Wang, J.A. A New Method For Dynamic Fracture Toughness Determination Using Torsion Hopkinson Pressure Bar. Conference Proceedings of the Society for Experimental Mechanics Series. doi: 10.1007/978-1-4614-4238-7, 2013.
15. Wang Jy-An, Kidane, A.: A new approach to determine the quasi-static and dynamic fracture toughness of engineering Materials, Dynamic Behavior of Materials, Proceedings of the Society for Experimental Mechanics Series, Vol. 1, 545–551, 2013
16. Naik, N.K., Asmelash, A., Kavala, V.R.: Interlaminar shear properties of polymer matrix composites: Strain rate effect. Mech. Mater. **39**(12), 1043–1052 (2007). doi:10.1016/j.mechmat.2007.05.003
17. Mallon, S., Koohbor, B., Kidane, A., Reynolds, A.P.: On the effect of microstructure on the torsional response of AA7050T-7651 at elevated strain rates. Mater. Sci. Eng. A **639**, 280–287 (2015)
18. Marc Andre, M.: Dynamic Behavior of Materials, p. 493. Wiley, New York (1994)
19. David, B.: Elementary Engineering Fracture Mechanics. Noordhoff International, Leyden (1974)
20. Bradley, W., Kobayashi, A.: An investagation of propagating crack by dynamics photoelasticity. Exp. Mech. **10**(3), 106–113 (1970)

Chapter 27
Which One Has More Influence on Fracture Strength of Ceramics: Pressure or Strain Rate?

M. Shafiq and G. Subhash

Abstract Both strain rate and confinement pressure have been known to have a strong influence on the failure strength of ceramics. However, the debate about which one has more influence has never been resolved. This manuscript aims to conclusively prove that confinement pressure has significantly higher influence than strain rate on the compressive strength of a ceramic. Normalized shear stress versus hydrostatic pressure plot on a variety of ceramics shows that the quasi-static and dynamic strength of brittle solids is a strong function of applied pressure and not the strain rate. The plots also revealed that despite the differences in material properties, test methods, and strain rate the data on failure strength fall in a narrow range and therefore a unified model, that extends the traditional Mohr-Coulomb criteria by adding an exponential term, can capture the overall deformation behavior of structural ceramics at high pressures.

Keywords JH-2 model • Brittle ceramics • Mohr-Coulomb • Ultrahigh pressures • HEL

27.1 Introduction

Brittle ceramics deformation behavior is known to be influenced by applied pressure and strain rate under controlled multi-axial loading conditions. One of the most extensively used criteria for characterizing brittle materials is Mohr-Coulomb (M-C) model due to its ability to account for the effect of hydrostatic pressure and Lode angle parameter. The traditional M-C model describes the failure strength (τ) as a function of applied pressure as [1]

$$|\tau| = \tau_0 + \alpha P \tag{27.1}$$

where $|\tau|$ is the shear strength under multiaxial loading, τ_0 corresponds to the material strength in pure shear (i.e. shear strength under no pressure), P is the hydrostatic pressure, and α represents the coefficient of internal friction or the proportionality constant. This criterion has been traditionally applied to materials which often exhibit elastic response followed by brittle fracture under unconfined or low-confinement pressure. With increasing confinement and/or high strain rates (e.g. shock loading) traditional brittle ceramics do exhibit large inelasticity, especially at stress levels beyond the Hugoniot elastic limit (HEL) or at large confinement pressures (e.g. Heard and Cline [2]). Under such multi-axial loading conditions the constitutive response may be comprised of micro-cracking, micro-plasticity, phase transformation, defect generation or some combination of the above. Several models have been proposed to describe the complex constitutive behavior of ceramic materials beyond the HEL. However, the failure strength under such multiaxial conditions may not be properly represented by the above linear dependence of pressure shown in Eq. (27.1). Also, the strain-rate dependence (if it exists) is not included in the model. Several models have been proposed to represent the complex constitutive behavior of ceramic materials and among the existing phenomenological models Johnson and Holmquist (JH) [3, 4] model has been applied to several ceramics with some success. The model incorporates a Drucker-Prager yield surface that evolves with damage through equivalent plastic fracture strain and represents the normalized intact strength σ_i^* of a ceramic as

$$\sigma_i^* = A\left(P^* + T^*\right)^N\left(1 + C\ln\dot{\varepsilon}^*\right) \tag{27.2}$$

where A and N are the intact material strength coefficient and exponent, respectively, and C is the strain rate coefficient. The pressure P and maximum hydrostatic tensile pressure T are normalized by P_{HEL} to yield P^* and T^*, respectively. The strain

M. Shafiq • G. Subhash (✉)
Department of Mechanical & Aerospace Engineering, University of Florida, Gainesville, FL 32611, USA
e-mail: subhash@ufl.edu

D. Casem et al. (eds.), *Dynamic Behavior of Materials, Volume 1*, Conference Proceedings of the Society for Experimental Mechanics Series, DOI 10.1007/978-3-319-41132-3_27

rate ($\dot{\varepsilon}$) is normalized using the reference strain rate of $\dot{\varepsilon}_0 = 1\ s^{-1}$ as $\dot{\varepsilon}^* = \dot{\varepsilon}/\dot{\varepsilon}_0$. However, the model shortcomings evolve from the inherent fashion in which the brittle ceramic response is formulated into pressure-sensitive regime (the term in the first parenthesis) and subsequent pressure insensitive regime (the term in the second parenthesis). Note from Eq. (27.2) that the pressure-dependence is modeled by a power law, while the rate dependence is modeled using logarithmic strain. Also, the model requires the curve to pass through the spall strength at one end and HEL of the material at the other end. The JH model, therefore, cannot always successfully capture both the low-pressure and the high pressure data simultaneously. Figure 27.1 illustrates the limitation of JH-2 model in capturing the pressure-sensitive response of several structural ceramics including boron carbide (B_4C) [5], silicon carbide (SiC) [8], aluminum nitride (AlN) [11], zirconium diboride-silicon carbide (ZrB_2-SiC) [14], float glass [15], and granite [16].

It can be observed from Fig. 27.1 that in case of AlN, for example, the procedural requirement of JH-2 model passing through the spall strength and HEL of material forces the curve to have a high stiffness in the beginning at low stress levels which gradually decreases as it progresses towards the HEL. The high stiffness at low hydrostatic pressure is contrary to normally expected behavior of lower specimen stiffness at low hydrostatic pressure and vice versa. Therefore, the model fails to provide a fit representative of full spectrum of the experimental data. Another limitation of the model is that Eq. (27.2) cannot be applied to post-HEL regime, where the behavior of a ceramic is dramatically different and is contrary to the model prediction of monotonic strength increase. At these high pressures, plasticity, phase transformation, defect generation, and cracking dominate the behavior, and the restriction described earlier will not be able to capture the experimental data effectively. Table 27.1 summarizes the relevant JH-2 model parameters reported in the literature for various ceramics.

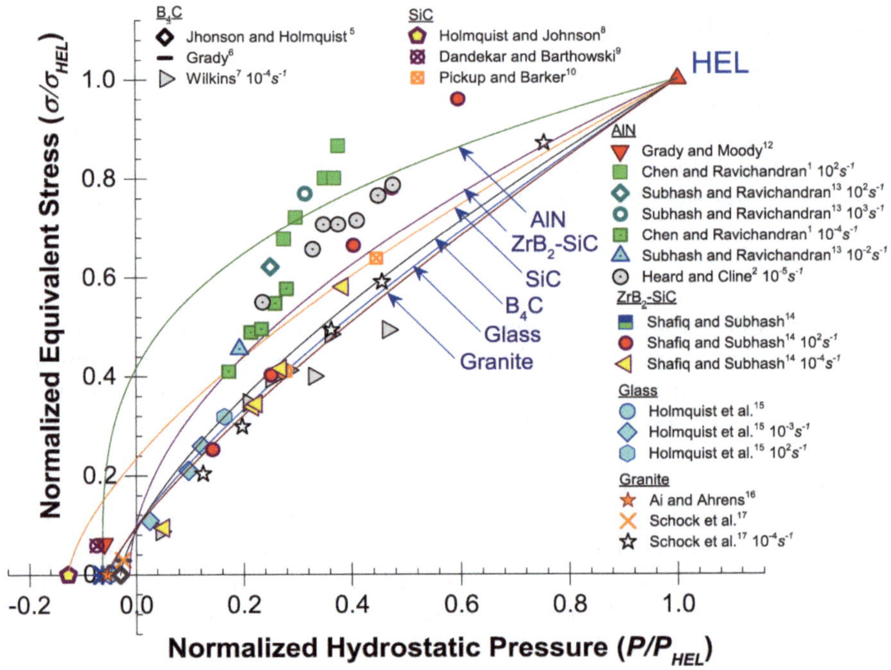

Fig. 27.1 Test data and JH-2 model fit for various ceramics including boron carbide (B_4C) [5], silicon carbide (SiC) [8], aluminum nitride (AlN) [11], zirconium diboride-silicon carbide (ZrB_2-SiC) [14], float glass [15] and granite [16]

Table 27.1 JH-2 model constants for B_4C [5], SiC [8], AlN [11], ZrB_2-SiC [14], glass [15] and granite [16]

Description	B_4C	SiC	AlN	ZrB_2-SiC	Glass	Granite
A	0.93	0.84	0.85	0.95	0.93	0.91
N	0.69	0.66	0.31	0.53	0.76	0.80
C	0.005	0.009	0.012	0.0038	0.003	0.005
HEL (GPa)	19	14.6	9	11.63	5.95	4.5
σ_{HEL} (GPa)	15.44	13	6	9.1	4.54	2.66
τ_{HEL} (GPa)	7.72	6.5	3	4.55	2.27	1.33
P_{HEL} (GPa)	8.71	5.9	5	5.4	2.92	2.73

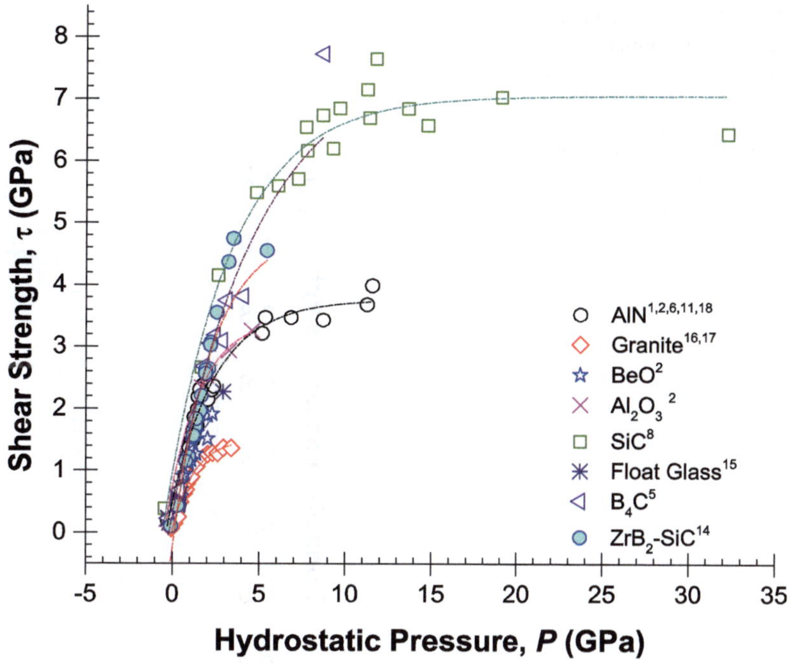

Fig. 27.2 Quasi-static and high strain rate data in hydrostatic pressure vs shear strength space for various structural ceramics [28]

27.2 Mechanisms of Deformation in Brittle Solids

The same data along with aluminum oxide (Al_2O_3), beryllium oxide (BeO) from Heard and Cline [2] and high pressure data for AlN [6, 18] and SiC [8] in the wake of HEL is re-plotted in hydrostatic pressure (P) vs shear strength (τ) space as presented in Fig. 27.2. Interestingly, similar trends are observed in the deformation behavior of all brittle ceramics. The above normalized ($P-\tau$) data for a series of structural ceramics reveals three distinct regimes: (i) macroscopic linear elastic response, which is highly pressure sensitive (ii) nonlinear pressure-dependent inelastic response, and (iii) pressure-insensitive response. In the linear elastic regime, the compressive strength is governed by several factors including initial microstructure, composition, defect distribution, etc. With increase in axial stress and confinement, a strongly linear pressure dependent behavior is observed. This regime evolves primarily due to the growth and interaction of large number of randomly distributed cracks emanating from pre-existing flaws within the material. As the pressure is further increased a new deformation mechanism (plasticity or phase transformation) comes into play and the material response transitions smoothly from crack dominated regime to plasticity. Beyond HEL the material tends to maintain its shear strength independent of pressure and dislocation-based failure mechanisms dominate the subsequent deformation behavior of the material until and unless a phase transformation occurs within the crystal structure [1, 19]. Among the material presented in Fig. 27.2, only SiC and AlN have data, which extend well beyond HEL and both show almost a pressure in-sensitive response. The data for B_4C is limited, but it has been well-known that it exhibits a strength weakening mechanism beyond HEL [6, 20]. This behavior has been attributed to the formation of nanometer-size amorphous bands [21] when the stress exceeds beyond HEL. Whether plasticity in the form of dislocations occurs at this stress level in this ceramic is unknown.

27.3 Proposed Model

Clearly the data for various ceramics, shown in Fig. 27.2, is different for different ceramics. However their overall behavior reveals similar trends. Therefore, we normalize the data with respect to their HEL values and plot them in shear stress and hydrostatic pressure space as shown in Fig. 27.3. Interestingly, the data collapses into a narrow range despite the differences in material properties, test methods, and strain rate. Since the data collapses into a single curve, a generalized constitutive model for brittle ceramics can now be constructed in the (P$-\tau$) plane as,

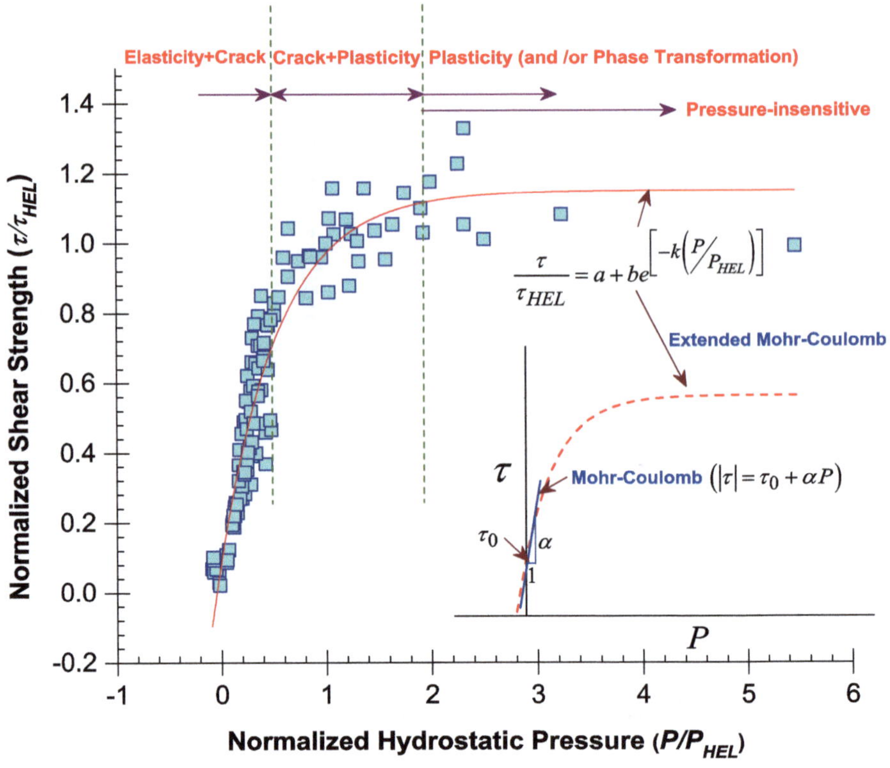

Fig. 27.3 Low and high strain rate data in normalized ($P-\tau$) space for several structural ceramics. The inset shows schematic representation of conventional Mohr-Coulomb model and extended Mohr-Coulomb model [28]

$$\tau = \tau_{HEL}\left(a + be^{-k\left(P/P_{HEL}\right)}\right) \tag{27.3}$$

Where a, b and k are constants and are 1.15, -1.06, and 1.78, respectively. Note that in this model, the strength is only pressure dependent and no strain rate dependence is required. The dominance of pressure on material behavior is expected because at high confining pressures (well beyond HEL), cracks can be suppressed and the ceramics yield by plastic deformation or phase transformation [6, 8, 18]. This behavior is well captured by the model beyond $\tau = \tau_{HEL}$, where the material has no pressure dependence because the stress remains constant (Fig. 27.3) due to activation of ductile deformation mechanisms with increasing confining pressure.

27.4 Influence of Pressure and Strain Rate

The lack of strain rate dependence is also well justified because even in JH-2 model [Eq. (27.2)], the strain rate coefficient is extremely small (0.003–0.012) and more than an order of magnitude smaller compared to the strength coefficient (A) and strength exponent (N) as shown in Table 27.1. These model constants are plotted in Fig. 27.4 for better comparison. Note that the range of these parameter values are reasonably close despite vast differences in their HEL values. For e.g., there is a six fold increase in equivalent stress at HEL (σ_{HEL}) or four fold increase in HEL value between boron carbide and granite, but the differences in the strength coefficient (A), strength exponent (N) and strain rate coefficient (C) is negligible. Figure 27.4 also reveals that the above parameter values remain almost constant for these ceramics indicating that the deformation characteristics of different structural ceramics are similar in nature. Equation (27.3) qualifies the earlier observations of Grady [6], Wilkins [7] and Johnson et al. [8] that pressure alone contributes to the low and high strain rate constitutive response of brittle ceramics and the influence of strain rate is not significant. A stronger evidence for reduced strain rate dependence of strength was provided by Anderson and Royal-Timmons [22] who conducted depth of penetration (DOP) experiments on alumina ceramic tiles as a function of confinement pressure and observed a strong coupling between the ballistic performance of the ceramic and impact velocity at low confinement pressures. However, with increase in confinement pressure, the ballistic performance of the ceramic exhibited less dependency on the impact velocity.

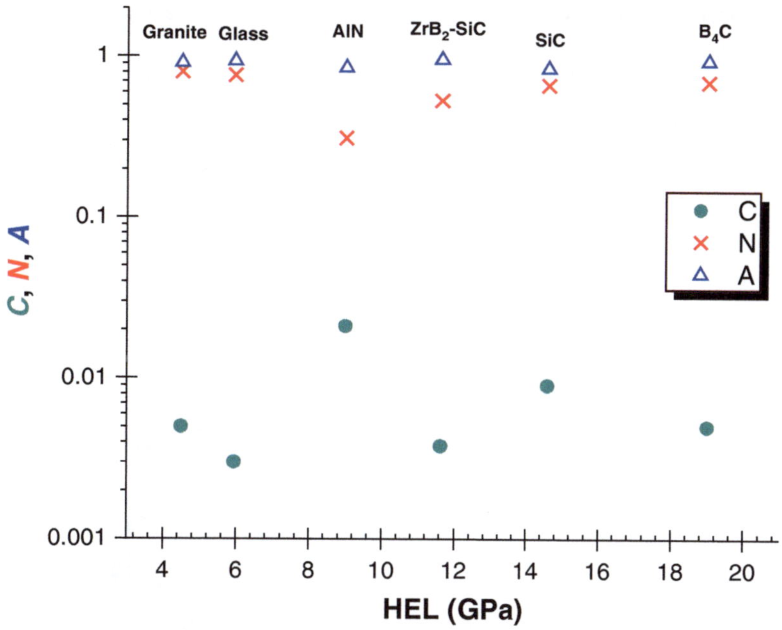

Fig. 27.4 Plot of strength coefficient (A), strength exponent (N) and strain rate coefficient (C) for a range of structural ceramics [14]

This result implies that the strength of a ceramic transitions from strain rate-dependent behavior to pressure-dependent behavior with increase in pressure i.e., from the crack inertia-dominated mechanism to crack shutdown mechanism and plasticity which activate in the moderate to high pressure regimes. Furthermore, considering the underlying low-pressure failure mechanisms, which are crack initiation, propagation, and interaction, it is perceivable that the newly formed crack surfaces slide against each other. Thus friction between the crack surfaces and crack growth accommodate the inelastic deformation. These frictional mechanisms and crack growth characteristics are in general strain rate dependent and hence contribute to the commonly observed strain rate-dependent failure strength at low or no confinement pressure [23]. The frictional resistance is further increased when external pressure is superposed which tends to close the crack, thus increasing the load bearing capacity of the ceramic and then resulting in higher failure strength. However, it is worth noting that ceramics exhibit an increase in strength from quasi-static to high strain loading even under unconfined conditions [1, 13, 23–26]. The apparent increase in strength under high strain rate is related to delay in the crack initiation or crack growth due to inertial effects. During the dynamic loading, the stress rises more rapidly than the ability of the crack to open and propagate due to material inertia. This behavior is reflected as an increase in the load bearing capacity of the ceramic. With continued increase in confinement pressure, the stress exceeds beyond the yield strength of the ceramic and other mechanisms such as plasticity and phase transformation come into play. In general, these mechanisms are known to occur at almost constant stress as noted in the high-pressure regime III (Figs. 27.2 and 27.3) and are insensitive to strain rate. Figure 27.3 also seems to provide an answer to one of the controversies on HEL, whether it represents the onset of widespread microcracking or widespread plastic deformation [27]. It can be seen in Fig. 27.3 that even before HEL is reached the curve starts to deviate from linearity. In the pre-HEL regime, microcracking is the dominant mechanism whereas plasticity can be sporadic and is a secondary deformation mechanism. At higher pressures, plasticity emerges as the primary deformation mechanism in the immediate post-HEL regime until a state is reached where the material flows freely without any further increase in resistance to deformation. Thus, at low confinement pressure brittle materials follow the traditional M-C criterion and at high pressure, the extended M-C criterion is more appropriate as schematically represented in the inset of Fig. 27.3. However, explicit determination of pressure component is not possible due to the strong coupling between pressure and strain rate and the inherent difficulties in building competent testing techniques.

27.5 Validation

To further validate the above model, let us consider the reported shear strength values in the literature from low and high strain rate confinement experiments on various ceramics. To obtain the failure strength τ_0 under pure shear (at zero confinement pressure), it is a common practice to conduct confinement tests at low pressure (typically below 500 MPa) by shrink-fitting a ceramic with a metal sleeve and then conducting experiments in uniaxial loading at different strain rates

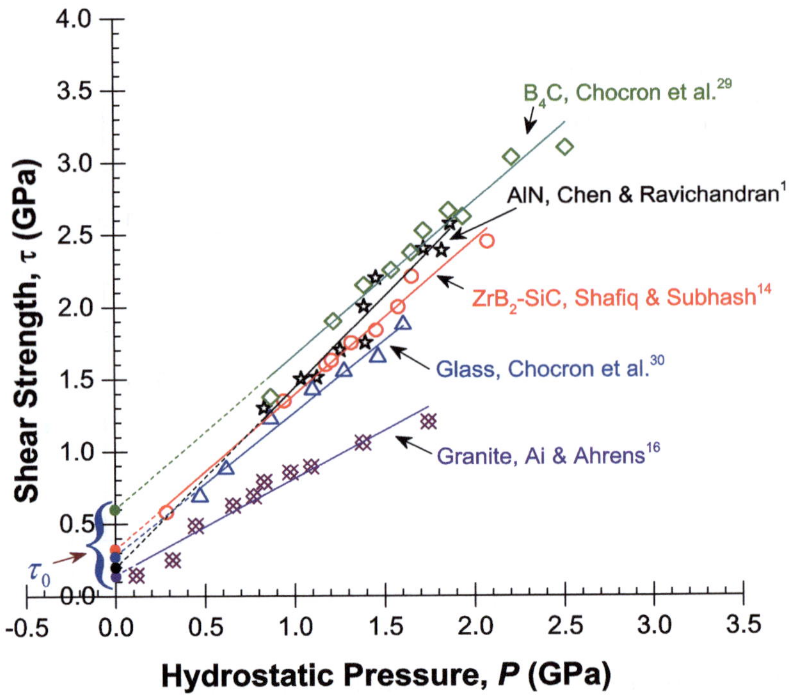

Fig. 27.5 Pressure versus shear strength plot to extract pure shear strength (τ_0) at zero confinement pressure and frictional slope (α) of various ceramics [28]

Table 27.2 Comparison of experimental and model predicted values of pure shear strength (τ_0) and frictional slope (α) for different ceramics [24].

Description		B_4C	SiC	AlN	ZrB_2-SiC	Glass	Granite
Pure Shear Strength (τ_0) GPa	Eq. (27.4)	0.69	0.60	0.18	0.40	0.20	0.12
	Experimental	0.59	0.86	0.28	0.32	0.29	0.14
Coefficient of friction (α)	Eq. (27.5)	1.67	2.08	1.13	1.58	1.47	0.92
	Experimental	1.07	1.21	1.26	1.07	0.82	0.67

[1, 14, 28]. Such data at low confinement pressure from literature for AlN, Granite, ZrB_2-SiC, and glass are plotted in Fig. 27.5 as shear stress ($\tau = [(\sigma_1 - \sigma_3)/2]$) vs hydrostatic pressure (P) plot for different levels of external confinement. The shear strength data for SiC at low confinement pressure and JH-2 model constants for Al_2O_3 and BeO are not available from literature and therefore they are not included in Fig. 27.5.

As data suggests, in this low pressure range, the linear Mohr-Coulomb criterion is well suited. For comparison, note that the extended Mohr-Coulomb model spans well into the GPa regime, see Figs. 27.2 and 27.3. Once again, the data for Fig. 27.5 are extracted from various sources in the literature as indicated. By extrapolating this data for each material to $P = 0$, the shear strength under pure shear (τ_0) can be determined. These values are given in Table 27.2 as experimental extrapolations. Interestingly, we can derive these τ_0 values from our extended Mohrs-Coulomb model (Eq. (27.3)) by setting $P = 0$ to obtain

$$\tau_0 = \tau_{HEL}(a + b) = 0.09{*}\tau_{HEL} \tag{27.4}$$

Knowing the value of τ_{HEL} given in Table 27.1, we can calculate the shear strength under pure shear for the above ceramics. Comparing these values, presented in Table 27.2, with those determined from Eq. (27.4) for various structural ceramics reveal a reasonably good agreement. Thus the proposed extended Mohr-Coulomb model with exponential pressure dependence of failure strength at high pressure indeed captures the low pressure dependent behavior of ceramics as well.

Additionally, Taylor series expansion of the exponential term in the extended model (Eq. (27.3)) also yields the frictional coefficient in Eq. (27.1) as

$$\alpha = -bk\tau_{HEL}/P_{HEL} \quad (27.5)$$

Note that α is also the slope of the experimental data in Fig. 27.4. From the values provided in Table 27.1 for τ_{HEL} and P_{HEL}, we can calculate the value of α in Eq. (27.5). A comparison of the two α values, shown in Table 27.2, for various ceramics reveals a larger discrepancy than for τ_0 values because the exponential model is intended to capture the data for the entire pressure range well beyond HEL and the α value only provides slope in the linear regime where the confinement pressure is low. Nevertheless, the proposed extended Mohr-Coulomb model with exponential pressure dependence of failure strength at high pressure indeed captures the overall behavior of ceramics reasonably well.

27.6 Conclusion

In conclusion, we showed that all structural ceramics considered here exhibit similar deformation behavior and contain three distinct deformation regimes: (i) elasticity and crack dominated regime, (ii) crack and plasticity dominated regime and finally (iii) plasticity and/or phase transformation dominated regime. A unified constitutive model representative of the deformation response of most structural ceramics was therefore constructed. The model showed that ceramics constitutive behavior is more influenced by applied pressure than strain rate.

References

1. Chen, W., Ravichandran, G.: Failure mode transition in ceramics under dynamic multiaxial compression. Int. J. Fract. **101**(1–2), 141–159 (2000)
2. Heard, H., Cline, C.: Mechanical behaviour of polycrystalline BeO, Al_2O_3 and AlN at high pressure. J. Mater. Sci. **15**(8), 1889–1897 (1980)
3. Johnson, G., Holmquist, T., Lankford, J., Anderson C., Walker, J.: A computational constitutive model and test data for ceramics subjected to large strains, high strain rates, and high pressures. Honeywell Report, DE-AC04-87AL-42550 (1990)
4. Johnson, G.R., Holmquist, T.J.: An improved computational constitutive model for brittle materials. AIP Conf. Proc. **309**(1), 981–984 (1994)
5. Johnson, G.R., Holmquist, T.J.: Response of boron carbide subjected to large strains, high strain rates, and high pressures. J. Appl. Phys. **85** (12), 8060–8073 (1999)
6. Grady, D.: Shock-wave compression of brittle solids. Mech. Mater. **29**(3), 181–203 (1998)
7. Wilkins, M., Cline C., Honodel, C.: Fourth Progress Report Of Light Armor Program (1969)
8. Holmquist, T.J., Johnson, G.R.: Response of silicon carbide to high velocity impact. J. Appl. Phys. **91**(9), 5858–5866 (2002)
9. Dandekar, D. P., Bartkowski, P. T.: Tensile strengths of silicon carbide (SiC) under shock loading (2001)
10. Pickup, I. Barker, A.: Measurement of the shear strength and damage kinetics of silicon carbide, 1 (1997)
11. Holmquist, T.J., Templeton, D.W., Bishnoi, K.D.: Constitutive modeling of aluminum nitride for large strain, high-strain rate, and high-pressure applications. Int. J. Impact Eng. **25**(3), 211–231 (2001)
12. Grady, D. Moody, R.: Shock compression profiles in ceramics (1996)
13. Subhash, G., Ravichandran, G.: Mechanical behaviour of a hot pressed aluminum nitride under uniaxial compression. J. Mater. Sci. **33**(7), 1933–1939 (1998)
14. Shafiq, M., Subhash, G.: Dynamic deformation characteristics of zirconium diboride–silicon carbide under multi-axial confinement. Int. J. Impact Eng. **91**, 158–169 (2016)
15. Holmquist, T. J., Johnson, G. R., Lopatin, C., Grady, D., Hertel, E. S. Jr.: High strain rate properties and constitutive modeling of glass (1995)
16. Ai, H., Ahrens, T.: Simulation of dynamic response of granite: A numerical approach of shock-induced damage beneath impact craters. Int J. Impact Eng. **33**(1), 1–10 (2006)
17. Schock, R., Heard, H., Stephens, D.: Stress-strain behavior of a granodiorite and two graywackes on compression to 20 kilobars. J. Geophys. Res. **78**(26), 5922–5941 (1973)
18. Rosenberg, Z., Brar, N., Bless, S.: Dynamic high-pressure properties of AlN ceramic as determined by flyer plate impact. J. Appl. Phys. **70**(1), 167–171 (1991)
19. Hu, G., Ramesh, K., Cao, B., McCauley, J.: The compressive failure of aluminum nitride considered as a model advanced ceramic. J. Mech. Phys. Solids **59**(5), 1076–1093 (2011)
20. Ghosh, D., Subhash, G., Zheng, J.Q., Halls, V.: Influence of stress state and strain rate on structural amorphization in boron carbide. J. Appl. Phys. **111**(6), 063523 (2012)
21. Chen, M., McCauley, J.W., Hemker, K.J.: Shock-induced localized amorphization in boron carbide. Science **299**(5612), 1563–1566 (2003)
22. Anderson, C.E., Royal-Timmons, S.A.: Ballistic performance of confined 99.5%-Al_2O_3 ceramic tiles. Int. J. Impact Eng. **19**(8), 703–713 (1997)
23. Huang, C., Subhash, G.: Influence of lateral confinement on dynamic damage evolution during uniaxial compressive response of brittle solids. J. Mech. Phys. Solids **51**(6), 1089–1105 (2003)
24. Lankford, J., Predebon, W., Staehler, J., Subhash, G., Pletka, B., Anderson, C.: The role of plasticity as a limiting factor in the compressive failure of high strength ceramics. Mech. Mater. **29**(3), 205–218 (1998)

25. Ravichandran, G., Subhash, G.: A micromechanical model for high strain rate behavior of ceramics. Int. J. Solids Struct. **32**(17–18), 2627–2646 (1995)
26. Subhash, G., Ravichandran, G.: Split Hopkinson pressure bar testing of ceramics. In: ASM Handbook, Mechanical Testing and Evaluation, vol. 8, pp. 497–504. ASM International, Materials Park (2000)
27. Chen, M., McCauley, J., Dandekar, D., Bourne, N.: Dynamic plasticity and failure of high-purity alumina under shock loading. Nat. Mater. **5** (8), 614–618 (2006)
28. Shafiq, M., Subhash, G.: An extended Mohr-Coulomb model for fracture strength of intact brittle materials under ultrahigh pressures. J. Am. Ceram. Soc. **99**(2), 627–630 (2016)
29. Chocron, S., Anderson, C.E., Dannemann, K.A., Nicholls, A.E., King, N.L.: Intact and predamaged boron carbide strength under moderate confinement pressures. J. Am. Ceram. Soc. **95**(1), 350–357 (2012)
30. Chocron, S., Anderson, C.E., Nicholls, A.E., Dannemann, K.A.: Characterization of confined intact and damaged borosilicate glass. J. Am. Ceram. Soc. **93**(10), 3390–3398 (2010)

Chapter 28
Dynamic Strength and Fragmentation Experiments on Brittle Materials Using Theta-Specimens

Jamie Kimberley and Antonio Garcia

Abstract Characterization of the strength and fragmentation response of brittle materials poses unique challenges related to specimen gripping and alignment. These challenges are often exacerbated when the characterization is to be conducted at elevated strain rates. Tensile strength of brittle samples are often characterized using the Brazilian disk testing geometry. While this ameliorates issues related to specimen alignment, the stress field in the specimen is not uniform, complicating the analysis of the test results. The theta specimen geometry was designed specifically to provide a uniform state of uniaxial stress in the specimen gage section when the exterior of the sample is subjected to compressive loading. Here we evaluate the use of the theta specimen geometry with a compressive Kolsky bar to measure the dynamic tensile strength and fragmentation response of a brittle polymer, Poly(Methyl methacrylate). Finite element simulations are used to investigate the effect of geometry and loading pulse shape on the ability to establish a state of uniaxial stress in the gage section. Particular attention is given the excitation of lateral vibrations in the gage section, which would perturb the desired uniaxial stress state.

Keywords Dynamic testing • High-rate • Tension • Kolsky bar • Split Hopkinson • Theta

28.1 Introduction

Measurement of the tensile strength using "direct" testing methods (e.g. dog-bone pull tests) presents significant difficulties with regard to specimen gripping and alignment when applied to the testing of brittle materials. The very act of gripping the specimen may introduce stress concentrations leading to preferential failure at the grip, and invalid testing results. Minor amounts of misalignment of the test specimen with respect to the loading axis can also introduce non-uniform stress state in the specimen, introducing error into the measurements. In an effort to reduce the occurrence of these errors in testing, so called "indirect" tensile test methods have been developed for determining the tensile strength of brittle materials. Most of these methods use specimen geometry to convert a compressive load into tensile stress, relying on the fact that the compressive strength of the material to be tested is significantly greater than the tensile strength.

One commonly used indirect tensile test method is the Brazilian disk test, in which a disk of material is subjected to compression by diametrically opposed forces. This configuration results in the development of a tensile state of stress acting normal to the compression axis. The value of tensile stress varies with distance from the center of the specimen, and achieves a maximum value $\sigma_t = -\frac{2P}{\pi D t}$ at the mid-plane of the sample, where P is the applied load (assumed positive in tension), D is the disk Diameter, and t is the thickness of the disk [1]. While the Brazilian disk test method eliminates issues with regard to alignment and gripping, it introduces a non-uniform, multi-axial stress state in the specimen. This added complexity adds difficulty to the interpretation of the measurements, and raises questions of under what state of stress and what volume of material was the strength was measured. The latter is a critical aspect of the development and validation of flaw based strength and fragmentation models for brittle materials. The Brazilian disk test has also been successfully applied to dynamic testing by compressing the specimen dynamically in a Kolsky (split Hopkinson) bar [2]. Here the indirect nature is of major benefit, because dynamic compressive loads are generally much easier to generate than dynamic tensile load, and specimen gripping strategies may adversely affect the wave propagation in the system.

J. Kimberley (✉)
Department of Mechanical Engineering, New Mexico Institute of Mining and Technology, 801 Leroy Place, Socorro, 87801, NM, USA

Energetic Materials Testing and Research Center, New Mexico Institute of Mining and Technology, 801 Leroy Place, Socorro, NM 87801, USA
e-mail: jamiek@nmt.edu

A. Garcia
Department of Mechanical Engineering, New Mexico Institute of Mining and Technology, 801 Leroy Place, Socorro, 87801, NM, USA

D. Casem et al. (eds.), *Dynamic Behavior of Materials, Volume 1*, Conference Proceedings of the Society for Experimental Mechanics Series, DOI 10.1007/978-3-319-41132-3_28

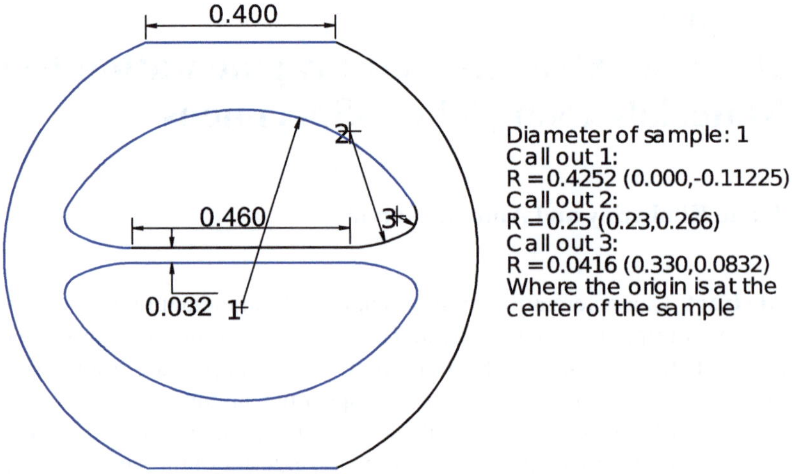

Fig. 28.1 Specimen geometry, scaled to an overall diameter of unity. The specimen studied here has a diameter of 10 mm. Specimen thickness is chosen to be half the diameter, but this is not a strict requirement

Another indirect tensile test method utilizes theta-shaped specimen geometries, such as that shown in Fig. 28.1. The application of compressive forces normal to the central web section results in the development of a uniform uniaxial tensile stress state in the central web. This test method was initially developed by [3] in the late 1960s, but never saw wide adoption due to the complex specimen geometry. More recently [4, 5] made modifications to the initial specimen geometry to perform strength measurements of Si wafer material on small-scale specimens. The stress state developed in the modified specimen geometry was analyzed using finite element analysis, and was shown to maintain a state uniform uniaxial stress in the gage section [5]. Similar the Brazilian disk test, the stress in the web is related to the applied load P and takes the form $\sigma = -K\frac{P}{Dt}$, where K is a constant that depends on the geometry of the sample. Durelli et al. [3] found $K = 13.8$ via photoelastic analysis, and Gaither et al. [5] found $K = 14.239$ using 3-D finite element analysis.

In this paper we aim to investigate the use of theta-shaped specimens in conjunction with a compressive Kolsky bar to measure the dynamic tensile strength of materials. A slightly modified version of the "arch" specimen geometry [4, 5] is analyzed via dynamic finite element simulations. The simulation results are used to evaluate the uniformity and the magnitude of the stress state in the central web for different loading pulses. Specific attention is paid to the excitation of transverse deflection of the web due to acceleration of the specimen during loading.

28.2 Methodology

The specimen geometry of the theta-shaped specimen studied here is shown in Fig. 28.1. Units of measure are scaled to a diameter of unity for ease of adaptation to various sizes or systems of measure. The internal geometry is identical to that used by [4, 5] but the external utilizes flats instead of the "top hat" geometry in the load bearing area. These flats aid with specimen alignment, and alleviate the potentially high contact stresses associated with a disk in contact with a flat plate. For the study presented here, the outside diameter was chosen to be 10 mm and the thickness 5 mm. The material of the specimen was poly-methyl methacrylate with Young's modulus $E_s = 3.2$ GPa, Poisson's ratio $\nu_s = 0.35$, and mass density $\rho_s = 1190\ \text{kg/m}^3$. The constant K in the load–stress relation has not yet been determined for the geometry studied here, but we expect it to be close to that for the geometry of Gaither et al. [5] because the internal geometries are so similar.

A Kolsky bar system with incident bar length of 1.0 m and transmitted bar length of 0.5 m was modeled to apply the loading to the specimen. All bars had diameter of 12.7 mm and were made of aluminum with Young's modulus $E_b = 69$ GPa, Poisson's ratio $\nu_b = 0.3$, and mass density $\rho_b = 2730\ \text{kg/m}^3$. All components were modeled in the finite element software Comsol Multiphysics version 5.0 as two-dimensional plane stress entities (note: the bars were modeled as having square cross section 12.7 mm $\times$ 10.0 mm resulting in cross sectional area equivalent to a 12.7 mm-diameter circular section bar). All components were meshed with linear elastic, 4-node quadrilateral elements, as shown in Fig. 28.2. The majority of the bar areas were meshed with a structured mesh with seven elements across and aspect ratio of 1. Near the specimen this mesh transitions to a free mesh to allow for better mapping of nodes between specimen and bar. The specimen mesh resulted in eight elements across the thickness of the central web and a free mesh outside of the web region. A convergence study was performed and further refinement of the mesh size in the specimen did not significantly affect the results.

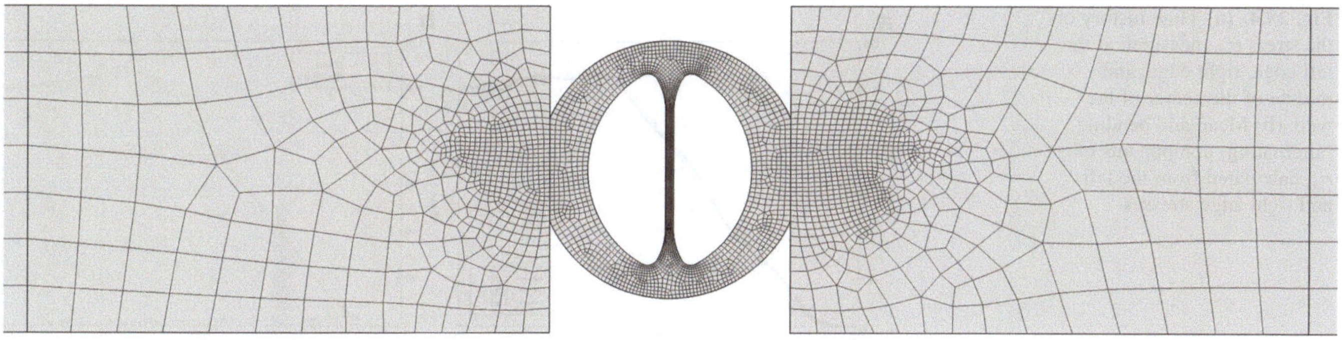

Fig. 28.2 Meshed specimen and bar geometries. Only portions of the bars are shown to illustrate detail of the mesh refinement near the specimen

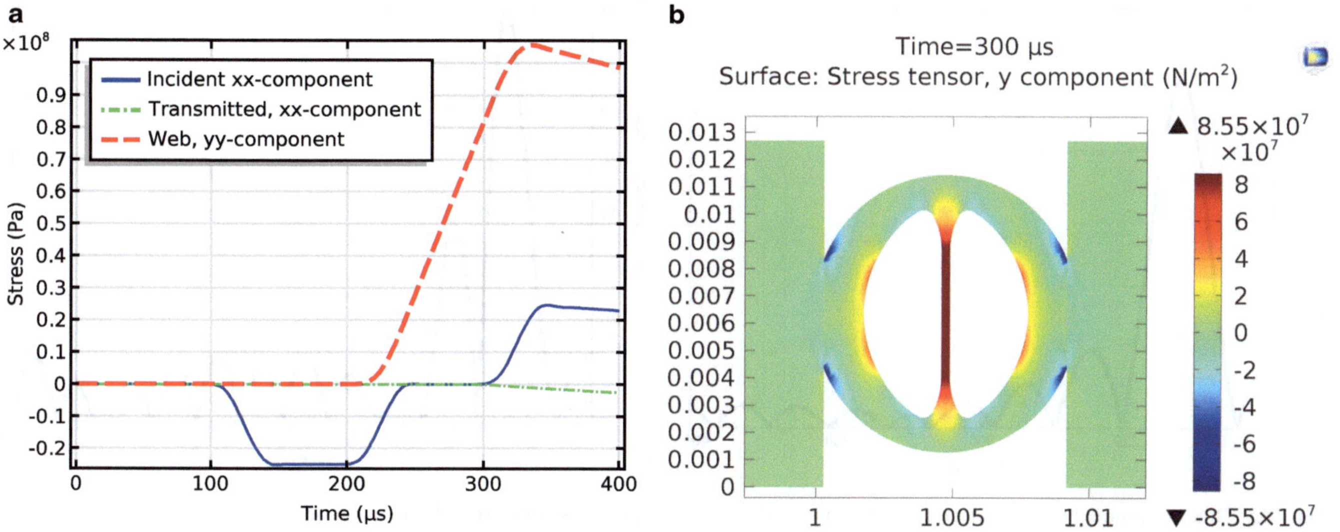

Fig. 28.3 (**a**) Stress-time histories taken at the midpoints of the incident bar, transmitted bar, and web. Stresses in the bars are the axial component σ_{xx}, stress in the web is the axial component σ_{yy}. (**b**) Contours of σ_{yy} at time = 300 µs, showing uniformity of stress in the web section

The incident pulse in the Kolsky bar was simulated by applying a pressure boundary condition to the impact end of the incident bar. This provides greater control over the input pulse shape, and removes the computational overhead required to model an impact of the striker on the incident bar. The incident pulse was a smoothed trapezoid with a compressive stress magnitude of 25 MPa and variable rise/fall and plateau times.

28.3 Results and Discussion

The stress-time histories taken at the midpoints of the incident bar, transmitted bar, and specimen web are shown in Fig. 28.3a. The incident pulse has rise and fall times of ~30 µs and a plateau region of ~60 µs. The stress in the web increases linearly after arrival of the incident stress pulse. At approximately 325 µs, the web stress decreases corresponding to the end of loading. Figure 28.3b shows contours of σ_{yy} in the specimen at $t = 300$ µs; note that the central web shows uniform stress values. The value of the stress in the web is higher than other locations in the specimen indicating that failure is likely to occur first in the web section.

While these results are encouraging, further investigation of the stress state in the web indicates that there is some non-uniformity due to vibrations induced in the web. The transverse vibrations are induced by the acceleration of the sample by the loading pulse and the effect of these vibrations can be seen by plotting the stress σ_{yy} at the left edge, right edge, and middle of the web as shown in Fig. 28.4. The stress at the left and right edges oscillate 180° out of phase, but with similar amplitude. The stress measured at the left σ_L and right σ_R edges can be used to express the observed behavior in terms of a

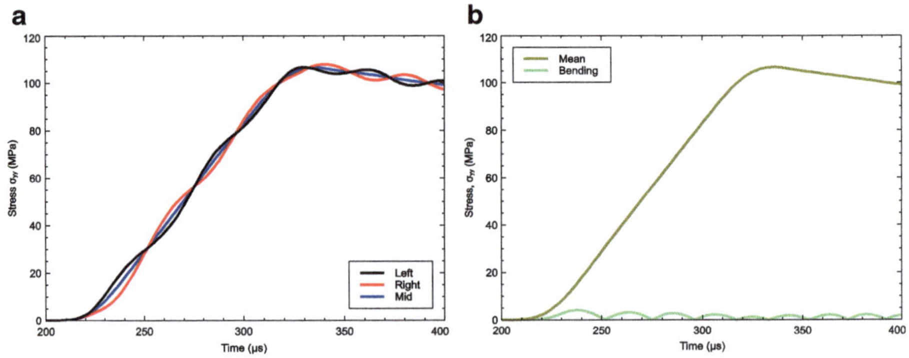

Fig. 28.4 (**a**) Time history of the stress σ_{yy} measured at the left edge, right edge, and middle of the center of the web. (**b**) Mean and bending (alternating) components of σ_{yy} calculated from the left and right edge stresses

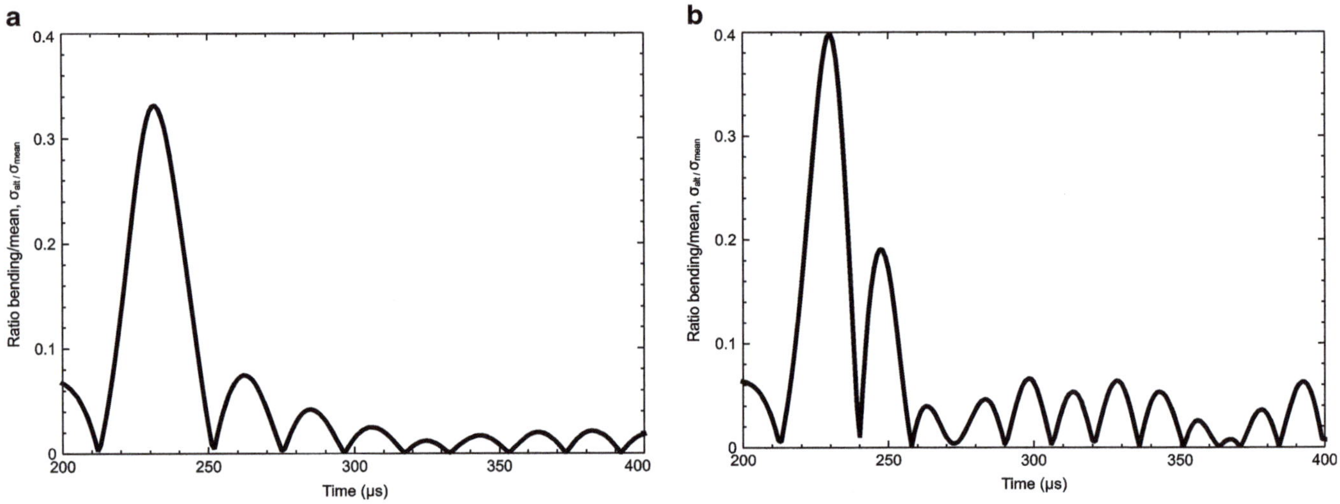

Fig. 28.5 Ratio of bending stress to mean stress for input pulse with (**a**) ~30 μs rise time, (**b**) ~15 μs rise time

mean stress $\sigma_{mean} = \frac{\sigma_L+\sigma_R}{2}$ corresponding to the axial stress and an alternating stress $\sigma_{alt} = \frac{|\sigma_L-\sigma_R|}{2}$ corresponding to bending. For this particular load case the stress induced by bending is only a few MPa, and represents a small portion of the maximum axial stress developed in the web.

The ratio of alternating stress to mean stress for a sample subjected to a loading pulse with 30 μs rise time is shown in Fig. 28.5a. Here we see that for early times the alternating stress due to bending is a large fraction of the mean (axial stress). As time progresses the sample deforms more, resulting in an increase in mean stress, and the ratio of alternating to mean stress diminishes. Thus, if we design the experiment so that the specimen fails near the end of the loading pulse, stress variations due to bending should be minimized. From one-dimensional wave theory, we know that the stress in the bar is proportional to the particle velocity [6], thus the nominally linear rise of our stress pulse corresponds to a constant acceleration with magnitude inversely proportional to the rise time of the pulse. By decreasing the rise time we expect larger amplitude oscillations to be induced resulting in an increased ratio of alternating to mean stress as shown in Fig. 28.5b (rise time of ~15 μs). This provides one experimental parameter that can be tailored through techniques like pulse shaping (e.g. [7]) to ensure that the variations in stress due to bending are kept to an acceptable level.

The input pulse amplitude is also a parameter that can be adjusted to control the amplitude of the acceleration, however this will affect the strain rate that the specimen is deformed at. Traditional Kolsky bar analysis [8] indicates that the strain rate in the sample is proportional to the amplitude of the pulse reflected from the sample. This holds true for the strain rate in the web section of the theta specimen because the predicted strain is proportional to the applied displacement. Since our sample has significantly lower impedance than the bar, the reflected pulse amplitude will be almost the same as the amplitude of the incident pulse. Thus, the strain rate in the sample will be proportional to the incident pulse amplitude. We expect that a more generalized analysis including sample parameters (diameter and material properties) will define limits for these parameters that can be applied to a wide range of brittle materials. An analytical analysis based on an Euler-Bernoulli beam subjected to constant base acceleration is currently under development. Experimental verification of the technique is also underway.

While the initial simulation results are promising, we expect there to be several challenges/limitations related to this technique. One particular challenge will be the machining of specimens, which will be difficult, and expensive if this technique is to be applied to high-strength ceramics. Other limitations will be related to material grain size and flaw populations that are required to have a representative response in the thin gage section. That said, this technique might prove useful for the development/validation of material models related to brittle failure. It may play a complimentary role to the easier to conduct Brazilian test, providing data under uniaxial stress conditions, as opposed to the multiaxial state of the Brazilian test. It may also prove useful for dynamic fragmentation, where many models are based upon a 1-D state of stress.

28.4 Conclusions

Finite element simulations have shown that theta-shaped specimens can be utilized in conjunction with a compressive Kolsky bar to perform dynamic measurements of tensile strength of brittle materials. The simulations show that transverse vibration modes are excited by the arrival of the loading pulse. These transverse deflections superpose a bending stress component, which disrupts the desired uniaxial stress state in the web. Appropriate selection of the input pulse properties (i.e. rise time and amplitude) can minimize the amplitude of the bending stress, leading to stress variations that are acceptable for testing purposes.

References

1. Mellor, M., Hawkes, I.: Measurement of tensile strength by diametral compression of discs and annuli. Eng. Geol. **5**(3), 173–225 (1971)
2. Kolsky, H.: An investigation of the mechanical properties of materials at very high rates of loading. Proc. Phys. Soc. B **62**, 676–700 (1949)
3. Durelli, A., Morse, S., Parks, V.: The theta specimen for determining tensile strength of brittle materials. Mater. Res. Stand **2**, 114–117 (1962)
4. Gaither, M.S., DelRio, F.W., Gates, R.S., Fuller, E.R. Jr., Cook, R.F.: Strength distribution of single-crystal silicon theta-like specimens. Scr. Mater. **63**(4), 422–425 (2010)
5. Gaither, M.S., DelRio, F.W., Gates, R.S., Cook, R.F.: Deformation and fracture of single-crystal silicon theta-like specimens. J. Mater. Res. **26** (20), 2575–2589 (2011)
6. Kolsky, H.: Stress waves in solids. J. Sound Vib. **1**(1), 88–110 (1963)
7. Frew, D.J., Forrestal, M.J., Chen, W.: Pulse shaping techniques for testing brittle materials with a split Hopkinson pressure bar. Exp. Mech. **42** (1), 93–106 (2002)
8. Ramesh, K.T.: High strain rate and impact experiments. In: Sharpe, W.N. (ed.) Springer Handbook of Experimental Solid Mechanics, Ch. 33, pp. 929–960. Springer, New York (2008)

Chapter 29
DTEM In Situ Mechanical Testing: Defects Motion at High Strain Rates

Thomas Voisin, Michael D. Grapes, Yong Zhang, Nicholas J. Lorenzo, Jonathan P. Ligda, Brian E. Schuster, Melissa K. Santala, Tian Li, Geoffrey H. Campbell, and Timothy P. Weihs

Abstract Defect nucleation and motion during high strain rate experiments has not been observed in situ at the nanoscale in metals. However, imaging dislocation and twin nucleation and propagation will enhance our understanding and ability to predict dynamic behavior and spall strength. In the experiments described here we use the Dynamic TEM at the Lawrence Livermore National Laboratory which is capable of recording pictures with a 20-ns time resolution in movie mode (a short multi-frames experiment), and we developed a new TEM holder capable of deforming samples at strain rates ranging from quasistatic to 10^4 s^{-1}. The holder uses two piezoelectric actuators that bend rapidly to load samples and TEM specimens with small gauge sections to obtain high strain rates. The TEM specimens and their narrow gauge sections are machined from bulk specimens using a femtosecond laser. The 50-μm wide gauge sections are ion milled to create electron transparent areas. We present high strain rate in situ mechanical test results for copper specimens.

Keywords In situ TEM • High strain rate • Dynamic TEM • Mechanical testing • Plasticity

29.1 Introduction

In situ Transmission Electron Microscopy (TEM) mechanical testing attracts significant interest as it allows one to image micro- to nano-scaled transient states of the microstructure during straining of samples. One can obtain information on defect velocities, nucleation mechanisms, defect transmission across multiple grains, grain boundary migration, etc. Unfortunately, all current in situ TEM techniques are limited to quasi-static strain rate experiments [1–3]. However, many investigations conducted at high rates have highlighted the strain rate sensitivity of flow stress [4–6], changes of the deformation and fracture mechanisms [7, 8], dislocation cell formation [9], dynamic recrystallization [10], and grains reorientation [11]. The TEM components of these studies, though, were conducted *post mortem*. The only current techniques that allow one to study microstructural evolution of samples in situ under high rate deformation are X-ray synchrotron methods [12]. With the objective of understanding and predicting the dynamic mechanical properties of metals at micro and nano-scales, we developed a novel in situ TEM technique enabling one to deform specimens rapidly. To achieve this, we used the Dynamic TEM (DTEM) at the Lawrence Livermore National Laboratory (LLNL) with a time resolution of 20 ns thanks to large electron pulses (greater than 10^9 electrons per pulse) generated by bombarding a photoelectron Ta cathode with a high intensity nanosecond laser [13]. We also designed and built the first in situ TEM holder capable of loading samples in tension at strain rates as high as 4×10^3 s^{-1} [14]. This system utilizes very small samples with a short gage length to reach such strain rates. In this paper, we present the general design of the TEM holder and then focus on sample preparation that combines laser machining and ion milling, as well as the imaging of defects in movie mode (a short multi-frames experiment) in pure Cu specimens.

T. Voisin (✉) • M.D. Grapes • T.P. Weihs
Department of Materials Science and Engineering, Johns Hopkins University, Baltimore, MD 21218, USA
e-mail: voisin.thomas31@gmail.com

Y. Zhang
Department of Mechanical Engineering, Johns Hopkins University, Baltimore, MD 21218, USA

N.J. Lorenzo • J.P. Ligda • B.E. Schuster
Weapons and Materials Research Directorate, Army Research Laboratory, Aberdeen Proving Ground, Aberdeen, MD 21005, USA

M.K. Santala • T. Li • G.H. Campbell
Materials Science Division, Lawrence Livermore National Laboratory, Livermore, CA 94550, USA

D. Casem et al. (eds.), *Dynamic Behavior of Materials, Volume 1*, Conference Proceedings of the Society for Experimental Mechanics Series, DOI 10.1007/978-3-319-41132-3_29

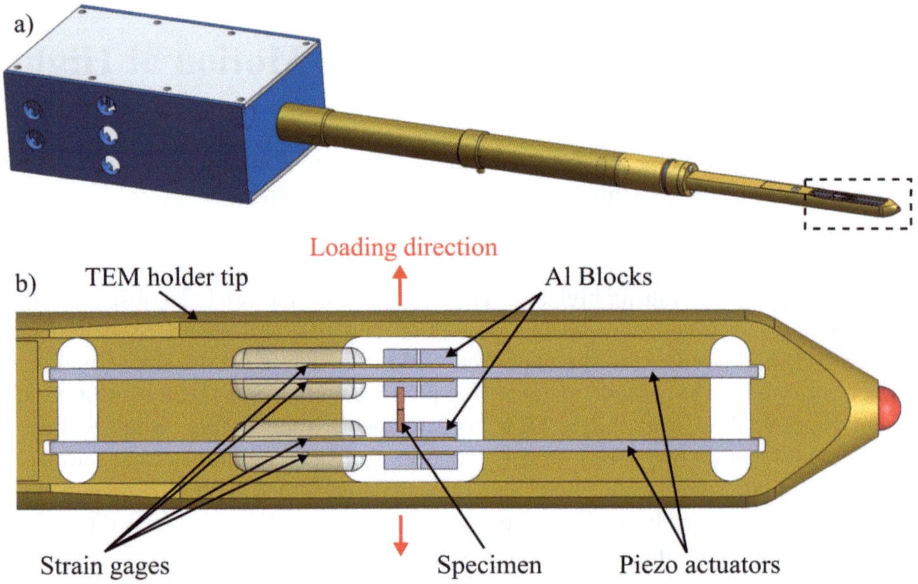

Fig. 29.1 High strain rate TEM holder: (**a**) General view and (**b**) zoomed in view of the sample area (highlighted in the Fig. 29.1a by the *dash line*)

29.2 High Strain Rate TEM Holder

Prior to the development of movie mode imaging in the DTEM, no TEM could image fast enough to record high strain rate experiments. Conventional frame rates are limited to approximately 30 frames per second [1]. Thus, existing in situ TEM holders have been designed to pull on samples at quasi-static strain rates. The TEM holder we designed to achieve high strain rates consists of two piezoelectric actuators, mounted in parallel, which bend in opposite directions when a voltage is applied (Fig. 29.1). The specimens are mounted perpendicular to the piezo actuators with their two ends glued onto two different aluminum blocks, each one being linked to a different piezo actuator (Fig. 29.1b). When these actuators bend, they load the specimen in tension. Thanks to four strain gages mounted on both sides of each piezo, displacement of the piezo actuators can be recorded during an experiment, and the force applied to the sample can be inferred from the resulting displacement data. This piezoelectric system offers the advantage of varying the strain rate from quasi-static to around $4 \times 10^3\ s^{-1}$, thereby enabling a complete investigation of a wide range of deformation rates. However, the actuators employed here are limited in displacement, force, and velocity to 100 μm, 200 mN, and 0.2 m s^{-1}, respectively.

29.3 Sample Preparation

In preparing TEM specimens for high strain rate experiments we sought to obtain several attributes:

- A narrow (25–50 μm) gage length to increase the strain rate.
- An effective cross-sectional area (width × thickness) that is small enough to minimize the resistance of samples during loading, but large enough so that specimens can be handled and mounted on the Al blocks without damage.
- A repeatable sample geometry to ensure consistent loading and imaging during experiments.
- A high throughput sample preparation process so that many samples may be fabricated at a moderate cost. Many samples are needed initially to synchronize loading and imaging and to optimize the new in situ technique.

Conventional techniques for preparing TEM samples did not provide suitable specimens [15]. Electro-polishing led to significant variations of the sample geometry, as well as samples that were simply too big. Preparation techniques that rely on Focused Ion Beam (FIB) methods are relatively slow and are not cost effective when producing many specimens. To obtain the attributes listed above, we used a femtosecond laser to machine 2 mm by 400 μm rectangles (Fig. 29.2a) from 10 μm thick rolled sheets of pure Cu and then milled a trench in the middle of the rectangles to serve as a narrow gauge section (Fig. 29.2b). The length, width, and thickness of the trench are 200, 25, and ≈ 3 μm, respectively. The laser beam remains perpendicular to the Cu sheet during machining at room temperature. Samples are ion milled at −180 °C with a

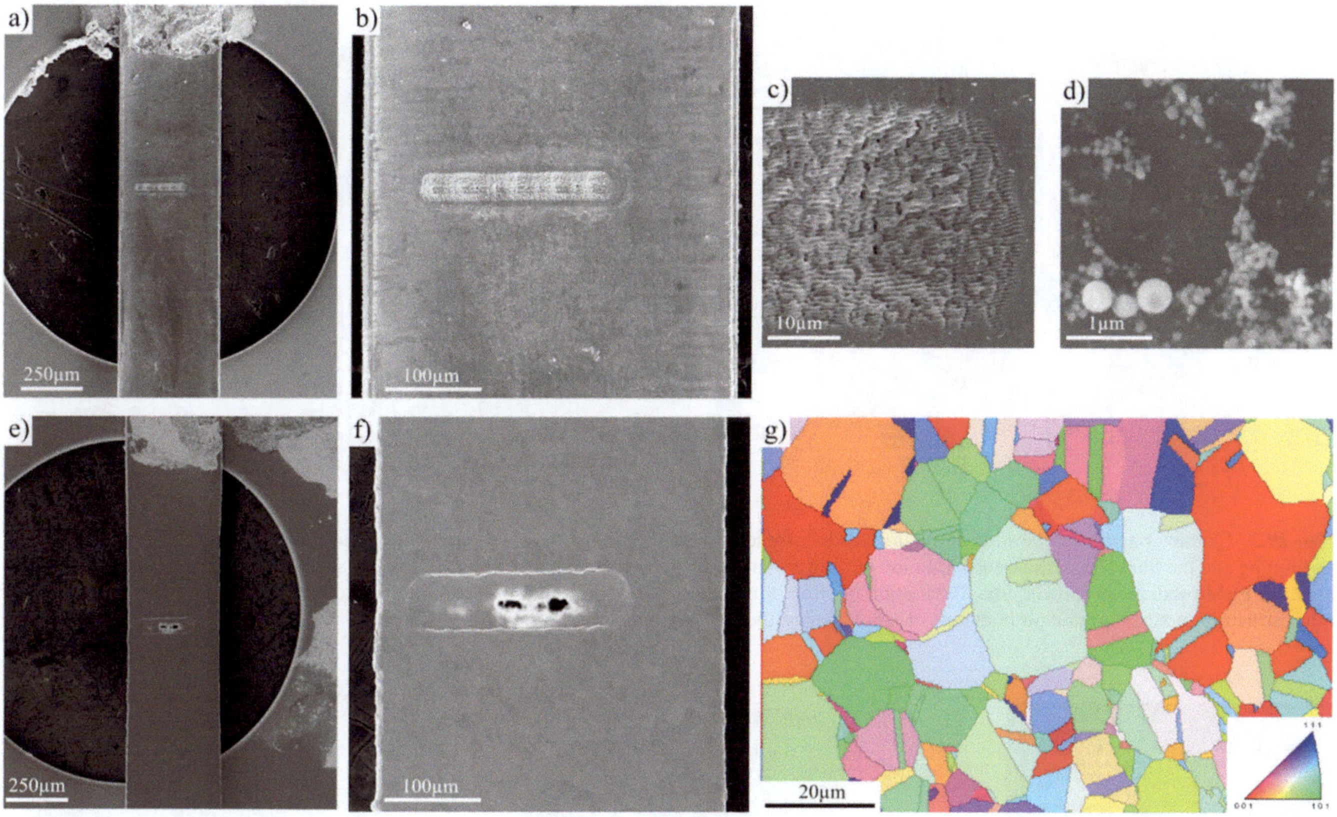

Fig. 29.2 Samples for high strain rate in situ DTEM experiments. (**a, b**) Sample after laser machining at two different magnifications. (**c**) Zoom in the right end of the trench of the same laser machined sample. (**d**) Zoom on small particles redeposited during the laser machining. (**e, f**) Same sample after ion milling with a PIPS at two different magnifications. (**g**) EBSD map in an area of the same sample far from the trench

Precision Ion Milling System II (PIPS II) produced by Gatan, following the recipe described previously [15]. Note that the thin, melted layer shown in Fig. 29.2c, reported to be about 100 nm deep by several authors [16–18], and the redeposited particles (Fig. 29.2d) are completely removed during the ion milling process. We optimized the laser machining parameters and found that a fluence of 2.9 J cm^{-2} presented a good balance between damage to the sample and efficiency of the machining. Nevertheless, as detailed in a previous paper [15], the laser machining still induces local deformation. While we attempt to minimize damage by optimizing the laser energy, we anneal samples after laser machining at 200 °C for 1 h in vacuum to recrystallize the damaged region and obtain a microstructure similar to the coarse grain microstructure found for the rest of the sample (Fig. 29.2g).

Figure 29.3 displays EBSD maps and TEM images of an area within the laser machined and ion-milled trench following annealing. The microstructure (Fig. 29.3a) and texture (Fig. 29.3b) are similar to those seen elsewhere in the specimens, away from the trench, and indicate removal of the laser damages during the 1 h anneal at 200 °C. The substantial broadening of the trench, due to ion milling, is marked by dashed lines in Fig. 29.3a. After ion milling, the electron transparent area is located near the holes. Figure 29.3c provides an example of FEM calculations showing the distribution of stresses around a hole in a sample as it is loaded in tension. The maximum stresses are found on the left and right sides of the hole when load is applied vertically. TEM pictures taken in such areas, prior to loading, show a clear microstructure virtually free of deformation. However, many growth twins are seen and are assumed to form during the annealing due to the low stacking fault energy of copper (≈78 J.m^{-2}).

29.4 Imaging in Movie Mode with the DTEM

When the DTEM is operated in movie mode, a train of nine electron pulses go through the sample one by one, spaced at time intervals as short as 70 ns, and are then deflected and distributed onto a single detector (Fig. 29.4a). This time resolution enables one to image transient states during ultrafast events such as phase transformations, solidifications, or chemical reactions that

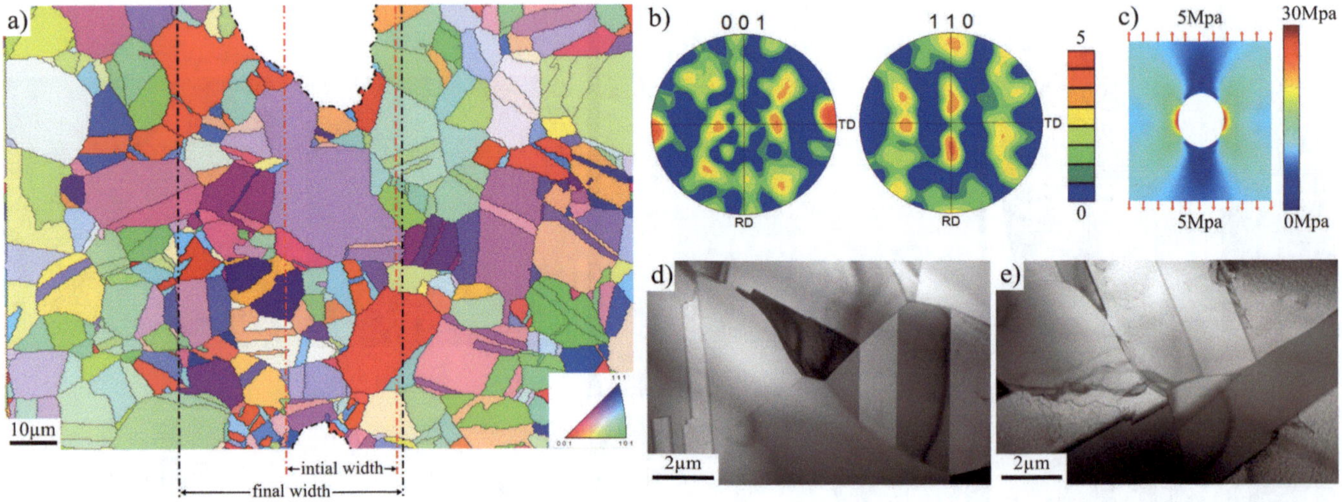

Fig. 29.3 Characterization of the microstructure inside a trench. (**a**) EBSD map across a trench where the initial size of the trench after laser machining and the final size of the trench after ion milling are highlighted by the *red* and *black dash lines*, respectively. (**b**) Pole figure of the same area. (**c**) FEM predictions showing the distribution of stress around a hole on a sample subject to a uniaxial loading. (**d, e**) TEM pictures of the microstructure in areas located on both side of a hole where the stress is highest

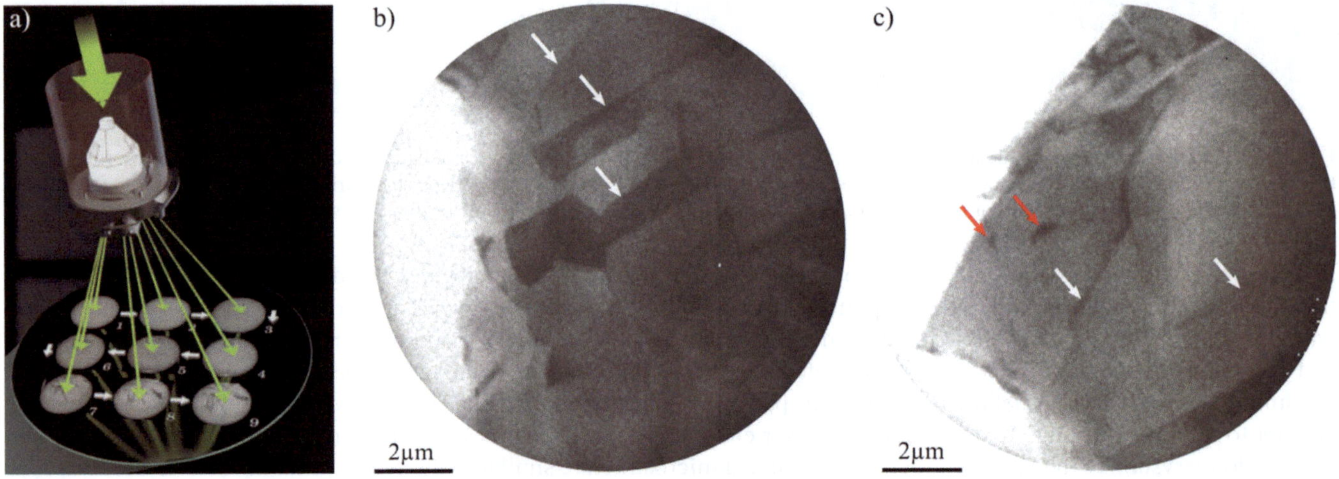

Fig. 29.4 Imaging with the DTEM in movie mode. (**a**) Nine electron pulses distributed on a same detector after crossing the sample [19]. (**b**) One of the nine pictures taken with the DTEM in movie mode with a 20 ns exposure time in a pure Cu sample. *White arrows* point out grain boundaries. (**c**) Picture taken in the same condition in another area of the sample. *White arrows* also point out grain boundaries and *red arrows* shows stacking faults

would be missed by other TEM techniques. Unfortunately, during such imaging spatial resolution is approximately 20 nm, much worse than conventional TEM which offer resolutions smaller than 0.2 nm. Thus, imaging nanoscale features such as dislocations, twins, or particles is challenging. Figure 29.4b, c show preliminary images of pure Cu samples taken in movie mode. Grain boundaries (including twin boundaries) are clearly distinguishable, as well as stacking faults spreading over more than 100 nm. We'll show in a forthcoming paper that dislocations and twins can be also imaged.

29.5 Conclusion

In this paper we have described our progress in studying the deformation of metallic samples at high strain rates using the Dynamic TEM. The TEM holder we designed enables one to deform samples at rates as high as $4 \times 10^3\ s^{-1}$ in tension, with control of displacement and monitoring of force. The limited maximum forces and displacements that are offered by the piezoelectric system required the development of non-conventional sample preparation techniques. By combining femtosecond laser micromachining and ion milling, we were able to produce small samples with narrow gauge lengths and electron transparent areas. Adjustment of laser parameters such as energy per pulse and the number of passes, as well as post-machining anneals, led to microstructures virtually defect free samples with a controlled grain size. Imaging of these samples in movie mode within the DTEM confirmed the possibility of capturing grain boundaries and defects in a rapid sequence of images during loading.

Acknowledgment Funding was provided by the U.S. Army through the Materials in Extreme Dynamic Environments CRA.

References

1. Legros, M.: In situ mechanical TEM: seeing and measuring under stress with electrons. C R Phys. **15**, 224–240 (2014)
2. Bernal, R.A., Ramachandramoorthy, R., Espinosa, H.D.: Ultramicroscopy **156**, 23–28 (2015)
3. Haque, M.A., Espinosa, H.D., Lee, H.J.: MRS Bull. **35**, 375–381 (2010)
4. Ulacia, I., Salisbury, C.P., Hurtado, I., Worswick, M.J.: J. Mater. Process. Technol. **211**, 830–839 (2011)
5. Murr, L. E., Staudhammer, K. P., Meyers, M.A.: Metallurgical Applications of Shock-Wave and High-Strain Rate Phenomena, CRC, Boca Raton (1986)
6. Follansbee, P.S., Frantz, C.: J. Eng. Mater. Technol. **105**, 61–66 (1983)
7. Ramesh, K.T.: Metall Mat. Trans. A **33**, 927–935 (2002)
8. Johari, O., Thomas, G.: Acta Metall. **12**, 1153–1159 (1964)
9. Nolder, R.L., Thomas, G.: Acta Metall. **12**, 227–240 (1964)
10. Andrade, U., Meyers, M.A., Vecchio, K.S., Chokshi, A.H.: Acta Metall. Mater. **42**, 3183–3195 (1994)
11. Bhattacharyya, A., Rittel, D., Ravichandran, G.: Scr. Mater. **52**, 657–661 (2005)
12. M. Majkut: Department of Mechanical and Materials Engineering, Queen's University, Kingston (2013)
13. Armstrong, M.R., Boyden, K., Browning, N.D., Campbell, G.H., Colvin, J.D., DeHope, W.J., Frank, A.M., Gibson, D.J., Hartemann, F., Kim, J.S., King, W.E., LaGrange, T.B., Pyke, B.J., Reed, B.W., Shuttlesworth, R.M., Stuart, B.C., Torralva, B.R.: Ultramicroscopy **107**, 356–367 (2007)
14. Grapes, M.D., Zhang, Y., Santala, M.K., Voisin, T., Campbell, G.H., Weihs, T.P.: In: **Shaw III**, G., Prorok, C.B., Starman, L. (eds.) MEMS and Nanotechnology, Proceedings of the 2015 Annual Conference on Experimental and Applied Mechanics, vol. 5, pp. 25–30. Springer International, Cham (2016)
15. Voisin, M., Grapes, Y., Zhang, N., Lorenzo, J., Ligda, B., Schuster, T., Weihs, T.P.: under review at Ultramicroscopy (2016)
16. Colombier, J.P., Combis, P., Stoian, R., Audouard, E.: Phy. Rev. B **75**, 104105 (2007)
17. Rethfeld, B., Sokolowski-Tinten, K., von der Linde, D., Anisimov, S.I.: Phys. Rev. B **65**, 092103 (2002)
18. Wang, S.Y., Ren, Y., Cheng, C.W., Chen, J.K., Tzou, D.Y.: Appl. Surf. Sci. **265**, 302–308 (2013)
19. Hansen, R.: Science and Technology Review, pp. 4–11. Lawrence Livermore National Laboratory, Livermore (2013)

Chapter 30
High-Strain-Rate Deformation of Ti-6Al-4V Through Compression Kolsky Bar at High Temperatures

S. Gangireddy and S.P. Mates

Abstract In this paper, we present our first results from the study of the constitutive response of a popular Titanium alloy, Ti-6Al-4V, using a variation of the compression Kolsky Bar technique that employs electrical pulses to achieve high temperatures. Experiments are conducted at temperatures ranging from room temperature to 1000 °C at a strain rate of about 2200 s^{-1} and a heating rate of about 1500 °C/s. The dynamic stress-strain results demonstrate significant thermal softening in the alloy that could be described by Johnson-Cook equation with $m = 0.8$ up to 650 °C. Above 650 °C the rate of change in the flow stresses was faster, which is attributed to allotropic transformation that results in a change in the phase fractions of the hcp and bcc phases present in the alloy. Evidence of transformation is observed in the microstructure of post-compression specimens, which showed an acicular morphology formed from the high temperature bcc phase on quenching.

Keywords Kolsky Bar • Ti-6Al-4V • High-temperature • Dynamic-response • Thermal-softening

30.1 Introduction

Methods for mechanical testing under rapidly applied loading are well established, such as the Split-Hopkinson Pressure Bar (SHPB) or Kolsky Bar technique. Variations of Kolsky Bar with additional components for achieving high temperatures exist as well. Typically miniature furnaces, induction coils, or radiation heating are used [1–3]. Heating times associated with these methods are typically on the order of minutes. At NIST, a new variation of Kolsky Bar system has been developed [4, 5] where electrical current is pulsed directly through the sample while it sits fixed between the incident and transmission bars. This technique gives us a unique advantage of rapid heating rates, up to 6000 K/s, where the heating time can be less than one second. Such rapid heating combined with rapid loading conditions creates a closer simulation to extreme physical processes such as high speed machining, explosive impact and other highly dynamic thermo-mechanical events. These results can give valuable insights that may lead to improved cutting processes or development of better blast-resistant structures.

Ti-6Al-4V is a widely used titanium alloy with aerospace applications owing to its excellent combination of high specific strength and corrosion resistance [6]. Pure titanium is allotropic and undergoes transformation from hcp (α) to bcc (β) crystal structure at elevated temperatures. The alloy Ti-6Al-4V contains both alpha stabilizing Aluminum (6 wt%) and beta stabilizing Vanadium (4 wt%) and consists of a mixture of $\alpha + \beta$ phases at room temperature [7]. At higher temperatures, the phase fraction of β increases, and above a temperature of about 995 °C the alloy is 100 % β [8]. This is known as the β-transus temperature. The mechanical response of this material can therefore be expected to be strongly dependent on temperature. The temperature sensitivity of the dynamic mechanical response of a commercial Ti-6Al-4V alloy was investigated using NIST's electrical pulse-heated Kolsky Bar system in the temperature range of 23–1008 °C under a strain rate of about 2200 s^{-1}.

30.2 Experimental Procedure

A commercial Ti-6Al-4V alloy of composition, 6.65 % Al, 4 % V, 0.02 % C, 0.23 % Fe, 0.007 % N, 0.197 % O, 0.003 % H, is studied in this investigation. The as-received material's microstructure consisted of globular α phase grains in a β matrix. The alloy, purchased in the form of a 2 mm thick plate, was cut using electrical discharge machining (EDM) to

S. Gangireddy (✉) • S.P. Mates
MML, NIST, Materials Science and Engineering Division, Gaithersburg, MD, USA
e-mail: sindhu.g.reddy@gmail.com

D. Casem et al. (eds.), *Dynamic Behavior of Materials, Volume 1*, Conference Proceedings of the Society for Experimental Mechanics Series, DOI 10.1007/978-3-319-41132-3_30

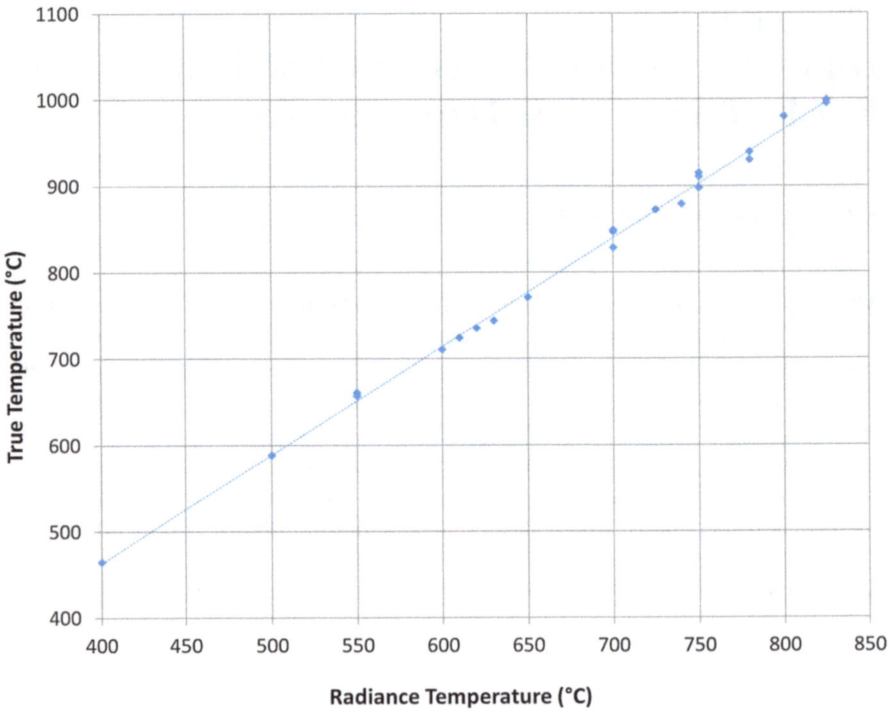

Fig. 30.1 Thermodynamic or true temperature measured by the thermocouple and the radiance temperature measured by the pyrometer at the time of impact show a strong correlation between 400 and 1000 °C

obtain the compression samples in cylindrical form of 4 mm diameter and 2 mm thickness. These samples were placed in between the incident and transmission bars of 150 mm diameter. Electrical current is conducted directly through the sample using the bar ends as electrodes. Owing to the large difference in the sample and bar cross-sectional areas and the very short heating times involved, the sample alone heats up while the bars themselves remain cool, with a less than 25 °C temperature rise even while the samples heated to 1000 °C [4]. In this series of experiments, the total heating time, which includes a transient heating period followed by a hold period, is limited to 3.5 s.

The temperature of the sample is monitored through three signals: a thermocouple spot welded onto the sample surface and two fast response infrared spot pyrometers focused on opposite sides of the sample surface. During heating by the electrical current, the thermocouple signal gets affected by electromagnetic interference. Hence one of the pyrometers is used as a feedback sensor to a specialized power supply that controls the sample temperature. The second pyrometer is used to monitor temperature uniformity. The pyrometer signal measures only the radiance temperature of the specimen. The thermodynamic, or true, temperature is determined from the thermocouple signal, which takes a few milliseconds to settle after the current is turned off. Hence compression pulse is timed to arrive about 20 ms after the current is switched off so a clean thermocouple signal can be obtained prior to impact. Just at the time of impact, the pyrometer signals could also be used to obtain radiance temperature, but after impact the specimen moves out of the view of pyrometers and the signals are lost. The radiance and true temperatures at impact, plotted in Fig. 30.1, are linearly correlated, indicating that even using the pyrometer as the feedback control signal, good true temperature control is possible.

The usual analysis method for deducing dynamic stress-strain from strain gauge signals is performed with a correction for the presence of graphite foils used for uniform contact conductance between the sample and the bars. The foil and the sample are treated as separately deforming elements and the foil response is deducted from the overall contraction between the compression bars to obtain the sample contraction. The detailed analysis for this deduction of dynamic stress-strain curves is presented elsewhere [4].

30.3 Results and Discussion

The dynamic true stress- true strain curves from compression tests conducted at temperatures ranging from 23 to 1008 °C under a strain rate of 2200 s^{-1} ($\pm$800 s^{-1}) are depicted in Fig. 30.2. The total time of heating is 3.5 s before the sample is impacted. The test temperature quoted is the true temperature measured using the thermocouple at the moment the first compression pulse hits the sample.

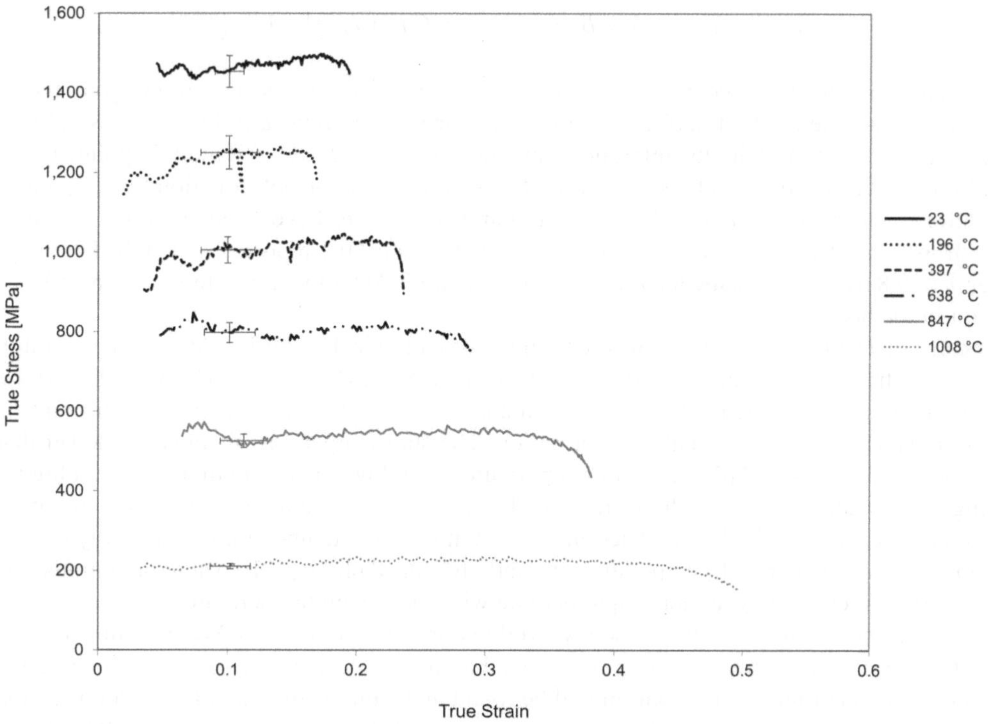

Fig. 30.2 Dynamic true stress—true strain graphs of Ti-6Al-4V alloy undergoing compression under a strain rate of 2200 s^{-1} ($\pm$800 s^{-1}) at temperatures between 23 and 1008 °C

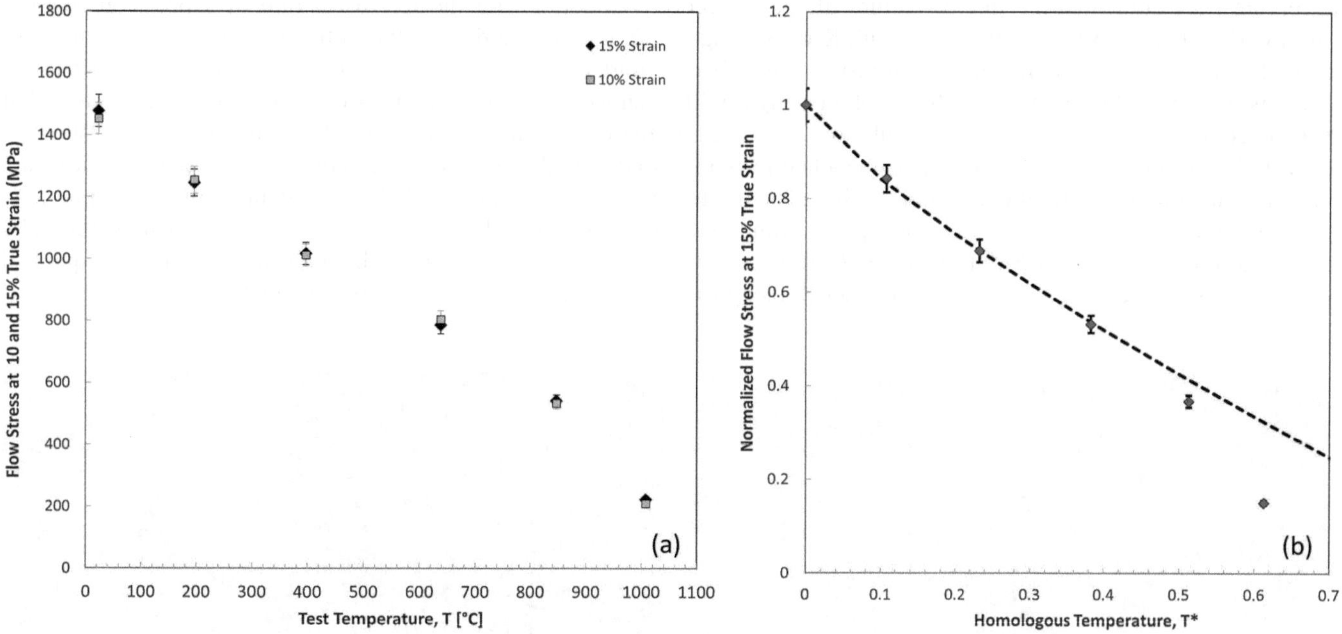

Fig. 30.3 (**a**) Flow stresses at 0.1 strain and 0.15 true strain plotted as a function of thermodynamic temperature. (**b**) Normalized flow stresses $\sigma(T)/\sigma(T_{ref})$ at 0.15 true strain plotted as a function of homologous temperature $T^* = (T-T_{ref})/(T_{melt}-T_{ref})$, with T_{ref} = 23 °C, T_{melt} = 1630 °C

Thermal softening is evident, as the flow stress decreases significantly with increase in temperatures. The flow stresses at 0.1 and 0.15 true strains are plotted as a function of test temperature in Fig. 30.3a. The Johnson-Cook empirical equation is often used to describe material response as a function of plastic strain, strain rate and temperature [9]:

$$\sigma\left(\varepsilon_p, \dot{\varepsilon}_p, T\right) = \left[A + B\left(\varepsilon_p\right)^n\right]\left[1 + C\,ln\left(\dot{\varepsilon}_p{}^*\right)\right]\left[1 - \left(T^*\right)^m\right] \tag{30.1}$$

where thermal softening is encompassed by the last term, $\left[1 - \left(T^*\right)^m\right]$. T* is the homologous temperature given by $(T - T_{ref})/(T_{melt} - T_{ref})$. In an attempt to describe our results similarly, the normalized flow stresses $\sigma(T)/\sigma(T_{ref})$ at 0.15 true strain are plotted as a function of T* with the reference temperature, $T_{ref} = 23$ °C, and melting point, $T_{melt} = 1630$ °C. This graph in Fig. 30.3b shows that the data could be well described by the Johnson-Cook equation for temperatures below 650 °C with a thermal softening parameter *m* of value 0.8. Similar values of *m* have been reported in other high strain rate investigations such as: Seo et al. [3] whose *m* was 0.7 while Johnson [10] reported $m = 0.8$. Dorogoy et al. [11] have reported a similar *m* of 0.8 from quasi-static loading conditions as well. At temperatures higher than 650 °C however, there is a steeper reduction in the flow stresses.

This behavior can be a result of the ongoing allotropic transformation in Ti-6Al-4V. At room temperature, pure Titanium exists in hcp (α) crystal structure and when heated to 882 °C, it changes into bcc (β) [7]. The alloy Ti-6Al-4V containing both α-stabilizing Aluminum and β-stabilizing Vanadium elements does not have a sharp transition point, instead there is a temperature range in which there is a gradual transformation. The start temperature α-transus is lower than room temperature, and the β-transus is about 995 °C [8]. At room temperature, the alloy contains both α and β. At higher temperatures, α starts transforming into β, increasing the β phase fraction. The increase in β volume fraction is insignificant below 600 °C, after which it starts to rise slowly initially and becoming rapid at elevated temperatures, especially closer to the β-transus. The α phase is known to show better high-temperature strength than the β phase [12], so at temperatures above 600 °C, it can be expected that the flow stresses show a more rapid decline with increasing temperature.

It has to be considered here that the heating time involved in our tests is only 3.5 s. Whether this time is sufficient for the $\alpha \rightarrow \beta$ transformation needs to be verified before concluding that the low flow stresses observed are a result of the allotropic transformation. Confirmation of this transformation will be found in the microstructure of post-compression specimens. The electrical-pulse heated Kolsky Bar allows rapid cooling rates of about 1000 °C/s owing to the swift heat transfer from the hot specimen to the cool bar ends once the heating current is turned off. If the test temperature is higher than the martensitic start temperature M_S, this quenching causes any new β produced by the $\alpha \rightarrow \beta$ transformation on heating to transform into martensitic acicular α′ [12].

Figure 30.4 compares the microstructure of the as-received material with the microstructure of a post-compression specimen, tested at 850 °C, both etched with Kroll's Reagent. Figure 30.4 (a depicts the initial/as-received microstructure consisting of globular α grains (bright contrast) in a matrix of β (dark contrast). At room temperature, the phase distributions consists of about 90 % α and 10 % β by volume. Figure 30.4b shows a region in the post-compression specimen tested at 850 °C consisting of a large area of acicular α′. According to the phase diagram [12], the phase composition at 850 °C is expected to be 66 % α and 34 % β. When quenched from this condition, the volume fraction of primary α remains the same, but the β starts to transform into acicular α'. Since the martensite finish temperature M_F is lower than room temperature, the transformation is incomplete, leaving some β remaining in the microstructure. The quenched microstructure can therefore be expected to contain about 66 % primary α, 24 % α′ and 10 % β. Acicular α' is easily distinguishable from the equiaxed primary α as can be seen in Fig. 30.4b, and hence its appearance confirms the allotropic transformation.

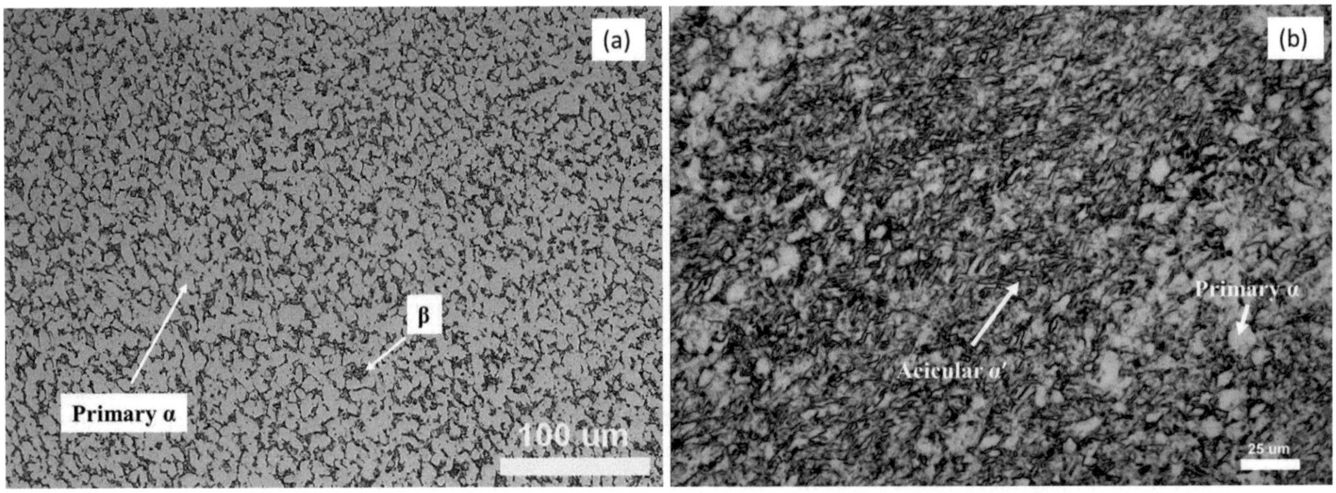

Fig. 30.4 (**a**) Microstructure of as received Ti-6Al-4V (**b**) microstructure of Ti-6Al-4V post-compression specimen tested at 850 °C

The initial results indicate that the constitutive response of Ti-6Al-4V from the pulse heated Kolsky Bar system is comparable to the behavior reported by other investigators [1, 3, 14] from room temperature to about 650 °C. However, at higher temperatures we observe the thermal softening could not be similarly described using Johnson-Cook's equation owing to the allotropic transformation, which has not been considered by most of the other studies [1, 3, 14]. Modifications to the Johnson-Cook equation are often made to account for phase transformation [12], but they are essentially step functions only suitable for well-defined transformation temperatures. But in a material such as Ti-6Al-4V, where there is a continual phase transformation in a two-phase field, they are not ideal. A new modification to the Johnson-Cook equation is therefore essential to describe thermal softening in Ti-6Al-4V, one that considers the changing phase fractions together with the difference in the high temperature strengths. We propose to conduct more experiments in the 700–1000 °C range with smaller temperature steps to more completely describe Ti-6Al-4V's thermal softening behavior through the transition region.

30.4 Conclusions

A compression Kolsky Bar system, which directly pulses electrical current through samples for achieving high temperature, is used to study the widely used Titanium alloy Ti-6Al-4V. The first results from this investigation where compression tests were conducted at about 2200 s^{-1} strain rate at temperatures ranging from 23 to 1008 °C are reported in this paper. The results show that the dynamic response of the material has a strong dependence on temperature. The thermal softening occurs more rapidly above 650 °C, and is attributed to allotropic transformation from hcp (α) to bcc (β) crystal structure. Evidence of this transformation is discovered in the microstructures of post-compression specimens, which show an acicular α' phase formed from quenching of β when the test temperatures are above martensite start, M_S. Further data is essential, especially in the 700–000 °C temperature range, to develop a new modification to the Johnson-Cook equation, which considers the changing phase fractions as a function of temperature to be able to describe the Ti-6Al-4V's mechanical response.

References

1. Lee, W.S., Lin, C.F.: Plastic deformation and fracture behavior of Ti-6Al-4V alloy loaded with high strain rate under various temperatures. Mater. Sci. Eng. **A241**, 48–59 (1998)
2. Seo, S., Min, O., Yang, H.: Constitutive equation for Ti-6Al-4V at high temperatures using the SHPB technique. Int. J. Impact Eng. **31**(6), 735–54 (2005)
3. Gray, G.T. III: Classic Split Hopkinson pressure bar testing. In: Kuhn, H., Medlin, D. (eds.) ASM Handbook, vol. 8, pp. 462–476. ASM International, Materials Park (2000)
4. Mates, S.P., Rhorer, R., Whitenton, E., Burns, T.J.: A pulse-heated Kolsky bar technique for measuring the flow stress of metals at high loading and heating rates. Exp. Mech. **48**, 799–807 (2008)
5. Basak, D., Yoon, H.W., Rhorer, R., Burns, T.J., Matsumoto, T.: Temperature control of pulse heated specimens in a Kolsky bar apparatus using microsecond time resolved pyrometry. Int. J. Thermophys. **25**(2), 561–574 (2004)
6. Wood, R.A.: Titanium Alloy Handbook. Metals and Ceramics Information Center. Battelle, Columbus. Publication No. MCIC-HB-02 (1972)
7. Froes, F.H.: Titanium—Physical Metallurgy, Processing and Applications. ASM International, Materials Park (2015)
8. Semiatin, S.L., Seetharaman, V., Weiss, I.: The thermomechanical processing of alpha/beta titanium alloys. JOM **49**, 33–39 (1997)
9. Johnson, G.R., Cook, W.H.: A constitutive model and data for metals subjected to large strains, high strain rates and high temperatures. In: Proceedings of the 7th International Symposium on Ballistics, vol. 21, The Hague. pp. 541–547(1983)
10. Johnson, G.R.: Strength and fracture characteristics of a titanium alloy (0.6Al, 0.4V) subjected to various strains, strain rates, temperatures and pressures. Technical report TR 86-144, Naval Surface Warfare Center, Potomac (1985)

11. Dorogoy, A., Rittel, D.: Determination of Johnson-Cook material parameters using the SCS specimen. Exp. Mech. **49**, 881 (2009)
12. Donachie, M.J.: Titanium: A Technical Guide. ASM International, Materials Park (2000)
13. Andrade, U., Meyers, M.A., Vecchio, K.S., Chokshi, A.H.: Dynamic recrystallization in high strain, high strain rate plastic deformation of copper. Acta Metall. Mater. **42**, 3183–95 (1994)
14. Nemat-Nasser, A., Guo, W.G., Nesterenko, V.F., Indrakanti, S.S., Gu, Y.B.: Dynamic response of conventional and hot isostatically pressed Ti-6Al-4V alloys: experiments and modeling. Mech. Mater. **33**(8), 425–39 (2001)

Chapter 31
Parametric Study of the Formation of Cone Cracks in Brittle Materials

Brady Aydelotte, Phillip Jannotti, Mark Andrews, and Brian Schuster

Abstract Brittle materials such as ceramics and oxide glasses are widely employed because they possess a variety of useful properties including hardness, strength, wear resistance and/or transparency. Normal and oblique impacts of spherical projectiles on brittle materials are similar to the classical Hertzian sphere indentation problem, yet different in significant ways. In the case of oblique impacts, these differences can result in the formation of unique damage patterns, including cone cracks, which are not axisymmetric, and they are distinct from partial cone cracks formed by sliding indentation. Some selected results of oblique sphere impacts on brittle targets are shown and discussed, identifying unique features. A parametric study of the factors contributing to the differences between normal and oblique impacts is reported. The effects of normal and tangential velocity, friction, and softening are investigated as factors influencing the peak principal stresses produced by oblique impacts. Friction and projectile softening were found to have a significant impact on peak maximum principal stress values. An obliquely impacting projectile's lateral motion and its interaction with the sides of the cone crack were also found to have a significant effect on the stress field and the shape of the resulting cone crack.

Keywords Cone crack • Impact • Indentation • Ceramic • Damage

31.1 Introduction

The causes of damage during oblique impacts is the subject of this paper. Sliding indentation, the obvious quasi-static analogue, has received considerable attention in mechanics literature. Hamilton and Goodman [1] developed an analytical solution for indentation with tangential motion for sliding friction. Lawn [2] combined Hamilton and Goodman's stress field solution and some concepts from fracture mechanics to deduce that the presence of a tangential load reduces the normal load required to pop in a cone crack by large factors of 100 or more depending on the indenter radius and the coefficient of friction. He also found that Auerbach's law ($P_{crit} \propto R$) still holds over a range of indenter sizes, but outside that range $P_{crit} \propto R^2$. Lawn [2], noting the large tensile region created at the back of a sliding indenter with a non-zero coefficient of friction concluded that partial cone cracks would form behind the indenter, and published images from sliding indentation that indeed suggest the formation of partial cone cracks in the wake of a sliding indenter with the cone axis oriented away from the indentation direction. Gilroy and Hirst [3] conducted sliding indentation studies in glass with a variety of indenter sizes. They found that indeed $P_{crit} \propto R$ over a range of R values then as R grew larger, $P_{crit} \propto R^2$. They did find that sliding indentation did cause a significant reduction in P_{crit}, but that the actual local tensile failure strength doesn't change; the tension region created by sliding indentation is simply more tensile resulting in a lower combined load to initiate fracture. They found a reduction in P_{crit} on the order or 5–10× less than Lawn's model predicted. Powell and Tabor's [4] results in titanium carbide essentially agree with Gilroy and Hirst [3]. Lawn et al. [5] used some results published by Roesler [6] to develop a model that predicts the orientation of the cone crack assuming a fixed cone geometry based on the coefficient of friction. They found this agreed reasonably well with their experimental data on glass. Laugier [7] reexamined Lawn's [2] model predicting the reduction in the critical normal load leading to initial fracture. Laugier concluded that Lawn's [2] model was using overestimates of the non-sliding indentation load to cause fracture. When adjusted for more accurate non-sliding critical loads, the reductions in sliding critical load were reasonably close to experimental values in alumina.

Damage due to normal impact on ceramics by spheres, long rods (length to diameter ratio $\geq$ 10) and short rods (length to diameter ratio $<$ 10) has been studied since at least the 1960s (see for example [8, 9, 10, 11, 12, 13]). Much less research has been published on the damage phenomenology that results from oblique impacts in ceramics, though a few papers exist. Salman et al. [14] studied the effects of oblique impacts on damage to ceramic particles, documenting a few different

B. Aydelotte (✉) • P. Jannotti • M. Andrews • B. Schuster
US Army Research Laboratory, 328 Hopkins Road RDRL-WML-H, Aberdeen Proving Ground MD 21005, USA
e-mail: brady.b.aydelotte.civ@mail.mil

D. Casem et al. (eds.), *Dynamic Behavior of Materials, Volume 1*, Conference Proceedings of the Society for Experimental Mechanics Series, DOI 10.1007/978-3-319-41132-3_31

damage modes. Sadanandan and Hethrington [15] and Hohler et al. [16] studied various simplified armor packages impacted at different obliquities and their resistance to single impacts as measured by metrics like V_{50}. There was little information published about impact-induced damage. Fawaz et al. [17] presented some results of modeling oblique and normal impacts on ceramic targets. Grant et al. [18] published a study of low velocity impacts of irregular projectiles on layered glass targets. Forde et al. [19] published a study of normal and oblique impacts of mild steel rods on borosilicate glass targets.

More closely related to this work, Wiederhorn et al. [20] studied the impact of small, irregular silicon carbide particles on glass targets. They observed elongated craters and postulated that the transverse particle motion might play some role in the crack-driving force. Chaudhri and Liangyi [21, 22] examined oblique sphere impacts on soda lime glass in detail. They found that oblique impacts led to cone cracks which were rotated relative to the impact surface normal, and that the cone crack actually appeared to change its included angle during the impact event. They also observed that cone crack orientation was essentially the opposite of that predicted and observed by Lawn [2] and Lawn et al. [5]. The main axis of the cone crack, rather than being rotated to point away from the indenter sliding direction, was actually rotated toward the direction of sliding. Chaudhri and Liangyi [21, 22] explained this by showing a dynamic impact experiment and a static indentation where they filmed the events with circularly polarized light. The resulting fringe contours didn't appear distorted by friction, leading them to conclude that there was very little friction. In turn, this lead Chaudhri and Liangyi [21, 22] to conclude that the shape of the cone crack resulted from the displacement of the applied load absent much friction. As the projectile slides across the surface, the leading edge of the cone crack aligns with steeper principal stress trajectories under the projectile while the trailing edge of the cone crack aligns with much shallower principal stress trajectories.

Chaudhri and Liangyi's [21, 22] work on characterizing oblique impacts was exclusively conducted on soda-lime glass. Chaudhri and Kurkjian [23] had shown that oxide glasses other than borosilicate glass and soda-lime glasses could have significantly different failure morphologies. Therefore, Aydelotte and Schuster [24] explored oblique impact induced damage morphology in polycrystalline ceramics. They used flash X-ray measurements to observe many of the same features reported by Chaudhri and Liangyi [21, 22] including cone cracks that were rotated with respect to the impact surface and cone cracks which were distorted due to the spatially and temporally varying stress field caused by the translating projectile.

Aydelotte et al. [25] took the recovered samples generated by the experiments reported in Aydelotte and Schuster [24] and conducted X-ray computed tomography (XCT) scans of them. They compared this data with XCT data generated from fused silica samples that had been subjected to oblique impacts to look for differences between the responses of fused silica glass and commercial polycrystalline boron carbide to oblique impacts. Aydelotte et al. [25] found that damage produced by normal and oblique impacts on fused silica and boron carbide targets were similar in many ways but also distinct in a few ways. Oblique impacts on boron carbide samples resulted in cone cracks with included angles very similar to those of normal impacts when compared on the basis of velocity normal to the impact surface. Cone cracks due to oblique impacts were often rotated relative to the surface normal of the target sample. The cone crack angle in boron carbide was consistently larger when measured on a plane which intersected the apex of the fracture conoid and contained the shot-line vector than when the cone angle was measured on a plane perpendicular to that containing the shot-line.

Aydelotte et al. [25] found that, similar to boron carbide samples, oblique impacts on fused silica samples produced curving fracture conoids which were rotated toward the direction the impacting projectile traveled. Fused silica samples did not show any consistent difference between measurements of the cone angle on the plane containing the shot-line vector or a perpendicular plane.

Chaudhri and Liangyi's [21, 22], Aydelotte and Schuster [24], Aydelotte et al. [25] all found that cone crack rotation was very different from what was predicted by Lawn [2] and Lawn et al. [5]. The purpose of this work is to explore the stress fields created by oblique impacts of various types onto different material descriptions numerically to gain additional insight into the formation of damage due to oblique impacts.

31.2 Selected Oblique Impact Experimental Results

Samples made of borosilicate glass were impacted with spheres at varying obliquities. The borosilicate glass samples, Schott Borofloat 33, were purchased cut and polished from Swift Glass Co Inc (Elmira Heights, NY). They were 101.6 mm (4.0 in.) × 101.6 mm (4.0 in.) × 12.7 mm (0.5 in.) or 25.4 mm (1.0 in.). A Shimadzu HPV-X high speed video camera or a Specialized Imaging Kirana high speed video camera were used to film the impact of projectiles against the glass targets. Flash bulbs or a Photogenix flash provided illumination. A mirror was used to film both the side profile of the impact event and the rear view of the impact and the resulting damage.

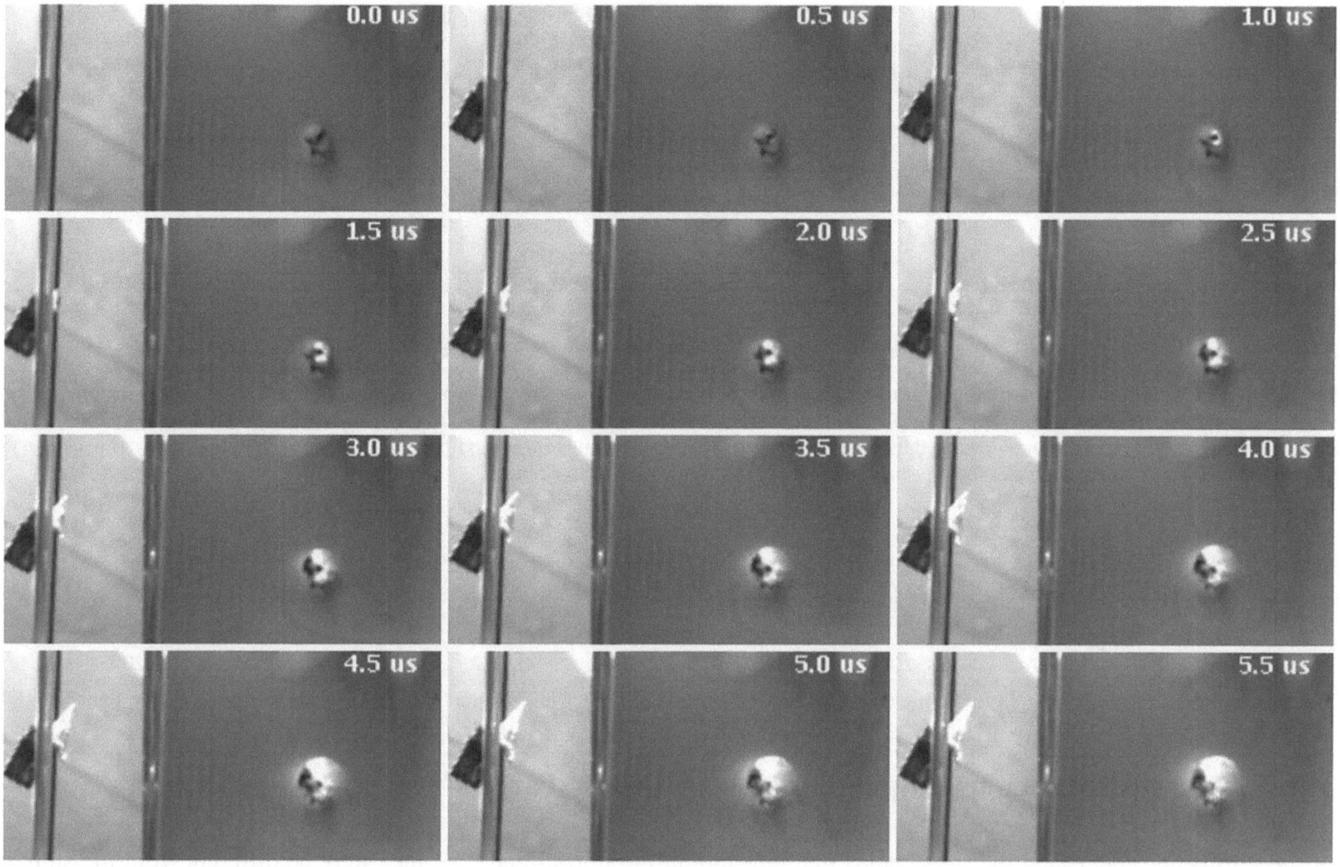

Fig. 31.1 Montage of images from an oblique impact on borosilicate glass at 179 m/s. The side view and the corresponding rear view are visible in each frame. Note that a crack is clearly visible by approximately 1.0 μs post impact and quickly grows. The crack doesn't appear to initiate in the wake of the translating projectile and the crack orientation is opposite what would be expected from [5]

The borosilcate glass samples were impacted with 6.35 mm diameter (0.25 in.) borosilicate glass spheres fired from a.30 caliber powder gun using plastic, non-discarding obturators. Break screens (from Whithner Screen Graphics Baltimore, MD) or orthogonal pairs of 150 kV flash X-rays were used to determine projectile velocities.

Hamilton and Goodman's [1] elastic stress field solution for a quasi-static sliding indenter shows that a large tensile region forms behind a sliding indenter in the presence of friction. Chaudhri and Liangyi[21, 22] believed that friction was not significant based on their observations. We would expect that if friction is significant, failure will initiate behind the impact sphere. If friction isn't significant we do not expect failure to initiate at the rear of the sliding sphere.

Figure 31.1 is a series of images from high speed video of a borosilicate glass target impacted at 179 m/s at 60° of obliquity. This sequence of images shows features which were typically observed during impact in all glass samples. The ring crack appears to form nearly simultaneously between one frame (0.0 μs) and the next (0.5 μs), but is very difficult to see the crack until 1.0 μs. There is no evidence that the ring forms in the rear of the sphere where the tension region would be expected or that the cone crack is oriented like those observed in [5]. This is consistent with the observations of Chaudhri and Liangyi's [21, 22].

As time advances in Fig. 31.1, the cone crack develops a pronounced forward tilt. That is, the cone axis is rotated toward the direction the projectile sphere is moving in as it translates across the surface of the glass target. It also appears that the leading edge of the cone crack is growing more than the trailing edge. We also observed that the trailing edge of the cone crack curves back toward the surface. Other high speed video of borosilicate glass targets struck by glass spheres have yielded similar observations.

Similar results were reported in Aydelotte et al. [25] in fused silica. A cross section of a fused silica sample impacted at 30° of obliquity is shown in Fig. 31.2. Note that the leading edge of the cone crack, the edge of the cone crack closest to the side toward which the projectile was traveling, is turned toward the bottom of the sample. The trailing edge is turned up toward the top surface. This is consistent with what is shown in Fig. 31.1 and what has been reported by other investigators [21, 22].

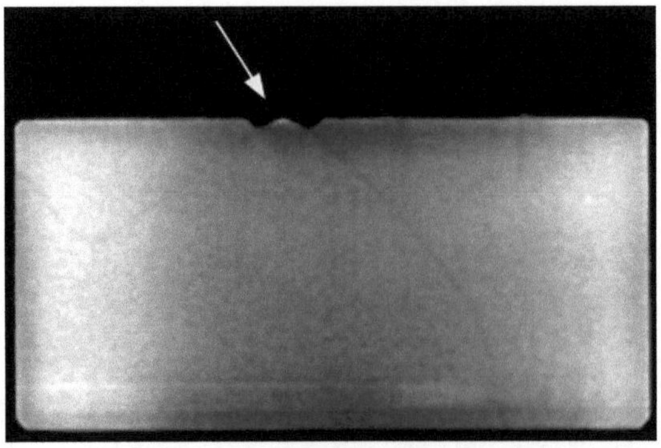

Fig. 31.2 A cross section of a fused silica sample impacted by a borosilicate glass sphere at 30° of obliquity and a velocity of 121 m/s. This sample was part of the experiments reported in Aydelotte et al. [25]. The *white arrow* in the image shows the approximate orientation of the projectile's velocity. The image was generated by XCT scans of the sample. The image was modified using Fiji [26], an image analysis package, to enhance the image contrast and make the presence of a cone crack more readily apparent

Table 31.1 Borosilicate and fused silica glasses were taken to have the following properties

Material	Youngs modulus (GPa)	Poisson's ratio	Density (kg/m^3)
Borosilicate glass	64	0.2	2200
Fused silica	73	0.17	2200

The same equation of state was used for both glasses, SESLAN table 7386. Further information about the equation of state can be found in [27]

31.3 Numerical Modeling of Sphere Impacts on Glass Targets

In order to better understand the stress fields that develop during impact and how they may be different from the sliding elastic indentation stress field solutions derived by Hamilton and Goodman [1], a series of computational models was used to evaluate the stress field during various impact conditions. The ALEGRA multimaterial hydrocode [27], developed and supported by Sandia National Laboratories, was used for all calculations. The models were run on the Army DSRC system Excalibur. Both Lagrangian and Eulerian calculations were used to explore the stress fields as a function of velocity normal to the surface, velocity tangential to the surface, and the coefficient of friction. Friction was modeled as a constant which was user specified as either 0 or 1.

The simulations, impacts of glass spheres onto glass targets, were all conducted with a mesh resolution of approximately 45 elements across the projectile diameter or 0.141 mm on a side. Most of the calculations utilized linear elastic constitutive models for either borosilicate glass or fused silica with properties assumed shown in Table 31.1. A few were run allowing material softening using the glass model developed by Richard Becker [28] and parameterized using data from Chocron et al. [29, 30] to explore the effect of softening and failure for both the target and the projectile on the peak value of the maximum principal stress field. The same model was studied previously by Aydelotte and Schuster [31].

Plots of the maximum principal stress at 1 μs post impact are shown in Fig. 31.3a, b. Reporting maximum principal stress values was done for two reasons: cone cracks were generally observed to have formed within that time period post impact and the stresses oscillated in time so that peak principal stress values didn't always occur at the same times. Reporting maximum principal stress values 1 μs post impact was one way of making a fair comparison between simulations.

In Fig. 31.3a, Lagrangian simulations were used to study the effects of friction, velocity, and obliquity for normal and oblique impacts. Normal impact velocity was 100 or 300 m/s and velocity tangential to the surface was between 0 and 300 m/s. Some selected Eulerian results are shown as well for comparison. All of the materials were elastic for the simulations showed in Fig. 31.3a. Allowing the coefficient of friction to increase from 0 to 1 increased the maximum principle stress σ_1 by a factor of around 10 in many cases. This indicates that, as shown by the static solution of Hamilton and Goodman[1], the effect of friction has huge impact on the tensile stresses that are developed during an oblique impact. Eulerian meshes also generated very high levels of σ_1 post impact. This is a result of ALEGRA's inability to support a velocity discontinuity in mixed material cells; the projectile and the target are effectively welded together.

Eulerian simulations showing the effects of material softening for normal impacts at 300 m/s and oblique impacts with a normal velocity of 300 m/s and a velocity tangential to the surface of 100 m/s are shown in Fig. 31.3b. The classical stress field solution of Hamilton and Goodman [1] is purely elastic, but during impact experiments the projectile often shatters.

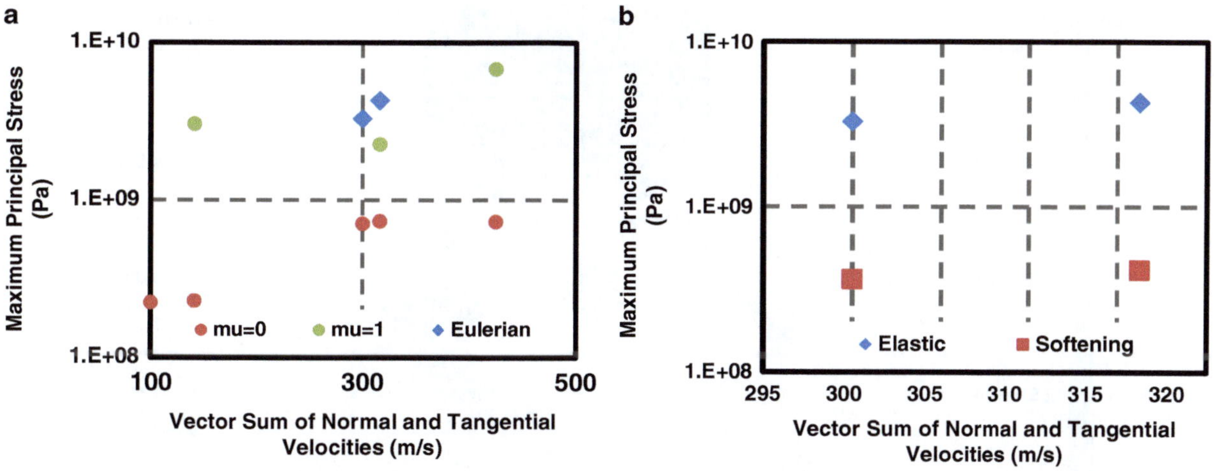

Fig. 31.3 Plots of the maximum principal stress versus the sphere impact velocity at 1 μs post impact. (**a**) Lagrangian simulations exploring the effects of friction, velocity, and obliquity. Some selected Eulerian results are shown as well. All of the materials were elastic. Transitioning from a coefficient of friction of 0–1 increased the maximum principle stress σ_1 by a factor of around 10 in many cases. Eulerian meshes also generated very high levels of σ_1 due to interface welding of mixed material cells. (**b**) Eulerian simulations showing the effects of material softening by examining what happens to the principal stress value when both the sphere and the target were elastic and when both are permitted to soften. Softening in both target and projectile reduced the maximum principle stress σ_1 by about as much as friction

This was certainly the case in the experiments discussed above. In these simulations the projectile sphere and target either both remained elastic or both were allowed to soften. Softening in both target and projectile reduced the maximum principle stress σ_1 by about as much as eliminating friction.

One calculation was performed to assess the effect of the projectile interacting with the fracture conoid after it forms by inserting an artificial cone crack and then impacting the top of the cone crack. The artificial cone crack was created by removing the frustum of a cone 3 elements thick from the solid model before meshing. The root of the artificial cone crack has a radius of 1.5 times the element size. A sphere impact at an obliquity of 60° and a velocity of 300 m/s was simulated, the same obliquity as that observed in Fig. 31.1.

The stress field Fig. 31.4 is color coded to show scalar intensity of the 1st principal stress. The lines indicate the orientation of the σ_3 stresses. These are perpendicular to σ_1 and show on a per element basis the approximate direction that an opening crack would grow due to σ_1. The σ_3 stress lines emanating the trailing edge of the cone crack show potential crack paths that are nearly horizontal and then curving upward toward the top surface. The potential crack paths emanating from the tip of the cone crack closest to the projectile are directed at a downward angle toward the bottom of the elastic cylinder. This is consistent with the experimental observations in Figs. 31.1 and 31.2. This appears to be due to the fact that the action of the σ_1 stresses is vertical under the sliding indenter and the angle becomes more shallow to either side of the indenter. As the sphere slides across the surface and applies pressure, the crack is influenced by σ_1 stresses under the indenter which act vertically. The result is that as the sphere translates across the surface of a cracking body, the cone crack on the leading edge re-orients itself to more be more and more vertical, often leading to concave down curvature of the leading edge of the cone crack.

We also see uneven stresses the roots of the cone crack in Fig. 31.4. As the sphere moves to the right, the stresses at the root of the cone crack closest to the sphere become more tensile. When the sphere impinges on the outside edge of the cone crack, the stress concentration at the root of the crack becomes even more severe. This appears to be why the leading edges-the cone crack edge closest to the sphere-of cone cracks which form during oblique impacts often grow more than the trailing edges of those same cone cracks during an oblique impact. This effect would probably be even more pronounced if the projectile had gouged into the surface, providing a more effective means for the sphere to apply a shear load to the surface. These results support one of the suppositions of Wiederhorn et al. [20], that the lateral motion of the projectile contributes to the crack driving force.

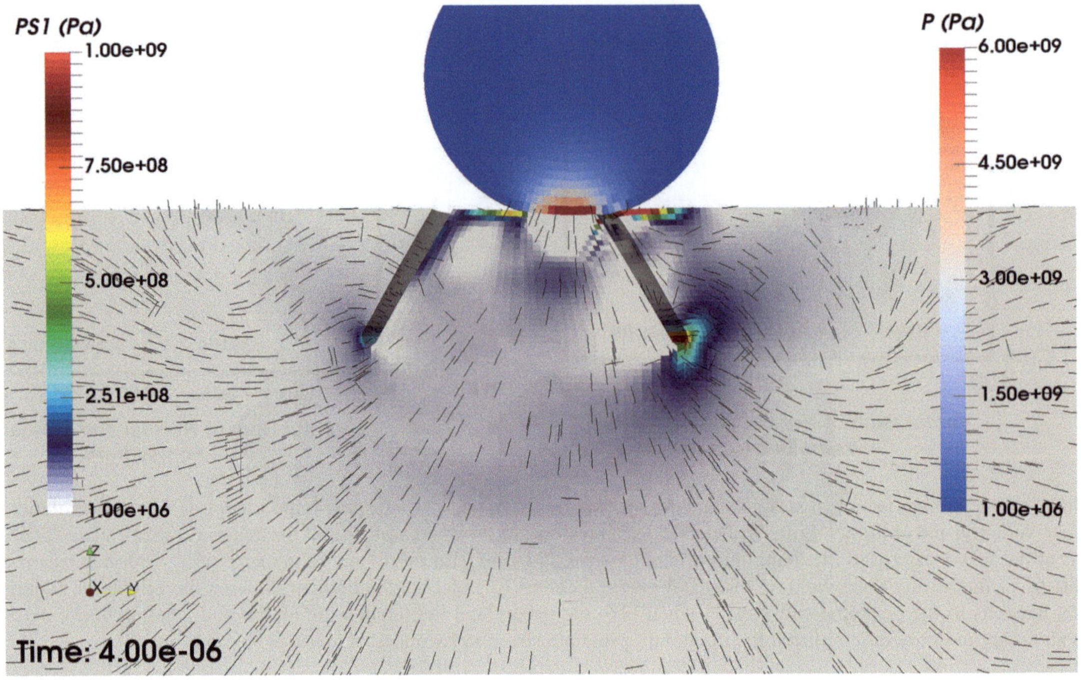

Fig. 31.4 Plot of σ_1 due to oblique impact by a sphere (moving from *left* to *right*) onto an elastic cylinder containing an approximation of a cone crack. The *lines* indicate the orientation of the σ_3 stresses. These are perpendicular to σ_1 and indicate approximate crack paths within the bulk of the target. Note that the root of the artificial crack away from the direction of travel shows potential crack paths that are horizontal and then curving upward toward the top surface. The potential crack paths emanating from the tip of the cone crack closest to the projectile are directed at a downward angle toward the *bottom* of the elastic cylinder. This is consistent with the experimental observations in Fig. 31.1

31.4 Conclusions

Static indentation is a very different process than dynamic impact of spherical projectiles. Though some similarities exist, the low levels of friction that appear to exist during dynamic impacts and, in this case, the softening projectiles used during the impact experiments, have a very different outcome. Rather than partial cone cracks which form rotated away from the direction of travel, complete cone cracks form which are rotated toward the direction of travel. There is no evidence that the cone crack forms in the wake of the sliding impacter.

Modeling and simulation of the impact event provide some insight into the failure process. Friction is able to have a dramatic effect on the tensile stress field, very similar to observations made by Goodman and Hamilton [1], however experimental observations do not suggest that much friction is present. Instead, it appears that the formation of cone cracks during oblique impacts is not so strongly influence by the friction as by the translation of the projectile across the target surface and the projectile's interaction with the sides of the cone crack.

Acknowledgements We gratefully acknowledge Todd Rumbagh of Hadland Imaging (San Francisco, CA) for the loan of the Shimadzu HPV-X high speed video camera and Brian Love (ARL) and Mark Foster (formerly ARL) for the loan of the Specialized Imaging Kirana high speed video camera and assistance in running the Kirana during experiments. We also acknowledge the support of the Sandia CUBIT team for assistance with generating suitable finite element meshes, and the range 309A technicians who assisted in these experiments. The computer simulations were made possible by a grant of computer time from the High Performance Computing Modernization Program (HPCMP) on the ARL HPC system Excalibur.

References

1. Hamilton, G.M., Goodman, L.E.: The stress field created by a circular sliding contact. J. Appl. Mech. **33**, 371–376 (1966)
2. Lawn, B.R.: Partial cone crack formation in a brittle material loaded with a sliding spherical indenter. Proc. R. Soc. Lond. A Math. Phys. Eng. Sci. **299**, 307–316 (1967)
3. Gilroy, D.R., Hirst, W.: Brittle fracture of glass under normal and sliding loads. J. Phys. D Appl. Phys. **2**(12), 1784 (1969)

4. Powell, B.D., Tabor, D.: The fracture of titanium carbide under static and sliding contact. J. Phys. D Appl. Phys. **3**(5), 783 (1970)
5. Lawn, B.R., Wiederhorn, S.M., Roberts, D.E.: Effect of sliding friction forces on the strength of brittle materials. J. Mater. Sci. **19**, 2561–2569 (1984)
6. Roesler, F.C.: Indentation hardness of glass as an energy scaling law. Proc. Phys. Soc. Sec. B **69**(1), 55 (1956)
7. Laugier, M.: The surface fracture of alumina under a sliding spherical indenter. J. Mater. Sci. Lett. **5**(3), 253–254 (1986)
8. Wilkins, M.L.: Mechanics of penetration and perforation. Special Issue: Penetration Mech. **16**(11), 793–807 (1978)
9. Shockey, D.A., Marchand, A., Skaggs, S., Cort, G., Burkett, M., Parker, R.: Failure phenomenology of confined ceramic targets and impacting rods. Int. J. Impact Eng. **9**(3), 263–275 (1990)
10. LaSalvia, J.C., Normandia, M.J., Miller, H.T., Mackenzie, D.E.: Sphere impact induced damage in ceramics: I. Armor-Grade SiC and TiB2. In: Advances in Ceramic Armor: A Collection of Papers Presented at the 29th International Conference on Advanced Ceramics and Composites, January 23–28, 2005, Cocoa Beach, Florida. Ceramic Engineering and Science Proceedings, pp. 170–181. Wiley (2008)
11. LaSalvia, J.C., Normandia, M.J., Miller, H.T., MacKenzie, D.E.: Sphere impact induced damage in ceramics: II. Armor-Grade B4c and WC. In: Advances in Ceramic Armor: A Collection of Papers Presented at the 29th International Conference on Advanced Ceramics and Composites, January 23–28, 2005, Cocoa Beach, Florida. Ceramic Engineering and Science Proceedings, pp. 183–192. Wiley (2008)
12. LaSalvia, J.C., Normandia, M.J., MacKenzie, D.E., Miller, H.T.: Sphere impact induced damage in ceramics: III. Analysis. In: Advances in Ceramic Armor: A Collection of Papers Presented at the 29th International Conference on Advanced Ceramics and Composites, January 23–28, 2005, Cocoa Beach, Florida. Ceramic Engineering and Science Proceedings, pp. 193–202. Wiley (2008)
13. LaSalvia, J., Leavy, R., Houskamp, J., Miller, H., MacKenzie, D., Campbell, J.: Ballistic impact damage observations in a hot-pressed boron carbide. In: *Advances in Ceramic Armor V: Ceramic Engineering and Science Proceedings*, vol. 30, issue 5, pp. 45–55. Wiley (2009)
14. Salman, A., Gorham, D., Verba, A.: A study of solid particle failure under normal and oblique impact. In: 8th International Conference on Erosion by Liquid and Solid Impact, vol. 186–187, Part 1, pp. 92–98 (1995)
15. Sadanandan, S., Hetherington, J.G.: Characterisation of ceramic/steel and ceramic/aluminium armours subjected to oblique impact. Int. J. Impact Eng. **19**, 811–819 (1997)
16. Hohler, V., Weber, K., Tham, R., James, B., Barker, A., Pickup, I.: Comparative analysis of oblique impact on ceramic composite systems. Int. J. Impact Eng. **26**, 333–344 (2001)
17. Fawaz, Z., Zheng, W., Behdinan, K.: Numerical simulation of normal and oblique ballistic impact on ceramic composite armours. Compos. Struct. **63**, 387–395 (2004)
18. Grant, P., Cantwell, W., McKenzie, H., Corkhill, P.: The damage threshold of laminated glass structures. Int. J. Impact Eng. **21**, 737–746 (1998)
19. Forde, L.C., Proud, W.G., Walley, S.M., Church, P.D., Cullis, I.G.: Ballistic impact studies of a borosilicate glass. Int. J. Impact Eng. **37**, 568–578 (2010)
20. Wiederhorn, S.M., Lawn, B.R., Hockey, B.J.: Effect of particle impact angle on strength degradation of glass. J. Am. Ceram. Soc. **62**, 639–640 (1979)
21. Liangyi, C., Chaudhri, M.M.: The influence of tangential stress on the impact damage of soda-lime glass with spherical projectiles. In: SPIE Proceedings, vol. 1032, pp. 955–964 (1989)
22. Chaudhri, M., Liangyi, C.: The orientation of the Hertzian cone crack in soda-lime glass formed by oblique dynamic and quasi-static loading with a hard sphere. J. Mater. Sci. **24**, 3441–3448 (1989)
23. Chaudhri, M., Kurkjian, C.: Impact of small steel spheres on the surfaces of "normal" and "anomalous" glasses. J. Am. Ceram. Soc. **69**, 404–410 (1986)
24. Aydelotte, B., Schuster, B.: Observation and modeling of cone cracks in ceramics. Conference Proceedings of the Society for Experimental Mechanics Series, vol. 1, pp. 19–23. Springer International Publishing (2016)
25. Aydelotte, B., Jannotti, P., Andrews, M., Schuster, B.: A Comparison of Damage in Glass and Ceramic Targets. Wiley, New York. Manuscript submitted for publication
26. Schindelin, J., Arganda-Carreras, I., Frise, E., Kaynig, V., Longair, M., Pietzsch, T., Preibisch, S., Rueden, C., Saalfeld, S., Schmid, B., et al.: Fiji: an open-source platform for biological-image analysis. Nat. Methods **9**, 676–682 (2012)
27. Robinson, A.C., Carroll, S.K., Drake, R.R., Hansen, G.A., Hensinger, D.M., Kramer, R., Labreche, D.A., Love, E., Luchini, C.B., Mosso, S.J., Petney, S.V., Sanchez, J.J., Siefert, C., Strack, O.E., Voth, T.E., Niederhaus, J.H.J., Ober, C.C., Rider, W.J., Weirs, V.G., Wong, M.K., Haill, M.K.: ALEGRA User Manual. Technical Report SAND2015-4970, Sandia National Labs (2015)
28. Becker, R.C.: A Glass Model Capturing High-Rate Fracture Observations. Technical Report ARL-TR-6086, Army Research Laboratory, Aberdeen Proving Ground, MD (2012)
29. Chocron, S., Anderson Jr. C.E., Nicholls, A.E., Dannemann, K.A., Bigger, R.P.: Characterization of Soda-Lime Glass by Confined Compression Testing With Numerical Validation. Technical Report 18.12544/021, South West Research Institute, San Antonio, Texas (2010)
30. Chocron, S., Anderson, C.E., Nicholls, A.E., Dannemann, K.A.: Characterization of confined intact and damaged borosilicate glass. J. Am. Ceram. Soc. **93**, 3390–3398 (2010)
31. Aydelotte, B., Schuster, B.: A Computational Study of the ARL Glass Model and Its Predictions of Ballistic Penetration and Fracture Conoid Development, pp. 1117–1128. International Ballistics Society (as PDF)

Chapter 32
Shockless Characterization of Ceramics

J.L. Zinszner, B. Erzar, and P. Forquin

Abstract The dynamic response of brittle materials like ceramics is usually studied by means of flyer plate impact experiments. In this work, original tests has been carried out on several ceramics (alumina and silicon carbides) using a high pulsed power generator named GEPI. This electromagnetic device allows obtaining a ramp loading, spread over 500 ns. This particularity is a great advantage to get data on brittle fragmentation through spalling tests in which strain-rate can be accurately determined. Several samples have been recovered partially damaged, giving an interesting insight into the fragmentation process. The ramp loading has also been used to study the compressive response of ceramics thanks to lagrangian analysis. Data have been gathered up to more than 15 GPa on a silicon carbide.

Keywords High-pulsed power • Dynamic characterizations • Lagrangian analysis • Spalling • Strain-rate effect

32.1 Introduction

For several decades, ceramic materials (such as alumina, silicon carbide or boron carbide) are one of the most used family of materials for protective systems. Indeed, besides their very high compressive strengths and hardnesses [1–3], ceramics also present a low density, ensuring a considerable weight benefit in comparison with monolithic steel armors [4]. However, their brittle nature combined with their relatively low tensile strength lead to an inevitable fragmentation of the target after an impact [5–7]. During the impact and the penetration of the projectile, the ceramic is subjected to a succession of three loading states which will have, each one of them, an influence on the performance of the armor system to stop a threat. The first loading state is a dynamic compressive state and lasts several hundreds of nanoseconds. During this period, very high compressive stresses are generated in the projectile and in the ceramic. The projectile damage begins during this first loading state. In the ceramic, the compressive stresses can exceed the Hugoniot Elastic Limit (HEL) of the material leading to the initiation of microdamage or microplasticity mechanisms [8, 9]. The second loading state is a dynamic tensile state. Tensile stresses exceeding the failure strength are generated, leading to the fragmentation of the target. Finally, during the last stage, lasting several tens of microseconds, the projectile or its fragments penetrates into the fragmented ceramic. In order to better understand the role played by the microstructure of the ceramic material on its ballistic efficiency, one of the methods consists in determining and understanding the behavior of the material in each one of these three stages.

Usually, plate impact experiments are performed to determine the dynamic behavior of materials at high strain-rates (higher than 10^3 s^{-1}) [1]. However, this method presents several drawbacks. Indeed, several tests at various loading stresses are needed to determine the dynamic compressive behavior of a material. Moreover, in the case of spalling tests, the shock induced by the impacting plate prevents from precisely determining the strain-rate at failure.

In this study, an electromagnetic generator based on high-pulsed power (HPP) technologies is used to investigate the compression and tensile responses of a SiC ceramic. For studying the dynamic compressive response, the symmetry of the load region of the machine allows generating the same pressure on two different specimens and thus using the lagrangian analysis method. Assuming a known volumetric response, the lagrangian analysis allows determining the whole deviatoric response with a unique test. Data have been gathered up to more than 15 GPa on a silicon carbide. For studying the dynamic tensile response, the tailored loading allows performing spalling tests in which strain-rate can be accurately determined. By

J.L. Zinszner (✉) • B. Erzar
CEA/DAM/GRAMAT, BP 80200, Gramat 46500, France
e-mail: jean-luc.zinszner@cea.fr

P. Forquin
Soils Solids Structures Risks (3SR) Laboratory, Grenoble-Alps University, Grenoble Cedex 9 38041, France

D. Casem et al. (eds.), *Dynamic Behavior of Materials, Volume 1*, Conference Proceedings of the Society for Experimental Mechanics Series, DOI 10.1007/978-3-319-41132-3_32

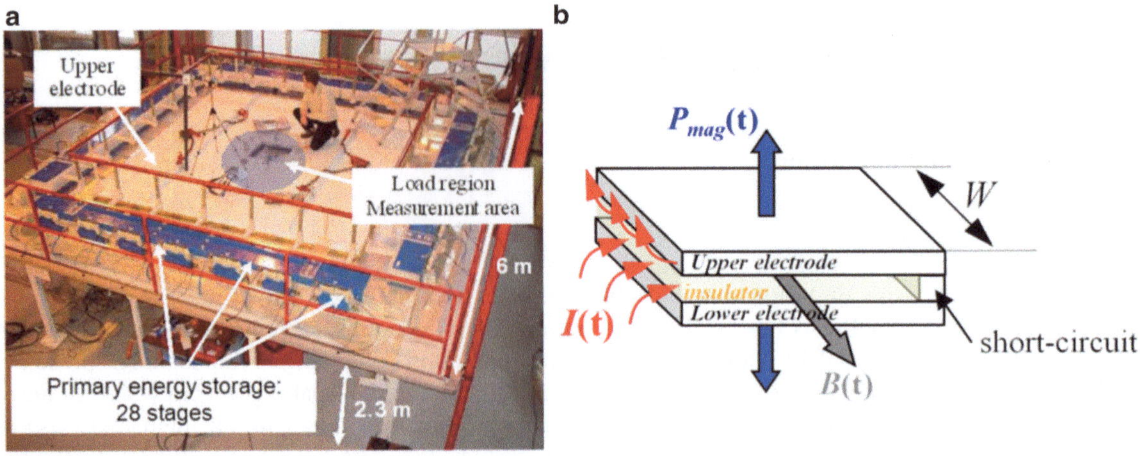

Fig. 32.1 (**a**) GEPI machine overview and (**b**) principle of the strip line design

varying the applied pressure, strain-rate effects on the dynamic behavior of materials can be investigated. Moreover, the absence of propulsive gas allows an easier recovering of specimens after tests in comparison with plate impact tests where a special casing is needed. Several samples have been recovered partially damaged, giving an interesting insight into the fragmentation process.

32.2 Principle of the GEPI Machine

The GEPI machine shown in Fig. 32.1a and settled in CEA Gramat (France) is a high-pulsed power generator based on the strip line concept [10] and devoted to perform quasi-isentropic experiments or hypervelocity impact by accelerating a flyer plate [11]. This machine is used to study the dynamic behavior of inert materials like ceramics, metals, composites and also mortar. The primary energy storage is composed of 28 stages and the total energy stored reaches 70 kJ from a charging voltage of 85 kV. In the central part of the machine, the energy storage stages are connected to the load by a strip line. As shown in Fig. 32.1b, the two electrodes are separated by an insulator (Mylar and Kapton dielectric foils) and locally connected by a short-circuit. Additional peaking capacitors have also been employed to smooth the temporal current profile and to push away the formation of a shock front when high pressures are generated. Using these peaking capacitors, the released current rises over 500 ns to a value about 3 MA.

The current combined with the intense magnetic field generate a compressive pulse applied to the internal skin of each electrode. Thanks to the symmetry of the central part, the same pressure is applied on each electrode. This pressure then propagates through the electrodes and through the materials to be characterized. The magnetic compressive pulse σ_{mag} applied on the electrodes can be easily assessed from the current signal using the relation

$$\sigma_{mag}(t) = k_p \frac{\mu_0}{2} \left(\frac{I(t)}{W} \right)^2, \tag{32.1}$$

where k_p is an edge effect coefficient, μ_0 is the magnetic permeability of free space, W is the width of the strip line and $I(t)$ is the current released measured by a calibrated Rogowski coil. By varying the current $I(t)$ and/or the width of the strip line W, a large range of pressure level (from several tens of MPa up to about several tens of GPa) can be obtained with the GEPI device.

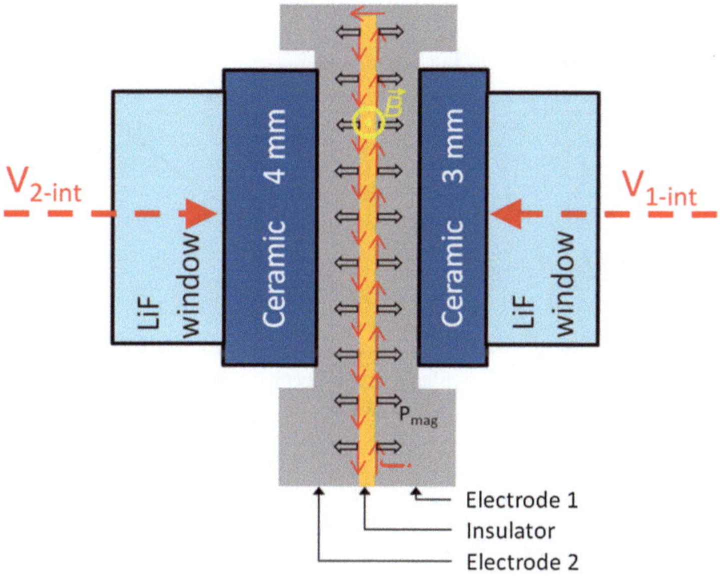

Fig. 32.2 Setup of GEPI compression tests in "lagrangian analysis" configuration

32.3 Quasi-Isentropic Compression Experiments

The Hugoniot Elastic Limit of ceramics (particularly for silicon carbides or boron carbides) being high (about 13–14 GPa for silicon carbides) [1], an intense pressure needs to be applied on the electrodes. To obtain sufficiently high pressure, the strip line width is set to 15 mm and the total energy stored is set to the maximum value. Thanks to the symmetry of the applied loading, one can test two samples in a unique test. When two samples of the same nature and having different thicknesses are tested with the same loading, one can perform a lagrangian analysis and then asses the whole deviatoric response of the material (knowing its volumetric behavior). An overview of the compression tests setup is presented in Fig. 32.2.

Laser interferometry is used to measure the velocity on the rear face of each ceramic specimen. This velocity is a key parameter in order to determine the stress level within the material. One can also observe that LiF windows are placed after each ceramic specimen. These windows, transparent to the laser wavelength, are used to trap the release waves which can perturb velocity signals and distort analyses. However, in order to guaranty the validity of analyses using velocity measurements at a given time, it is necessary to ensure a uniaxial strain state (absence of release waves coming from lateral surfaces) at the measurement locations. For SiC ceramics, preliminary numerical simulations have shown that, for a specimen diameter of 15 mm, thicknesses of 3 and 4 mm ensure a uniaxial strain-state in the axis of revolution of each specimen during the GEPI compressive loading.

The lagrangian analysis is useful for characterizing the dynamic behavior of materials [12, 13]. Indeed, it allows determining the whole loading path of the material with a unique test whereas only one point on this loading path can be obtained after one plate impact experiment (cf. Fig. 32.3a). This method is based on the integration of the equations of conservation of mass, momentum and energy. Moreover, no assumptions on the behavior of the material are needed. Its principle is based on a comparison of velocity signals obtained on two different thicknesses of a material. For each interface velocity u_{int}, an associated wave velocity $C_l(u_{int})$ can be calculated from times Δt (cf. Fig. 32.3b). In a first time, the in situ or particular velocity is calculated from this wave velocity and considering the ceramic and the window impedances. Then, this wave velocity (also noted $C_l(u_p)$) allows to compute increments of longitudinal stress σ_x, longitudinal strain ε_x or specific volume v associated to each increment of particular velocity following:

$$d\sigma_x = -\rho_0 C_L(u_p) du_p \tag{32.2}$$

$$dv = -\frac{1}{\rho_0 C_L(u_p)} du_p \tag{32.3}$$

$$\varepsilon_x = \ln\frac{v}{v_0} \tag{32.4}$$

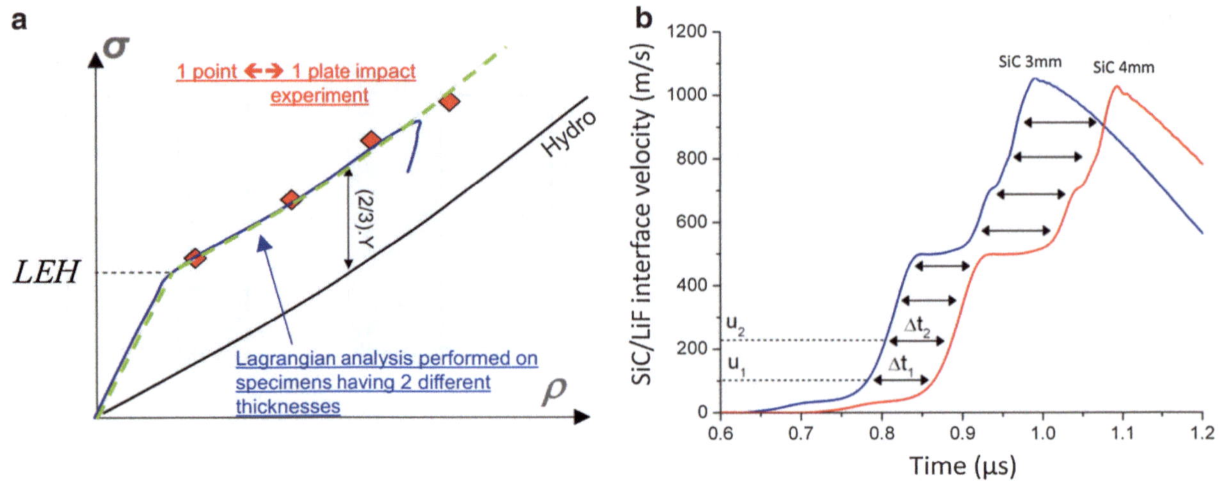

Fig. 32.3 Lagrangian analysis method: (**a**) interest in comparison with plate impact experiments, (**b**) determination of the wave velocity $C_l(u_{int})$ associated to each particular interface velocity u_{int}

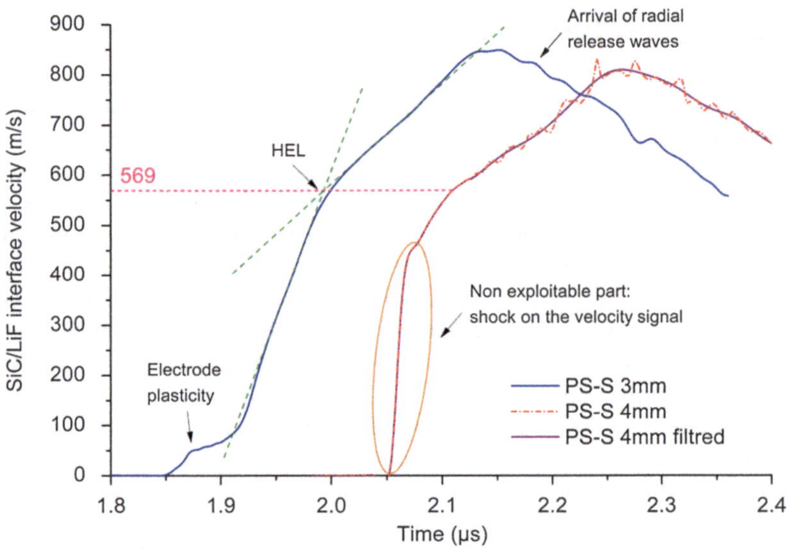

Fig. 32.4 Velocity signals obtained after a GEPI compression test in "lagrangian analysis" configuration on the PS-S SiC grade

This lagrangian analysis allows in the first time to determine easily the Hugoniot Elastic Limit of ceramics. Errors or uncertainties can be done when we determine the HEL just using the velocity at which a change of the slope of the velocity signal is observed. Indeed, following the tested material, a strong work-hardening can be observed after the elastic limit inducing a small change of the slope of the velocity signal.

Figure 32.4 presents velocity signals obtained during a GEPI compression test on a silicon carbide grade, sintered in solid state using pressureless sintering process. This grade, called PS-S in this paper, presents a density greater than 98 % of the theoretical density of the SiC and a fine microstructure (grain size in the range 4–10 μm). One can observe the presence of a shock on the first part of the velocity signal measured on the 4 mm width ceramic. This shock is due to presence of voids within the glue and present at the interface between the electrode and the ceramic specimen. Numerical simulations have shown that this shock has no influence on the velocity signal on the exploitable part. Thus, a lagrangian analysis has been performed with the two velocity signal for interfaces velocities greater than 500 m/s.

Using the velocity at which a change of slope of velocity signals is observed (SiC/LiF interface velocity of 569 m/s), a HEL of 14.9 GPa is obtained. By performing a lagrangian analysis and plotting the longitudinal stress/longitudinal strain curve, it is observed that the elastic/inelastic transition begins at 15.1 GPa. These two HEL values are similar but a better confidence is given to results obtained with the lagrangian analysis. In all cases, the uncertainty is evaluated at ±0.2 GPa.

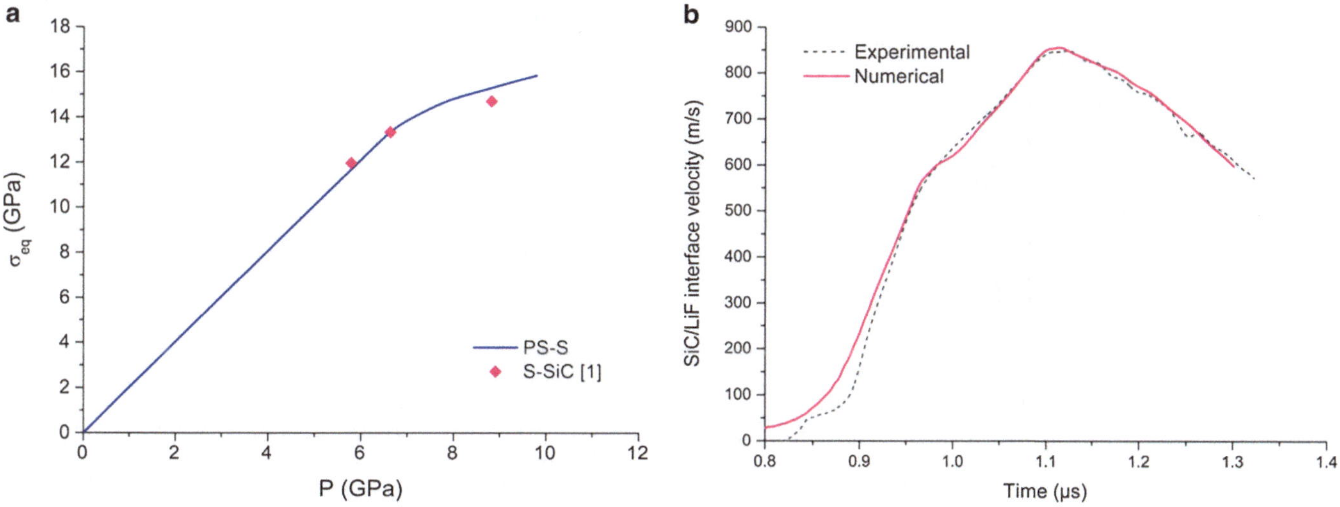

Fig. 32.5 (**a**) Equivalent stress as function of the hydrostatic pressure obtained from lagrangian analysis results on PS-S SiC grade and comparison with plate impact experiment results obtained by [1], (**b**) comparison between experimental interface velocity and numerical simulation by imposing the determined deviatoric response

The longitudinal stress obtained with the lagrangian analysis is necessary but not sufficient to determine the equivalent stress σ_{eq}. However, during a uniaxial deformation, one can write:

$$\sigma_{eq} = \frac{3}{2}(-\sigma_x - P), \tag{32.5}$$

where P denotes the hydrostatic pressure. Thus, knowing the equation of state for the ceramic, one can evaluate the evolution of equivalent stress. During the elastic compression of the PS-S grade and other SiC grades, a low variation of the $K+(4/3)G$ modulus is observed. A constant bulk modulus K is assumed, meaning that the pressure is known during the whole experiment. Thus, using Eq. (32.5), one can deduce the equivalent stress in the ceramic specimen during the compressive loading. The equivalent stress as a function of the hydrostatic pressure is presented in Fig. 32.5a. In this figure is also plotted experimental values obtained by Bourne et al. [1] by performing plate impact experiments on a SiC grade presenting a microstructure similar than the one of the PS-S grade. One can observe that lagrangian analysis data obtained after a GEPI test on PS-S grade are consistent with experimental data obtained by plate impact experiments.

Knowing the pressure applied on each electrode (cf. Eq. (32.1)), one can perform a numerical simulation by imposing to the ceramic the deviatoric response presented Fig. 32.5a. This numerical simulation has been performed with the Abaqus/Explicit finite element code. It can be observed that the interface velocity signals obtained experimentally and by the numerical simulation are similar (cf. Fig. 32.5b); validating the plasticity model identified using the lagrangian analysis.

32.4 Spalling Experiments

The GEPI machine also presents a great advantage for performing spalling experiments in comparison with plate impact technique. Indeed, when performing a spalling test by plate impact, the strain-rate at failure cannot be determined due to the applied shock loading and the spall strength can only be compared to the level of applied compressive stress. Thus, strain-rate effects on the dynamic behavior of materials cannot be investigated. By using the GEPI machine, the ramp pressure transmitted to the specimens allows determining accurately, by means of numerical simulations, the strain-rate everywhere and at every time within the specimen.

For performing a spalling test with the GEPI machine, the experimental setup slightly defers from the compressive tests setup. Indeed, in order to generate tensile stresses within the specimen, the windows have to be removed (cf. Fig. 32.6a). In this case, the crossing of the incident pulse and the pulse reflected on the free-end produces a rapid increase of the tensile stress, initiating the fragmentation process. As for compressive tests, two samples can be tested in a unique experiment. However, it has been chosen to test only one ceramic specimen per experiment. One of the electrodes is let free (without specimen).

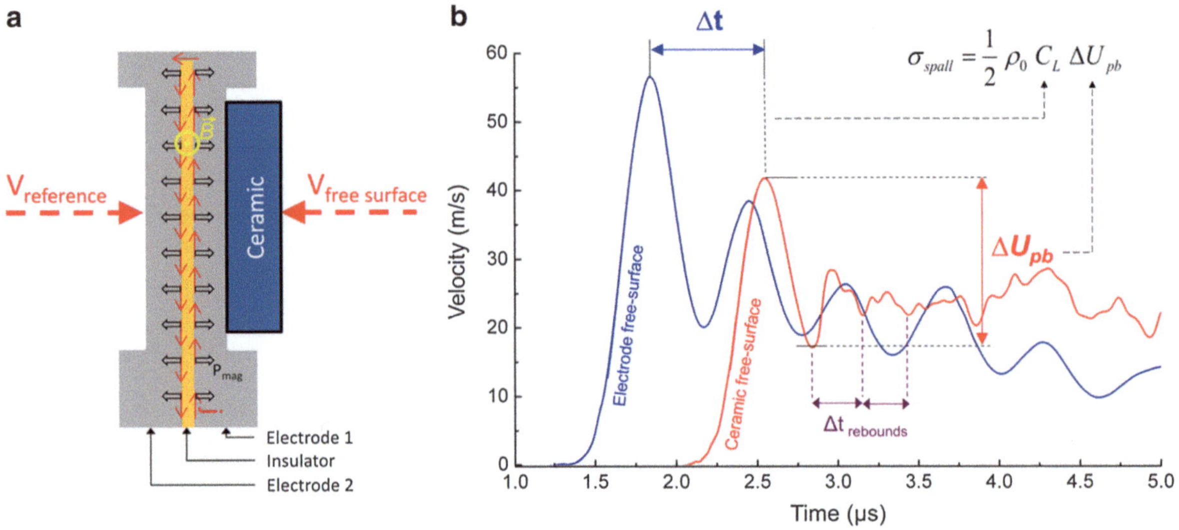

Fig. 32.6 GEPI spalling tests: (**a**) experimental setup and (**b**) data processing for determining the spall strength

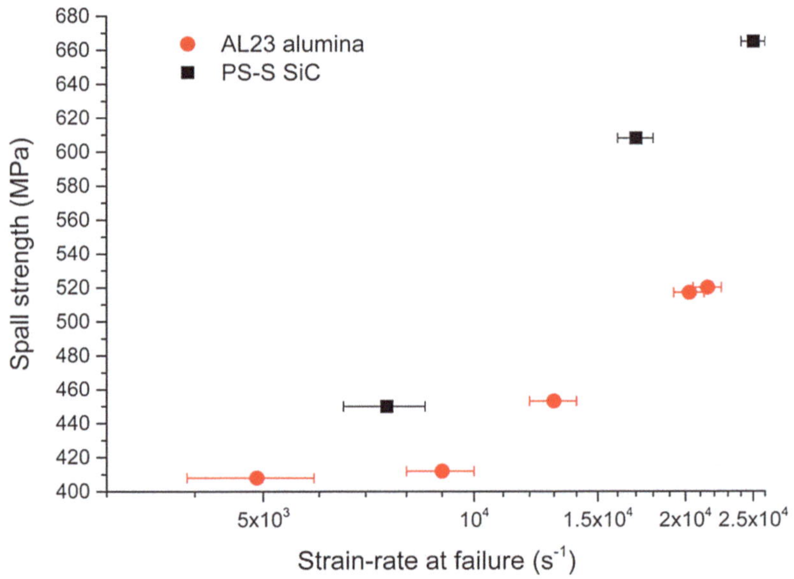

Fig. 32.7 Spall strength of AL23 alumina and PS-S as function of the strain-rate at failure

The velocity is measured at the free-surface of the ceramic and at the free-surface of the electrode (cf. Fig. 32.6b). The pullback velocity ΔU_{pb} observed at the ceramic free-surface velocity profile is the most important parameter for determining the spalling strength. Indeed, when based on the free-surface velocity, the spalling strength is expressed by Ref. [14]:

$$\sigma_{spalling} = \frac{1}{2}\rho_0 C_L \Delta U_{pb}, \tag{32.6}$$

where ρ_0 is the initial density and C_L the velocity of the elastic waves in the ceramic. The velocity measured at the free-surface of the second electrode is used as a reference. It allows determining directly the elastic waves velocity in the ceramic (cf. Fig. 32.6b). Rebounds are observable after the pullback velocity on the ceramic free-surface velocity. Their frequency directly depends on the distance between the free-surface and the spalling fracture zone.

The rising time of the pressure ramp being always the same, the strain-rate is directly influenced by the level of the compressive pulse. Thus, by changing the pressure level between experiments, strain-rate effects can be investigated.

Experiments have been performed on alumina [15] and silicon carbide [16]. For each test presented in this work, the pressure is less than the HEL of the tested material to prevent any damage in the ceramic due to the applied compressive loading. Figure 32.7 presents the spall strength of a coarse grain alumina (AL23 alumina) and of the PS-S SiC as a function of the strain-rate.

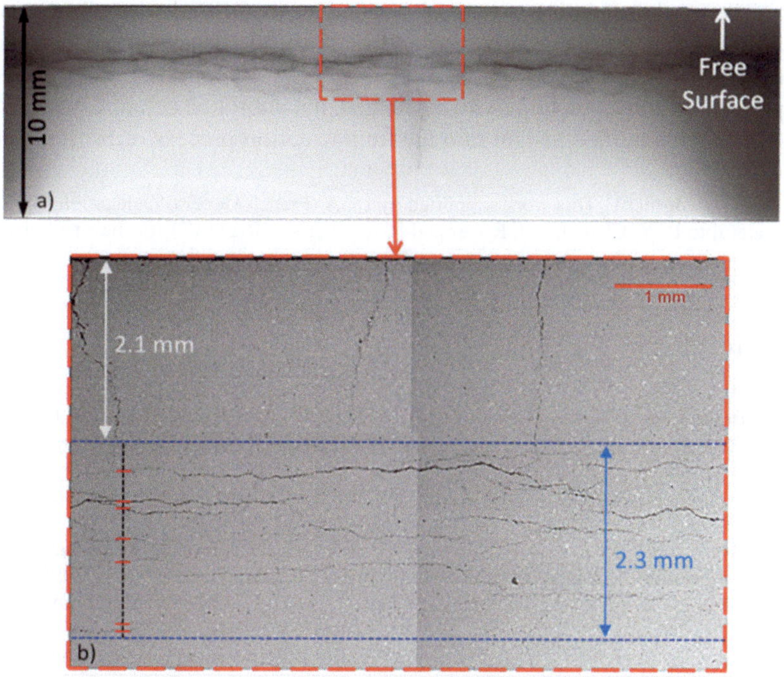

Fig. 32.8 Post-test analysis of the alumina specimen tested at the spalling limit: (**a**) Optical and (**b**) scanning electron microscopy images of the inner damage provoked by the tensile loading

One can observe that the two tested ceramics present strong strain-rate sensitivity. Moreover, this strain-rate effect depends on the ceramic. For alumina, spall strength increases from 410 to 520 MPa for strain-rates between 4900 and 21,500 s^{-1}, whereas it increases from 450 to 665 MPa for strain-rates between 7500 and 25,000 s^{-1}. In all cases, the uncertainty on the strain-rate at failure is about $\pm 1000\ s^{-1}$.

Another great advantage of GEPI spalling tests in comparison with plate impact technique is the capacity to recover damaged specimens. Indeed, using plate impact technique, a special setup is needed to recover the specimen after test. During GEPI tests, due to the absence of propulsive gases and striking projectile, no additional system is necessary. The size of recovered parts after test depends on the applied pressure. For tests realized at high strain-rates, the specimen is totally fragmented due to the amplitude of release waves. For tests performed at strain-rates about $10^4\ s^{-1}$, specimens are cut in two parts by the spalling fracture plane. After the test performed at 4900 s^{-1} on alumina ceramic, the specimen was recovered in one piece without any visible damage at the macroscopic scale. However, after cutting the specimen, microscopic analysis clearly indicates damage within the material. During this test, the tensile loading dropped just after the initiation of cracks, stopping their propagation. In this case, the specimen was tested at the spalling limit. Thus, the microscopic analysis gives a unique insight into the fragmentation process of brittle materials. Figure 32.8 presents a post-test analysis of the alumina specimen tested at the spalling limit.

One can observe the presence of horizontal and vertical cracks. Vertical cracks (parallel to the sample axis) are the result of lateral release waves crossing each other. However, these cracks were initiated after the spalling damage explaining that they do not cross the damaged zone composed of horizontal cracks. The spalling fracture zone is 2.3 mm width and is located at 2.1 mm from the free-surface. This distance can be easily obtained from rebounds observed in the ceramic free-surface velocity (Fig. 32.6b).

32.5 Conclusion

In this work, an electromagnetic device (GEPI) is used to characterize the dynamic behavior of ceramics. Thanks to the ramp loading applied on specimens to characterize and thanks to the symmetrical configuration, this machine presents great advantages in comparison with plate impact experiments generally performed. For characterizing the dynamic compressive behavior of ceramics, very high pressures can be applied, exceeding the Hugoniot Elastic Limit of ceramics. By performing a test on two specimens having different thicknesses, the lagrangian analysis method allows determining the whole loading path of the material. By assuming a known equation of state, the evolution of equivalent stress has been determined. Good comparisons are observed with plate impact experiment results obtained on a similar ceramic. To characterize the dynamic tensile behavior, the ramp pressure applied by the GEPI device allows performing spalling test in which the strain-rate can be accurately determined. Strong strain-rate sensitivities have been observed on an alumina and a SiC ceramic. Furthermore, in comparison with plate impacts, the recovery of damaged samples is easier and do not need any special setup. After a test performed at the spalling limit, a damaged but unbroken ceramic has been recovered. A microscopic analysis of the damaged zone gave a unique insight into the fragmentation process of brittle materials.

Acknowledgment This work supported by DGA (French General Delegation for Armament, French Ministry of Defense). The authors are also grateful to P.-Y. Chanal, J.-J. Royer and P. Rey (CEA GRAMAT) for their technical contribution.

References

1. Bourne, N., Millet, J., Pickup, I.: Delayed failure in shocked silicon carbide. J. Appl. Phys. **81**, 6019–6023 (1997)
2. Feng, R., Raiser, G.R., Gupta, Y.M.: Material strength and inelastic deformation of silicon carbide under shock wave compression. J. Appl. Phys. **83**(1), 79–86 (1998)
3. Forquin, P., Denoual, C., Cottenot, C.E., Hild, F.: Experiments and modelling of the compressive behavior of two SiC ceramics. Mech. Mater. **35**, 987–1002 (2003)
4. Roberson, C.J.: Ceramic materials and their use in lightweight armour systems. In: Proceedings of the Lightweight Armour System Symposium. Royal Military College of Science, Cranfield, 28–30 June 1995
5. Forquin, P., Tran, L., Louvigné, P.-F., Rota, L., Hild, F.: Effect of aluminum reinforcement on the dynamic fragmentation of SiC ceramics. Int. J. Impact Eng. **28**, 1061–1076 (2003)
6. Riou, P., Denoual, C., Cottenot, C.E.: Visualization of the damage evolution in impacted silicon carbide ceramics. Int. J. Impact Eng. **21**(4), 225–235 (1998)
7. Zinszner, J.L., Forquin, P., Rossiquet, G.: Experimental and numerical analysis of the dynamic fragmentation in a SiC ceramic under impact. Int. J. Impact Eng. **76**, 9–19 (2015)
8. Yeshurun, Y., Brandon, D.G., Rosenberg, Z.: Impact damage mechanisms in debased alumina. In: Chiem, C.Y., Kunze, H.-D., Meyer, L.W. (eds.) Impact Loading and Dynamic Behaviour of Materials, vol. 1, pp. 399–405. DGM Informationsgesellschaft-Verlag, Oberursel (1987)
9. Mukhopadhyay, A.K., Joshi, K.D., Dey, A., Chakraborty, R., Mandal, A.K., Rav, A., Ghosh, J., Bysakh, S., Biswas, S.K., Gupta, S.C.: Electron microscopy of shock deformation in alumina. Ceram. Int. **37**, 2365–2376 (2011)
10. Mangeant, C., Lassalle, F., L'Eplattenier, P., Héreil, P.L., Bergues, D., Avrillaud, G., Syrinx project: HPP generators devoted to isentropic compression experiments. In: Furnish, M.D., Chhabildas, L.C., Hixson, R.S. (eds.) AIP Conference Proceedings 620, Shock Compression of Condensed Matter 2001, pp. 1173–1176. American Institute of Physics, Melville (2002)
11. Héreil, P.L., Lassalle, F., Avrillaud, G.: GEPI: An ICE generator for dynamic material characterization and hypervelocity impact. In: AIP Conference Proceeding 706, Shock Compression of Condensed Matter 2003, Portland, 20–25 July. pp. 1209–1212 (2004)
12. Fowles, R., Williams, R.F.: Plane stress wave propagation in solids. J. Appl. Phys. **41**(1), 360–363 (1970)
13. Cowperthwaite, M., Williams, R.F.: Determination of constitutive relationships with multiple gauges in nondivergent waves. J. Appl. Phys. **42** (1), 456–462 (1971)
14. Novikov, S.A., Divnov, I.I., Ivanov, A.G.: The study of fracture of steel, aluminium and copper under explosive loading. Fizika Metallov I Metallovedeniye. **21** (1966)
15. Zinszner, J.L., Erzar, B., Forquin, P., Buzaud, E.: Dynamic fragmentation of an alumina ceramic subjected to shockless spalling: an experimental and numerical study. J. Mech. Phys. Solids **85**, 112–127 (2015)
16. Zinszner, J.L., Erzar, B., Forquin, P., Rossiquet, G., Experimental and numerical analysis of the tensile behaviour of two silicon carbide grades subjected to spalling, In: Proceedings of Personal Armour System Symposium 2014, Cambridge

Chapter 33
Dynamic Hyper Elastic Behavior of Compression Shock Loaded Vibration Dampers

V.B.S. Rajendra Prasad and G. Venkata Rao

Abstract A variety of rubber dampers are available, these having been designed over the years through approximate methods of analyses and based on the experience drawn from their operation over the years. Behavior of rubbers, polymers and elastomers is highly non-linear posing difficulties in the analyses. Although the behavior of rubber blends have been researched in the past decades, its behavior in high strain range has attracted the attention of researchers only in the recent past. In the present work, the characterization of natural rubber and high performance SB rubber under severe dynamic compression loading is carried out based on the material response curves provided by recent researchers using material models. With a view to examining the feasibility of using these rubber formulations, these material models are then applied to the case of dynamically compression-loaded dampers in the high strain rate regimes. The discrepancies in the results obtained by utilizing the different material models are discussed in detail.

Keywords Hyperelasticity • Strains • Non linerity

33.1 Introduction

Elastomers, more precisely Rubber like materials usually have long molecular chains, which can with stand high strains and does not undergo permanent deformation or fail due to fracture. Rubbers have typical physical properties like extensibility, resilience and durability, which generally are not found in other materials. These materials generally have a complicated behavior that generally exceed linear elastic theory and contain large deformations, plastic and visco—elastic properties and softening due to stresses.

The paper utilizes the results of experimental characterization of SB rubber utilizing expensive and controlled experiments on rectangular specimens under high strain rates.

The behavior of rubber exhibits three states of behavior vizz. Elastic behavior at small strains and deformations, visco-elastic behavior during transition and finally hyper-elastic behavior at large strain rates and large deformations.

Various mathematical models have been proposed and used earlier in the analysis of hyper-elastic material and for the design of vibration dampers. Hyper elastic models are based on phenomological aspects (i.e. only the physical behavior is represented by the relationship and not the microscopic aspects).

This General behavior of rubber is reflected in Rivlin's theory, it exhibits a isotropic behavior in elastic behavior and is nearly incompressible.

Elastic properties of Rubber can be expressed in the form of stress invariants $\mathbf{I_1}$, I_2,I_3 and this will enable solving the problem as independent of microscopic properties.

The general consideration for Strain energy density and I_1 I_2 I_3 can be given as $\sum C_{ijkl}(I_1 - 3)^I\, I_2 - 3)^j.\ (I_3 - 1)^k$ where $i + j + k = 1$ and when the volume is considered to be unchanged during this behavior the third term will become unity and the expression reduces to $\sum C_{ijkl}(I_1 - 3)^I(I_2 - 3)^j$.

33.1.1 Features of the Behavior of a Solid Rubber

The material is close to ideally elastic. i.e. (i) when deformed at constant temperature or adiabatically, stress is a function only of current strain and independent of history or rate of loading, (ii) the behavior is reversible, no net work is done on the

V.B.S.R. Prasad (✉) • G. Venkata Rao
Vasavi College of Engineering, Ibrahimbagh, Hyderabad, Telangana, India
e-mail: vbsrprasad@gmail.com

D. Casem et al. (eds.), *Dynamic Behavior of Materials, Volume 1*, Conference Proceedings of the Society for Experimental Mechanics Series, DOI 10.1007/978-3-319-41132-3_33

solid when subjected to a closed cycle of strain under adiabatic or isothermal conditions. The material strongly resists volume changes. The bulk modulus (the ratio of volume change to hydrostatic component of stress) is comparable to that of metals or covalently bonded solids; the material is very compliant in shear and shear modulus is of the order of 10^{-5} times that of most metals; the material is isotropic, its stress-strain response is independent of material orientation. The shear modulus is temperature dependent: the material becomes stiffer as it is heated, in sharp contrast to metals; when stretched, the material gives off heat.

Different hyper elastic models [7] of different series of elements will be used to predict the hyper elastic behavior, however the choice of the model depends on its application, and only the few described below can be better used to describe the complete behavior of these materials, for different loading conditions, they are: Mooney, Arruda–Boyce, Polynomial, Yeoh, Neo Hookean, Gent and Ogden.

33.1.1.1 Generalized Neo-Hookean Solid

$$\overline{U} = \frac{\mu_1}{2}(\overline{I}_1 - 3) + \frac{K_1}{2}(J - 1^2)$$

where μ_1 and K_1 are material properties (for small deformations, μ_1 and K_1 are the shear modulus and bulk modulus of the solid, respectively). Elementary statistical mechanics treatments predict that $\mu_1 = NKT$, where N is the number of polymer chains per unit volume, k is the Boltzmann constant, and T is temperature. This is a rubber elasticity model, for rubbers with very limited compressibility, and should be used with $K_1 >> \mu_1$. The stress-strain relation follows as

$$\sigma_{ij=\frac{\mu_i}{j^{\frac{5}{3}}}\left(B_{ij}-\frac{1}{3}B_{ij\delta_{ij}}\right)+k_1(j-1)\delta_{ij}}$$

The fully incompressible limit can be obtained by setting $K_1(J - 1) = \frac{p}{3}$ in the stress-strain law.

33.1.1.2 Generalized Mooney-Rivlin Solid

Deformations, the shear modulus and bulk modulus of the solid are $\mu = \mu_1 + \mu_2$ and $K = K_{1.}$ This is a rubber elasticity model, and should be used with $K_1 >> \mu_1$. The stress strain relation follows as:

$$\overline{U} = \frac{\mu_1}{2}(\overline{I}_1 - 3) + \frac{\mu_2}{2}(\overline{I}_2 - 3) + \frac{K_1}{2}(J - 1)^2$$

33.1.1.3 Generalized Polynomial Rubber Elasticity Potential

$$\overline{U} = \sum_{i+j=1}^{N} C_{ij}(\overline{I}_1 - 3)^i(\overline{I}_2 - 3)^j + \sum_{i=1}^{N}\frac{k_i}{2}(J - 1)^{2i}$$

where C_{ij} and k_i are material properties. For small strains the shear modulus and bulk modulus follow as $\mu = 2(c_{01} + c_{10})$, $K = 2K_1$,This model is implemented in many finite element codes. Both the neo-Hookean solid and the Mooney-Rivlin solid are special cases of the law (with $N = 1$ and appropriate choices of C_{ij}). Values $N > 2$ are rarely used, because it is difficult to fit such a large number of material properties to experimental data.

33.1.1.4 Ogden Model

$$\sum_{i=1}^{N} \frac{2\mu_i}{\alpha_i^2} \left(\overline{\lambda}_1^{\alpha_i} + \overline{\lambda}_2^{\alpha_i} + \overline{\lambda}_3^{\alpha_i} - 3 \right) + \frac{K_1}{2} (J-1)^2$$

where, $\overline{\lambda}_i = \frac{\lambda_i}{J^{1/3}}$ and $\mu_{i,}\ \alpha_{i,}\ K$ are material properties. For small strains the shear modulus and bulk modulus follow as $\mu = \sum_{i=1}^{N} \mu_i, K = K_{1.}$ This is a rubber elasticity model, and is intended to be used with The stress can be computed using the formulas but are too lengthy to write out in full here.

33.1.1.5 Arruda–Boyce Model

$$\overline{U} = \mu \left\{ \frac{1}{2} \left(\overline{I}_1 - 3 \right) - \frac{1}{20\beta^2} \left(\overline{I}_1^2 - 9 \right) + \frac{11}{1050\beta^4} \left(\overline{I}_1^3 - 27 \right) + \ldots \right\} + \frac{K(J-1)^2}{2}$$

where μ, β, K are material properties. For small deformations μ, K are the shear and bulk modulus, respectively. This is a rubber elasticity model, so $K >> \mu$ The potential was derived by calculating the entropy of a simple network of long-chain molecules, and the series is the result of a Taylor expansion of an inverse Langevin function. The reference provided lists more terms if you need them. The stress-strain law is

$$\sigma_{ij} = \frac{\mu}{J^{\frac{5}{3}}} \left(1 + \frac{B_{kk}}{5J^{\frac{2}{3}}\beta^2} + \frac{33(B_{kk})^2}{525\beta^4 J^{\frac{4}{3}}} + \ldots \right) \left(B_{ij} - \frac{B_{kk}}{3}\delta_{ij} \right) + K(J-1)\delta_{ij}$$

33.1.2 Experimentation

A rectangular speicmen [3] of styrene butadiene rubber is considered in the experiment, with one end fixed and the other end is loaded, so the specimen will remain in uni-axial Tension and compression. The experimental values of the stress—strain behavior of the same material given in the references is utilized for a relative comparison of the experimental data with different models available to access the hyper-elastic behavior. The general comparison is done for all the models individually. The cumulative results of comparison of the graphs and their matching positions with the experimental behavior are tabulated and the best fit criteria is taken for the analysis in static and transient conditions of model as shown in Table 33.1.

33.1.3 Finite Element Modeling

Utilizing the same experimental results [1] on SB rubber provided by earlier investigators referred to above, finite element analysis is carried out to characterize the rubber in order to choose the best fit mathematical model to the experimental data. For this purpose, finite element code in ANSYS 14.5 has these models incorporated. It must be noted that the results provided are for high rates of strain, which implies sudden loading.

33.2 Results and Discussions

The experimental data for the stress-strain behaviour of the SB rubber is obtained from the earlier works done and quoted in the reference [3]. The comparison of the work is done by the available models for characterization of Hyper-elastic behaviour, which are provided in the Ansys 14.5 version.

A detailed comparison of the available data is done with various models like, Mooney-Rivlin, Ogden, Polynomial, Neo-hookean, Arruda–Boyce, Yeoh, Gent .etc, comparison of the experimental stress- strain behaviour and the models

Table 33.1 Comparison of graphs

S no:	Model	Number of points match-ing with experimental curve	Remarks	Positive deviation (mpa)	Negative deviation (mpa)	Cumulative deviation (mpa)
	Arruda–Boyece	7		4	2	6
2	Mooney (two parameters)	9		3	3	6
3	Mooney (three parameters)	11	*	3	3	6
4	Mooney (five parameters)	11	*	3	4	7
5	Mooney (nine parameters)	10		2	3	5
6	Neo-hookean	7		4	2	6
7	Ogden (first order)	12	*	3	3	6
8	Ogden (second order)	11		3	3	6
9	Ogden (third order)	11		3	3	6
10	Polynomial (first order)	9		3	3	6
11	Polynomial (second order)	12	*	2	3	5
12	Polynomial (third order)	9		2	3	5
13	Yeoh (first order)	8		4	1	5
14	Yeoh (second order)	11		4	2	6
15	Yeoh (third order)	12	*	3	4	7
16	Gent	2		–	Maximum	

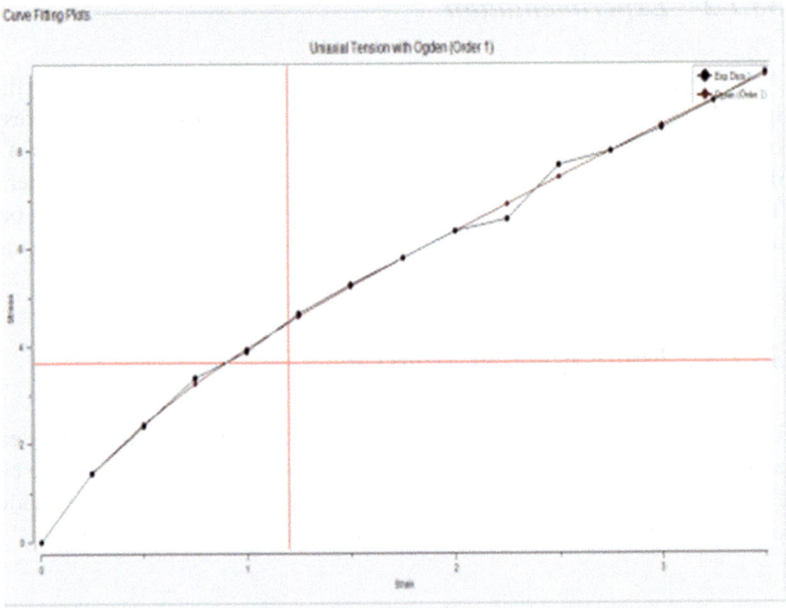

Fig. 33.1 Experimental vs. Ogden

shown in the Table 33.1 and the better fit models are identified based upon the curve-fitting details as listed. Amongst the total 16 iterations considered the picked are Mooney 3 parameter with 11 points coinciding with the actual curve, Ogden first order with 12 points coinciding, polynomial second order 12 . Upon considering the deviations from the actual curve, Arruda–Boyce model was picked as the deviation obtained is minimum, 6 Mpa. The variations of the presented models with the actual stress-strain behaviour shown in Fig. 33.1, Mooney-Rivlin, Fig. 33.2 for Ogden, and Fig.33.3 Polynomial and Fig.33.4 for Arruda–Boyce.

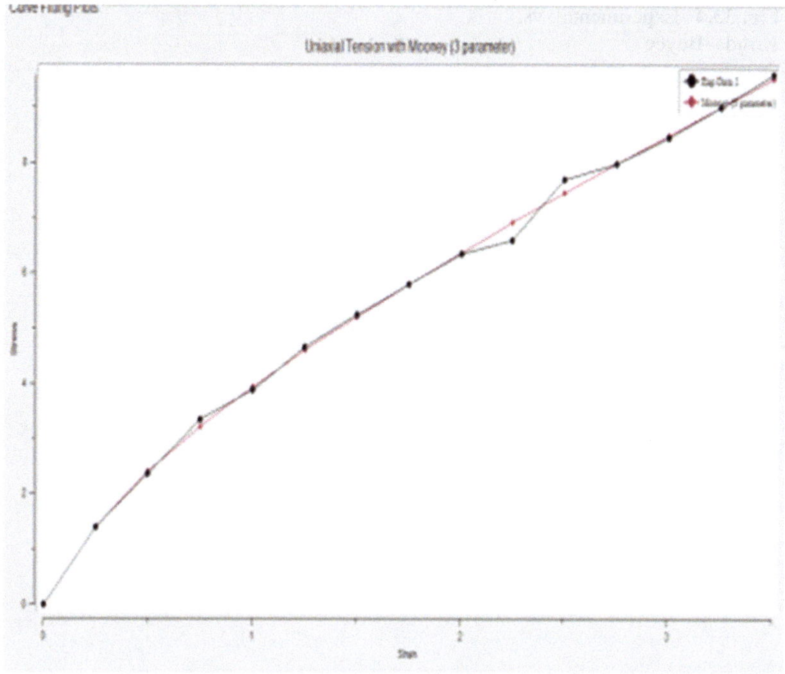

Fig. 33.2 Experimental. vs. Mooney

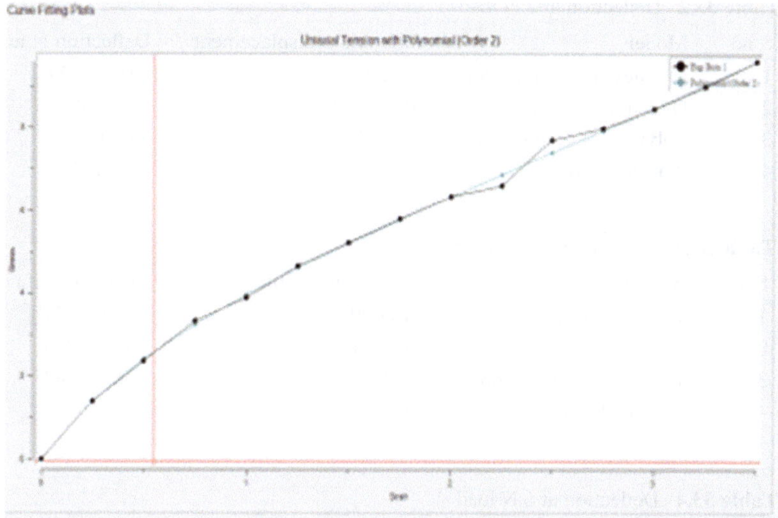

Fig. 33.3 Experimental vs. polynomial order 3

The deflection in static behavior for the selected models were listed in Table 33.2 for 2 N load, and in Table 33.3 for 4 N load, in Table 33.4 for 6 N load, Table 33.5 for 8 N load and Table 33.6 for 10 N load . The deflections are noted down for node number 35, 78 and 168, which were picked at salient positions on the FE model of the sample analyzed. The relative comparison of the load and displacement characteristics were shown in figures from Figs. 33.5, 33.6, 33.7 and to 33.8.

Transient analysis for the model was done to predict the response of the test sample after the release of the load which is acting for a very brief period. Initially a loading of 2 N to 10 N is considered in steps of 2 N to obtain the response in transient and the corresponding deflections are considered for the various models at the nodes mentioned earlier and were tabulated in the Table 33.7. The stress and strains for various loadings in static behaviour were tabulated in Tables 33.8, 33.9 and 33.10, the max. stress experienced and min. stress experienced is observed. However in the present paper the data for load set of 10 N is only presented. The velocity and acceleration of the transient response were also plotted in Figs. 33.9 and 33.10 through which the variation in the velocity and acceleration over the time step can be observed.

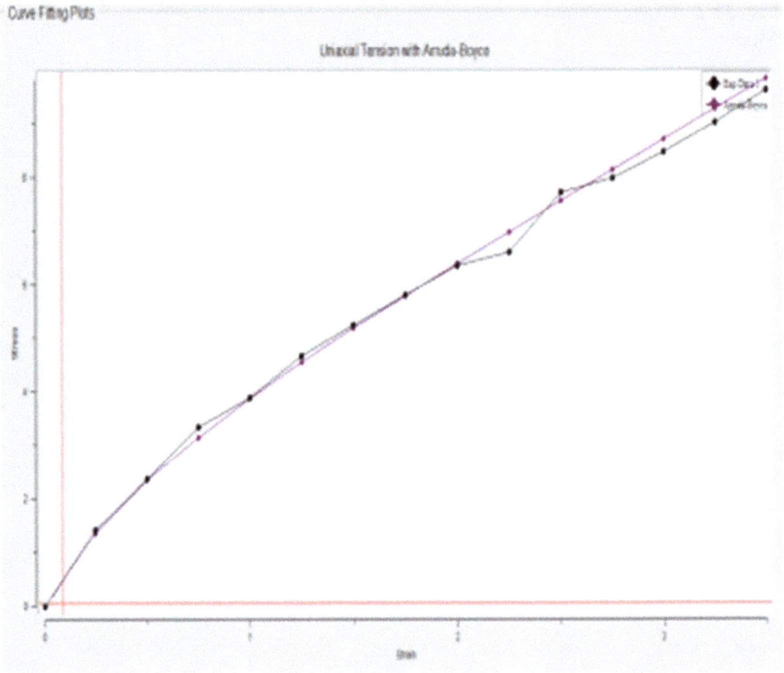

Fig. 33.4 Experimental. vs. Arruda–Boyce

Table 33.2 Deflection at 2 N load

S no	Model	Maximum displacement	Deflection at node 35	Deflection at node 78	Deflection at node 168
1	Mooney third parameter	0.529269	−0.265657	−0.428494	−0.403862
2	Ogden first order	0.529943	−0.265998	−0.42904	−0.404381
3	Polynomial second order	0.522554	−0.262271	−0.423122	−0.398711
4	Arruda–Boyce	0.554	−0.27818	−0.448747	−0.422909

Table 33.3 Deflection at 4 N load

S no	Model	Maximum displacement	Deflection at node 35	Deflection at node 78	Deflection at node 168
1	Mooney third parameter	1.06993	−0.541321	−0.875604	−0.823341
2	Ogden first order	1.0711	−0.541924	−0.87656	−0.824259
3	Polynomial second order	1.05949	−0.535984	−0.867288	−0.815233
4	Arruda–Boyce	1.11992	−0.566997	−0.91731	−0.862426

Table 33.4 Deflection at 6 N load

S no	Model	Maximum displacement	Deflection at node 35	Deflection at node 78	Deflection at node 168
1	Mooney third parameter	1.6242	−0.827346	−1.3419	−1.25884
2	Ogden first order	1.6257	−0.828126	−1.34313	−1.26002
3	Polynomial second order	1.61258	−0.82136	−1.33288	−1.24977
4	Arruda–Boyce	1.70016	−0.8668	−1.40624	−1.31892

Table 33.5 Deflection at 8 N load

S no	Model	Maximum displacement	Deflection at node 35	Deflection at node 78	Deflection at node 168
1	Mooney third parameter	2.19402	−1.12423	−1.828	−1.71104
2	Ogden first order	2.19565	−1.1251	−1.8293	−1.71235
3	Polynomial second order	2.1834	−1.11876	−1.8203	−1.70279
4	Arruda–Boyce	2.29677	−1.17809	−1.9161	−1.79308

Table 33.6 Deflection at 10 N load

S no	Model	Maximum displacement	Deflection at node 35	Deflection at node 78	Deflection at node 168
1	Mooney third parameter	2.78073	−1.43725	−2.3341	−2.18026
2	Ogden first order	2.78272	−1.4331	−2.3354	−2.18156
3	Polynomial second order	2.77288	−1.42824	−2.32947	−2.17434
4	Arruda–Boyce	2.91112	−1.50112	−2.4472	−2.28517

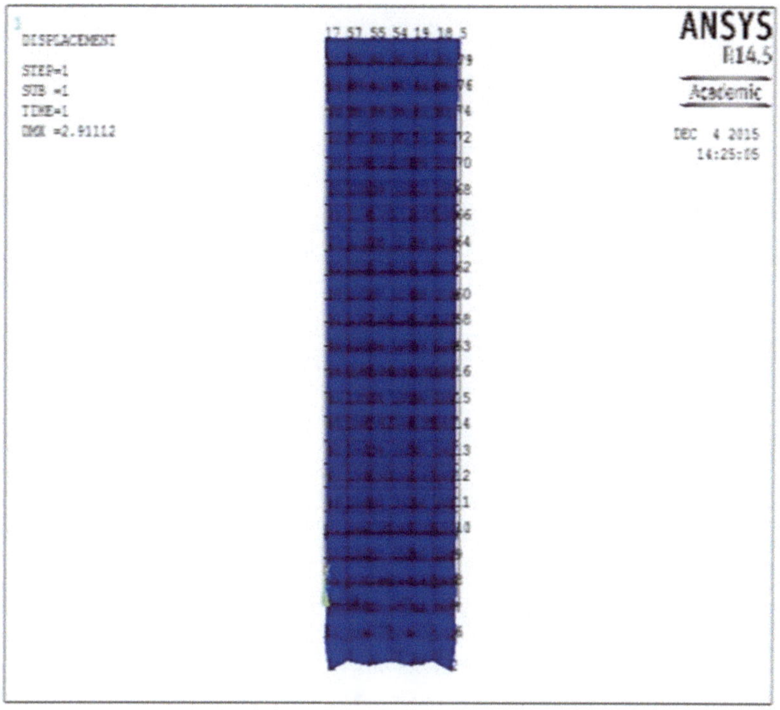

Fig. 33.5 Mooney static—deflection in (mm) on YY axis

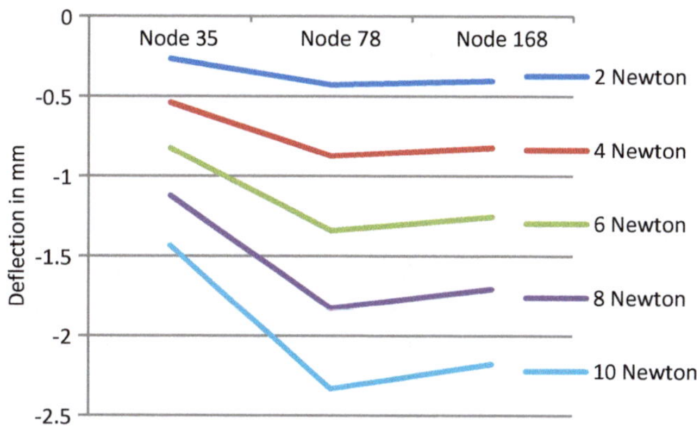

Fig. 33.6 Ogden—static—deflection. (mm) on YY axis

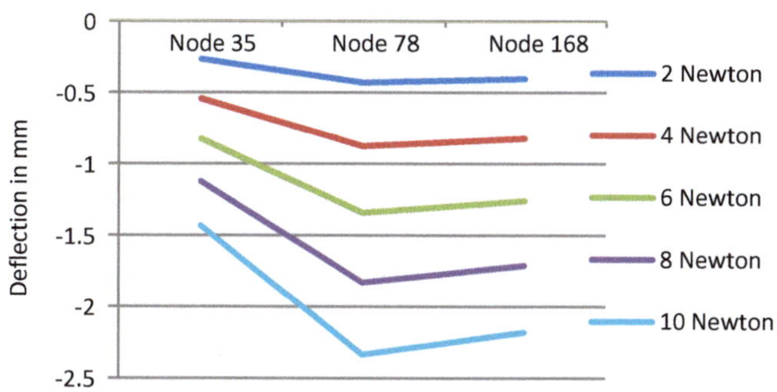

Fig. 33.7 Polynomial second order static—deflection (mm) on YY axis

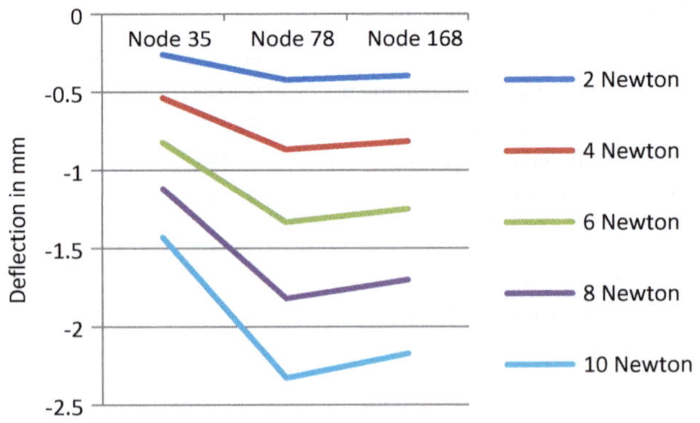

Fig. 33.8 Arruda–Boyce staic—deflection (mm) on YY axis

Table 33.7

S no	Model	Maximum displacement	Deflection at node 35	Deflection at node 78	Deflection at node 168
1	Mooney third parameter	0.01902	0.000330249	−0.0165898	0.010078
2	Ogden first order	0.019051	0.00033151	−0.0166176	0.0100978
3	Polynomial second order	0.018789	0.000316669	−0.0163584	0.00991376
4	Arruda–Boyce	0.020758	0.0003802	−0.0181224	0.0110526

Table 33.8

S no	Node number	Model	Stress	Strain
1	35	Mooney third parameter	0.680948	0.0936537
2	35	Ogden first order	0.680983	0.0937044
3	35	Polynomial second order	0.6808	0.0934518
4	35	Arruda–Boyce	0.683867	0.0979328

Table 33.9

S no	Node number	Model	Stress	Strain
1	78	Mooney third parameter	0.629548	0.0922082
2	78	Ogden first order	0.629548	0.0922082
3	78	Polynomial second order	0.629548	0.0920552
4	78	Arruda–Boyce	0.632017	0.0962871

Table 33.10

S no	Node	Model	Stress	Strain
1	168	Mooney third parameter	0.637138	0.0855128
2	168	Ogden first order	0.637125	0.0855625
3	168	Polynomial second order	0.637767	0.0853081
4	168	Arruda–Boyce	0.640355	0.089587

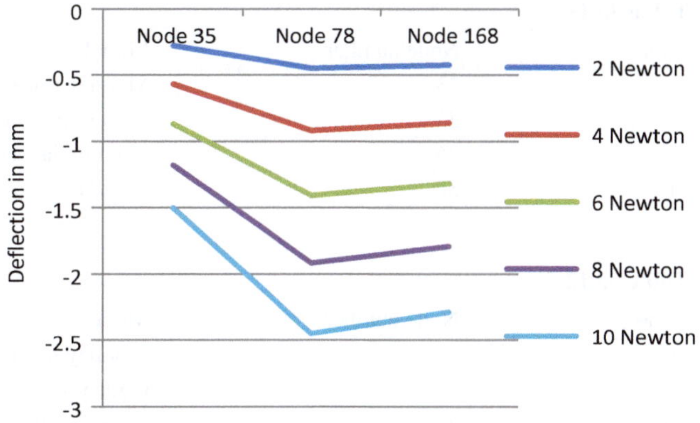

Fig. 33.9 Arruda–Boyce transient analysis—velocity

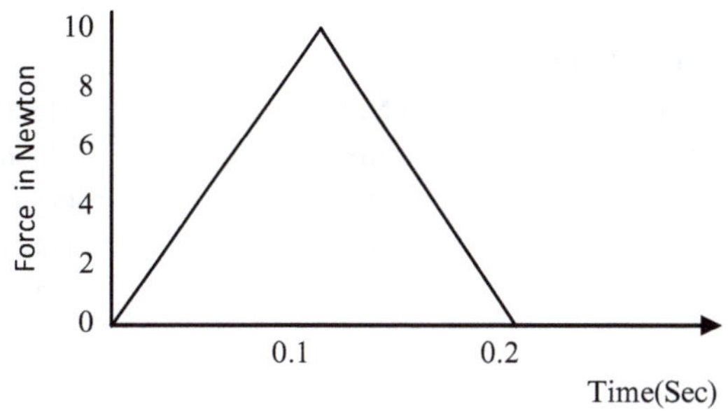

Fig. 33.10 Arruda–Boyce transient analysis—acceleration

33.3 Curve Fits

33.3.1 Static Behavior of Rubber Under Varying Loads

- All deflections can be considered in mm.

33.4 Transient Analysis

Dampers for reducing vibration of mechanical systems have been based on springs, dashpots, a combination of these and hydraulic device. The advantages of rubber dampers are that they do not require may maintenance, less expensive and less messy due to no oil being present. In some of the situations, they are preferred like in automobiles, aircraft and civil bridges *etc*. The same tensile specimen is subjected to transient loading at high strain rates as per the loading diagram given below (Tables 33.11, 33.12 and 33.13; Figs.33.11, 33.12, 33.13 and 33.14).

Table 33.11

S.no	Node number	Model	Stress	Strain
1	35	Mooney third parameter	$8.59532e^{-006}$	$-2.37126e^{-006}$
2	35	Ogden first order	$8.6037e^{-006}$	$-2.37317e^{-006}$
3	35	Polynomial second order	$8.50105e^{-006}$	$-2.37938e^{-006}$
4	35	Arruda–Boyce	$9.01099e^{-006}$	$-2.58605e^{-006}$

Table 33.12

S.no	Node number	Model	Stress	Strain
1	78	Mooney third parameter	−0.0036632	−0.000487846
2	78	Ogden first order	−0.00366185	−0.000488272
3	78	Polynomial second order	−0.00369833	−0.000485963
4	78	Arruda–Boyce	−0.00381438	−0.000529619

Table 33.13

S no	Node	Model	Stress	Strain
1	168	Mooney third parameter	−0.00041681	$-6.08635e^{-005}$
2	168	Ogden first order	−0.000416795	$-6.10726e^{-005}$
3	168	Polynomial second order	−0.00041944	$-5.83329e^{-005}$
4	168	Arruda–Boyce	−0.00043754	$-6.79535e^{-005}$

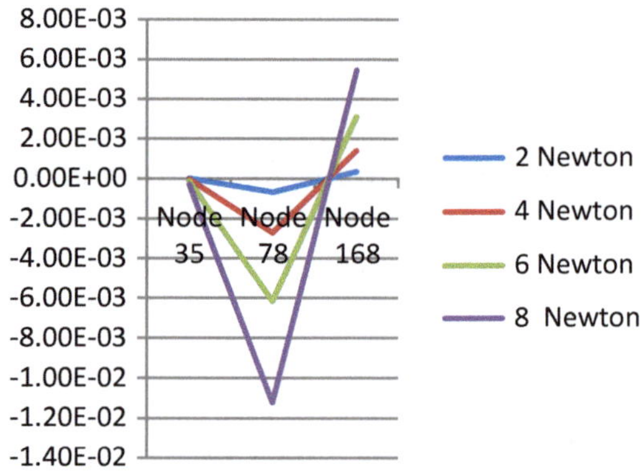

Fig. 33.11 General model with mesh and deformation, under static loading

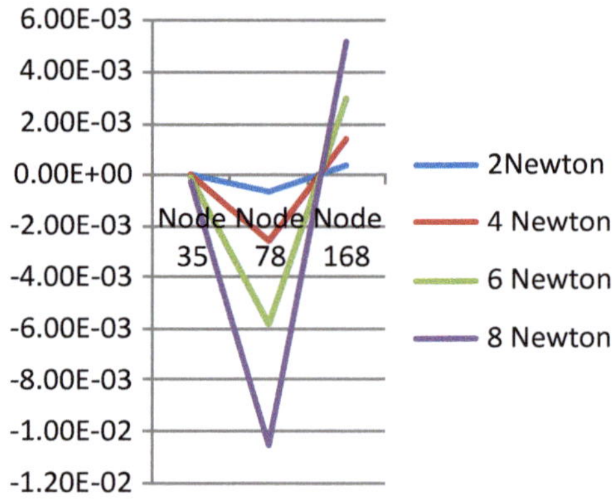

Fig. 33.12 Arruda–Boyce static-FE model

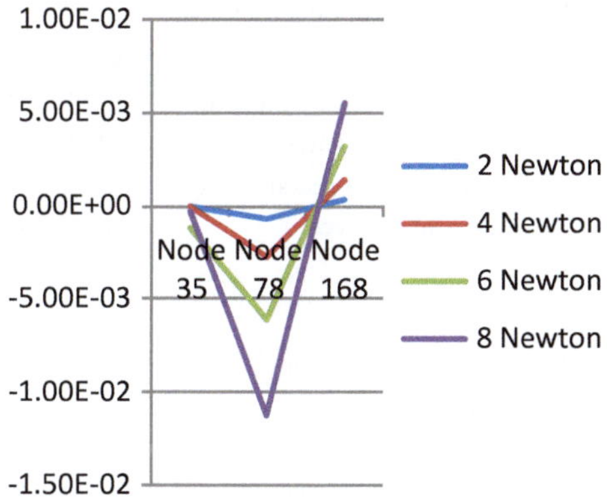

Fig. 33.13 Mooney static—deflection in (mm) on YY axis

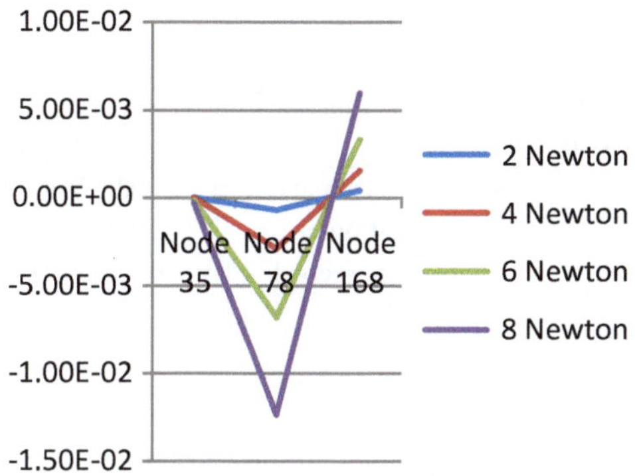

Fig. 33.14 Polynomial static—deflection in (mm) on YY axis

33.5 Transient Analysis for Compression

33.5.1 2 Newtons

Table 33.14

33.5.2 4 Newtons

Table 33.15

33.5.3 6 Newtons

Table 33.16

33.5.4 8 Newtons

Table 33.17

Table 33.14 Deflection at 2 N load

S no	Model	Maximum displacement	Deflection at node 35	Deflection at node 78	Deflection at node 168
1	Mooney third parameter	7.80E−04	−5.45E−06	−0.00066	0.000358
2	Ogden first order	7.82E−04	−5.48E−06	−0.00067	0.000359
3	Polynomial second order	7.50E−04	−5.20E−06	−0.00064	0.000345
4	Arruda–Boyce	8.55E−04	−6.16E−06	−0.00073	0.000392

Table 33.15 Deflection at 4 N load

S no	Model	Maximum displacement	Deflection at node 35	Deflection at node 78	Deflection at node 168
1	Mooney third parameter	0.003174	−3.60E−05	−0.00269	0.001409
2	Ogden first order	0.003185	−3.62E−05	−0.0027	0.001413
3	Polynomial second order	0.003032	−3.40E−05	−0.00258	0.001352
4	Arruda–Boyce	0.00349	−4.10E−05	−0.00296	0.001543

Table 33.16 Deflection at 6 N load

S no	Model	Maximum displacement	Deflection at node 35	Deflection at node 78	Deflection at node 168
1	Mooney third parameter	0.007312	−0.000115916	−0.00616	0.003116
2	Ogden first order	0.007342	−0.000116505	−0.00618	0.003127
3	Polynomial second order	0.006919	−0.000108778	−0.00585	0.002975
4	Arruda–Boyce	0.008065	−0.00013	−0.00678	0.003413

Table 33.17 Deflection at 8 N load

S no	Model	Maximum displacement	Deflection at node 35	Deflection at node 78	Deflection at node 168
1	Mooney third parameter	0.01341	−0.00027581	−0.0112112255	0.005457
2	Ogden first order	0.13491	−0.000277413	−0.01127	0.005478
3	Polynomial second order	0.012551	−0.00025684	−0.0105589	0.005175

33.6 Stresses and Strain of Nodes

33.6.1 2 Newtons

Tables 33.18, 33.19 and 33.20

33.6.2 4 Newton

Tables 33.21, 33.22 and 33.23

33.6.3 6 Newtons

Tables 33.24, 33.25 and 33.26

Table 33.18 Stresses and strain of node 35

S no	Model	Node number	Stress	Strain
1	Mooney third parameter	35	1.63E−07	−4.67E−08
2	Ogden first order	35	1.64E−07	−4.69E−08
3	Polynomial second order	35	1.60E−07	−4.48E−08
4	Arruda–Boyce	35	1.71E−07	−5.14E−08

Table 33.19 Stresses and strain of node 78

S no	Model	Node number	Stress	Strain
1	Mooney third parameter	78	−6.91E−05	−9.65E−06
2	Ogden first order	78	−6.92E−05	−9.6864E−06
3	Polynomial second order	78	−6.72E−05	−9.23E−06
4	Arruda–Boyce	78	−7.25E−05	−1.06E−05

Table 33.20 Stresses and strain of node 168

S no	Model	Node number	Stress	Strain
1	Mooney third parameter	168	−6.73E−06	−9.71E−07
2	Ogden first order	168	−6.24E−06	−9.74E−07
3	Polynomial second order	168	−6.60E−06	−9.41E−07
4	Arruda–Boyce	168	−7.05E−06	−1.07E−06

Table 33.21 Stresses and strain of node 35

S no	Model	Node number	Stress	Strain
1	Mooney third parameter	35	6.80E−07	−1.93E−07
2	Ogden first order	35	6.81E−07	−1.94E−07
3	Polynomial second order	35	6.66E−07	−1.83E−07
4	Arruda–Boyce	35	7.14E−07	−2.13E−07

Table 33.22 Stresses and strain of node 78

S no	Model	Node number	Stress	Strain
1	Mooney third parameter	78	−0.00028	−3.98E−05
2	Ogden first order	78	−0.00028	−3.99E−05
3	Polynomial second order	78	−0.00027	−3.75E−05
4	Arruda–Boyce	78	−0.0003	−4.39E−05

Table 33.23 Stresses and strain of node 168

S no	Model	Node number	Stress	Strain
1	Mooney third parameter	168	−2.72E−05	−4.18E−06
2	Ogden first order	168	−2.73E−05	−4.19E−06
3	Polynomial second order	168	−2.66E−05	−4.05E−06
4	Arruda–Boyce	168	−2.85E−05	−4.61E−06

Table 33.24 Stresses and strain of node 35

S no	Model	Node number	Stress	Strain
1	Mooney third parameter	35	1.61E−06	−4.49E−07
2	Ogden first order	35	1.61E−06	−4.51E−07
3	Polynomial second order	35	1.57E−06	−4.20E−07
4	Arruda–Boyce	35	1.61E−06	−4.96E−07

Table 33.25 Stresses and strain of node 78

S no	Model	Node number	Stress	Strain
1	Mooney third parameter	78	−0.00066	−9.31E−05
2	Ogden first order	78	−0.00066	−9.36E−05
3	Polynomial second order	78	−0.00062	−5.61E−05
4	Arruda–Boyce	78	−0.0007	−1.03E−04

Table 33.26 Stresses and strain of node 168

S no	Model	Node number	Stress	Strain
1	Mooney third parameter	168	−6.32E−05	−1.05E−05
2	Ogden first order	168	−6.34E−05	−1.05E−05
3	Polynomial second order	168	−6.16E−05	−1.01E−05
4	Arruda–Boyce	168	−6.68E−05	−1.17E−05

Table 33.27 Stresses and strain of node 35

S no	Model	Node number	Stress	Strain
1	Mooney third parameter	35	3.06E−06	−8.20E−07
2	Ogden first order	35	3.07E−06	−8.24E−07
3	Polynomial second order	35	2.96E−06	−7.58E−07
4	Arruda–Boyce	35	3.26E−06	−9.07E−07

Table 33.28 Stresses and strain of node 78

S no	Model	Node number	Stress	Strain
1	Mooney third parameter	78	−0.00123	−0.00017
2	Ogden first order	78	−0.00123	−0.00018
3	Polynomial second order	78	−0.00113	−0.00016
4	Arruda–Boyce	78	−0.0013	−0.00019

Table 33.29 Stresses and strain of node 168

S no	Model	Node number	Stress	Strain
1	Mooney third parameter	168	−0.00012	−2.19E−05
2	Ogden first order	168	−0.00012	−2.20E−05
3	Polynomial second order	168	−0.00012	−2.11E−05
4	Arruda–Boyce	168	−0.00013	−2.49E−05

33.6.4 8 Newtons

Tables 33.27, 33.28 and 33.29

33.7 Conclusions

The best suited model to model the SBR is Arruda–Boyce when max, deflection is the criteria under static analysis based on the results obtained and tabulated from Tables 33.2 to 33.6 shown above, and for the Transient analysis Polynomial of second order is preferred, based on the tabulated results in the Tables 33.13, 33.14, 33.15 and 33.16 for deflection and stresses and strains considering various intensities of loading in both tension and compression. The same data and procedure can be applied for studying the dynamic characteristics of rubber dampers.

References

1. Song, B., Chen, W.: One dimensional dynamic compressive behaviour of EPDM rubbber. J. Eng. Mat. Technol. **125**, 294–301 (2003)
2. Ombes, B., Thuillier, S., Pilvin, P., Grohens, Y., Gillet, S.: Effective properties of carbon black filled natural rubber: experiments and modeling. J. Compos. **39**, 1141–1149 (2008)
3. Michelle, S., Fatt, H., Ouyang, X.: Integral based constitutive equation for rubber at high strain rates. Int. J. Solid Struct. **44**, 6491–6506 (2007)
4. Ali, A., Hosseini, M., Sahari, B.B.: A review and comparison on some rubber elasticity models. J. Sci. Ind. Res. **69**, 495–500 (2009)
5. Boyce, M.C., Arruda, E.M.: Constitutive models for rubber elasticity: a review. Rubber Chem. Technol. **73**, 504–523 (2000)
6. Bauman, J.T.: Fatigue, Stress, and Strain of Rubber Components—a Guide for Design Engineers. Hanser Publishers, Munich (2008)
7. Bower, A.F.: Applied Mechanics of Solids. CRC Press, Boca Raton (2009)
8. Varanasi, K.K., Nayfeh, S.A.: Damping of flexural vibration by low density foams and granular materials. In: ASME Design Engineering Technical Conferences, Chicago, 2–6 September, 2003
9. Treloar, L.R.G.: Stress-strain data for vulcanized rubber under various types of deformations. Trans. Faraday Soc. **40**, 59–70 (1944)
10. Mooney, M.: A theory of large elastic deformation. J. Appl. Phys. **11**, 582–596 (1940)
11. Zehil, G.P., Gavin, H.P.: Unified constitutive modeling of rubber like materials under diverse loading conditions. Int. J. Eng. Sci. **62**, 90–105 (2013)

Chapter 34
Specimen Size Effect on Stress-Strain Response of Foams Under Direct-Impact

Behrad Koohbor, Addis Kidane, Wei-Yang Lu, and Ronak Patel

Abstract Stress-strain response of rigid closed-cell polymeric foam under direct impact loading conditions is investigated, focusing on the specimen size effects. Cylindrical specimens with two different length-to-diameter ratios are impacted at different projectile velocities. Stereovision high speed photography in conjunction with 3D digital image correlation is utilized to study the full-field deformation of specimens under impact loading. A non-parametric analysis is then conducted to extract the local stress-strain curves within specimens. The proposed analytical method takes into account the concurrent influences of inertia stresses and material compressibility. The inertia stress within the area of interest is evaluated using the full-field distributions of acceleration and density. The calculated inertia stress is then superimposed with the stress measured at the boundary to enable the determination of full-field stress distribution over the entire gauge area. The results obtained in this work confirm that the effects of inertia stresses become more pronounced as specimen length-to-diameter ratio increases; whereas the degree of strain and strain rate variability is also elevated.

Keywords Polymeric foam • Dynamic loading • Digital image correlation • Shock tube • Direct impact

34.1 Introduction

Rigid polymeric foams have gained extensive attention in engineering applications due to their excellent energy absorption and low structural weight. Owing to their superior damping performance, polymeric foam structures are generally utilized in applications in which dynamic loading conditions are dominant. Therefore, knowledge on the constitutive response of the material at dynamic loading conditions is an essential factor in the design of structures fabricated from polymeric foams. One particular challenge faced in the study of dynamic constitutive response of polymeric foams is due to their low mechanical impedance. The low impedance nature of these materials results in a belated state of quasi-static equilibrium and makes the conventional boundary measurement techniques quite inaccurate. Several solutions have been proposed to compensate for the non-equilibrium stress state present during the dynamic deformation of foams. Among those, application of pulse shaping and the use of polymeric bars have been successfully implemented and confirmed to result in more accurate evaluation of the constitutive response of foams at high strain rate loading conditions [1–3]. These two solutions are only applicable in Hopkinson pressure bar (SHPB).

A more general solution to compensate for the non-equilibrium condition in dynamic testing of low impedance materials is the application of full-field measurements to include the effect of inertia loading into the analysis. This has successfully been achieved in recent years through the use of Virtual Fields Method [4], as well as a direct non-parametric methodology proposed by Othman et al. [5] to identify the constitutive response of low impedance materials in dynamic conditions. The non-parametric approach has successfully been adopted in the study of constitutive response of various polymeric foams subjected to direct impact and up to strain rates of 2500 s^{-1} [6, 7]. The objective in the present work is to extend the previous studies further by investigating the effects of specimen size (length-to-diameter ratio) on the non-parametric solution proposed in Ref. [7]. In addition, strain rate conditions of up to 5500 s^{-1} are developed in the specimens, and the influence of spatial variability of strain rate along the specimen axis is discussed in detail.

B. Koohbor (✉) • A. Kidane • R. Patel
Department of Mechanical Engineering, University of South Carolina, 300 Main Street, Columbia, SC 29208, USA
e-mail: koohbor@email.sc.edu

W.-Y. Lu
Sandia National Laboratories, Livermore, CA, USA

D. Casem et al. (eds.), *Dynamic Behavior of Materials, Volume 1*, Conference Proceedings of the Society for Experimental Mechanics Series, DOI 10.1007/978-3-319-41132-3_34

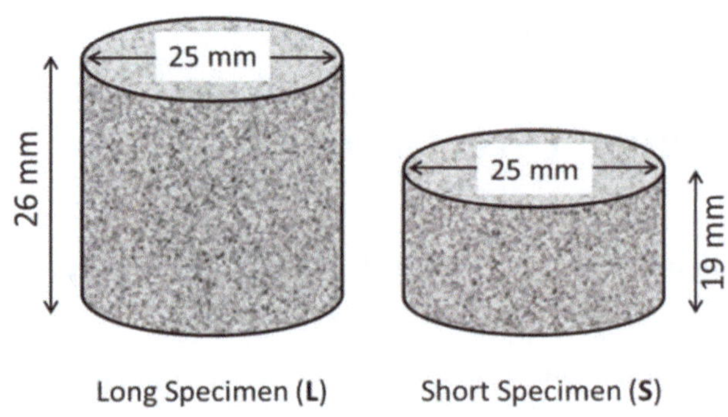

Fig. 34.1 Cylindrical specimens with different initial dimensions used for direct impact experiments

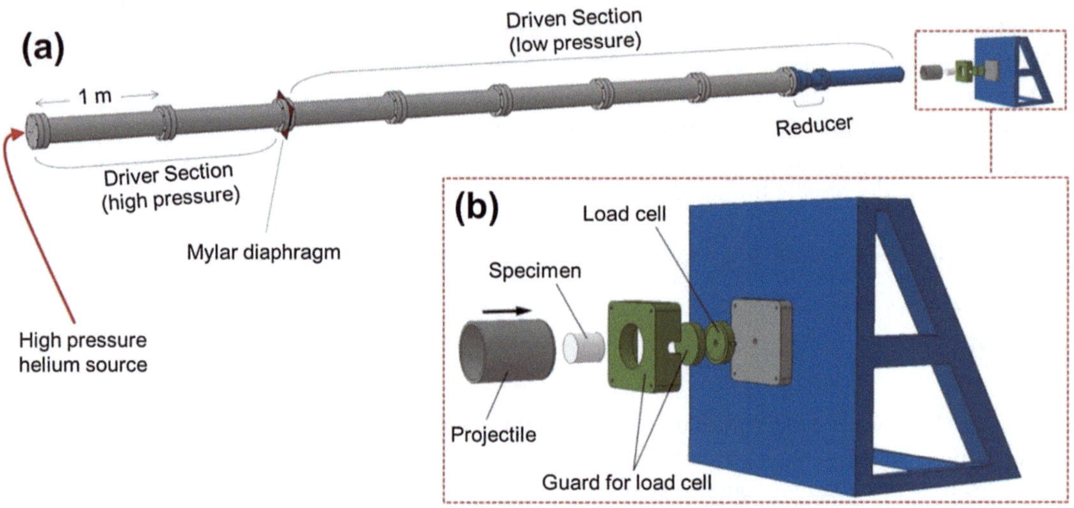

Fig. 34.2 (**a**) Schematic of the shock tube with (**b**) a close up view of the load fixture

34.2 Materials and Methods

Rigid closed-cell polyurethane foam under the commercial name TufFoam35 [8] is examined in this work. Nominal density of the as-received foam is measured as 560 kg/m^3. Longitudinal elastic wave velocity of the material is calculated as $c_l = 1180$ m/s, using its quasi-static elastic modulus ($E = 780$ MPa). Cylindrical specimens with two different length-to-diameter ratios are extracted from the as-received billets, as shown schematically in Fig. 34.1. High contrast black and white speckle patterns are applied on the lateral surface of each specimen for image correlation purposes.

Controlled direct impact loading is applied on the specimens using a shock tube apparatus. Design factors of the utilized shock tube can be found in detail elsewhere [7, 9]. A 70 gr projectile made from high strength 7068 aluminum alloy is directly shot at the specimen to increase the momentum transfer (see Fig. 34.2). The impact velocity of the projectile can be manipulated by varying the number and/or thickness of the plastic diaphragms (Mylar sheets) separating the driver and the driven sections of the tube. In the present work, to maintain the value of velocity to specimen length ratio, impactor velocities of 162 m/s and 123 m/s are applied on long and short specimens, respectively.

34.3 High Speed Imaging and Digital Image Correlation

High speed stereo imaging in conjunction with 3D digital image correlation is used to study the dynamic deformation response of specimens. The stereo camera system is comprised of a pair of Photron SA-X2 high speed cameras each equipped with a 100 mm macro lens. Images are acquired at a rate of 2×10^5 fps at full-field resolution of 256×152 pixel^2. The exposure time used here is 293 ns and the spatial resolution is measured as 0.19 mm/pixel.

Images acquired during the impact experiments are analyzed using the image correlation software Vic-3D. In this software, subset size and step size of 19 × 19 pixel^2 and 1 pixel are used for correlation process, respectively, to extract the full-field displacement distribution. Strain field is obtained by differentiating the displacement fields, and smoothed using Gaussian filtering with a filter size of 13. 'Normalized squared differences' is used as the matching criterion. Optimized 8-tap interpolation with affine shape functions are also utilized in the image correlation process.

34.4 Data Analysis

Let us consider the general dynamic stress equilibrium as:

$$\sigma_{ij,j} + \rho b_i = \rho a_i \tag{34.1}$$

where σ_{ij} denotes Cauchy stress tensor; ρ is mass density, b_i is body force in i-direction and a_i is the acceleration in i-direction. In the case of uniaxial compression of a cylindrical specimen, as schematically shown in Fig. 34.3, and in the absence of body forces Eq. (34.1) is reduced to:

$$\frac{\partial \sigma_z}{\partial z} = \rho a_z \tag{34.2}$$

where subscript z denotes the axial direction. The solution to Eq. (34.2) is given by:

$$\sigma_z(L,t) = \sigma_z(0,t) + \int_{\xi=0}^{\xi=L} \rho(\xi,t) a_z(\xi,t) d\xi \tag{34.3}$$

Equation (34.3) indicates that the axial stress at any given position z $= L$ along the specimen axis can be reconstructed from the stress magnitude measured on the boundary (i.e. $z = 0$) and the integral term on the right hand-side of Eq. (34.3), which represents inertia stress.

As can be seen in Eq. (34.3), the inertia stress is expressed as a function of the spatial variations of axial acceleration and mass density. The density term could be assumed constant if the specimen was made of an incompressible material, e.g. metal. However, for cellular materials and particularly polymeric foams, the assumption of constant density is not valid, since mass density itself varies with the amount of deformation applied.

Compressibility and change in density of the material during dynamic deformation is taken into account in this work using the following equation [7]:

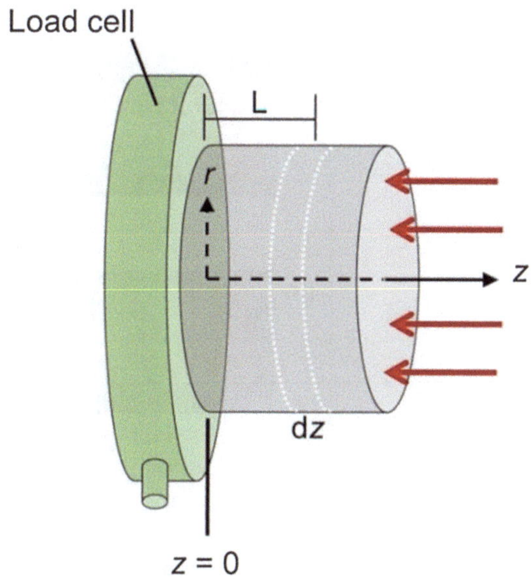

Fig. 34.3 Schematic of a cylindrical specimen subjected to direct impact

$$\rho(z,t) = \rho_0\left[\exp(\varepsilon_z(z,t))\right]^{2\upsilon(z,t)-1} \tag{34.4}$$

This equation is proposed assuming the principle of mass conservation, and allows for the calculation of the local mass density, $\rho(z, t)$, as a function of initial density, ρ_0, local axial compressive strain, $\varepsilon_z(z, t)$, and local Poisson's ratio, $\upsilon(z, t)$. Local Poisson's ratio can also be evaluated using the local radial and axial strain values as:

$$\upsilon(z,t) = -\frac{d\varepsilon_r(z,t)}{d\varepsilon_z(z,t)} \tag{34.5}$$

Further details on the compressibility model used in the present work can be found in Ref. [7].

34.5 Full-Filed Deformation Response

Considering the longitudinal elastic wave speed in the material ($c_l = 1180$ m/s), specimen characteristic time, i.e. time for a single stress wave traverse along the axis, is calculated as 16 μs for the short specimen and 22 μs for the long specimen. It is well-established that at least three consecutive stress wave reverberations are required before the quasi-static equilibrium is achieved in a dynamically deformed material [10]. Accordingly, minimum durations of 48 μs and 66 μs are required for the stress to equilibrate in the short and long specimens, respectively. Failure in the long specimen initiates well before stress equilibrium. Such a condition makes the conventional way of determining the constitutive response based on boundary measurements quite inaccurate, thus necessitating the inclusion of inertia loading effects during the transient stress state. Although less stress variability along the axis of the short specimen is expected due to the short time required to reach equilibration, it will still be necessary to take into account the development of inertia stresses to acquire more meaningful and more accurate constitutive information for this specimen.

Axial strain distribution over the gauge area is determined using the displacement fields obtained from DIC. Figure 34.4 shows typical axial strain maps obtained from DIC. Considerable spatial variability in the axial strain is observed in both specimens, while the degree of this spatial variability is lower in the short specimen. In both specimens, the highest magnitude of strain is developed at the middle sections of each specimen. To explain this observation, note that the applied impact produces two distinct waves: (1) an elastic wave traveling at the speed of sound in the specimen, and (2) a plastic wave moving at a significantly slower velocity. The elastic wave created at the instant of impact does not plastically deform the material. However, the plastic wave traverses over the length of the specimen reducing its velocity to zero while plastically deforming the material. The stress developed behind the plastic wave deforms the material axially, as well as laterally and forms the mushroom configuration at the impacted side of the specimen, as clearly shown in Fig. 34.4. Furthermore, the highest strain magnitudes are also developed at the location of the plastic wave front.

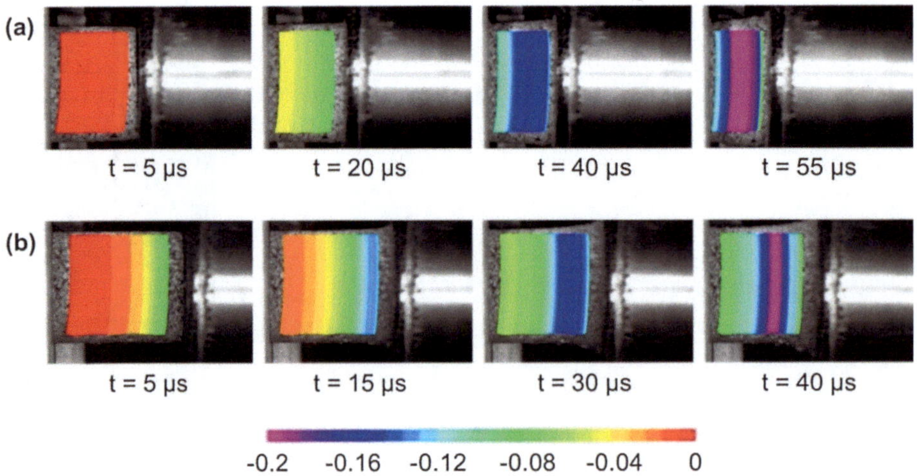

Fig. 34.4 Full-field axial strain maps obtained for (**a**) short and (**b**) long specimens at different times after the impact

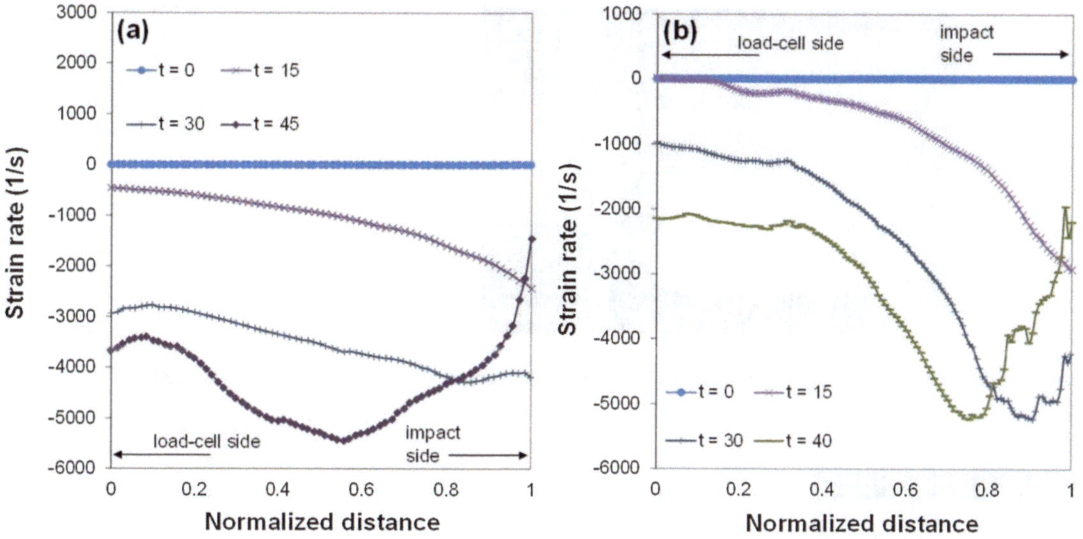

Fig. 34.5 Variation of axial strain rate along the specimen axis for (**a**) short and (**b**) long specimens

Evolution of full-field strain rate is also expected to show similar patterns. Axial strain rate fields are calculated from the axial strain by applying a central difference scheme as:

$$\dot{\varepsilon}_z(t) = \frac{\varepsilon_z(t+\Delta t) - \varepsilon_z(t-\Delta t)}{2\Delta t} \tag{34.6}$$

where $\dot{\varepsilon}_z$ denotes axial strain rate and Δt is the inter-frame time (=5 μs). Variation of axial strain rate along the specimen axis has been plotted in Fig. 34.5. It is seen in Fig. 34.5a that the highest strain rate values are developed at the impacted side of the short specimen at $t < 30$ μs. The location of the highest strain rate gradually moves towards the middle sections of the specimen after $t = 30$ μs. Similar response is also observed in Fig. 34.5b for the long specimen; whereas the strain rate variability is indicated to be more substantial for the case of the long specimen. This behavior can be explained through the plastic wave traverse in the specimen. As the plastic wave passes along the specimen from the impacted side, it reduces its velocity to zero, while plastically deforming the portions of the specimen it has already passed through [11]. Had the specimen been free on its left side, the strain rate behind the plastic wave would have been very close to zero. However, the reaction force imposed on the left side of the specimen gives rise to a compressive wave travelling in the opposite direction—from left to right—elevating the magnitude of the compressive strain rate over the regions already undergone plastic deformation. An interesting point here is the rapid increase in the value of strain rate in the short specimen after $t = 30$ μs. This is due to the reversion of the axial acceleration in the short specimen, which results in a change in sign of the inertia stress. Similar behavior is observed for long specimen, as well. More explanations on this are provided in later sections.

Next, change of density of the material during dynamic deformation is evaluated using the compressibility model described earlier in Eq. (34.4). The first step here is to measure the Poisson's ratio of the material. Poisson's ratios of the specimens in this work are measured following the procedure detailed in Refs. [7, 12]. Due to the formation of an irregular mushroom shape after the impact, each specimen is divided into two halves—load-cell half and projectile half, as shown in Fig. 34.6—and the average Poisson's ratios are measured within each half. Poisson's ratios of 0.35 and 0.26 are determined in projectile and load-cell halves of the short specimen, respectively. For the long specimen, Poisson's ratios determined in projectile and load-cell halves are 0.25 and 0.20, respectively. For both specimen geometries, the average Poisson's ratio values are higher over the regions more closely associated with the impacted side. This observation is consistent with the fact that both specimens undergo relatively higher radial expansions at their impacted side. The change in density during deformation is then calculated for each specimen using the axial strain fields and the Poisson's ratios determined earlier. Accordingly, density increase of up to 10 % was determined for the specimens in this work.

Axial acceleration is calculated from the axial displacement by applying a central difference scheme as:

$$a_z(t) = \frac{u_z(t+\Delta t) - 2u_z(t) + u_z(t-\Delta t)}{\Delta t^2} \tag{34.7}$$

where u_z is axial displacement component. Figure 34.7 illustrates the variation of axial acceleration in time extracted from

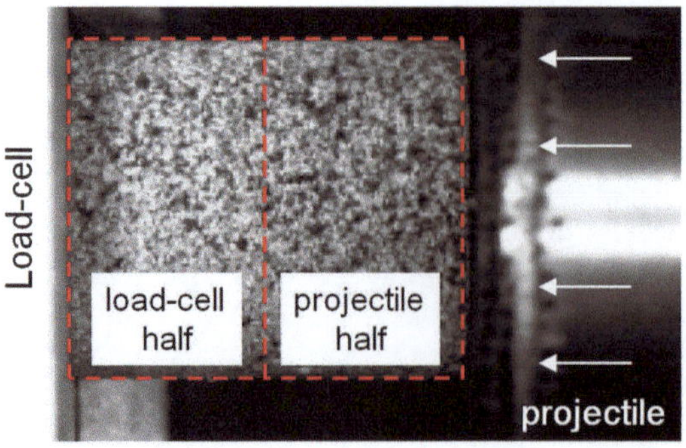

	Specimen	
	Short	Long
Projectile half	0.35	0.25
Load-cell half	0.26	0.20

Fig. 34.6 Illustration of the load-cell and projectile halves over which the average Poisson's ratios are evaluated

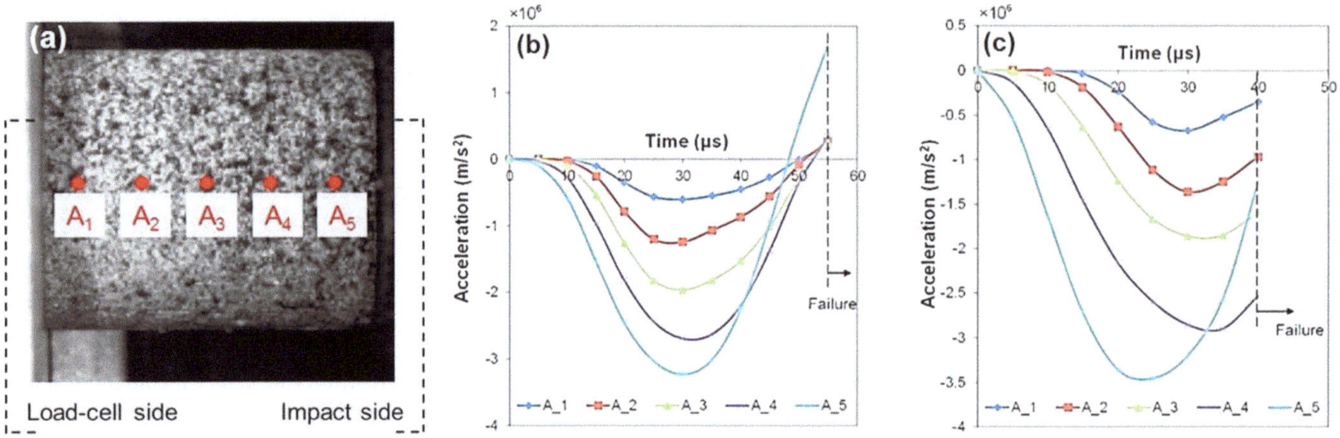

Fig. 34.7 Variation of axial acceleration in time extracted from five representative points shown in (**a**), for (**b**) short and (**c**) long specimen

the representative points shown in Fig. 34.7a. Acceleration curves shown in Fig. 34.7 exhibit significant axial variability. This is particularly evident during the earlier stages of deformation for both specimens. In both specimen geometries, the magnitude of axial acceleration indicates a significant and rapid increase during the first 25–30 μs after the impact. The large magnitudes of acceleration calculated in early stages of impact results in the development of significant inertia stresses. In the current work, owing to the substantially high impact velocity and the resultant applied strain rate, the long specimen fails before acceleration damping and quasi-static stress equilibration. This is why the acceleration curves do not return back to zero before failure. In addition, although deformation time is longer for the short specimen compared with that of the long specimen, complete damping does not take place in the short specimen either.

34.6 Inertia Stress and the Local Constitutive Curves

Inertia stress distribution over the entire area of interest can be evaluated using the full-field density and acceleration presented earlier. Inertia stresses are calculated in this work by numerically integrating the product of density and acceleration in axial direction, following the procedure detailed in Ref. [7, 12]. Variation of axial inertia stress along the specimen axis for both specimens is shown in Fig. 34.8. It is clearly shown that at any given time, the inertia stress takes larger magnitudes at the impact side of the specimen. This is due to the higher acceleration values developed within the areas closer to the impacted side of both specimens (see Fig. 34.7). It is also evident that the magnitude of the inertia stress increases progressively during the first 30 μs after the impact in the short specimen. This trend is reversed at $t \geq 30$ μs such that the compressive inertia stresses become smaller in magnitude. A similar response is also observed in the long specimen after $t = 30$ μs.

Using the inertia stress values shown in Fig. 34.8 and the boundary stresses measured at the load-cell side, distribution of axial stress along the axis of the specimen is determined following Eq. (34.3). Variation of total (inertia plus boundary

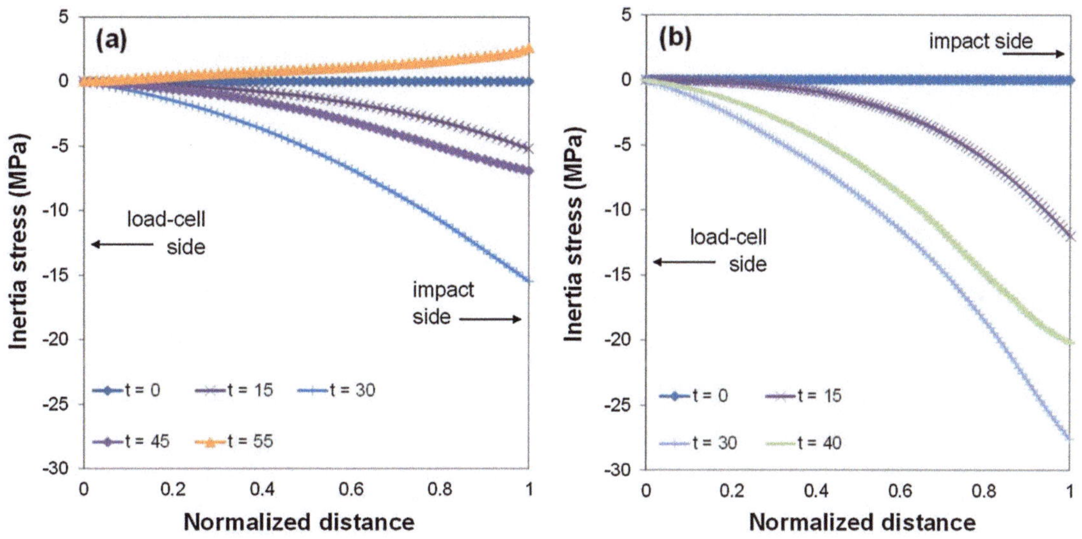

Fig. 34.8 Variation of inertia stress along the specimen axis for (**a**) short and (**b**) long specimens

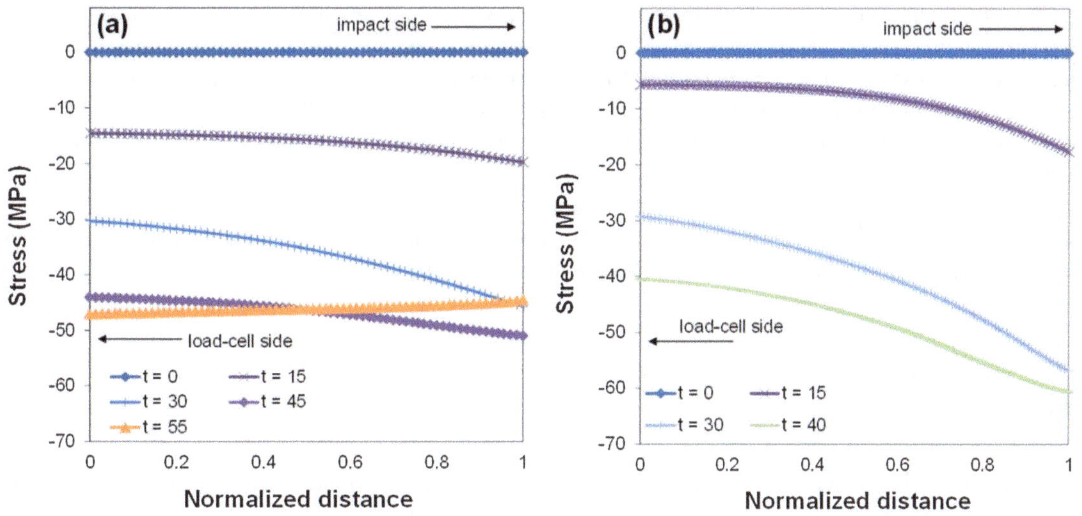

Fig. 34.9 Variation of total axial stress along the specimen axis for (**a**) short and (**b**) long specimens, at different times after the impact

measured) axial stress along the specimen axis is plotted in Fig. 34.9. The following remarks must be highlighted in this regard:

1. Total axial stress values indicate considerable length-wise variability, the degree of which is higher for the long specimen. This substantial degree of axial variability in stress for the long specimen is attributed to the more significant gradient of strain rate over the length of the specimen, as shown and discussed earlier in Fig. 34.5.
2. Magnitude of the total axial stress is larger at the impact side of the specimen, over the entire deformation time in the long specimen, as well as the first 45 μs after the impact in the short specimen.

Having obtained the full-field strain, strain rate and stress distributions over the entire area of interest, local stress-strain response of the material can be extracted at any given time and on any given location within the area of interest. Accordingly, the local stress-strain curves extracted from three locations along the specimen axis are plotted in Fig. 34.10 along with their associated strain rate curves. Note that in this figure, compressive strain, strain rate and stress are all plotted as positive value for convenience.

The stress-strain curves shown in Fig. 34.10 represent the local constitutive response for the non-equilibrium stress conditions. The discrepancy observed in the curves is due to the local variability of strain rate. Note that owing to the strong

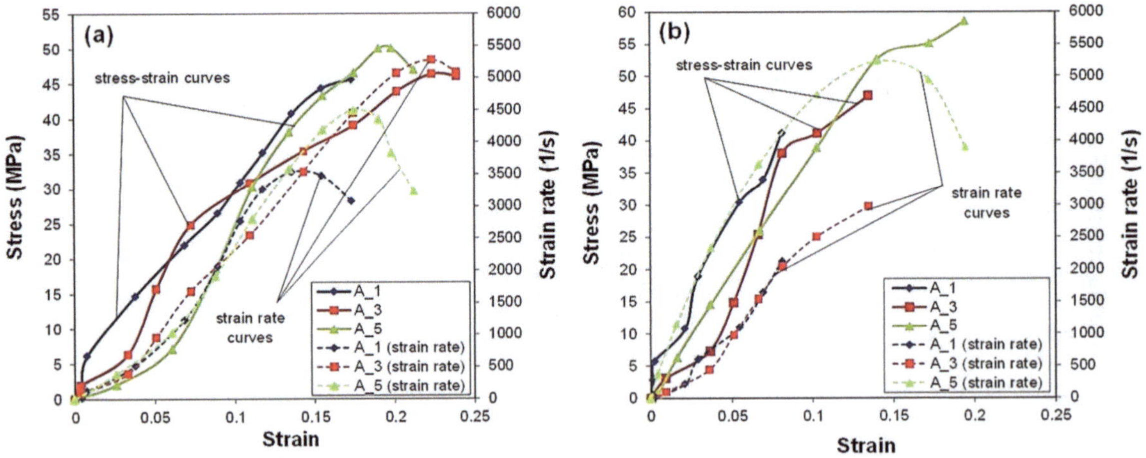

Fig. 34.10 Local stress-strain curves in (**a**) short and (**b**) long specimen, extracted for the representative locations indicated in Fig. 34.7a

rate sensitivity of the examined material and the fact that there is a significant axial gradient in the strain rate values, local variation of stress-strain response was actually expected in this work. In addition, the initial part of the stress-strain curves obtained here is different from what was previously observed for the same material tested at lower strain rates [7]. This could be due to the effect of local strain rate, which in this case is almost three times higher than the strain rates applied on the specimen in our previous research. Additional investigation is needed to verify the observations made in the present work.

Results presented in this study clearly do not allow for an accurate assessment of the constitutive response of the material, since local flow curves are not plotted at constant strain rates. This issue is more challenging in the case of the long specimen, in which stronger strain rate gradients are developed. However, the results obtained in this work can be further analyzed via various numerical algorithms to interpolate the flow curves at constant strain rate conditions. Attempts are currently being made by the authors to utilize numerical tools such as simple scattered data interpolation and artificial neural networks to identify the trends hidden within the scattered strain-strain rate-stress data obtained in this work, and finally extract the constitutive response of the material at constant strain rates ranging from 1000 to 5000 s^{-1}.

34.7 Summary

Constitutive response of rigid closed-cell polymeric foam subjected to direct impact is investigated in this work. The effects of material compressibility and inertia loading during the non-equilibrium state of deformation are concurrently taken into account using appropriate mathematical models, along with the results obtained from full-field DIC measurements. It is clearly indicated that the inertia stresses developed in the transient stress duration have a significant effect on the local constitutive response of the material; whereas a larger length-to-diameter ratio promotes the influence of the inertia stresses. Local constitutive curves within the specimens are determined from the full-field stress-strain distributions. Substantial discrepancy observed between the local constitutive curves obtained for each specimen is attributed to the strong strain rate gradients along the axis of the specimen; while such strain rate gradients are shown to be more significant in the long specimen geometry. The procedure described in this work can be used to identify rate-sensitive constitutive curves over a wide range of strain rate conditions using a minimum number of tests.

Acknowledgment Sandia National Laboratories is a multi-program laboratory managed and operated by Sandia Corporation, a wholly owned subsidiary of Lockheed Martin Corporation, for the U.S. Department of Energy's National Nuclear Security Administration under contract DE-AC04-94AL85000.

References

1. Zhao, H.: Testing of polymeric foams at high and medium strain rates. Polym. Test. **16**, 507–516 (1997). doi:10.1016/S0142-9418(97)00012-3

2. Chen, W., Lu, F., Winfree, N.: High strain rate compressive behavior of a rigid polyurethane foam with various densities. Exp. Mech. **42**, 65–73 (2002). doi:10.1007/BF02411053
3. Song, B., Chen, W., Frew, D.J.: Dynamic compressive response and failure behavior of an epoxy syntactic foam. J. Compos. Mater. **38**, 915–936 (2004). doi:10.1177/0021998304040552
4. Moulart, R., Pierron, F., Hallett, S.R., Wisnom, M.R.: Full-field strain measurement and identification of composites moduli at high strain rate with the virtual fields method. Exp. Mech. **51**, 509–536 (2011). doi:10.1007/s11340-010-9433-4
5. Othman, R., Aloui, S., Poitou, A.: Identifications of non-homogeneous stress fields in dynamic experiments with a non-parametric method. Polym. Test. **29**, 616 (2010). doi:10.1016/j.polymertesting.2010.03.013
6. Koohbor, B., Mallon, S., Kidane, A., Lu, W.Y.: The deformation and failure response of closed-cell PMDI foams subjected to dynamic impact loading. Polym. Test. **44**, 112–124 (2015). doi:10.1016/j.polymertesting.2015.03.016
7. Koohbor, B., Kidane, A., Lu, W.Y., Sutton, M.A.: Investigation of the dynamic stress-strain response of compressible polymeric foam using a non-parametric analysis. Int. J. Impact Eng. **91**, 170–182 (2016). doi:10.1016/j.ijimpeng.2016.01.007
8. Lu, W.Y.: Mechanical characterization of rigid polyurethane foams. Sandia Report, SAND2014-20708 (2014)
9. Mallon, S., Koohbor, B., Kidane, A., Sutton, M.A.: Fracture behavior of prestressed composites subjected to shock loading: a DIC-based study. Exp. Mech. **55**, 211–225 (2015). doi:10.1007/s11340-014-9936-5
10. Davies, E.D.H., Hunter, S.C.: The dynamic compression testing of solids by the method of the split Hopkinson pressure bar. J. Mech. Phys. Solid **11**, 155–179 (1963). doi:10.1016/0022-5096(63)90050-4
11. Nicholas, T., Recht, R.F.: Introduction to impact phenomena. In: Zukas, J.A. (ed.) High Velocity Impact Dynamics, pp. 1–63. Wiley, New York (1990)
12. Koohbor, B., Kidane, A., Lu, W.Y.: Dynamic flow response of rigid polymer foam subjected to direct impact. In: Song, B., Lamberson, L., Casem, D., Kimberley, J. (eds.) Conference Proceedings of the Society for Experimental Mechanics Series, Dynamic Behavior of Materials, vol. 1, pp. 163–170. Springer, New York (2016). doi:10.1007/978-3-319-22452-7_23

Chapter 35
Texture Evolution of a Fine-Grained Mg Alloy at Dynamic Strain Rates

Christopher S. Meredith and Jeffrey T. Lloyd

Abstract AMX602 (Mg-6%Al-0.5%Mn-2%Ca) is a high strength Mg alloy that was manufactured by the spinning water atomization process (SWAP) and extruded into a plate geometry. The processing produces an alloy with a weak rolling texture (for Mg) and grains between 0.5 and 5 μm. Quasi-static and dynamic compression experiments were carried out to probe the material's mechanical behavior in the three processing directions. The tested plate showed deformation mechanism induced anisotropy consistent with what has been observed for other Mg alloys. The texture evolution was measured after loading in the three directions using X-ray diffraction at the Cornell High Energy Synchrotron Source (CHESS). A computationally efficient crystal plasticity model that demarcates twinning, basal slip, and non-basal slip mechanisms was utilized to predict the texture evolution and compared to the experimental texture. The model was able to predict the reorientation of grains associated with twin dominated yielding in the extrusion and transverse directions, and strengthening of the texture associated with slip dominated deformation in the normal direction.

Keywords AMX602 • Dynamic strain rates • Texture evolution • Twinning • Crystal plasticity

35.1 Introduction

Magnesium is the lightest metal for structural applications with a specific strength similar to or exceeding aluminum and titanium [1]. However, commercial magnesium alloys have relatively low strength, limited ductility, anisotropic mechanical behavior, and relatively poor corrosion resistance. The strength and ductility can be optimized using alloy development, texture modification and/or grain refinement [3, 4]. For example, microalloying by adding trace amounts of certain elements which creates finely dispersed precipitates recently has been utilized to enhance precipitate hardening [2]. Also, a more random texture increases ductility versus typical wrought Mg alloys because stable plastic deformation is promoted by an increased strain hardening rate [5–7]. Finally, reducing the grain size is an effective way to increase the strength of a material through the well-known Hall–Petch relationship. When twinning dominates yielding the Hall–Petch has the same empirical form as for slip [8], thus twinning is suppressed as the grain size is reduced and is manifested as a higher Hall–Petch constant [8, 9]. Typical processing methods–like rolling and extrusion–limit the ability to reduce the grain size due to Mg's low ductility, and these methods produce sharp textures associated with increasing mechanical anisotropy.

Severe plastic deformation processing methods are effective at refining grain size, but they also produce sharp textures [10–13]. The sharp textures generally result in highly anisotropic mechanical behavior, tension/compression asymmetry and low ductility. Recent work has shown that equal channel angular pressing (ECAP) is able to reduce the grain size and tailor the texture for property improvement. Several authors [14–16] have performed ECAP on AZ31B and showed increased strength and reduced anisotropy while preserving ductility using specific processing routes and temperatures. The resulting basal texture is generally stronger than before, and is generally rotated with respect to a principal processing direction. Therefore, when determining the subsequent strength and ductility, the loading directions were at an angle with respect to the basal texture. When conventional rolled AZ31B is loaded along the processing directions, the loading direction coincides with the principal texture components, which maximizes the measured the anisotropic response. The reduced anisotropic behavior and negligible change in ductility of ECAPed AZ31B is partially a consequence of the chosen loading directions, and not an intrinsic material property. When samples are cut along principle material directions, the measured yield strength differences in different directions can still vary by a factor of 2 or more, and limited ductility is observed for particular loading conditions [14, 15]. In applications with complex loading conditions, where the full anisotropic material response is

C.S. Meredith (✉) • J.T. Lloyd
Weapons and Materials Research Directorate, Army Research Lab, Aberdeen Proving Ground, Aberdeen, MD 21005, USA
e-mail: christopher.s.meredith3.civ@mail.mil

D. Casem et al. (eds.), *Dynamic Behavior of Materials, Volume 1*, Conference Proceedings of the Society for Experimental Mechanics Series, DOI 10.1007/978-3-319-41132-3_35

probed, it is unclear whether this ECAPed AZ31B will exhibit its measured strength and ductility improvement. Because the sharp texture is thought to produce the measured anisotropy, a reduction in peak texture is seen as a promising route for overall performance enhancement.

Rapid solidification is another route for concurrent grain refinement and texture modification. Several authors [17–20] have used rapid solidification processes to prepare metal powders such as Fe, Mg, Al and Cu and their alloys. The Spinning Water Atomization Process (SWAP) is capable of producing metal powders with a fine-grained microstructure [21–24]. Recently, SWAP was applied to magnesium alloys and it has shown promise in producing relatively large volumes with a fine-grained microstructure and a comparatively weak texture [25–28]. This process consists of rapidly cooling molten material with water, which produces a coarse powder. The powder is consolidated and extruded to its bulk shape at its theoretical maximum density. The resulting high strength Mg alloys, AMX602 (Mg-6%Al-0.5%Mn-2%Ca) and ZAXE1711 (Mg-1%Zn-7%Al-1%Ca-1%La), have exhibited improved ballistic performance compared with conventionally processed AZ31B [29]. An initial batch of AMX602 specimens was produced in extruded bars with a width-to-thickness ratio of 1.5:1. These specimens had reduced peak texture, measured yield strengths of approximately 250–350 MPa, and strain-to-failures of about 10–20 %, depending on the loading direction, strain rate and sense (tension or compression) [26, 27, 30].

In this study AMX602 was prepared by SWAP and the coarse powder was consolidated and extruded into a plate (6:1 width-to-thickness ratio). We performed experiments to determine the material's mechanical behavior in compression at quasi-static and dynamic strain rates in each of the three processing directions–extrusion (ED), transverse (TD) and normal (ND) directions. We measured the texture at different amounts of compressive strain using X-ray diffraction at the Cornell High Energy Synchrotron Source (CHESS). Additionally, we recently developed and used a HCP crystal plasticity model [23, 34] to predict the material's texture evolution subjected to the aforementioned loading conditions.

35.2 Experimental Procedures

AMX602 coarse powder was prepared via SWAP. The entire process is performed in an inert atmosphere. The molten AMX602 is directly streamed into the chamber where water is spun around the interior surface. Impinging water droplets rapidly cool the molten stream, which locks in a fine-grained microstructure. The powder is consolidated into a green compact and extruded into the final shape. An initial batch of this material has been investigated by others [26, 27, 30]. The plate geometry investigated here is the latest batch to be produced. The powder was consolidated and extruded at 370 °C into a cross section of 152 mm × 25.4 mm. In this paper, AMX602 refers to the final extruded material. The response of the compact and extruded powder were not considered in this work.

We loaded the material to failure in compression at quasi-static and dynamic loading rates along the extrusion (ED), transverse (TD) and normal directions (ND). Quasi-static compression experiments were performed at a constant displacement rate corresponding to a strain rate of 10^{-4} s^{-1}. Two strain gages, with their output averaged, mounted on two orthogonal surfaces measured strain of the plate geometry specimens. The specimens measured 8.5 mm × 8.5 mm × 12.7 mm. We performed dynamic compression experiments using a split-Hopkinson pressure bar (SHPB) with 19 mm diameter maraging steel incident and reflected bars. The samples were 11.4 mm cubes. The stress-strain history within the sample was calculated from the incident, reflected and transmitted strain gage signals mounted on the input and output bars [31]. The strain rate was approximately 1000 s^{-1}. We cut all samples from the center of the plate to eliminate any surface effects. Experiments where samples were "recovered" after ~3 and 10 % true strain were performed to measure the texture of each.

We used synchrotron X-ray diffraction (SXRD) to measure the texture of AMX602 samples, which has much higher energies than reflection XRD. The X-ray beam passes through the sample so the measured diffraction patterns are characteristic of the bulk. We used SXRD at the Cornell High Energy Synchrotron Source (CHESS) using beamline F2. For each sample, a Laue pattern was recorded at several different orientations. We analyzed the raw patterns using the MAUD program [32]. MAUD calculates the texture that best matches the experimental diffraction line profile. We measured and calculated the texture of an as-received sample, and after subsequent compressive loading at both quasi-static and dynamic strain rates. All of the experimental pole figures are normalized by the condition with the strongest texture (ED loading direction, ~10 % true strain).

35.3 Modeling Framework

We used a combined crystal- and macro-plasticity framework to model the dynamic response of AMX602 specimens. Elsewhere, Becker and Lloyd [34] and Meredith et al. [35] describe details of the model so herein we restrict ourselves to essential model features and assumptions. The reduced-order model is based on the model by Graff et al. [33] where simplifications are made for increased computational efficiency and stability, thereby enabling the framework to simulate large dynamic problems of interest. The model uses a pseudo-slip description for extension twinning accompanied by basal <a> slip, prismatic <a> slip, and pyramidal <c+a> slip. Reorientation due to twinning is neglected for strength calculations; however, reorientation is accounted for during post-processing as to mimic extension twin reorientation affects. The model approximates the combined effect of the three basal <a> slip by assuming basal slip occurs on one aggregate system that glides in the direction of maximum shear stress on the basal plane. Because prismatic and pyramidal have similar strengths, and their combined effect forms an approximately closed yield surface in biaxial stress space [34], they are approximated using as a single isotropic, rate-dependent proportional plastic deformation mechanism. Both twinning and basal slip are assumed to be rate-independent, whereas rate-dependent isotropic slip is approximated using a phenomenological Johnson-Cook hardening relation.

35.4 Results and Discussion

We performed compression experiments at 10^{-4} and $\sim 1000\ s^{-1}$ in the ED, TD and ND to determine the influence of loading rate on the anisotropy of the mechanical response. All experiments resulted in the macroscopic failure of the sample, which is reflected in the stress-strain curves as a quick or sudden drop in the flow stress at large strains. Multiple experiments at each geometry, loading direction and strain rate were performed and showed great consistency. A single test is shown for each parameter for clarity.

Figure 35.1 shows the stress-strain response for the three loading directions of the plate at quasi-static and dynamic strain rates. Compression along the ED results in a characteristic s-shaped hardening curve, which is indicative of $\{10\bar{1}2\}$ extension twinning [36]. After yielding the material exhibits a stress plateau until strains of approximately 4 %. Additionally, despite seven orders of magnitude difference in the strain rate, there is no appreciable difference in the yield strength (~280 MPa) and little difference in the post-yield stress-strain response. The dynamic strain rate has a strain to failure that is 4 % larger than for the quasi-static strain rate. Compression along the TD, Fig. 35.1b, exhibits a similar response as compression in the ED, but the plateau region is not as pronounced, the yield strength is lower (~240 MPa), and the work hardening rate is slightly lower. Therefore, although the strain to failure is similar for compression along the ED and TD, the peak stress at failure is lower for compression along the TD than the ED. Figure 35.1c shows the stress-strain response when compressed along the ND. The main differences between the response along the ND and the other two directions are the lack of a well-defined yield, a concave-down, instead of an s-shaped, hardening curve, and greater strain rate sensitivity in both the initial yield strength and subsequent hardening behavior. The yield strengths for quasi-static and dynamic strain rates are approximately 250 and 280 MPa, respectively. Another takeaway is the anisotropy in the yield strength is relatively low versus conventially processed Mg alloys which is likely due to the reduced texture observed, which is discussed next.

Figure 35.2 displays the as-received experimental texture and a pole figure of randomly sampled orientations used as input for the model. The data for Fig. 35.2b was an electron backscatter diffraction (EBSD) scan [35], and shows the basal texture primarily in the ND. This is a typical rolling texture for an HCP material. The SXRD data shows generally the same texture orientation but with bifurcated peaks rotated a few degrees away from the ND toward the TD. The peak intensity of 5.2 MRD (multiples of a random distribution) is weak for typical Mg and its alloys. We are not sure why there is a difference in the peak intensity location. One explanation is the bifurcated peaks are an artifact of the imperfect fit between the experimental diffraction patterns and Rietveld analysis that MAUD uses to calculate the texture. A second possibility is the EBSD scan was not representative of the full volume of material. The synchrotron beam was ~2 mm × 0.1 mm and passed through about 9 mm of the material. The EBSD scans revealed most grains were between 0.5 and 5 μm but there were larger grains tens of micrometers in size. It is possible the larger grains had texture that was missed with the EBSD scans but are showing up in the SXRD data. Mg alloys can show double-peak character but they are mostly radially symmetric about the ND [35, 36]. Note that the color shading for all the pole figures share the same scale of between 0 and 10 MRD.

Plastic loading changes the texture through grain and subgrain rotation, which is indicative of the deformation mechanisms which are active. ED loading is perpendicular to the dominant c-axis basal texture. Compressing perpendicular to the c-axis, the extension along the c-axis is accommodated by $\{10\bar{1}2\}$ twins. This twin system reorients the basal pole by

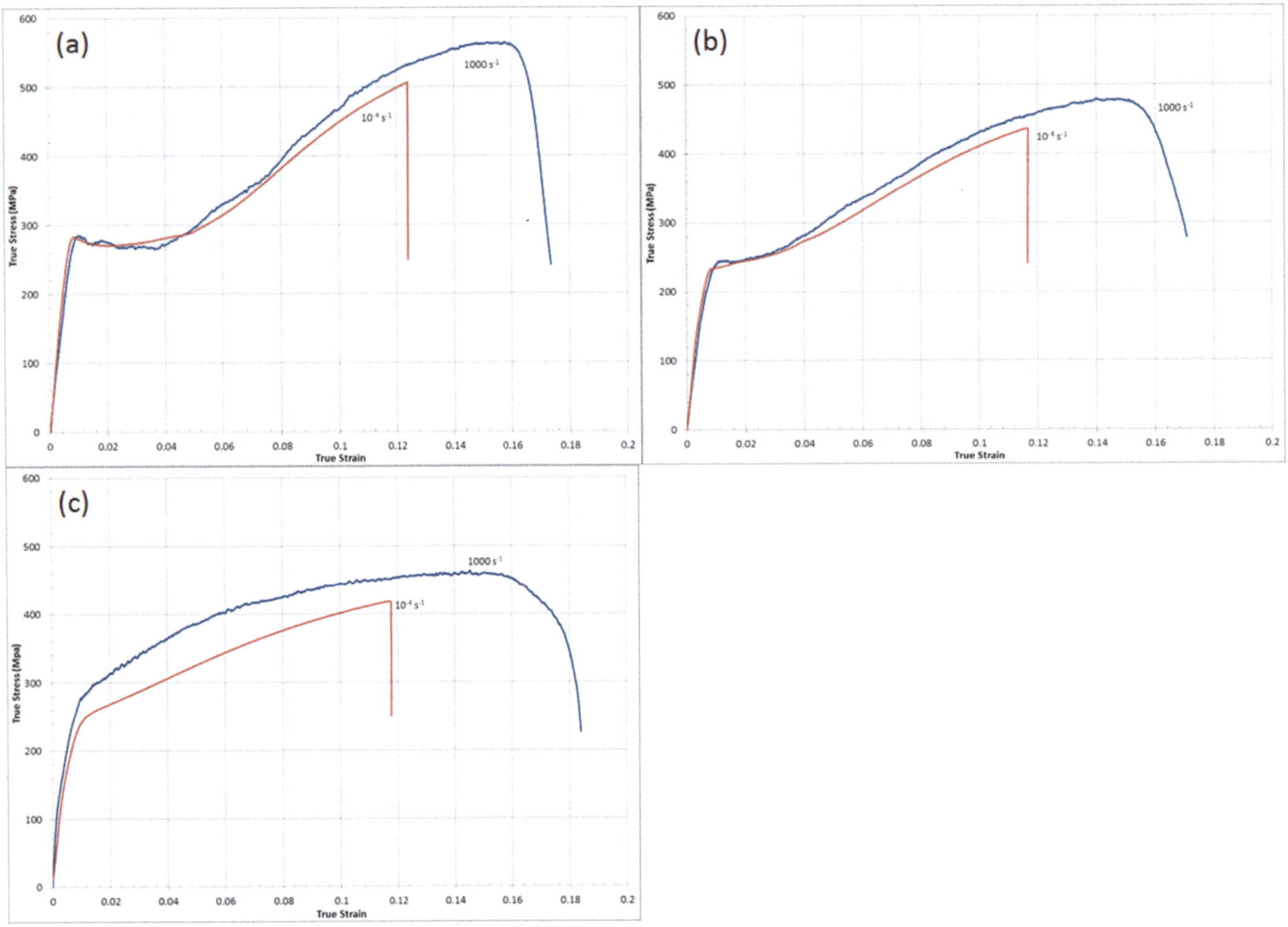

Fig. 35.1 Mechanical behavior of AMX602 in the (**a**) ED, (**b**) TD and (**c**) ND

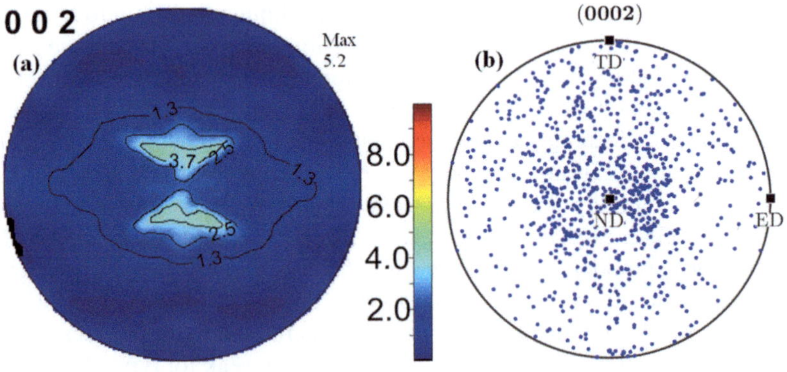

Fig. 35.2 As received (**a**) experimentally measured texture and (**b**) randomly sampled orientations used as input for the model. Both pole figures have the same orientation

86.6° in Mg so that it is nearly parallel to the loading direction [37, 38]. This is shown in Fig. 35.3. Note the center of the pole figures are the ED (loading direction), whereas the ND was at the center in Fig. 35.2. The experimental texture shows no residual peak in the ND at 3 % strain (Fig. 35.3a), whereas in the predicted texture from the model it does and an intensity peak along the ED has not formed. After 10 % true strain (Fig. 35.3b) the experimental texture is strongly aligned in the loading direction. The model still predicts a residual peak in the ND but one has formed in the ED. By 15 % compression (Fig. 35.3c) the model predicts texture to be only be aligned in the ED. The intensities of the model are weaker than the actual, which is likely due to it under predicting the extent of twinning. This was also reflected in the simulated stress-strain curves from the model [35]. The model does show the correct trend in the texture reorientation associated with twinning. We are not showing the pole figures subjected to quasi-static compression because the differences were minimal—they only had

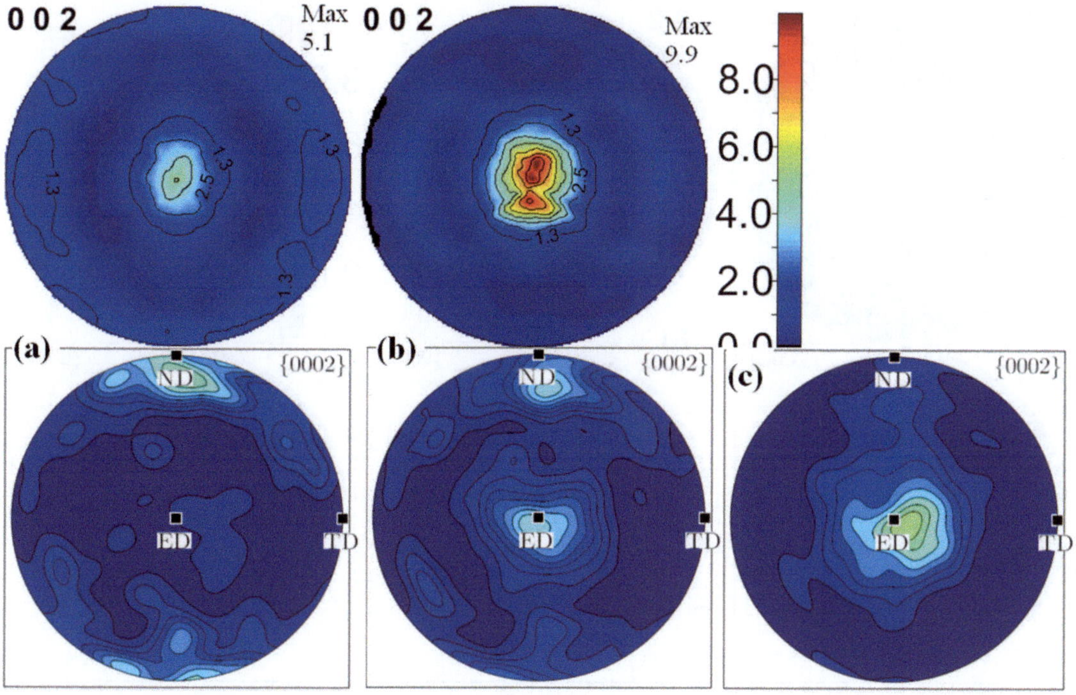

Fig. 35.3 Experimentally measured (*top*) and model predicted (*bottom*) texture of AMX602 subjected to (**a**) ~3 %, (**b**) ~10 % (**c**) ~15 % dynamic compression in the ED. All pole figures have the same orientation and scale

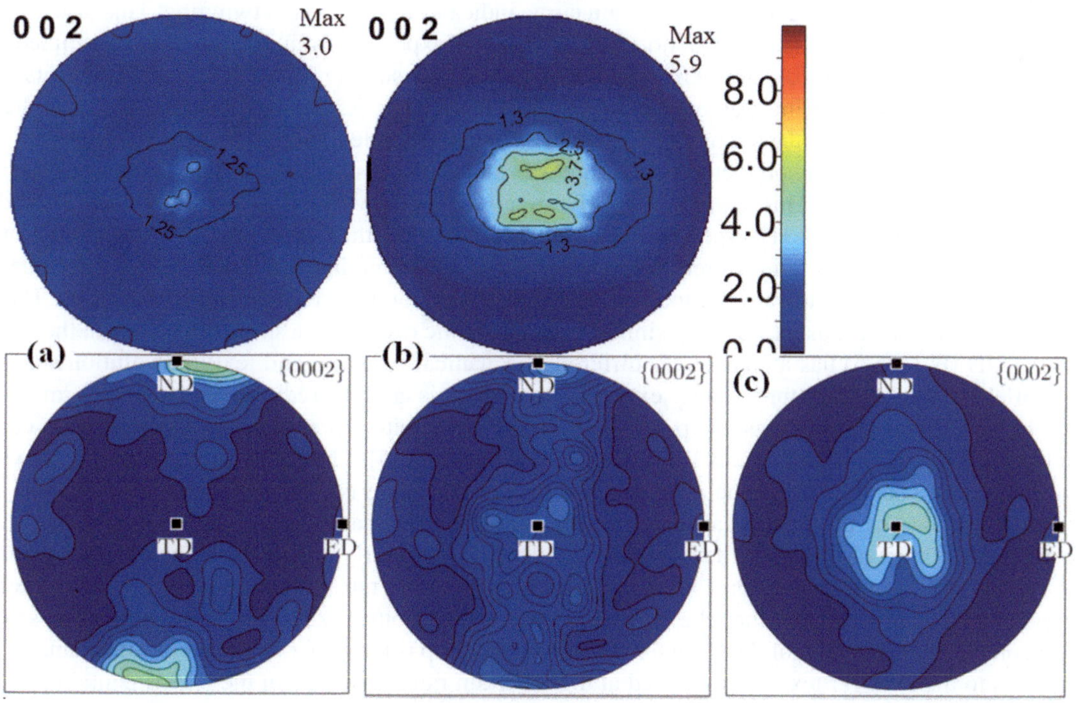

Fig. 35.4 Experimentally measured (*top*) and model predicted (*bottom*) texture of AMX602 subjected to (**a**) ~3 %, (**b**) ~10 % (**c**) ~15 % compression in the TD. All pole figures have the same orientation and scale

a small reduction in the intensity relative to the dynamic ones. This makes sense based on the small differences observed in the stress-strain response.

Figure 35.4 displays the texture after dynamic loading in the TD. Note, the TD is at the center of the pole figures. The same general trend is applicable in this loading direction, too. The loading path is still generally perpendicular to the basal

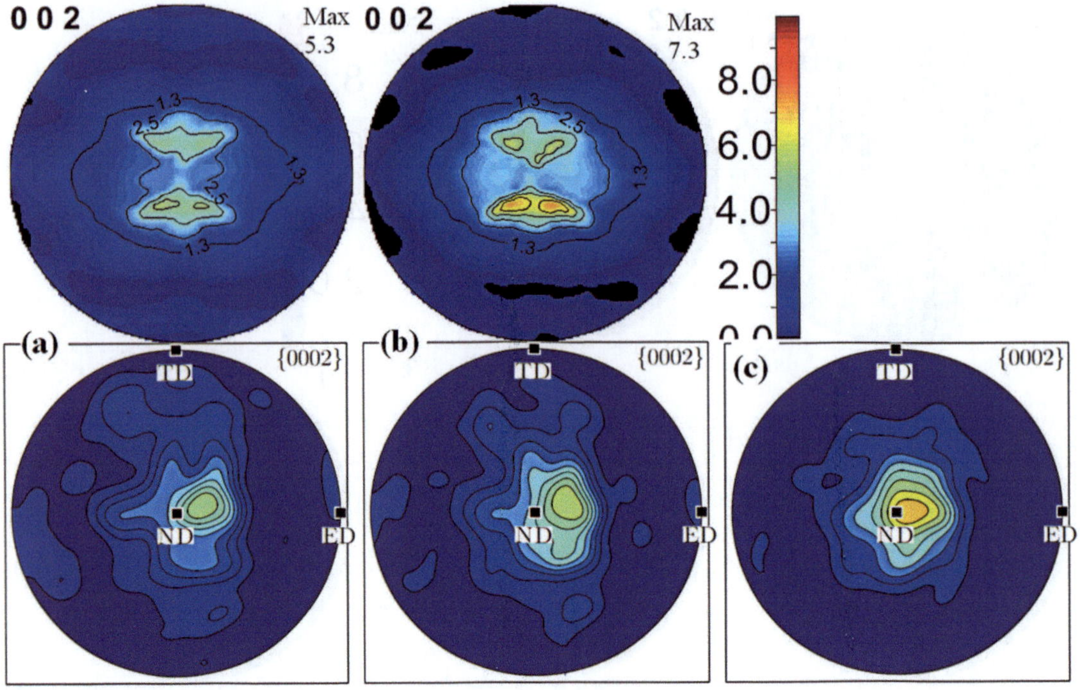

Fig. 35.5 Experimentally measured (*top*) and model predicted (*bottom*) texture of AMX602 subjected to (**a**) ~3 %, (**b**) ~10 % (**c**) ~15 % compression in the ND. All pole figures have the same orientation

texture, but the less pronounced plateau in the stress-strain curve indicates a reduction in twinning. This is also evident by the weaker reorientation of the texture. The reduction in twinning is expected from the randomly sampled orientations (Fig. 35.2b) because more of the c-axes are perpendicular to the ED versus the TD. The model lags in its prediction of the reorientation, as before. The experimental texture does show bifurcated peaks after 10 % strain similar to the as-received AMX602 but with a small increase in the intensity. As with the as-received pole figure, it is unclear whether this is real or an artifact of the MAUD calculation. The predicted pole figure does not show this. If it is real then it could be a holdover from the initial texture having this feature. This requires more investigation, but Brown et al. [37] and Yi et al. [38] did not see this when compressing coarse grained AZ31 perpendicular to the basal texture (their initial texture was typical of rolling and extrusion). Our material is, of course, a different alloy with much smaller grains. Notionally, the model does predict the correct trend of the texture rotating toward the loading direction as the compressive true strain increases. The quasi-static texture evolution is not shown because it was very similar to the dynamic except for a small difference in the intensity value.

Loading in the ND (Fig. 35.5) has a completely different mechanical response and texture evolution since the loading direction is parallel to the basal texture. Thus, extension twinning is greatly reduced and the deformation is mostly accommodated by slip. Basal and non-basal slip are predicted to contribute more or less equally above a couple percent true strain, and mostly non-basal slip at strains less than a couple percent [35]. The 90° reorientation of the texture does not occur so the c-axis texture stays in the same orientation and intensifies (the ND is at the center of the pole figure). The experimental texture maintains the same peak shape throughout the deformation which is a remnant of the as-received material. The predicted texture also stays in the ND while intensifying. Similarly as before, the pole figures from quasi-static loading in the ND are not shown because they were close to the high strain rate ones except for a weaker intensity. In this case the difference was somewhat larger than the other two loading directions. Likely the greater strain rate sensitivity does have some effect on the intensification of the basal texture. Non-basal slip is the only rate dependent mechanism in the model and this contributes to the stronger texture developed at higher strain rates observed in the experimental texture data.

35.5 Conclusions

In this study a high strength Mg alloy, AMX602 (Mg-6%Al-0.5%Mn-2%Ca), was produced by the spinning water atomization process (SWAP) and subsequently extruded into a plate. The microstructure was fine-grained (the vast majority of grains between 0.5 and 5 μm). Quasi-static and dynamic compressive loading showed AMX602 possesses high strength, reasonable ductility and low anisotropy in the principal processing directions. The initial texture of AMX602 was a weak rolling texture which was input into a computationally efficient plasticity model. The mechanical behavior, experimental texture and predicted texture all pointed to profuse extension twinning along the ED and TD, whereas slip dominated the response for compression along the ND. The model was able to correctly predict the reorientation of the texture toward the loading direction in the ED and TD, and only intensifying when along the ND. The model did under predict the texture intensity changes but the trends were correct.

Acknowledgement Thanks to Tyrone Jones (ARL) for providing the material used in this investigation and David Gray (ARL) for helping with the quasi-static tests. CHESS is supported by the NSF & NIH/NIGMS via NSF award DMR-1332208, and the MacCHESS resource is supported by NIGMS award GM-103485.

References

1. Mordike, B.L., Ebert, T.: Magnesium properties—applications—potential. Mater. Sci. Eng. A **302**, 37–45 (2001)
2. Hono, K., Mendis, C.L., Sasaki, T.T., Oh-ishi, K.: Towards the development of heat-treatable high-strength wrought Mg alloys. Scripta Mat. **63**, 710–715 (2010)
3. Yukutake, E., Kaneko, J., Sugamata, M.: Anisotropy and non-uniformity in plastic behavior of AZ31 magnesium alloy plates. Mater. Trans. **44**, 452–457 (2003)
4. Stanford, N., Barnett, M.: The origin of "rare earth" texture development in extruded Mg-based alloys and its effect on tensile ductility. Mater. Sci. Eng. A **496**, 399–408 (2008)
5. Agnew, S.R., Duygulu, O: Plastic anisotropy and the role of non-basal slip in magnesium alloy AZ31B. Int. J. Plasticity **21**(6), 1161–1193 (2005)
6. Bohlen, J., Nurnberg, M.R., Senn, J.W., Letzig, D., Agnew, S.R.: The texture and anisotropy of magnesium-zinc-rare earth alloy sheets. Acta Mater. **55**, 2101–2112 (2007)
7. Yukutake, E., Kaneko, J., Sugamata, M.: Anisotropy and non-uniformity in plastic behavior of AZ31 magnesium alloy plates. Mater. Trans. **44**, 452–457 (2003)
8. Barnett, M.: A rationale for the strong dependence of mechanical twinning on grain size. Scripta Mater. **59**, 696–698 (2008)
9. Barnett, M., Keshavarz, Z., Beer, A., Atwell, D.: Influence of grain size on the compressive deformation of wrought Mg-3Al-1Zn. Acta Mater. **52**, 5093–5103 (2004)
10. Chen, Y., Wang, Q., Peng, J., Zhai, C., Ding, W.: Effects of extrusion ratio on the microstructure and mechanical properties of AZ31 Mg alloy. J. Mater. Process Tech. **182**, 281–285 (2007)
11. Iwahashi, Y., Wang, J.T., Horita, Z., Langdon, T.G.: Principle of equal-channel angular pressing for the processing of ultra-fine grained materials. Scripta Mater. **35**, 143–146 (1996)
12. Kondoh, K., Aizawa, T.: Environmentally benign fabricating process of magnesium alloy by cyclical plastic working in solid-state. Mater. Trans. **44**, 1276–1283 (2003)
13. Kondoh, K., Kawabata, K., Oginuma, H.: Mechanical properties and texture of hot extruded magnesium alloys via RCP process in using coarse raw powder magnesium technology 2007. TMS. pp. 433–436 (2007)
14. Al-Maharbi, M., Karaman, I., Beyerlein, I., Foley, D., Hartwig, K., Kecskes, L., Mauthaudhu, S.: Microstructure, crystallographic texture, and plastic anisotropy evolution in an Mg alloy during equal channel angular extrusion processing. Mater. Sci. Eng. A **528**, 7616–7627 (2011)
15. Foley, D., Al-Maharbi, M., Hartwig, K., Karaman, I., Kecskes, L., Mathaudhu, S.: Grain refinement vs. crystallographic texture: mechanical anisotropy in a magnesium alloy. Scripta Mate. **64**(2), 193–196 (2011)
16. Hammond, V., Mauthaudhu, S., Doherty, K., Walsh, S., Vargas, L., Placzankis, B., Labukas, J., Pepi, M., Trexler, M., Barnett, B., Jones, T., Kecskes, L.: Ultrahigh-strength magnesium alloys for the future force: a final report on the 5-year mission program, 2009–2013. ARL-TR-6788. pp. 1–58 (2014)
17. Kusy, M., Grgac, P., Behulova, M., Vyrostkova, A., Miglierini, M.: Morphological variants of carbides of solidification origin in the rapidly solidified powder particles of hypereutectic iron alloy. Mater. Sci. Eng. A **375–377**, 599–603 (2004)
18. Rawers, J., Sauer, W., German, R.: Planar solidification of rapidly solidified powders. J. Mater. Sci. Lett. **16**, 1327–1329 (1993)
19. Wright, R.N., Anderson, I.E.: Age-hardening behavior of dynamically consolidated rapidly solidified Cu–2%Zr powder. Mater. Sci. Eng. A **114**, 167–172 (1989)
20. Kawamura, Y., Hayashi, K., Inoue, A., Masumoto, T.: Rapidly solidified powder metallurgy Mg97Zn1Y2 alloys with excellent tensile yield strength above 600 MPa. Mater. Trans. **42**, 1172–1176 (2001)
21. Endo, I., Otsuka, I., Okuno, R., Shintani, A., Yoshino, M., Yagi, M.: Fe-based amorphous soft-magnetic powder produced by spinning water atomization process (SWAP). IEEE Trans. Magnet. **35**(5), 3385–3387 (1999)

22. Yamauchi, I., Ohnaka, I., Kawamoto, S., Fukusako, T.: Hot extrusion of rapidly solidified Al-Si alloy powder by rotating-water-atomization process. Trans. Japan Inst. Met. **27**(3), 195–203 (1986)
23. Ohnaka, I., Yamauchi, I., Kawamoto, S., Fukusako, T.: Production and properties of rapidly solidified Al-4.5%Cu alloy powder by the rotating-water-atomization process. J. Mat. Sci. **20**, 2148–2158 (1985)
24. Matsuura, K., Kudoh, M., Kinoshita, H., Takahishi, H.: Precipitation of Si particles in a super-rapidly solidified Al-Si hypereutectic alloy. Mater. Chem. Phys. **81**(1–2), 393–395 (2003)
25. Kondoh, K., Hamada, E.-S.A., Imai, H., Umeda, J., Jones, T.: Microstructures and mechanical responses of powder metallurgy non-conbustive magnesium extruded alloy by rapid solidification process in mass production. Mater. Design **31**, 1540–1546 (2010)
26. Elsayed, A., Kondoh, K., Imai, H., Umeda, J.: Microstructure and mechanical properties of hot extruded Mg-Al-Mn-Ca alloy produced by rapid solidification powder metallurgy. Mater. Design **31**, 2444–2453 (2010)
27. Elsayed, A., Umeda, J., Kondoh, K.: The texture and anisotropy of hot extruded magnesium alloys fabricated via rapid solidification powder metallurgy. Mater. Design **32**, 4590–4597 (2011)
28. Ayman, E., Junko, U., Katsuyoshi, K.: Application of rapid solidification powder metallurgy to the fabrication of high-strength, high-ductility Mg-Al-Zn-Ca-La alloy through hot extrusion. Acta Mater. **59**, 273–282 (2011)
29. Jones, T., Labukas, J., Placzankis, B., Kondoh, K.: Ballistic and corrosion analysis of new military-grade magnesium alloys AMX602 and ZAXE1711 for armor applications. ARL-TR-5931. pp. 1–50 (2012)
30. Shen, J., Kondoh, K., Jones, T., Mathaudhu, S., Kecskes, L., Wei, Q.: Effect of strain rate on the mechanical properties of magnesium alloy AMX602. Mater. Sci. Eng. A **649**, 338–348 (2016)
31. ASM Metals Handbook Volume 8: Mechanical Testing and Evaluation)
32. Wenk, H.R., Lutterotti, L., Vogel, S.: Texture analysis with the new HIPPO TOF diffractometer. Nucl. Instrum. Meth. Phys. Res. Sect. A **515** (3), 575–588 (2003)
33. Graff, S., Brocks, W., Steglich, D.: Yielding of magnesium: from single crystal to polycrystalline aggregates. Int. J. Plast. **23**, 1957–1978 (2007)
34. Becker, R., Lloyd, J.: A reduce-order crystal model for HCP metals: application to Mg. Mech. Mater. **98**, 98–110 (2016)
35. Meredith, C.S., Lloyd, J.T., Sano, T.: The quasi-static and dynamic response of fine-grained Mg alloy AMX602: An experimental and computational study. Mater. Sci. Eng. A **673**, 73–82 (2016)
36. Jain, A., Agnew, S.R.: Modeling the temperature dependent effect of twinning on the behavior of magnesium alloy AZ31B sheet. Mater. Sci. Eng. A **462**, 29–36 (2007)
37. Brown, D.W., Agnew, S.R., Bourke, M.A.M., Holden, T.M., Vogel, S.C., Tomé, C.N.: Internal strain and texture evolution during deformation twinning in magnesium. Mater. Sci. Eng. A **399**, 1–12 (2005)
38. Yi, S.B., Davies, C.H.J., Brokmeier, H.G., Bolmaro, R.E., Kainer, K.U., Homeyer, J.: Deformation and texture evolution in AZ31 magnesium alloy during uniaxial loading. Acta Mater. **54**, 549–562 (2006)

Chapter 36
Failure Processes Governing High Rate Impact Resistance of Epoxy Resins Filled with Core Shell Rubber Nanoparticles

Erich D. Bain, Daniel B. Knorr Jr., Adam D. Richardson, Kevin A. Masser, Jian Yu, and Joseph L. Lenhart

Abstract Epoxy resins are classically toughened by rubber additives, but the effectiveness of rubber toughening tends to diminish with increasing strain rate, decreasing temperature, and decreasing matrix ductility. In this study we demonstrate that low loadings of 100–200 nm core-shell rubber (CSR) particulate additives can improve high strain rate (10^4–10^5 s^{-1}) impact resistance by nearly 200 % for epoxy resins with glass transition temperatures T_g in a range between 60 and 110 °C, without large reductions in T_g or stiffness. Size and surface chemistry of the CSR particles influence the ballistic response, with 200 nm diameter, weakly bound, poorly dispersed CSR particles providing the greatest toughening performance at low filler loadings and high rates. Impact resistance for a systematic series of CSR modified epoxies covers a transition from brittle to tough behavior, where the failure mechanism changes with effective fracture resistance. For brittle resins, failure is dominated by initiation of Hertzian cone fracture which depends strongly on fracture toughness K_{IC}, while for tough resins, failure is dominated by plastic yield at the impact site and is independent of fracture toughness above a minimum K_{IC} value of approximately 1.2–1.5 MPa-m$^{1/2}$. Interestingly, quasistatic mechanical properties are reasonably effective qualitative predictors of high rate impact resistance, suggesting that the toughening mechanisms of CSR particles are similar over the rates studied here. The insights gained from this study are valuable for design of next generation adhesives, polymers, and polymer composite matrices for lightweight protective applications.

Keywords Rubber toughened epoxies • High rate impact resistance • Failure mechanisms • Composites • Fracture

36.1 Introduction

Cross-linked epoxy polymers are major components in technologically important material packages such as fiber reinforced polymer composites (FRPCs), adhesives, coatings, and bulk plaques for protective systems servicing a wide range of military, aerospace, transportation, and construction industries [1–3]. In FRPCs, where epoxy resin acts as the matrix, the relationship between epoxy toughness and composite high rate impact resistance is complex. Tensile failure of primary yarns and interlaminar delamination tend to be two primary dissipative mechanisms, where the former is dominated by the fiber properties [4], while the latter is at least partially dependent on the matrix [5]. For example, Kinloch, Sprenger, and coworkers have shown that interlaminar fracture energy G_c for a range of fiber and weave types systematically increases with epoxy matrix fracture toughness, when the matrix is toughened with rubbery particles, silica nanoparticles, or both [6, 7]. These authors additionally demonstrated improved high rate impact resistance of carbon fiber composites formulated with a "hybrid" epoxy matrix containing both rubber and silica toughening particles [7]. While optimization of fibers and laminates is well advanced [8], less work has been done to understand and optimize high rate impact resistance of the epoxy matrix itself [9–12].

Extensive research has focused on improving fracture toughness of epoxies through hard and soft particulate additives [13–19], as well as the manipulation of the epoxy chemistry and chain structure [20, 21]. Rubber toughening is a common approach for improving the quasi-static fracture toughness in epoxy thermosets [22, 23], and is capable of improving fracture resistance by a factor of 2 or more through incorporation of phase-separated liquid rubber particles [13–15, 24, 25], block copolymer micelles [26–29], or pre-formed core-shell rubber (CSR) particles [19, 30–32]. A primary mechanism of thermoset toughening by particulate fillers is particle cavitation or debonding of the matrix from the particles, both of which relieve the triaxial constraint ahead of a propagating crack and initiate plastic void growth and shear band formation in

E.D. Bain (✉) • D.B. Knorr Jr. • A.D. Richardson • K.A. Masser • J. Yu • J.L. Lenhart
U.S. Army Research Laboratory, 4600 Deer Creek Loop, Aberdeen Proving Ground, Aberdeen, MD 21005, USA
e-mail: erich.d.bain.civ@mail.mil

D. Casem et al. (eds.), *Dynamic Behavior of Materials, Volume 1*, Conference Proceedings of the Society for Experimental Mechanics Series, DOI 10.1007/978-3-319-41132-3_36

matrix ligaments between voids [24, 33, 34]. However, particle cavitation does not guarantee the beginning of matrix shear deformation, but must produce a stress concentration sufficiently close to the matrix yield stress to induce local shear deformation, making the shear strength (i.e., ductility) of the matrix an important parameter [19, 35].

More ductile resins are generally more effectively toughened at quasistatic rates via the matrix void growth and matrix shear banding mechanism described above, with the molecular weight between crosslinks (M_c) being a predictor of toughenability [25, 36]. This has resulted in poor additive toughening of highly crosslinked materials over a wide temperature range relative to glass transition temperature T_g, while readily tough enable resins have showed improved properties even far below T_g [37]. Furthermore, while quasi-static fracture has been intensively investigated, high rate performance is less well understood. The effectiveness of rubber toughening tends to decrease as strain rate increases or temperature decreases [14, 15, 27, 38, 39]. Therefore, a key question for protection applications is, "Does rubber toughening work at ballistic impact strain rates over the broad temperature ranges required for practical applications?"

Our previous work [10–12] has shown that improvements in ballistic performance of crosslinked polymer networks tend to come at the expense of reductions in modulus and/or T_g, which are undesirable for many structural applications. Therefore, the present work [40] is motivated by a need to improve high strain rate impact resistance within an epoxy matrix, over a broad range of operational temperatures, while simultaneously maintaining a relatively high glass transition temperature and glassy modulus. Our model system is composed of low loadings of core-shell rubber (CSR) nanoparticles integrated into diamine cured DGEBA resins, which exhibit a broad range of glass transition temperatures. The epoxy used in this study, DGEBA, is commonly found in many applications, and the T_g and M_c of the network were tuned by varying the diamine hardener.

36.2 Experimental

All materials used in this experiment were purchased from Sigma-Aldrich unless noted otherwise. Diglycidyl ether of bisphenol A (DGEBA) was purchased from Hexion Inc. under the product name EPON 825. PACM (4,4′-methylenebis (cyclohexylamine)) was purchased from Air Products. Samples of polyetheramines, trademarked as Jeffamine D-230 and D-400, were supplied by Huntsman Chemical. Kane Ace MX-125, MX-156, and MX-257 are core-shell rubber (CSR) particle formulations pre-dispersed in DGEBA, and were provided by Kaneka, USA. A listing of the CSR particle characteristics is given in Table 36.1. MX-125 and MX-156 contain a small percentage of epoxy groups in the PMMA shell, while MX-257 has no epoxy or other reactive groups in the PMMA shell. The chemical structures of all monomers used are given in Fig. 36.1. As all samples contain DGEBA, they will be referred to throughout the paper by the amine hardener and what weight percent, if any, of CSR they contain.

The preparation for both quasistatic and high rate specimens was identical. Stoichiometric amounts of epoxy (with the desired amount of CSR filler) and diamine were heated to 60 °C. They were then mixed using an overhead mixer for 5 min. The mixture was degassed, poured into stainless steel molds (6 in. × 6 in. × 0.25 in.) that had been polished, treated with Frekote 55NC release agent, and pre-heated to 60 °C, and degassed once more. All formulations were cured under a nitrogen purge with a cure cycle of 80 °C for 2 h, 150 °C for 8 h, 200 °C for 2 h.

Ballistic tests [41] were carried out using a 0.22 caliber gas gun. The projectile was a 5.56 mm diameter stainless steel ball bearing (Type 302, 0.69 g) and its speed was measured by Doppler radar (BR-3502, Infinition Inc.). The polymer specimen dimensions were 62.5 mm × 62.5 mm × 6.4 mm. Each specimen was sandwiched between two aluminum plates with a 50.8 mm diameter circular aperture. An aluminum foil witness plate (0.0108 mm thick) was placed 50.8 mm behind the target. Ballistic failure of the epoxy was assigned when the projectile or a fragment of the epoxy target penetrated the witness

Table 36.1 MX series core-shell rubber (CSR) nanoparticle properties

Product	Loading in DGEBA (%wt)	Nominal particle size (nm)	Measured particle size (nm)[a]	Core composition	Core T_g (°C)[b]	Shell composition	Shell functionalization
MX-125	25	100	104 ± 10	PS/PB	−50	PMMA	Some epoxy
MX-156	25	100	104 ± 7	PB	−75	PMMA	Some epoxy
MX-257	37	200	167 ± 8	PB	−71	PMMA	None

[a]Number average diameters measured from SEM images of as-received particle mixtures spincast from toluene and cured with PACM
[b]Taken from low temperature peaks in DMA tan δ curves of particle mixtures cured with D230

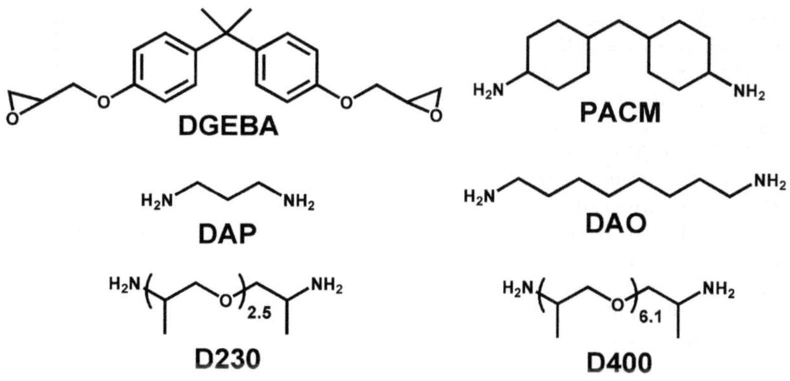

Fig. 36.1 Structures of epoxy (DGEBA) and diamine curing agents used in this study

plate. Twelve targets were used of each formulation. The V_{50} ballistic limit was calculated by taking the arithmetic mean of the three highest non-failed and the three lowest failed impact velocities. The kinetic energy associated with V_{50}, i.e., KE_{50}, was obtained from the V_{50} value and the mass of the projectile m, as $KE_{50} = 0.5\,\mathrm{m}V_{50}^{2}$. Selected impact events were imaged at 180,000 frames per second using a FASTCAM SA1.1 model 675 K-M1 high speed video camera (Photron USA, Inc.) mounted behind the specimen back face, with specimens illuminated from the front.

Dynamic mechanical analysis (DMA) was conducted to determine T_g using a TA Instruments Q800 on samples that were nominally 35 mm × 12 mm × 3.25 mm for the rubbery region and 35 mm × 12 mm × 1.5 mm for the glassy region. The quasistatic mode I fracture toughness of epoxy resins was determined according to ASTM D5045-99 using the single-edge-notch bending (SENB) geometry with dimensions of 6.35 mm × 12.7 mm × 63 mm ($B \times W \times L$) and a cross head speed of 10 mm/min. The pre-crack was initiated by tapping with a cryogenically frozen razorblade. Tensile specimens were molded to the dimensions of ASTM D638-10, Type IV. Cross head speed was 5 mm/min and digital image correlation (DIC) was used to obtain strain values. Compression samples 12 mm in diameter and 6 mm thick were prepared and tested using ASTM D695-10 as a guide, with cross head speed of 1.3 mm/min. Selected fracture surfaces of tested specimens were sputtered with an approximately 5 nm thick film of gold-palladium and were imaged in a Hitachi S-4700 scanning electron microscope using an accelerating voltage of 3 kV.

36.3 Results and Discussion

36.3.1 Effect of CSR Particles on Ballistic and Quasistatic Performance as a Function of Glass Transition Temperature

We produced a series of DGEBA resins cured with the diamine hardeners shown in Fig. 36.1, both neat and filled with MX-125 such that the overall CSR nanoparticle concentrations are 1, 3, and 5 wt%. Results of ballistic impact tests for this series of resins are plotted in Fig. 36.2 in the form of KE_{50}, the projectile kinetic energy associated with a 50 % probability of target failure, as a function of $T-T_g$, where T is the measurement temperature of 22 °C. For clarity, the results are normalized using the KE_{50} value of unfilled PACM/DGEBA as a standard. The"x" symbols shown in Fig. 36.2 represent data for previously studied [10] diamine cured DGEBA resins measured at 22 °C without any rubber fillers, to provide context for the present measurements. The step change in KE_{50} values for unmodified resins at about 25 °C below T_g is similar to that observed for other materials such as polyurethane coatings [42–45].

T_g values of all resins (measured by DMA) do not change significantly upon addition of the CSR particles. KE_{50} values of the highest T_g resins, including those cured with PACM, DAP, and DAO, do not show significant improvement with CSR nanoparticle modification up to 5 wt% over the relatively poor performance of the unfilled resins. In contrast, KE_{50} of DGEBA cured with D230 more than doubles for the resins filled with 3 and 5 wt% nanoparticles, while the 1 wt% sample shows a modest improvement. Interestingly, the modified D400 cured resin does not show significant improvement over the already high normalized KE_{50} value of about 3.5.

Although mode I fracture toughness is not strictly analogous to impact resistance, the two quantities can be related via the complex set of fracture processes involved in impact failure [46–48]. In Fig. 36.3 we compare literature values of quasistatic mode I fracture toughness (K_{IC}) for various epoxies cured with different monomers and diamine hardeners, covering a wide range of $T-T_g$ values, next to quasistatic K_{IC} values measured for the neat and CSR-modified samples in this study. The

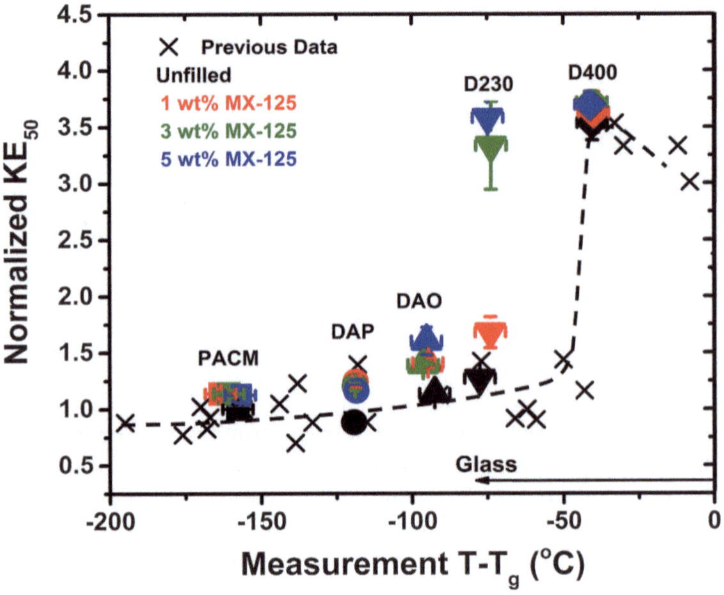

Fig. 36.2 Normalized high rate impact resistance KE_{50} of MX-125 core-shell rubber modified DGEBA cured with various diamines as a function of $T - T_g$: PACM (*filled square*), DAP (*filled circle*), DAO (*filled up-pointing triangle*), D-230 (*filled down-pointing triangle*), and D-400 (*filled diamond*). Previous data for unfilled resins from [10] are represented by (x). The *dotted line* is intended as a guide to the eye for the trend in KE_{50} of the unfilled resins

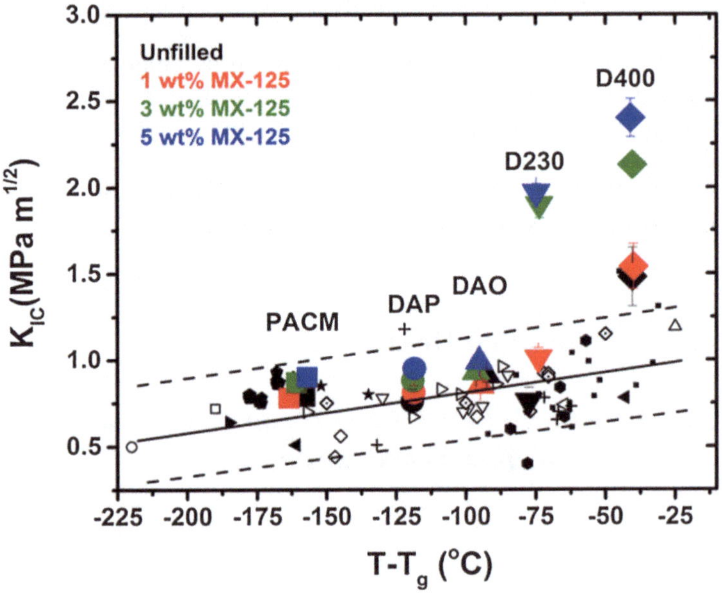

Fig. 36.3 Quasistatic fracture toughness K_{IC} as a function of $T - T_g$ for CSR-modified epoxies (large symbols same as Fig. 36.2). *Small black* symbols represent unmodified epoxy/diamine systems from the literature [40]. The *solid line* is a guide for the eye illustrating the approximate trend in the literature data, while the region within the *dotted lines* encompasses ~95 % of the data

results agree well with general trends in the literature, showing that matrix toughenability due to rubber additives tends to increase with increasing $T - T_g$ [19, 28, 29]. Higher T_g resins including PACM, DAP, and DAO cured DGEBA exhibit little toughening effect in quasistatic K_{IC} as a result of CSR addition. D230 shows a minimal toughening effect at 1 wt% but a high degree of toughening at 3 wt% and 5 wt% modifier, reminiscent of the trends observed for KE_{50} of this resin. However, quasistatic K_{IC} of D400 also increases dramatically with increasing loading of CSR particles, in contrast to the negligible effect caused by the same filler on KE_{50} of this resin.

36.3.2 *Effect of CSR Particle Size and Surface Functionality on Ballistic and Quasistatic Performance of DGEBA-D230 Resin*

Because the DGEBA/D230 system exhibits large increases in both quasistatic fracture toughness and ballistic impact resistance with the addition of MX-125, we have investigated this resin further with additional CSR nanoparticle types.

These particle types include MX-156, which is similar to MX-125 in size (nominally 100 nm) and shell functionality (lightly epoxidized), but whose polybutadiene core has a lower T_g than the polystyrene-butadiene core of MX-125, and MX-257, which is larger (nominally 200 nm) and features an unfunctionalized PMMA shell and soft polybutadiene core. These attributes are summarized in Table 36.1.

Normalized room temperature KE_{50} data for DGEBA/D230 modified with each of the three CSR nanoparticle types is shown as a function of particle loading in Fig. 36.4. The performance of the two 100 nm particle additives (MX-125 and MX-156) is very similar: KE_{50} increases with loading up to 3 wt%, where a plateau KE_{50} value of about 3.5 is reached. In the case of D230 modified with the larger MX-257 particles, a similar plateau in KE_{50} is reached but for loadings as low as 0.5 wt %, suggesting that either particle size, surface functionalization, or both play important roles in the impact resistance.

Quasistatic fracture toughness K_{IC} of D230 as a function of filler loading for the three CSR particle types are shown in Fig. 36.5. The effects of MX-125 and MX-156 are similar over most of the loading range, reaching a plateau in quasistatic K_{IC} at a loading of about 5 wt%. The behavior of D230 containing MX-257 particles follows a similar trend, with slightly higher K_{IC} values. Unlike KE_{50}, which increase sharply at loadings of about 2 wt% for MX-125 and MX-156 and 0.5 wt% for MX-257, the increase in quasistatic fracture toughness with loading is more gradual for all three particle types.

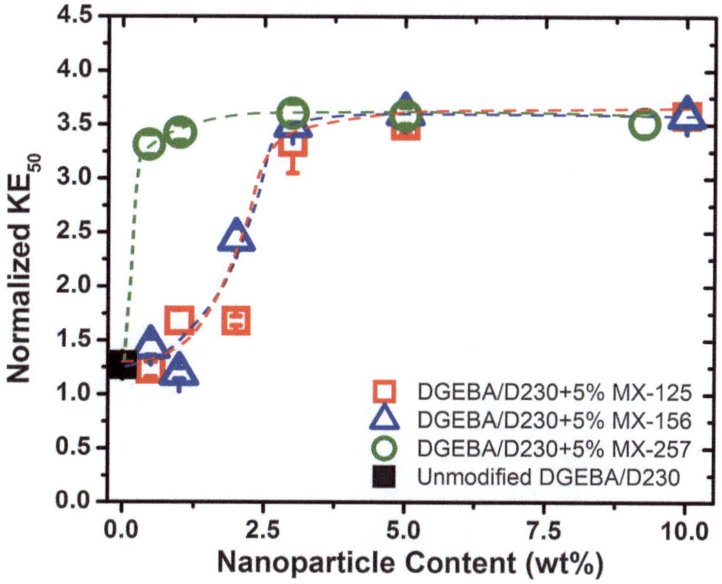

Fig. 36.4 Normalized high rate impact resistance KE_{50} as a function of CSR nanoparticle content for modified DGEBA/D230. The *dotted lines* are intended as guides for the eye

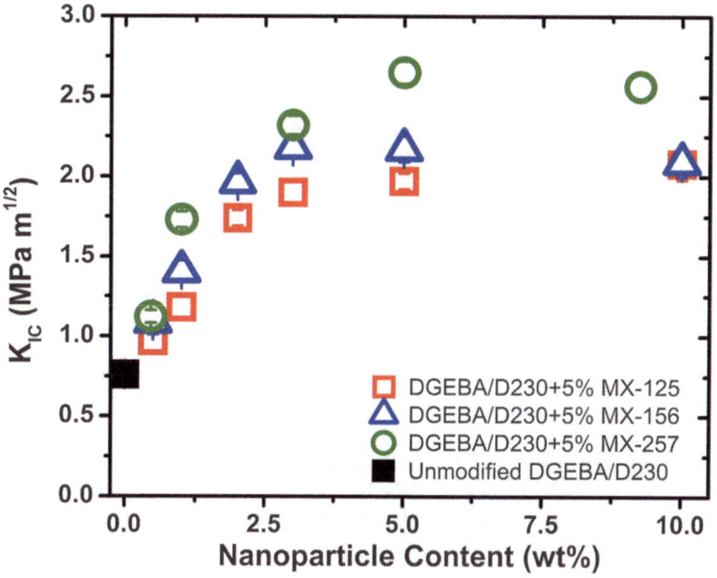

Fig. 36.5 Quasistatic fracture toughness K_{IC} as a function of CSR nanoparticle content for modified DGEBA/D230

36.3.3 *Damage Types Associated with Ballistic Impact*

In order to better understand the complex relationship between high rate impact resistance and mode I fracture toughness, it is necessary to understand the sequence of failure processes involved in ballistic impact events and the governing parameters for each type of damage. Photographs and schematics illustrating various damage types in ballistically impacted epoxy plates are shown in Fig. 36.6. Three types of ballistic damage will be considered in our treatment: radial cracking, cone cracking, and plastic deformation at the impact site.

Upon impact with steel spheres during V_{50} tests, all materials in this study formed radial cracks similar to those shown in Fig. 36.6a, b. Radial cracks are known to initiate at the target back face due to bending stress ahead of the impact site [46, 49]. In this study neat and modified epoxy tiles impacted at velocities well below the failure velocity V_{50} tended to form radial cracks only, whereas targets impacted at higher velocities formed both radial cracks and cone cracks. Radial cracks tended to originate from the $r = 0$ point immediately ahead of the impact site on the back face, propagating toward the sample edges and passing smoothly through the orthogonal cone cracks, if present, as seen for example in Fig. 36.6b. From these observations we hypothesize that radial cracks initiate prior to cone crack initiation, consistent with predictions for targets whose thickness is on the order of the projectile diameter [46, 50]. Furthermore, these observations suggest that radial cracking is a necessary, but not sufficient condition for target failure according to the V_{50} test.

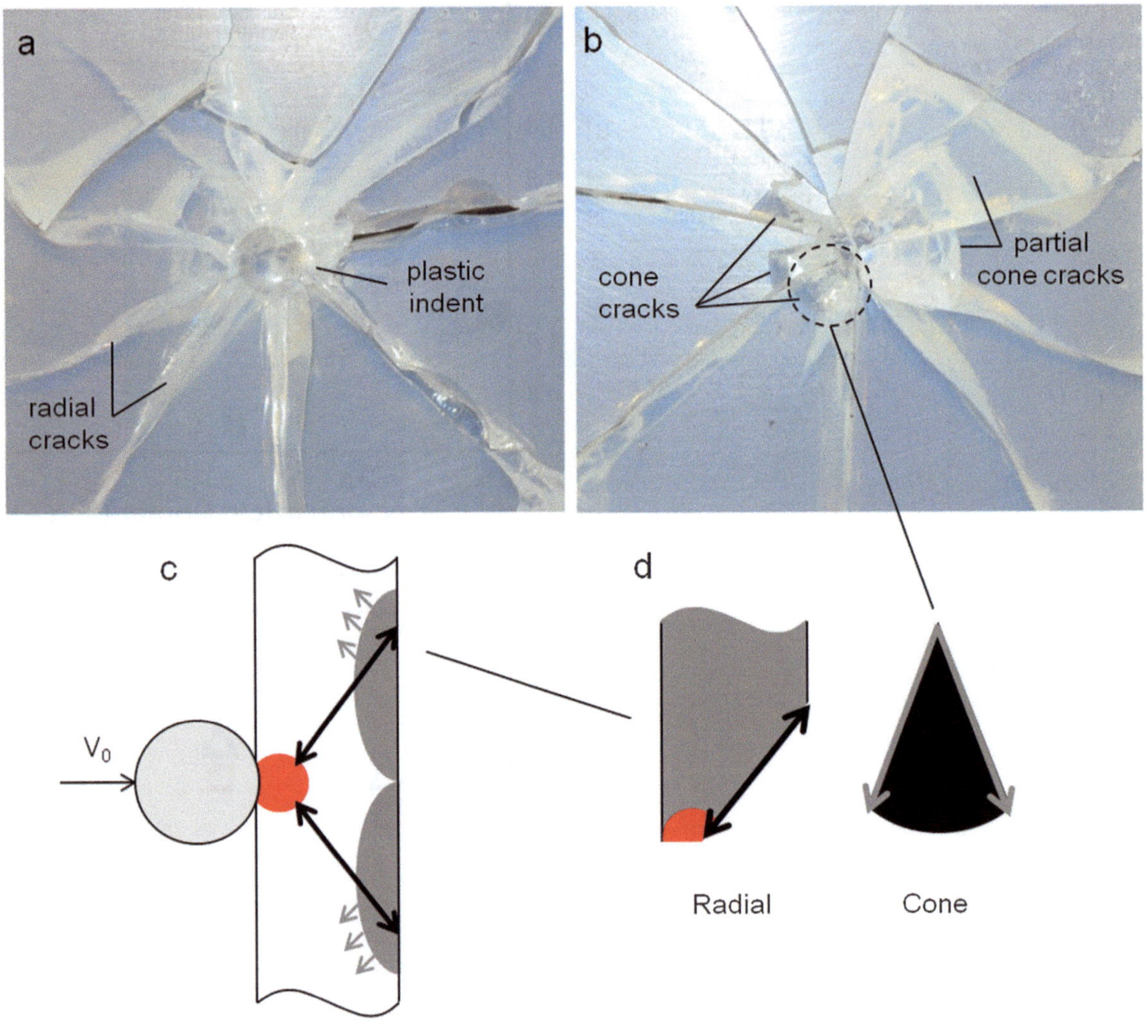

Fig. 36.6 (**a**) Front face of ballistically impacted DGEBA-D230 containing 5 wt% MX-125 CSR particles, showing plastic indentation from projectile and radial cracks. (**b**) Back face of same specimen showing cone cracks and radial cracks. (**c**) Schematic illustrating damage types during ballistic impact: plastic zone (*red*), cone cracks (*black*), and radial cracks (*gray*). (**d**) Schematic of fracture surface sections taken for SEM imaging

If the stress intensity at a flaw near the impact site exceeds the fracture toughness K_{IC} of the target, a ring crack will initiate. This crack tends to propagate along a cone-shaped trajectory opening from the front face toward the back. For ideally elastic (Hertzian) contact, the critical impact velocity for cone crack initiation V_c can be estimated from analytical relationships [46, 48]:

$$V_C \approx \frac{950 K_{IC}^{5/3}}{R \rho_1^{1/2} E^{*7/6}} \tag{36.1}$$

where K_{IC} is the fracture toughness of the target, R is the projectile diameter, ρ_1 is the projectile density, E^* is a reduced contact modulus [46], and the subscripts 1 and 2 designate the projectile and target, respectively.

The specimen shown in Fig. 36.6a features a circular indentation at the impact point on the front face, indicative of plastic deformation. Similar plastic indentations were observed for all targets with normalized KE_{50} between 3 and 4, whereas the most brittle targets with low ballistic limits had little or no indentation when impacted at V_{50}. Compton et al. [46] have used finite element simulations and experiments to investigate the critical velocity for plastic deformation (yielding) of ceramic tiles impacted with metal spheres. For projectiles in a high velocity regime, yield initiation was found to follow

$$V_Y \approx \frac{3\sigma_{Y,2}}{(\rho c)'} \tag{36.2}$$

Where $\sigma_{Y,2}$ is the yield stress of the target and $(\rho c)'$ is a reduced acoustic impedance of contact [46]. It is important to note that Eq. (36.2) was derived from simulation results for ceramic targets, whose yield stress is 5–10 times that of the metal projectile. In that case, the projectile deforms plastically prior to target yielding at intermediate and high velocities, causing significant deviations from the ideal Hertzian contact radius and pressure distribution. In our system, deviations from Hertzian contact also occur at high velocities, but it is due solely to plastic deformation of the epoxy targets, whose yield stresses are an order of magnitude lower than that of the steel projectile. Hence, Eq. (36.2) may not quantitatively describe the yield initiation threshold of our system. However, qualitative insight from Eq. (36.2) is valuable, and an essential prediction of these equations is that the critical velocity for yield initiation is a relatively simple function of target yield stress.

36.3.4 High Speed Videography of Ballistic Failure Events

Failure processes occurring during ballistic impact of selected specimens were imaged by a high speed video camera at 180,000 frames per second, as shown in Fig. 36.7. These images demonstrate that radial and cone cracks initiate within 10's

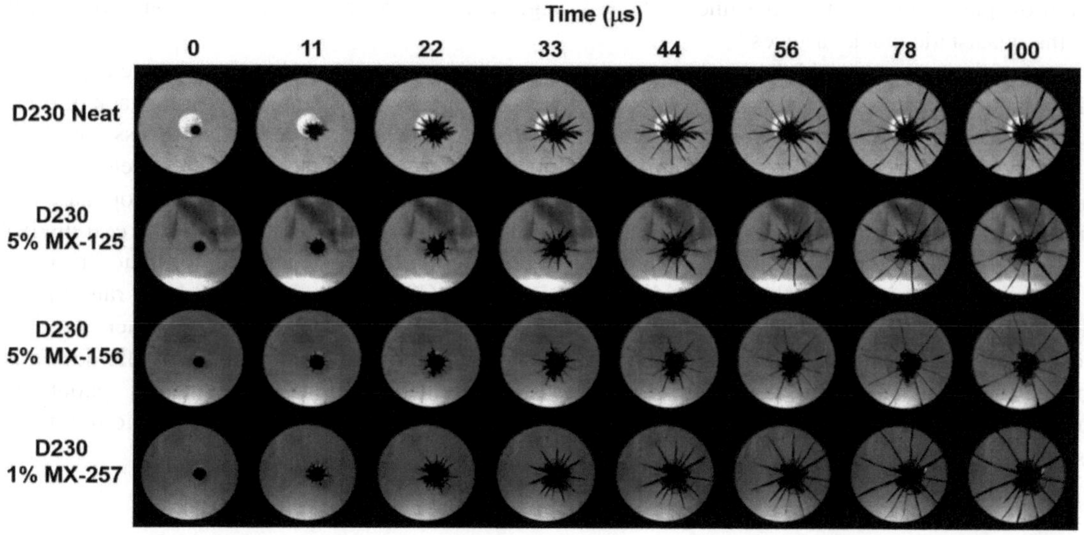

Fig. 36.7 High-speed camera images taken from the back face of samples impacted near V_{50}, including neat D230 (*top row*) and D230 containing 5 wt% MX-125 (*row 2*), 5 wt% MX-156 (*row 3*), and 1 wt% MX-257 (*row 4*), arranged as a function of time

of microseconds following impact, suggesting minimal time is available for plastic deformation processes associated with rubber toughening. These images provide visual evidence that radial cracks initiate prior to cone cracks during high rate impact. Material sections formed by cone cracks ahead of the impact site were observed to eject from samples at relatively long times (>2 ms, images not shown) and penetrate the witness foil, implying that cone cracking is a primary mechanism of specimen failure according to the standard V_{50} test for the materials in this study.

36.3.5 Fracture Surface Analysis of DGEBA-D230 Modified with Various CSR Particle Types

We now turn our focus to morphological evidence for the role played by the CSR additives in the failure processes described above, using scanning electron microscopy (SEM) images of filled and unfilled D230 fracture surfaces.

SEM images at high magnification of quasistatic K_{IC} process zones in rubber-modified DGEBA-D230 are shown in the top row of Fig. 36.8. Detailed analysis of void sizes formed on these fracture surfaces [40], suggest that MX-125 and MX-156 CSR particles form voids via cavitation of the rubber cores. On the other hand, results suggest voids around MX-257 particles are formed via matrix debonding around weakly bonded particles, rather than solely cavitation of the particle cores. Both of these void formation mechanisms are consistent with classical rubber toughening via plastic void growth and shear banding [24, 33]. The thermodynamically unfavorable particle-matrix interface causes MX-257 to form aggregates of 5–10 particles during curing, whereas MX-125 and MX-156 particles are well dispersed in the D230 matrix. Greater toughening effectiveness of aggregated, weakly bonded rubber particles has been previously observed due to an apparent tendency to depress yield stress and increase the size of the crack-tip plastic zone [31, 51]. Bright lines on the MX-257 filled fracture surface display clear evidence of significant matrix plastic deformation, consistent with high fracture resistance. Neat D230 shows no evidence of deformation or toughening mechanisms in high magnification SEMs [40].

The second row of Fig. 36.8 shows the unstable crack propagation regions (far ahead of the precrack tip [40]) of the quasistatic K_{IC} fracture surfaces. In this region of MX-125 and MX-156 filled systems there is little evidence of particle cavitation or matrix plastic deformation, consistent with rapid, unstable propagation of the crack front after reaching K_{IC}. On the other hand, the unstable fracture region of MX-257 filled D230 features matrix-particle debonding as well as lines indicating matrix deformation. These results suggesting that the toughening effect of MX-257 may be less rate-sensitive than those of MX-125 and MX-156, since crack propagation is known to speed up after becoming unstable.

The third row of high magnification SEM images in Fig. 36.8 are taken from cone crack surfaces near the ballistic impact sites. The fourth row of Fig. 36.8 contains SEM images taken near the back face on the same cone cracks. Particle morphologies on these cone cracks are similar to those seen in unstable fracture regions of quasistatic K_{IC} samples (row 2), but with a greater degree of contrast in the interior of MX-125 and MX-156 particles, suggesting the particle cores have cavitated to form shallow voids. Voids around MX-257 particles are similar to or slightly smaller than those in the unstable fracture region of quasistatic K_{IC} test, but lines indicating significant matrix plastic deformation appear, similar to those observed on the quasistatic crack surfaces.

No regions of greater average void diameter, analogous to process zones observed for quasistatic K_{IC} samples, could be found on any ballistic fracture surface specimens, including near the impact site. The reason for the lack of observed process zones is likely due to the time available for stable crack growth; in quasistatic K_{IC} tests the process zone develops over a period of 1–4 s, during which load increases steadily as a function of displacement until K_{IC} is met and unstable fracture begins. Conversely, as shown in Fig. 36.7, in the KE_{50} test radial and cone cracks initiate and propagate on the order of microseconds. Therefore any corresponding process zone would be vanishingly small due to the extremely limited time available for stable crack growth, as well as for cavitation/debonding and matrix plastic deformation processes. Since no observable process zone is formed, the toughening mechanism of CSR particles during high rate impact cannot be determined from these results. Two reasonable hypotheses are that (1) CSR particles do resist crack initiation during the initial microseconds of high rate impact via stable fracture, in which void formation and matrix plastic deformation occur in a small process zone that may be lost or destroyed during subsequent failure processes such as spallation, or (2) CSR particles continuously resist crack growth via a limited extent of void formation and matrix plastic deformation over the entire course of crack propagation [40].

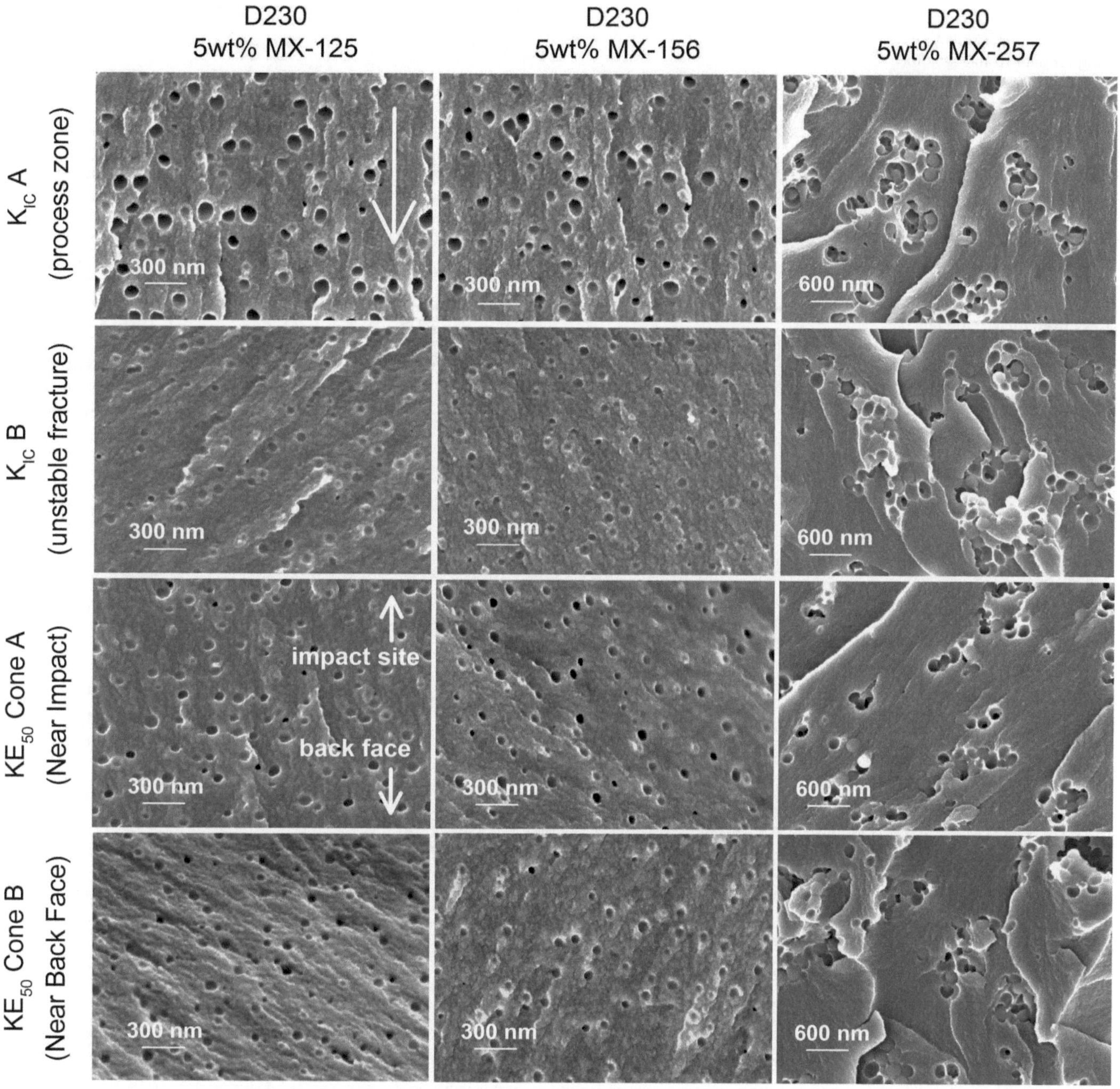

Fig. 36.8 High magnification scanning electron microscopy (SEM) images including fracture surfaces of DGEBA-D230 modified with 5 wt% each of MX-125 (*left column*), MX-156 (*center column*), and MX-257 (*right column*). Process zone (*row 1*) and unstable fracture (*row 2*) images are from quasistatic *KIC* fracture surfaces. Near impact (*row 3*) and near back face (*row 4*) images are from ballistic *KE50* cone crack surfaces. *Arrows* indicate the direction of crack propagation for *KIC* images, and the orientation relative to impact site and back face for cone crack images

36.3.6 Correlations Between Quasistatic and Ballistic Performance

Given the preceding investigations of impact failure processes and fracture surface morphology, we are now in a better position to understand the relationships between ballistic impact resistance and quasistatic mechanical properties of the rubber toughened epoxies. To this end, Fig. 36.9 shows a plot of KE_{50} measurements from Figs. 36.2 to 36.4 as a function of quasistatic K_{IC} measurements from Figs. 36.3 to 36.5. Plotted in this way, most of the data fall into two groups: brittle resins with $K_{IC} \leq 1$ MPa-m$^{1/2}$ and $KE_{50} < 2$, and tough resins with $K_{IC} > 1$ MPa-m$^{1/2}$ and $KE_{50} \approx 3.5$. The red dashed curve in Fig. 36.9 is a least squares fit of Eq. 36.1 using different values of Young's modulus for selected filled and unfilled resins

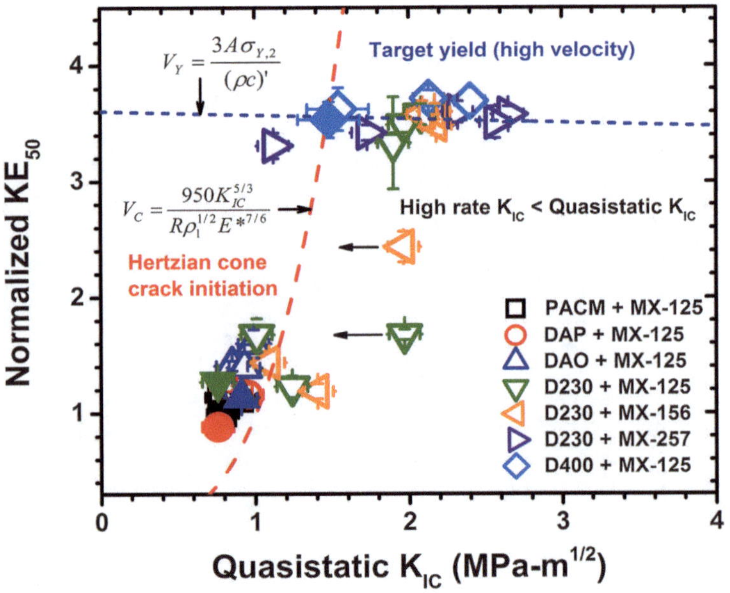

Fig. 36.9 Plot of KE_{50} as a function of quasistatic K_{IC} using all data from Figs. 36.2 to 36.5. *Closed symbols* represent unmodified resins while *open symbols* correspond to resins modified with CSR particles, with symbol shapes indicated in the legend. The lines are best fits to Eqs. (36.1) (*red dotted line*) and 36.3 (*blue dashed line*), using experimental values of Young's modulus and yield stress from quasistatic tensile tests, summarized in Table 36.2

Table 36.2 Quasistatic mechanical properties measured for selected filled and unfilled DGEBA resins at room temperature

Diamine	Modifier	Young's modulus (MPa)	Tensile yield stress (MPa)	Compressive yield stress (MPa)	K_{IC} (MPa m$^{1/2}$)	G_{IC} (J/m^2)	Ballistic failure type
PACM	None	2180 ± 140	–	116 ± 3	0.8 ± 0.1	240 ± 20	Brittle
DAP	None	2510 ± 70	–	106 ± 1	0.8 ± 0.1	240 ± 50	Brittle
DAO	None	2000 ± 50	–	81 ± 1	0.9 ± 0.1	390 ± 30	Brittle
D230	None	3070 ± 140	70 ± 5	87 ± 1	0.8 ± 0.1	180 ± 30	Brittle
D230	1 wt% MX-125	3010 ± 260	66 ± 1	84 ± 1	1.0 ± 0.1	430 ± 60	Brittle
D230	5 wt% MX-125	2830 ± 160	59 ± 1	79 ± 1	2.0 ± 0.1	1530 ± 150	Tough
D230	1 wt% MX-156	2970 ± 120	67 ± 1	83 ± 1	1.4 ± 0.1	580 ± 100	Mixed
D230	5 wt% MX-156	2830 ± 180	62 ± 3	74 ± 1	2.2 ± 0.1	1700 ± 150	Tough
D230	1 wt% MX-257	3215 ± 240	62 ± 2	82 ± 1	1.7 ± 0.1	1000 ± 150	Tough
D230	5 wt% MX-257	2670 ± 330	57 ± 1	75 ± 1	2.6 ± 0.1	2500 ± 200	Tough
D400	None	2556 ± 170	57 ± 4	68 ± 1	1.5 ± 0.2	760 ± 200	Tough

obtained from quasistatic tensile tests, listed in Table 36.2. The curve representing Eq. (36.1) follows the trend for the brittle group of data surprisingly well. The ballistic limit of the brittle resins is underpredicted by Eq. (36.1) by a consistent margin of about 0.5 normalized KE_{50}. Such a gap is not surprising because Eq. (36.1) predicts the critical impact velocity to initiate a cone crack, whereas the V_{50} test measures penetration of a witness foil behind the specimen by the projectile or an ejected fragment of the specimen. The gap of 0.5 normalized KE_{50} may represent the additional impact energy required to eject a cone section fragment from a specimen after initiating conical cracking. As shown in Fig. 36.7 and discussed above, radial cracking occurs prior to cone crack initiation, allowing conical fragments to dislodge relatively easily after forming. It is remarkable that the trend for brittle samples is well described by Eq. (36.1) with no adjustable parameters, since this equation was derived for ideal elastic, quasistatic Hertzian contact. Compton et al. [46] have presented finite element simulations suggesting that variations in contact radius due to plastic deformation cause dramatic deviations from the cone initiation threshold predicted by Eq. (36.1). Therefore, our result implies that plastic deformation does not influence the contact radius for brittle resins at failure velocity.

As seen in Fig. 36.9, data with $KE_{50} > 3$ are not a strong function of quasistatic K_{IC}, suggesting they may be best described by a relationship with the general form of Eq. (36.2). To test this hypothesis, we multiply Eq. (36.2) by an arbitrary constant A:

$$V_Y = \frac{3A\sigma_{Y,2}}{(\rho c)'} \tag{36.3}$$

The blue dotted curve in Fig. 36.9 represents a fit to Eq. (36.3) using $A = 3.8$ and yield stress and Young's modulus values from quasistatic tensile tests of selected filled and unfilled resins in the tough group, listed in Table 36.2. In reality, stresses in the plastic zone ahead of impact are a combination of tensile and compressive, and thus the fit using tensile yield stress is an approximation. Use of measured compressive yield stress values (also included in Table 36.2) for Eq. (36.3) results in a similarly good fit to the data, using a value of $A = 3$ rather than 3.8.

The value of A was chosen simply to give a good fit to the experimental KE_{50} data for tough samples, but it may represent a combination of physical phenomena. First, as discussed above, Eq. (36.2) was derived for ceramic targets with high yield stress, in which the contact radius is determined by plastic deformation of the projectile prior to yield of the target at high velocities. In our system it is the target rather than the projectile that first deforms plastically; thus a portion the prefactor A in Eq. 36.3 may account for different contact geometry. Secondly, Eq. 36.2 was derived as an approximate limit for *initiation* of yield at the target front face, whereas the KE_{50} data are measurements of complete target failure characterized by ejection of fragments or the projectile from the specimen's back face. Hence, a portion of A may constitute the additional energy required for full target failure after initiating plasticity, analogous to the gap of 0.5 between Eq. 36.1 and KE_{50} of the brittle samples. Thirdly, it is well know that mechanical properties of glassy polymers are highly rate dependent [52]. Quasistatic yield stress and modulus values used to construct the blue curve in Fig. 36.9 were measured at a nominal strain rate of 3×10^{-3} s^{-1}, while the ballistic impact events occur at effective strain rates between 10^4 and 10^5 s^{-1}. Therefore a significant portion of the prefactor A in Eq. 36.3 may account for the increase in yield stress over the quasistatic value due to the high rate of ballistic impact. The combination of these reasonable physical considerations, as well as the good agreement between Eq. 36.3 and KE_{50} of tough resins reported here, suggest that ballistic failure may be dominated by plastic yielding of the target for materials with fracture toughness above a certain threshold defined by the intersection of the two curves in Fig. 36.9.

Approximately five data points in Fig. 36.9 do not fit the trends described by Eqs. (36.1) or (36.3), comprising D230 modified with 0.5–2 wt% of either MX-125 or MX-156. These materials have relatively high quasistatic K_{IC} values between 1.1 and 2.0 MPa-m$^{1/2}$, yet their KE_{50} is lower than that predicted by Eq. (36.3) for the high K_{IC}, plastic yield dominated regime. A possible explanation for these outliers is that the toughening effects of MX-125 and MX-156 have a greater rate dependence than that of MX-257, as suggested above in the analysis of Fig. 36.8. In this case high rate K_{IC} would be much less than quasistatic K_{IC}, shifting these data to the left where they may fall along the trend defined by the red dashed curve (Eq. 36.1). On the other hand MX-257 modified D230 behaves like a tough material with $KE_{50} \approx 3.5$ for all compositions from 0.5 to 9 wt%. This contrast is particularly striking when considering the case of D230 modified with 0.5 wt% MX-257 ($K_{IC} = 1.1$ MPa-m$^{1/2}$, $KE_{50} = 3.3$), which approximately fits both trends and is thus apparently near the transition point between brittle and tough regimes.

36.4 Conclusions

We have demonstrated that core-shell rubber (CSR) nanoparticle modifiers are capable in some cases of improving ballistic performance of amine-cured epoxy resins without significantly lowering T_g or stiffness. The CSR particles impart dramatic improvements to ballistic impact resistance of resins within a relatively narrow range of T_g between 60 and 110 °C, whereas resins above or below this T_g range show little ballistic improvement with addition of CSR particles. While brittle resins with low molecular weight between crosslinks tend to exhibit modestly increasing KE_{50} as a function of fracture toughness K_{IC}, a limiting value of about 3.5 normalized KE_{50} was found for tougher formulations, irrespective of measured quasistatic K_{IC}. These trends are well described by two different failure mechanisms: for brittle materials ballistic failure is dominated by the critical velocity for Hertzian cone crack initiation, while for tough materials ballistic limit is related to the critical velocity for target plastic yield at the impact site. These insights will inform the design of next generation adhesives and matrix materials for fiber-reinforced composites used in a variety of lightweight protective applications.

References

1. Grossman, E., Gouzman, I.: Space environment effects on polymers in low earth orbit. Nucl. Instrum. Methods Phys. Res. B **208**, 48–57 (2003)
2. Committee on Opportunities in Protection Materials Science Technology for Future Army Applications and National Research Council: Opportunities in Protection Materials Science and Technology for Future Army Applications. The National Academies Press, Washington (2011)
3. Crawford, E., Lesser, A.J.: The effect of network architecture on the thermal and mechanical behavior of epoxy resins. J. Polym. Sci. Polym. Phys. **36**(8), 1371–1382 (1998)
4. Naik, N.K., Shrirao, P.: Composite structures under ballistic impact. Compos. Struct. **66**(1-4), 579–590 (2004)
5. Carrillo, J.G., et al.: Ballistic performance of thermoplastic composite laminates made from aramid woven fabric and polypropylene matrix. Polym. Test. **31**(4), 512–519 (2012)
6. Hsieh, T.H., et al.: The toughness of epoxy polymers and fibre composites modified with rubber microparticles and silica nanoparticles. J. Mater. Sci. **45**(5), 1193–1210 (2009)
7. Sprenger, S.: Fiber-reinforced composites based on epoxy resins modified with elastomers and surface-modified silica nanoparticles. J. Mater. Sci. **49**(6), 2391–2402 (2013)
8. David, N.V., Gao, X.L., Zheng, J.Q.: Ballistic resistant body armor: contemporary and prospective materials and related protection mechanisms. Appl. Mech. Rev. **62**(5), 050802 (2009)
9. Naik, N.K., et al.: High strain rate mechanical behavior of epoxy under compressive loading: experimental and modeling studies. Mater. Sci. Eng. A **528**(3), 846–854 (2011)
10. Knorr Jr., D.B., et al.: Glass transition dependence of ultrahigh strain rate response in amine cured epoxy resins. Polymer **53**(25), 5917–5923 (2012)
11. Masser, K.A., et al.: Relating structure and chain dynamics to ballistic performance in transparent epoxy networks exhibiting nanometer scale heterogeneity. Polymer **58**, 96–106 (2015)
12. Knorr, D.B., et al.: Overcoming the structural versus energy dissipation trade-off in highly crosslinked polymer networks: ultrahigh strain rate response in polydicyclopentadiene. Compos. Sci. Technol. **114**, 17–25 (2015)
13. McGarry, F.J.: Building design with fibre reinforced materials. Proc. R Soc. Lond. A Math. Phys. Sci. **319**(1536), 59–68 (1970)
14. Kinloch, A.J., et al.: Deformation and fracture behaviour of a rubber-toughened epoxy: 1. Microstructure and fracture studies. Polymer **24**, 1341–1354 (1983)
15. Yee, A.F., Pearson, R.A.: Toughening mechanisms in elastomer-modified epoxies Part 1 mechanical studies. J. Mater. Sci. **21**, 2462–2474 (1986)
16. Spanoudakis, J., Young, R.J.: Crack propagation in a glass particle-filled epoxy resin Part 1—effect of particle volume fraction and size. J. Mater. Sci. **19**, 473–486 (1984)
17. Hsieh, T.H., et al.: The mechanisms and mechanics of the toughening of epoxy polymers modified with silica nanoparticles. Polymer **51**(26), 6284–6294 (2010)
18. McGrath, L.M., et al.: Investigation of the thermal, mechanical, and fracture properties of alumina-epoxy composites. Polymer **49**(4), 999–1014 (2008)
19. Sue, H.J., et al.: Optimization of mode-I fracture toughness of high-performance epoxies by using designed core-shell rubber particles. Adv. Chem. Ser. **233**, 259–291 (1993)
20. Blanco, M., et al.: Thermoplastic-modified epoxy resins cured with different functionalities amine mixtures: morphology, thermal behavior, and mechanical properties. J. Appl. Polym. Sci. **114**(3), 1753–1760 (2009)
21. Fischer, M.: Properties and failure of polymers with tailored distances between cross-links. Adv. Polym. Sci. **100**, 313–355 (1992)
22. Garg, A.C., Mai, Y.W.: Failure mechanisms in toughened epoxy resins—a review. Compos. Sci. Technol. **31**(3), 179–223 (1988)
23. Bagheri, R., Marouf, B.T., Pearson, R.A.: Rubber-toughened epoxies: a critical review. Polym. Rev. **49**(3), 201–225 (2009)
24. Pearson, R.A., Yee, A.F.: Toughening mechanisms in elastomer-modified epoxies. Part 2—Microscopy studies. J. Mater. Sci. **21**, 2475–2488 (1986)
25. Pearson, R.A., Yee, A.F.: Toughening mechanisms in elastomer-modified epoxies 3. The effect of cross-link density. J. Mater. Sci. **24**(7), 2571–2580 (1989)
26. Dean, J.M., et al.: Micellar structure and mechanical properties of block copolymer-modified epoxies. J. Polym. Sci. Polym. Phys. **39**, 2996–3010 (2001)
27. Liu, J., et al.: Strain rate effect on toughening of nano-sized PEP-PEO block copolymer modified epoxy. Acta Mater. **57**(9), 2691–2701 (2009)
28. Liu, J., et al.: Nanocavitation in self-assembled amphiphilic block copolymer-modified epoxy. Macromolecules **41**, 7616–7624 (2008)
29. Thompson, Z.J., et al.: Block copolymer toughened epoxy—role of crosslink density. Macromolecules **42**, 2333–2335 (2009)
30. Sue, H.J., et al.: Fracture mechanisms in rigid core shell particle modified high performance epoxies. Colloid Polym. Sci. **274**(4), 342–349 (1996)
31. Bagheri, R., Pearson, R.A.: Role of blend morphology in rubber-toughened polymers. J. Mater. Sci. **31**, 3945–3954 (1996)
32. Lu, F., et al.: Toughening mechanisms in modified epoxy resins with different crosslink densities. Polym. Bull. **37**(3), 399–406 (1996)
33. Huang, Y., Kinloch, A.J.: Modeling of the toughening mechanisms in rubber-modified epoxy polymers Part I. Finite element analysis studies. J. Mater. Sci. **27,** 2753–2762 (1992)
34. Huang, Y., Kinloch, A.J.: Modelling of the toughening mechanisms in rubber-modified epoxy polymers Part II—A quantitative description of the microstructure-fracture property relationships. J. Mater. Sci. **27**, 2763–2769 (1992)
35. Arias, M.L., Frontini, P.M., Williams, R.J.J.: Analysis of the damage zone around the crack tip for two rubber-modified epoxy matrices exhibiting different toughenability. Polymer **44**(5), 1537–1546 (2003)
36. Aizpurua, B., et al.: Chemorheology and ultimate behavior of epoxy-amine mixtures modified with a liquid oligomer. J. Appl. Polym. Sci. **76** (8), 1269–1279 (2000)

37. Bradley, W.L., et al.: The synergistic effect of cross-link density and rubber additions on the fracture toughness of polymers. Adv. Chem. Ser. **233**, 317–334 (1993)
38. Raghavan, D., et al.: Strain rate dependence of fracture in a rubber-toughened epoxy system. J. Adhes. **78**(8), 723–739 (2002)
39. Cardwell, B.J., Yee, A.F.: Rate and temperature effects on the fracture toughness of a rubber-modified epoxy. Polymer **34**(8), 1695–1701 (1993)
40. Bain, E.D., et al.: Failure processes governing high-rate impact resistance of epoxy resins filled with core–shell rubber nanoparticles. J. Mater. Sci. **51**(5), 2347–2370 (2015)
41. US Department of Defense: V50 ballistic test for armor MIL-STD-662F (1997)
42. Bogoslovov, R.B., Roland, C.M., Gamache, R.M.: Impact-induced glass transition in elastomeric coatings. Appl. Phys. Lett. **90**(22), 221910 (2007)
43. Roland, C.M.: Mechanical behavior of rubber at high strain rates. Rubber Chem. Technol. **79**(3), 429–459 (2006)
44. Roland, C.M., Fragiadakis, D., Gamache, R.M.: Elastomer-steel laminate armor. Compos. Struct. **92**(5), 1059–1064 (2010)
45. Roland, C.M., et al.: Factors influencing the ballistic impact resistance of elastomer-coated metal substrates. Philos. Mag. **93**(5), 468–477 (2013)
46. Compton, B.G., Gamble, E.A., Zok, F.W.: Failure initiation during impact of metal spheres onto ceramic targets. Int. J. Impact Eng. **55**, 11–23 (2013)
47. Seagraves, A.N., Radovitzky, R.A.: An analytical theory for radial crack propagation: application to spherical indentation. J. Appl. Mech. **80**, 041018 (2013)
48. Lawn, B.R.: Indentation of ceramics with spheres: a century after Hertz. J. Am. Ceram. Soc. **81**, 1977–1994 (1998)
49. Sherman, D., Brandon, D.G.: The ballistic failure mechanisms and sequence in semi-infinite supported alumina tiles. J. Mater. Res. **12**, 1335–1343 (1997)
50. Sherman, D.: Impact failure mechanisms in alumina tiles on finite thickness support and the effect of confinement. Int. J. Impact Eng. **24**, 313–328 (2000)
51. Qian, J.Y., et al.: The role of dispersed phase morphology on toughening of epoxies. Polymer **38**(1), 21–30 (1997)
52. Iwamoto, T., Nagai, T., Sawa, T.: Experimental and computational investigations on strain rate sensitivity and deformation behavior of bulk materials made of epoxy resin structural adhesive. Int. J. Solids Struct. **47**, 175–185 (2010)

Chapter 37
Ballistic Response of Polydicyclopentadiene vs. Epoxy Resins and Effects of Crosslinking

Tyler R. Long, Daniel B. Knorr Jr., Kevin A. Masser, Robert M. Elder, Timothy W. Sirk, Mark D. Hindenlang, Jian H. Yu, Adam D. Richardson, Steven E. Boyd, William A. Spurgeon, and Joseph L. Lenhart

Abstract The ballistic performance of polydicyclopentadiene (pDCPD) was investigated and compared to two epoxy resins that a have similar glass transition temperature (T_g) to pDCPD. The ballistic performance of these materials (at an effective stain rate of 10^4–10^5 s^{-1}) was characterized by determining the kinetic energy of the projectile where there is a 50 % probability that the projectile will penetrate a witness foil behind the sample (KE_{50}). The ballistic performance of pDCPD showed a 300–400 % improvement over the structural epoxy resins. Typical, highly crosslinked epoxy networks become brittle at low temperatures, but pDCPD has a superior ballistic performance over a broad temperature range from (−55 to 75 °C), despite having a glass transition temperature of 142 °C, which characteristic of structural resins. pDCPD also exhibited a room temperature glassy storage modulus of 1.7 GPa, making pDCPD a potential structural resin that can overcome the structural vs. energy dissipation trade-off that commonly exists with some conventional crosslinked polymers. Quasi-static measurements of pDCPD when compared to epoxy resins suggested that the performance of pDCPD relates to higher fracture toughness and lower yield stress relative to typical epoxies, while molecular dynamics simulations comparing pDCPD to epoxy resins suggest that the performance of pDCPD is due to the lack of strong non-covalent interactions and the facile formation of nanoscale voids.

Keywords Polymer • Amorphous materials • Polydicyclopentadiene • Fracture toughness • Impact behavior

37.1 Introduction

The high strain rate performance of materials are becoming more and more important in both military and commercial applications. Crosslinked polymers are more widely used for many applications where they serve both a structural and protective role. [1–7] Crosslinked polydicyclopentadiene (pDCPD) has shown to have unique and impressive properties under high strain rate reached during a ballistic event, and is being investigated for use as a resin in fiber reinforced polymer composites (FRPC). [1] FRPC are becoming an increasingly important part of light weight protective equipment utilized by the military, aerospace, transportation, and construction industries. The polymer matrix in these FRPCs can be responsible for between 20 and 35 % of the total energy dissipation capability. [8] Typically, crosslinked epoxies are uses as the resins in these FRPC because of their high stiffness and strength which are requirement for structural applications, but epoxies are known to be brittle and tend to fracture during a ballistic event.

The performance of a polymer composite during a high strain rate impact is dependent on several complex energy dissipation mechanisms. [6, 7] In order to characterize the energy dissipation capability of these polymers at these high strain rates high velocity projectiles are used with an effective strain rate of 10^4–10^5 s^{-1}. This energy dissipation capability can be quantified by determining the materials KE_{50}, which is the kinetic energy of the projectile where there is a 50 % probability of penetration of a witness foil behind a sample. This method has proven to be a useful way the determine the energy dissipation capability of polymers, [2, 3] metals, [9] and polymeric composites. [10]

Previous work on a variety of epoxies shown that their energy dissipation capability is highly dependent on the measurement temperature compared to the T_g of the epoxy ($T-T_g$), which is worse at large $T-T_g$ or colder temperature. [2] pDCPD, on the other hand, has better energy dissipation capability than epoxy, and pDCPD energy dissipation capability increases at larger $T-T_g$. [1] The other advantage pDCPD has is its ductile failure mechanism during a ballistic event, which results in a small damage area. By comparison, epoxies exhibit a comparatively large damage area as a result of a brittle failure mechanism that creates both radial and cone fractures [11].

T.R. Long (✉) • D.B. Knorr Jr. • K.A. Masser • R.M. Elder • T.W. Sirk • M.D. Hindenlang •
J.H. Yu • A.D. Richardson • S.E. Boyd • W.A. Spurgeon • J.L. Lenhart
U.S. Army Research Laboratory, Aberdeen Proving Ground, MD 21005, USA
e-mail: tyler.r.long8.civ@mail.mil

D. Casem et al. (eds.), *Dynamic Behavior of Materials, Volume 1*, Conference Proceedings of the Society for Experimental Mechanics Series, DOI 10.1007/978-3-319-41132-3_37

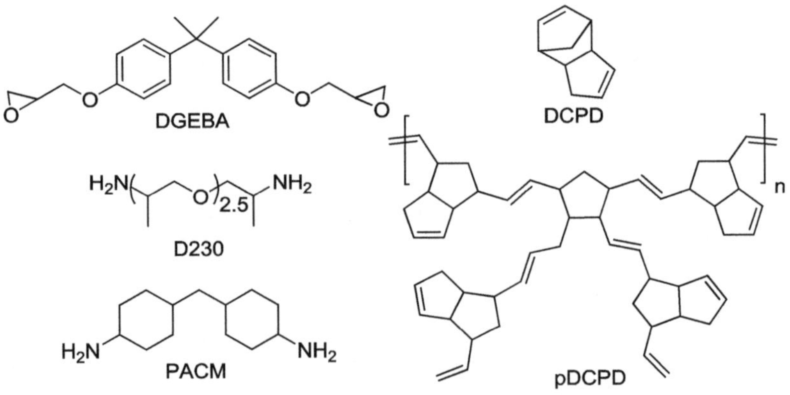

Fig. 37.1 Epoxy resins diglycidyl ether of bisphenol A (DGEBA), curing agents 4,4′-methylenebis (cyclohexylamine) (PACM), Jeffamine D-230, DCPD, and pDCPD with a single ring-opened cyclopentyl to form crosslink

The M_c for the polymers investigated were estimated using the theory of rubbery elasticity Eq. (37.1) [1, 12]. This theory relates the molecular weight between crosslinks or entanglements (M_c) to the inverse of the minimum in the rubbery region of storage modulus obtained from dynamic mechanical analysis (DMA) measurements. In Eq. (37.1), E'_r is the minimum of the storage modulus, T is the absolute temperature where the minimum of the storage modulus is reached, ρ is the density at temperature T, and R is the gas constant (8.314 J K^{-1} mol^{-1}). The density change due to thermal expansion was considered, and the room temperature density was adjusted using the coefficient of liner thermal expansion (COTE) data from the TMA measurement, assuming the material expansion was isotropic.

$$M_C = \frac{3\rho RT}{E'_r} \tag{37.1}$$

In this work, the ballistic performance, mechanical properties, and molecular dynamics simulation of pDCPD were compared for two epoxy resins: DGEBA cured with PACM and DGEBA cured with D230, the structures of which are shown in (Fig. 37.1). DGEBA/PACM and DGEBA/D230 both form highly crosslinked networks that have T_g values similar to pDCPD. DCPD polymerizes via ring-opening metathesis polymerization using a Grubbs first generation catalyst. This also creates a crosslinked networked when the less strained cyclopentyl ring is opened after the initial linear propagation from the opening of the highly strained norbornene ring [13].

37.2 Results and Discussion

Figure 37.2 compares normalized KE_{50}, or KE_{50} normalized to the room temperature KE_{50} of a stoichiometric mixture of DGEBA/PACM, to the experiment temperature relative to the materials T_g. pDCPD has a unique ballistic performance when it has been compared to epoxy resins. DGEBA/PACM has poor performance at all temperature investigated and DGEBA/D230 exhibit a strong temperature dependence, similar to previously reported systems [2]. These epoxies exhibit a significant decrease in ballistic performance at temperatures well below T_g and the performance is reduced in the range of interest to the army (−55 to 75 °C). pDCPD shows a steady increase in normalized KE_{50} as temperature is reduced, even deep into the glassy state and shows a 300–400 % improvement over the epoxies studied. pDCPD also exhibits no radial cracking near the impact site, instead exhibiting a shear-induced mechanism of failure. Unlike pDCPD, DGEBA/PACM and DGEBA/D230 exhibited a brittle failure mechanism manifesting itself as a significant amount of cracking and spalling, spreading out from the impact site.

A DMA was used to compare T_g, room temperature stiffness, as well as the stiffness at T_g–50 °C of pDCPD to DGEDA/PACM and DGEBA/D230. The T_g was determined by peak in the tan δ at 1Hz on hearing at 1 °C/min. The minimum in the storage modulus from the DMA was combined with density data from a pycnometer. The density was adjusted with the COTE data from the TMA, and was used to estimate the M_c of these polymers using Eq. (37.1). We found that pDCPD has a lower storage modulus than DGEDA/PACM and DGEBA/D230, which indicates that pDCPD is less crosslinked than the epoxy networks (Table 37.1).

To determine what quasi-static property could potentially be an indicator of improved high rate mechanical response, quasi-static compression, tensile and fracture toughness measurements were performed at room temperature and at T_g−50 °C. The DGEBA/PACM system was brittle at room temperature, with failure being dominated by flaws such that yielding was not observed. DGEBA/D230 and pDCPD both yield prior to tensile failure. DGEBA/D230 exhibits a higher

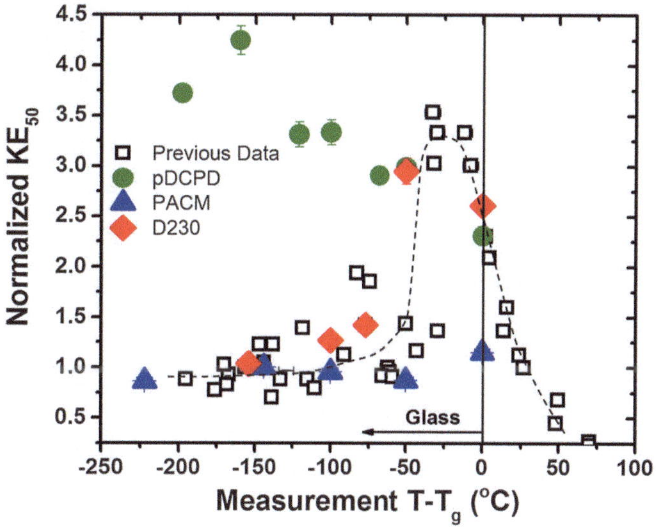

Fig. 37.2 Plot of the normalized KE_{50} of pDCPD, DGBEDA/PACM, and DGBEDA/D230 the measurement temperature as a function of $T-T_g$.[1] Previous data taken from [2]

Table 37.1 Summary of DMA, TMA and density results[1]

	T_g (°C)	E′ at 25 °C (MPa)	E′ at T_g−50 °C (MPa)	E′ min (MPa)	Glassy COTE m/(m · °C)	Rubbery COTE m/(m · °C)	Density at RT (g/cm³)	Density at T_g + 50 °C (g/cm³)	M_c (g/mol)
DGEBA/PACM	164 ± 3	2470 ± 120	1590 ± 90	36 ± 1	72 ± 3	183 ± 2	1.158 ± 0.004	1.09 ± 0.02	370 ± 20
DGEBA/D230	96 ± 3	2780 ± 150	2270 ± 110	19 ± 1	76 ± 5	202 ± 2	1.156 ± 0.002	1.10 ± 0.02	550 ± 30
pDCPD	142 ± 3	1700 ± 90	1500 ± 80	17 ± 1	86 ± 1	201 ± 6	1.041 ± 0.003	0.98 ± 0.02	670 ± 40

tensile yield stress than pDCPD, and both exhibit similar apparent tensile toughness. pDCPD shows a 30 % higher strain at break than DGEBA/D230 (Table 37.2). pDCPD tensile results agreed with literature values [14].

The tensile test at 114 °C (T_g−50 °C) for the DGEBA/PACM system showed that yield occurred due to the increased temperature with a yield stress of 46.4 MPa, which falls between the yield stress of pDCPD at 38.6 MPa and DGEBA/D230 at 55.4 MPa. pDCPD exhibited a lower tensile and compressive yield stress both at room temperature and T_g−50 °C when compared to the epoxies. pDCPD also exhibited necking beyond the yield point, which was not observed in either epoxy resin, leading to higher elongation to break. The necking of pDCPD demonstrates its ability to plastically deform, which DGEBA/D230 and DGEBA/PACM lack. pDCPD also exhibits a lower compressive yield stress than either epoxy resin. pDCPD room temperature fracture toughness of (2.19 MPa $m^{1/2}$) compared to the DGEBA/D230 of (0.76 MPa $m^{1/2}$) and DGEBA/PACM (0.69 MPa $m^{1/2}$) is due to the substantial tearing type behavior pDCPD exhibited compared to the conventional brittle failure mechanism of the epoxies. This demonstrated that pDCPD a much more ductile then the typical epoxy resin.

Molecular dynamics simulations were performed by comparing a pDCPD network containing six un-crosslinked monomers between crosslinked points to DGEBA/PACM and DGEBA/D230 networks. First, we investigated the work need to overcome the electrostatic inter-chain interactions during deformation by comparing the electrostatic potential energy to the strain [4]. As shown in Fig. 37.3a, hydrogen-bonds and the dipole-dipole interactions in the epoxy result in a substantial increase in the electrostatic energy during strain. pDCPD, on the other hand, which lacks the same strong intermolecular interaction, shows almost no increase in the electrostatic energy when strained. The ductility of thermoset systems may be related to the behavior of nanoscale voids formed during strain as recent research suggests [15]. As shown in Fig. 37.3b, DGEBA/PACM and DGEBA/D230 exhibit a larger distribution of nanovoid size compared to pDCPD. The nanovoid fraction verses strain shows pDCPD requires less void formation to accommodate strain. Figure 37.3c shows the energy difference between atoms near the void interface and the atoms in the bulk of the network. This indicates that epoxies require more energy to form the void than pDCPD, which is close to zero. These molecular dynamics simulations of epoxy and pDCPD suggest that the higher molecular weight between crosslinks (M_c) of pDCPD yields lower nanovoid volume at a given strain, which may be needed to accommodate strain. Void formation in pDCPD was also energetically neutral due to

Table 37.2 Mechanical properties of thermosets measured at 22 °C and T_g−50 °C.[1]

	T(C)	KE_{50}	Compressive yield strength (MPa)	Ultimate tensile strength (MPa)	Tensile yield strength (MPa)	Strain at break	Tensile modulus (MPa)	Apparent tensile toughness (MPa)	K_{IC} (MPa m$^{1/2}$)	G_{IC} (kJ/m^2)
pDCPD	22 ± 3	3.31 ± 0.12	73.2 ± 0.8	35.3 ± 2.5	52.4 ± 2.7	0.16 ± 0.03	1770 ± 90	6.1 ± 1.3	2.19 ± 0.15	2.07 ± 0.28
DGEBA/D230	22 ± 3	1.42 ± 0.03	93.6 ± 0.5	56.7 ± 4.4	70.2 ± 5.8	0.11 ± 0.02	3060 ± 170	6.2 ± 1.3	0.76 ± 0.08	0.20 ± 0.02
DGEBA/PACM	22 ± 3	1.00 ± 0.05	115.6 ± 2.3	40.5 ± 3.9	N/A	0.026 ± 0.004	2140 ± 70	0.61 ± 0.16	0.69 ± 0.05	0.19 ± 0.03
pDCPD	T_g−50	2.99 ± 0.07	49.3 ± 1.6	23.6 ± 1.6	38.6 ± 2.5	0.13 ± 0.02	1750 ± 170	3.4 ± 0.4	2.31 ± 0.18[a]	4.57 ± 0.4[a]
DGEBA/D230	T_g−50	2.95 ± 0.13	70.2 ± 1.6	37.8 ± 2.2	55.4 ± 3.0	0.09 ± 0.02	2630 ± 200	3.6 ± 0.6	0.98 ± 0.22	0.50 ± 0.2
DGEBA/PACM	T_g−50	0.87 ± 0.04	58.1 ± 1.0	42.7 ± 1.7	46.4 ± 0.9	0.078 ± 0.014	1760 ± 110	2.9 ± 0.6	0.91 ± 0.05	0.41 ± 0.03

[a]pDCPD samples tore during fracture toughness test and are not strictly valid based on ASTM D5045-99

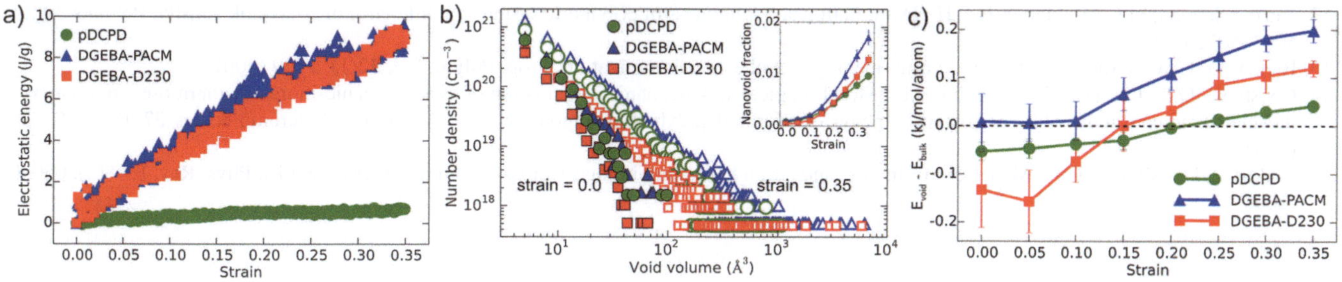

Fig. 37.3 (**a**) Change in total electrostatic energy, (**b**) void volume distribution and (*inset*) void fraction, and (**c**) difference in non-bonded interaction energy between atoms near voids and atoms in the bulk polymer for the simulated pDCPD, DGEBA/PACM, and DGEBA/D230 networks as a function of engineering strain at 150 K. *Solid* and *open* symbols in (**b**) are at 0 and 35 % strain, respectively[1]

the lack of strong electrostatic interactions. The combination of high M_c and a lack of strong electrostatic interactions in pDCPD suggests that less work is required for nanovoid formation, and this allows for more plastic deformation compared to epoxy networks with hydrogen bonds and a low M_c.

37.3 Conclusion

pDCPD's high T_g (142 °C) and room temperature storage modulus of 1.7 GPa make it suitable for uses as a structural resin. pDCPD's excellent ballistic performance and ductile failure mechanism make it unique when compared to typical highly crosslinked, high T_g epoxy networks. The comparison to DGEBA/PACM and DGEBA/D230 give some insight into what makes pDCPD unique. The high fracture toughness and low yield strength from the quasi-static test and the low crosslink density as determined by Eq. (37.1) may be general indicators of improved high rate mechanical response. Molecular dynamics simulations suggest that the weak intermolecular interactions and low crosslink density of pDCPD results in less nanovoid formation during strain and the nanovoid formation requires less energy, allowing for more plastic deformation. This combination of structural and ballistic performance make pDCPD a great candidate for integration into FRPCS.

Acknowledgement This research was supported in part by an appointment to the Postgraduate Research Participation Program at the U.S. Army Research Laboratory administered by the Oak Ridge Institute for Science and Education through an interagency agreement between the U.S. Department of Energy and USARL.

References

1. Knorr Jr., D.B., et al.: Overcoming the structural versus energy dissipation trade-off in highly crosslinked polymer networks: ultrahigh strain rate response in polydicyclopentadiene. Compos. Sci. Technol. **114**, 17–25 (2015)
2. Knorr Jr., D.B., et al.: Glass transition dependence of ultrahigh strain rate response in amine cured epoxy resins. Polymer **53**(25), 5917–5923 (2012)
3. Masser, K.A., et al.: Relating structure and chain dynamics to ballistic performance in transparent epoxy networks exhibiting nanometer scale heterogeneity. Polymer **58**, 96–106 (2015)
4. Elder, R.M., Andzelm, J.W., Sirk, T.W.: A molecular simulation study of the glass transition of cross-linked poly(dicyclopentadiene) networks. Chem. Phys. Lett. **637**, 103–109 (2015)
5. McGrath, L.M., et al.: Investigation of the thermal, mechanical, and fracture properties of alumina–epoxy composites. Polymer **49**(4), 999–1014 (2008)
6. Morye, S.S., et al.: Modelling of the energy absorption by polymer composites upon ballistic impact. Compos. Sci. Technol. **60**(14), 2631–2642 (2000)
7. Deka, L.J., Bartus, S.D., Vaidya, U.K.: Damage evolution and energy absorption of E-glass/polypropylene laminates subjected to ballistic impact. J. Mater. Sci. **43**(13), 4399–4410 (2008)
8. Zee, R.H., Hsieh, C.Y.: Energy absorption processes in fibrous composites. Mat. Sci. Eng. A Struct **246**(1–2), 161–168 (1998)
9. Czarnecki, G.J.: Estimation of the V-50 using semi-empirical (1-point) procedures. Compos. Part B Eng. **29**(3), 321–329 (1998)
10. Gellert, E.P., Cimpoeru, S.J., Woodward, R.L.: A study of the effect of target thickness on the ballistic perforation of glass-fibre-reinforced plastic composites. Int. J. Impact Eng. **24**(5), 445–456 (2000)
11. Bain, E.D., et al.: Failure processes governing high-rate impact resistance of epoxy resins filled with core–shell rubber nanoparticles. J. Mater. Sci. **51**(5), 2347–2370 (2015)

12. van der Sanden, M.C.M., Meijer, H.E.H.: Deformation and toughness of polymeric systems: 3. Influence of cross-link density. Polymer **34**(24), 5063–5072 (1993)
13. Bielawski, C.W., Grubbs, R.H.: Living ring-opening metathesis polymerization. Prog. Polym. Sci. **32**(1), 1–29 (2007)
14. Constable, G.S., Lesser, A.J., Coughlin, E.B.: Morphological and mechanical evaluation of hybrid organic-inorganic thermoset copolymers of dicyclopentadiene and mono- or tris(norbornenyl)-substituted polyhedral oligomeric silsesquioxanes. Macromolecules **37**(4), 1276–1282 (2004)
15. Mukherji, D., Abrams, C.F.: Microvoid formation and strain hardening in highly cross-linked polymer networks. Phys. Rev. E **78**(5), 050801 (2008)

MIX
Papier aus verantwortungsvollen Quellen
Paper from responsible sources
FSC® C105338

If you have any concerns about our products,
you can contact us on
ProductSafety@springernature.com

In case Publisher is established outside the EU,
the EU authorized representative is:
Springer Nature Customer Service Center GmbH
Europaplatz 3, 69115 Heidelberg, Germany

Printed by Libri Plureos GmbH
in Hamburg, Germany